FUNCTIONAL POLYMERS

Design, Synthesis, and Applications

FUNCTIONAL POLYMERS

Design, Synthesis, and Applications

Edited by
Raja Shunmugam

Apple Academic Press Inc.	Apple Academic Press Inc.
3333 Mistwell Crescent	9 Spinnaker Way
Oakville, ON L6L 0A2	Waretown, NJ 08758
Canada	USA

First issued in paperback 2021

No claim to original U.S. Government works

ISBN-13: 978-1-77463-592-6 (pbk)
ISBN-13: 978-1-77188-296-5 (hbk)

Library and Archives Canada Cataloguing in Publication

Functional polymers : design, synthesis, and applications/edited by Raja Shunmugam.
Includes bibliographical references and index.
Issued in print and electronic formats.
ISBN 978-1-77188-296-5 (hardcover).--ISBN 978-1-77188-297-2 (pdf)

1. Polymers. 2. Polymer engineering. 3. Polymerization. 4. Polymers--Industrial applications. I. Shunmugam, Raja, author, editor

| TP1087.F85 2016 | 668.9 | C2016-906164-7 | C2016-906165-5 |

CIP data on file with US Library of Congress

Apple Academic Press also publishes its books in a variety of electronic formats. Some content that appears in print may not be available in electronic format. For information about Apple Academic Press products, visit our website at **www.appleacademicpress.com** and the CRC Press website at **www.crcpress.com**

ABOUT THE EDITOR

Raja Shunmugam, PhD

Raja Shunmugam, PhD, is an Associate Professor and Ramanujan Fellow at the Polymer Research Centre, Department of Chemical Sciences, at the Indian Institute of Science Education and Research Kolkata, West Bengal, India. He was formerly an Assistant Professor, Indian Institute of Science Education and Research, Kolkata, India. Dr. Shunmugam has written a number of articles published in international journals and holds several patents. He has been an invited presenter at several international conferences. He performed his postdoctoral work at the Polymer Science and Engineering Department at UMass Amherst, Massachusetts, USA.

Dedicated to
IISER Kolkata, India

CONTENTS

LIST OF CONTRIBUTORS

Mehmet Arslan
Bogazici University, Istanbul, Turkey

Ramakrishnan Ayothi
JSR Micro Inc., Sunnyvale, CA 95136, USA E-mail: rayothi@jsrmicro.com

Sanjib Banerjee
IISER Kolkata, India

Dipankar Basak
IACS, Kolkata, India

Anjan Bedi
IISER Kolkata, India

Sourav Bhattacharya
IISER Kolkata

Xiaorui Chen
Gustaf Carlson School of Chemistry and Biochemistry, Clark University, Worcester, Massachusetts, USA

Sumana Roy Chowdhury
NCL Pune, India

Priyadarsi De
Polymer Research Centre, Department of Chemical Sciences, IISER Kolkata, Mohanpur 741246, Nadia, India, E-mail: p_de@iiserkol.ac.in

Prashant Deshmukh
Polymer Program, Institute of Materials Science, University of Connecticut, Storrs, Connecticut, USA

Raghavachari Dhamodharan
Department of Chemistry, IIT Madras, Chennai 600 036, India E-mail: damo@iitm.ac.in

Sergio Granados-Focil
Gustaf Carlson School of Chemistry and Biochemistry, Clark University, Worcester, Massachusetts, USA

Tugce Nihal Gevrek
Bogazici University, Istanbul, Turkey

Suhrit Ghosh
Polymer Science Unit, Indian Association for the Cultivation of Science, Kolkata 700032, India E-mail: psusg2@iacs.res.in

Ujjal Haldar
IISER Kolkata, India

Badri Nath Jha
IISER Kolkata, India

Mukundamurthy Kannan
IIT Madras, India

Rajeswari M. Kasi
Department of Chemistry, University of Connecticut, Storrs, CT 06269, USA E-mail: Kasi@ims.uconn.edu

Jayant Kumar
UMASS, Lowell, Massachusetts, USA

Rajeev Kumar
IISER Kolkata, India

Binoy Maiti
IISER Kolkata, India

Ramaswamy Nagarajan
Department of Plastics Engineering, University of Massachusetts Lowell, Lowell, MA 01854, USA E-mail: Ramaswamy_Nagarajan@uml.edu

Subhalakshmi Nagarajan
UMASS, Lowell, Massachusetts, USA

Chi Thanh Nguyen
Polymer Program, Institute of Materials Science, University of Connecticut, Storrs, Connecticut, USA

Brian Osborn
JSR Micro Inc., Sunnyvale, CA 94089, USA

Santu Sarkar
IISER Kolkata

Swaminathan Sivaram
Polymers and Advanced Materials Laboratory, National Chemical Laboratory, Pune 411008, India E-mail: s.sivaram@ncl.res.in

Amitav Sanyal
Department of Chemistry, Bogazici University, Bebek 34342, Istanbul, Turkey E-mail: amitav.sanyal@boun.edu.tr

Raja Shunmugam
Polymer Research Centre, Department of Chemistry, IISER-Kolkata, Mohanpur 741252, Nadia, India E-mail: sraja@iiserkol.ac.in

Venkatanarasimhan Swarnalatha
IIT Madras, India

Sanjio S. Zade
Department of Chemical Sciences, Indian Institute of Science Education and Research–Kolkata, Mohanpur 741252, Nadia, India E-mail: sanjiozade@iiserkol.ac.in

LIST OF ABBREVIATIONS AND SYMBOLS

ε	molar absorptivity
ε_o	permittivity of free space
ε_r	dielectric constant of the polymer
μ	field-effect mobility
μ_h	hole mobility
λ	wavelength
ω	frequency
σ	ionic conductivity
η	viscosity
χ	Flory-Huggins interaction parameters
μM	micromolar
$	dollar
°C	degree Celsius
°C/min	degree Celsius per minute
deg	degree
rads^{-1}	radian per second
2-PAM	2-pyridinealdoxime
3-MP	3-methyl pyrrole
A-co-B	copolymer of A and B
AAU	Anodic aluminium oxide
ABTS	2,2'-azino-bis(3-ethylbenzthiazoline-6-sulfonic acid) diammonium salt
ADMET	acyclic diene metathesis
AGET-ATRP	activators generated by electron transfer ATRP
AIBN	azobisisobutyronitrile
AIE	aggregation-induced emission
Al-OH	aluminol
AMM	activated monomer mechanism
AMPs	2-acrylamido-2-methylpropane sulfonate
ArF	argon fluoride laser
ARGET-ATRP	activators regenerated by electron transfer ATRP
ATP	attapulgite
ATR	attenuated total reflectance
ATRP	atom transfer radical polymerization

ATRPP	atom transfer radical precipitation polymerization
AuCs	gold clusters
AzVE	2-azidoethyl vinyl ether
BCD	block copolymer
BDT	5,6-bis(octyloxy)benzo-2,1,3-thiophene
BHJ	bulk-heteroconjugation
bipy	2,2′-bipyridyl
BLG-NCA	γ-benzyl-L-glutamate-N-carboxyanhydride
BMIM	1-butyl-3-methylimidazolium hexafluorophosphate
Bn-Glu	γ-benzyl-L-glutamate
Boc	*tert*-butyloxycarbonyl
BOD	bilirubin oxidase
BODIPY	boron-dipyrromethane
BPMODA	N,N-bis(2-pyridylmethyl)octadecylamide
BPPC	2,2-bis((2-bromo-2-methyl)propionatomethyl)propionyl chloride
bpy	5-(4-methoxystyryl)-5′-methyl-2,2′-bipyridine
BSA	bovine serum abunim
CA	chemically amplified
CCL	cone cross-linked
CD	circular dichroism
$CdCl_2$	cadmium chloride
$CDCl_3$	chloroform-d
CDP	4-cyano-4-(dodecylsulfanylthiocarbonyl)sulfanyl pentanoic acid
CdSe nps	cadmium selenide nanoparticles
CEVE	2-chloroethyl vinyl ether
CFRP	conventional free radical polymerization
CG	contact holes
CH_2Cl_2	dichloromethane
$CHCl_3$	chloroform
CHEF	chelation enhanced fluorescence
CL	ε-caprolactone
CNTs	carbon nanotubes
CO_2	carbon dioxide
COD	1,5-cyclooctadiene
CPs	conjugated polymers
CROP	cationic ring opening polymerization
CRP	controlled/living radical polymerization
CSA	10-camphosulfonic acid

CSOY	conjugated soybean oil
CTA	chain transfer agent
CWAs	chemical warfare agents
d	diameter
D	diffusivity
D/A	donar/acceptor
DBU	1,8-diazobicyclo[5.4.0]undec-7-ene
DCC	N,N′-Dicyclohexylcarbodiimide
DCCBMP	3-[3,5-bis(1-chloro-1-methylethyl)phenyl]-3-methylbutyl 2-bromo-2-methylpropionate
DCD	dicyclodiazo
DCE	5-*tert*-butyl-1,3-bis(1-methoxy-1-methylethyl)benzene
DCM	dichloromethane
DCP	dicyclohexyl fluorophosphates
DDS	drug delivery system
DECD	diethylchlorophosphate
DFP	diisopropyl fluorophosphates
DFT	density functional theory
DGEBA	bisphenol A diglycidyl ether
DLS	dynamic light scattering
DMAEMA	N,N-(dimethylamino)ethyl methacrylate
DMF	dimethylformamide
DMSO	dimethyl sulfoxide
DNA	deoxyribonucleic acid
DNQ	diazonaphthoquinone
DoF/DOF	depth of focus
DOX	doxorubicin hydrochloride
DP	double patterning
DPA	diphenylamine
DPCP	diphenylchlorophosphate
DPE	1,4-bis(1-phenylethenyl)benzene
DRS	dual reactive surfactant
DSA	directed self-assembly
DSC	differential scanning calorimetry
DSSC	dye sensitized solar cells
DTBP	2,6-di-tert-butylpyridine
DTE	1,1,-ditolylethylene
DTT	dithiothreitol
DUV	deep ultraviolet
DVB	divinylbenzene

E°	redox potential
E_g^{opt}	optical band gap
Ea	activation energy
e-beam	electron beam
eATRP	electrochemically mediated ATRP
EBDW	electron beam direct write
EEA	1-ethoxyethyl acrylate
EGDMA	ethylene glycol dimethacrylate
EHMA	2-ethylhexyl methacrylate
ELBL	electrostatic layer-by-layer
EOE	ethylidene bis(oxy-2,1-ethanediyl)ester
EP-PIB-EP	epoxy telechelic PIBs
EQE	external quantum efficiency
ESCAP	environmentally stable chemical amplified photoresist
ESIPT	excited-state intramolecular proton transfer
EtOH	ethyl alcohol
EUV	extreme ultraviolet
eV	electron volt
EVE	ethyl vinyl ether
F	quantum yield
FF	fill factor
FITC	fluorescein isothiocyanate
Fmoc	[(9-fluorenylmethyl)oxy]carbonyl
FPLC	fast protein liquid chromatography
FRET	fluorescence resonance energy transfer
FT-IR	Fourier transform infrared
G'	elastic modulus
G"	viscous modulus
GMA	glycidyl methacrylate
GPC	gel permeation chromatography
GSH	glutathione
GTP	group transfer polymerization
H_2O_2	hydrogen peroxide
Hb	hemoglobin
HCl	hydrochloric acid
HEMA	2-hydroxyethyl methacrylate
HEPES	(4-(2-hydroxyethyl)-1-piperazineethane sulfonic acid)
HFA	hexafluoro alcohol
HIF	high index fluids
HMDS	hexamethyldisilazane

HNT	halloysite nanotubes
HOES	4-(2-hydroxyethyl)styrene
HOMO	highly occupied molecular orbital
HPLC	high performance liquid chromatography
HRP	horseradish peroxidase
HSMMs	human skeletal muscle myoblast cells
HSQ	hydrogen silesquioxane
HUVECs	human umbilical vein endothelial cells
HVM	high volume manufacturing
IB	isobutylene
IC	integrated circuit
ICAR-ATRP	initiators for continuous activator regeneration ATRP
ICP-MS	inductively coupled mass spectroscopy
ICT	intermolecular charge transfer/internal charge transfer
ID	indene
IL	ionic liquid
ISC	intersystem crossing
ITO	indium tin oxide
IUPAC	International Union of Applied and Pure Chemistry
J	current density
J_{mpp}	minimum power point current
J_{sc}	short circuit current
KCl	potassium chloride
kDa	kilo Dalton
KeV	kilo electro volt
KIO_3	potassium iodate
L	thickness of the polymer
L-Ala	L-alanine
LAMA	lactobionamidoethyl methacrylate
L-Asp	L-aspartate
LBG	low band gap
LBL	layer-by-layer
LC	liquid crystalline
LCBBCs	liquid crystalline brush block copolymers
LCBCPs	liquid crystalline block copolymers
LCST	lower critical solution temperature
L_D	diffusion length
LE	litho etch
LED	light emission diode
LELE	litho-etch-litho-etch

LER	line edge roughness
LFLE	litho-freeze-litho-etch
L-Glu	L-Glutamic acid
LGS	liguin sulfonate
LLE	litho-litho-etch
L-Leu	L-Leucine
LMCT	ligand-to-metal charge transfer
LoD	load of detection
LOD	lactate oxidase
L-Phe	L-Phenylalanine
LPLO	litho-process-litho-etch
LS	low scattering/line-space
LUMO	lowest unoccupied molecular orbital
M	molar/monomer
M*	radical cation of the monomer
$[M]_0$	monomer concentration at time t = 0
$[M]$	monomer concentration at time t = t
$M^{-1}cm^{-1}$	per molar per milliliter
MAA	methylmethacrylate
MALDI-TOF	MS matrix-assisted laser desorption/ionization time-of-flight mass spectrometry
MAPPER	multiple picture pixel-by-pixel enhancement of resolution
MBA	N,N'-methylene bisacrylamide
MC	metal-centered
MDI	4,4'-methylenebis(phenylisocyanate)
MeCl	methyl chloride
MeOH	methanol
MEBDW	multiple e-beam direct write
MEEF	mask error enhancement factor
MEO_2MA	2-(2-methoxy ethoxy)ethyl methacrylate
$mgml^{-1}$	milligram per milliliter
MJLCP	mesogen-jacketed liquid crystalline polymer
MLCT	metal-to-ligand charge transfer
M_n	number average molecular weight
MNs	model networks
MPCS	2,5-bis[(4-methoxyphenyl)oxycarbonyl]styrene
MPEG	poly(ethylene glycol) monomethyl ether
MRI	magnetic resonance imaging
MTAB	tetradecyltrimethylammonium bromide
mW/cm^2	megawatt per centimeter square

M_w	weight average molecular weight
MWCNT	multi-walled carbon nanotubes
MWD	molecular weight distribution
N	degree of polymerization
N_2	nitrogen
N8HQ	norbornene functionalized 8-hydroxyquinoline
NA	numerical aperture of the optical system
NACE	nonaqueous capillary electrophoresis
NaCl	sodium chloride
NAM	normal amine mechanism
NAP	Naproxen
n-BuLi	n-butyllithium
$\text{n-Bu}_4\text{NCl}$	quaternary ammonium halide salt
NBCh9	norbornene bearing cholesterol with nine methylene spacer
NBCOOH	5-norbornerne-2-carboxylic acid
NBMPEG	α-methoxy-ω-norbornenyl-poly ethylene glycol
NBS	N-bromosuccinimide
NC	nanocapsules
NCA	N-carboxyanhydrides
NDI	diamino substituted naphthaimide
NDT	norbornene attached terpyridine
NDTH	NDT homopolymer
NHDFs	normal human dermal fibroplasts
NIL	nanoimprint lithography
NIPAM	N-isopropylacrylamide
NIR	near infrared
nm	nanometer
NMP	nitroxide-mediated polymerization/N-methyl-2-pyrrolidone
NMR	nuclear magnetic resonance
Nor-NHS	norbornene based n-hydroxy succinamide
Nor-Rh	norbornene derived rhodamine
Nor-Th	Thiol based norbornene
NPs	nanoparticles
NTD	negative tone develop
NWAs	nanowire arrays
NVS	1-(4-vinylstyryl)naphthalene
OAMA	oligo (2-aminoethyl methacrylate hydrochloride)
OEGMA	di(ethylene glycol) methacrylate polymer
OFETs	organic field-effect transistors
OLEDs	organic light emitting diodes

OoB	out-of-band
OPVs	organic photovoltaics
P3HT	poly(3-hexylthiophene)
P4VP	poly(4-vinylpyridine)
Pa	pascal
PA	polyacrylate
PAcOSt	p-acetoxy styrene
PAGE	polyacrylamide gel electrophoresis
PAGs	photoacid generators
PAM	polyacrylamide
PAN	poly acrylonitrile
PANI	poly aniline
P[Asp(DET)]	poly([N-(2-aminoethyl)-2-aminoethyl]-γ,ω-aspartamide
PAzEMA	poly(azidoethyl methacrylate)
PBLG	poly(γ-benzyl-L-glutamate)
PBnMA	poly(benzyl methacrylate)
PC$_{61}$BM	[6,6]-phenyl-C$_{61}$-butyric acid methyl ester
PCEM	poly(carbazole ethyl methacrylate)
PCL	poly(ε-Caprolactone)
PCMSt	poly(4-chloromethyl styrene)
PCT	photo-induced charge transfer
PDAC	poly(dimethyl diallylammonium chloride)
PDEGMMA	poly{methoxydi(ethylene glycol)methacrylate}
PDI	poly dispersive index
PDPA	poly(2-diisopropylamino)ethyl methacrylate
Pd(PPH$_3$)$_4$	Tetrakis(triphenylphosphine)palladium(0)
PDVB	poly divinylbenzene
PEDOT	poly(3,4-ethylenedioxy thiophene)
PEDOT:PSS	poly(3,4-ethylenedioxythiophene):poly(styrene slfonate)
PEEP	poly(ethyl-ethylene phosphate)
PEG	polyethylene glycol
PEGMA	poly ethylene glycol methyl ether methacrylate
PEHA	poly-2-ethylhexyl acrylate
PEO	poly(ethylene oxide)
PET	photo-induced electron transfer
PETI-ATRP	Pickering emulsion template interfacial atomic transfer radical polymerization
PFETs	polymer field-effect transistors
PF	polyfluorene
PFMA	poly(furfuryl methacrylate)

PFOP	poly{1,4-fullerene-alt-[1,4-dimethylene-2,5-bis(cyclohexylmethyl ether) phenylene]}
PFPNa	poly(9,9-bis(30-phosphatepropyl)fluorene-alt-1,4-phenylene) sodium salt
PFS	poly(ferrocenyldimethylsilane)
PGlyMA	poly(glycerol monomethacrylate)
PHEAA	poly(N-(2-hydroxyethyl)acrylamide)
PHEMA	poly(hydroxyethyl methacrylate)
pHOSt	p-hydroxy Styrene
PHPA	poly(N-3-hydroxypropyl)aspartamide
PHS	poly(4-hydroxystyrene)
PIB	polyisobutylene
PIB-A	polyisobutylene acrylate
PIB-Allyl-Cl	chloroallyl chain end-functionalized PIB
PIB-MA	polyisobutylene methacrylate
PIB-VE	polyisobutylene vinyl ether
PIB-b-PMMA	poly(isobutylene-b-methylmethacrylate)
PIB-b-PtVE	PIB-b-poly(tert-butyl vinyl ether)
PIB-b-PVA	PIB-b-poly(vinyl alcohol)
PIL	poly(ionic liquid)
pKa	acid dissociation constant
PLA	poly (DL-lactide)
PLG	poly (L-glutamate)
PLGA	poly(L-glutamic acid)
PLLA	poly(L-lactide)
PLLys	poly(L-lysine)
PMMA	poly(methyl methacrylate)
pMOSt	p-methoxy Styrene
PMPC	poly(2-methacryloyloxy)ethyl phosphoryl choline
PMPS	poly(γ-methacryloxypropyl trimethoxysilane)
PN8HQ	polyN8HQ
PNIPAM-SH	thiol terminated PNIPAM
PNVK	poly(N-vinyl carbazole)
POEGMA	poly(oxyethyleneglycol methacrylate)
POEM	poly(oxyethylene methacrylate)
POSS	polyhedral oligomeric silesquioxane
PPEGMEMA	poly[poly (ethylene glycol) methyl ether methacrylate]
PPEs	poly(phenyleneethylene)s
PPO	Poly propylene oxide
PPors	porphyrin-incorporated polymers

PPy   polypyrrole
PPV   P-phenylene vinylene
PRE   persistent radical effect
ps   picoseconds
PS   polystyrene
PSCs   polymer solar cells
PSEMA   poly(2-(succinyloxy)ethyl methacrylate)
PSPMA   poly(3-sulfopropyl methacrylate)
PS-POM   polystyrene-polyoxometallate
PSSA   polystyrene sulfonic acid
PSSNa   poly(styrene-4-sulfonate)
PT   polythiophene
PtBMA   poly(*tert*-butyl methacrylate)
PTBOCST   poly(tert-butoxycarbonyloxystyrene)
ptBOSt   p-tert-butoxy styrene
PTD   positive tone develop
PTEGMMA   poly{methoxytri(ethylene glycol)methacrylate}
pTSA   para-toluene sulfonic acid
PVAC   poly(vinylbenzydimethylhydroxyethyl ammonium
                chloride)
PVBA   poly(4-vinylbenzoic acid)
PVCs   photovoltaic cells
PVDM   poly(2-vinyl-4,4-dimethylazalactone)
PVFc   poly(vinylferrocene)s
PVP   poly(vinylphosphonic acid)
QCM   quartz crystal microbalance
QDs   quantum dots
q_{peg} peg crystalline temperature
r, nm   radius in nanometer
R   resolution
RAFT   reverse-addition fragmentation chain transfer
RC   regenerated cellulose
RCM   ring closing metathesis
rDA   retro-Diels–Alder
RDRP   reversible–deactivation radical polymerization
REBL   relective e-beam lithography
RI   refractive index
RITC-Dx   rhodamine B isothiocyanate–dextran
RMSt   R-methyl styrene
RNA   ribonucleic acid

ROMP	ring opening metathesis polymerization
ROP	ring opening polymerization
RT	room temperature
RT-PCR	reverse transcriptase polymerase chain reaction
RuO_4	ruthenium tetroxide
SADP	self-aligned double patterning
SAS	Sarin surrogate
SAv	streptavidin
SARA-ATRP	supplemental activators and reducing agents ATRP
SAW	surface acoustic wave
SAXS	small angle x-ray scattering
SBMA	sulfobetaine methacrylate
SBP	soybean peroxidase
SCATRP	surface confined ATRP
SCLC	space-charge-limited current
Scm^{-1}	Siemens per centimeter
SCO_2	supercritical carbon dioxide
sCT	salmon calcitonin
SCVPCP	self-condensing vinyl precipitation copolymerization
SDS	sodium dodecyl sulfate
SEC	size exclusion chromatography
SET-LRP	single-electron transfer living radical polymerization
SET-RAFT	single-electron transfer reversible addition fragmentation chain transfer polymerization
SI-ATRP	surface initiated ATRP
SI-CRP	surface initiated controlled radical polymerization
SI-NMP	surface initiated nitroxide mediated polymerization
SI-PIMP	surface initiated photo-initiated mediated polymerization
SI-RAFT	surface initiated RAFT
SmA	smectic A
SMCs	smooth muscle cells
SiNWAs	silicon nanowire arrays
SiNPS	silicon nanoparticles
SIPA	poly(styrene-*co*-isopropenyl acetate)
SPDP	N-succinimidyl 3-(2-pyridyldithio)propionate
SPR	surface plasmon resonance
SSS	sodium-4-styrene sulfonate
St	styrene
T	lifetime
tBA	tert-butylacrylate

TBDMES	4-[2-(*tert*-butyldimethylsiloxy)ethyl]styrene
TBOC	tert-butylcarbonate
t-BuOS	*t*-butoxystyrene
T_c	crystalline temperature
TCEP	tris(2-carboxyethyl)phosphine
TEAOH	tetraethylammonium hydroxide
TEM	transmission electron microscope
Tg	glass transition temperature
THF	tetrahydrofuran
TMAH	tetramethylammonium hydroxide
TMAPMA	3-(trimethoxysilyl)propyl methacrylate
TMS-CBM	trimethylsilyl carbamate
TPEs	thermoplastic elastomers
TPT	thiophene-phenylene-thiophene
TSA	toluenesulfonic acid
UCST	upper critical solution temperature
UV	ultraviolet
V	voltage
V_G	voltage applied at gate electrode
V_{mpp}	minimum power point voltage
V_{oc}	open circuit voltage
V_{th}	threshold voltage
VE	vinyl ether
VTT	volume thermal transistors
VUV	vaccum ultraviolet
WAXS	wide angle x-ray scattering
WGA	wheat germ agglutinin
XPS	x-ray photoelectron spectroscopy
XRD	x-ray differaction
$Zn(CH_3)_2$	dimethylzinc

PREFACE

In recent years, a large number of polymerization techniques have been reported. Though some functional monomers give direct polymerization, it is always a challenge to polymerize the different functional monomers.

Nowadays, the most challenging task for the researcher is not only to synthesize polymers in laboratories using different techniques of functional groups but also to sythesize those polymers on an industrial scale with their application. This book provides the solution to the problem cited above with a team of international experts addressing a wide range of topics with new synthetic techniques and their applications.

This book will be useful for chemists with no background of polymer science and will also function as a classroom text at the undergraduate as well as graduate level. The recent cutting-edge research will be valuable for chemists, chemical engineers, and many others.

—**Raja Shunmugam**

PART I
Controlled/Living Polymerizations of Functional Monomers

CHAPTER 1

SYNTHESIS OF FUNCTIONAL POLYMERS OF POLAR AND NONPOLAR MONOMERS BY LIVING AND/OR CONTROLLED POLYMERIZATION

SUMANA ROY CHOWDHURY and SWAMINATHAN SIVARAM

CONTENTS

1.1 SYNTHESIS OF FUNCTIONAL POLYMERS VIA LIVING AND/ OR CONTROLLED POLYMERIZATION

A living or controlled polymerization is defined as one that proceeds in the absence of or reduced incidence of chain transfer and termination reactions.[1] Well-defined polymers with low degrees of compositional heterogeneity can be prepared by this method. Ever since the first report of living polymerization by Szwarc,[1] living/controlled polymerization techniques such as anionic,[2] cationic,[3] group transfer polymerization (GTP),[4] and living/controlled free radical polymerization methods (atom–transfer polymerization (ATRP),[5] stable free radical polymerization (SFRP),[6] reversible addition fragmentation transfer (RAFT),[7] and nitroxide mediated polymerization (NMP)[8]) have emerged as powerful tools for the controlled synthesis of macromolecules. These methods of polymer synthesis have been well described in several recent books and compendiums.[9,10] One of the major goals of synthetic polymer chemistry is to prepare polymers with well-defined structures as well as adequate control over properties such as molecular weight (MW), molecular weight distribution (MWD), copolymer composition, microstructure, tacticity, chain-end functionality, and branching. Living/controlled polymerization is an attractive technique for the synthesis of useful macromolecules such as graft, block, star, α,ω-functional polymers, and macromonomers.[11] They find wide applications as interfacial agents in polymer blends, additives for oil industry, general-purpose resins, adhesives, impact modifiers, and as processing aids in textiles, advanced fibers, and so forth.[12]

1.2 LIVING ANIONIC POLYMERIZATION (LAP) OF NONPOLAR MONOMERS

One of the distinguishing features of a living anionic polymerization (LAP) is that after the monomer has been completely consumed, the polymer chains retain their activity. Further addition of monomer results in continued polymerization. It is, therefore, possible to synthesize well-defined block copolymers and introduce suitable functional groups at the end of the polymer chain by use of a suitable terminating agent. Although the efficiency of these reactions is often not always 100%, it still provides a useful methodology for the synthesis of chain-end functionalized polymers.[13,14]

LAP is best suited for vinyl/diene type monomers where polymerization takes place at the double bond and cyclic monomers, where, the polymerization proceeds by ring-opening. For anionic polymerization any substitution on the double bond should be electron withdrawing and inert to a carbanionic chain-end. If the substituents contain active hydrogen atoms, it is necessary to protect these groups suitably.[15] There exists a range of monomers that can be polymerized anionically without any transfer or termination reactions such as styrenes, dienes, epoxides, cyclic siloxanes, and lactones. The reactivity of monomers in anionic polymerization depends upon the pK_a of the conjugate acids of their propagating anions. The stability of the corresponding carbanionic chain-ends and the choice of a suitable initiator are also important. Usually, a good initiator for a monomer is one that has a pK_a value for its conjugate acid that is similar to the propagating carbanionic chain-ends.

Choice of an appropriate solvent is very important for LAP. Polymerization of styrene and diene monomers can be performed in either a hydrocarbon solvent, like benzene or toluene, or in a polar solvent, like tetrahydrofuran (THF). In a hydrocarbon solvent, both aggregated [(RMt)n] and un-aggregated (RMt) species can exist. On the other hand, in a polar solvent, in addition to the aggregated and un-aggregated species, free ions ($R^+ + Mt^-$), contact ion pairs (R^+,Mt^-), and solvent separated ion pairs ($R^+//Mt$) can also exists as shown in Scheme 1-1 (Winstein spectrum).[16,17]

$$(RMt)_n \rightleftharpoons n\ RMt \rightleftharpoons R^-Mt^+ \rightleftharpoons R\text{-}//\ Mt^+ \rightleftharpoons R^- + Mt^+$$

SCHEME 1-1   Winstein spectrum.

Each species in the Winstein spectrum can participate in the propagation. It must also be recognized that each species can polymerize with different rates of propagation. The major species that exists depends on the structure of the carbanion, nature of the solvent, temperature, and counterion. Additionally, when polar solvents are used there is a possibility of chain transfer reactions of the active initiator or the propagating species to the solvent. Therefore, polymerizations in polar solvents are carried out at low temperatures. In contrast, there are usually no side reactions at room temperature in hydrocarbon solvents (except for toluene).[17] However, since the carbanionic chain-ends remain aggregated in hydrocarbon solvents, addition of a small amount of Lewis base helps to dissociate the aggregates and makes the reaction more efficient, since the C–Li bond is more active in the un-aggregated species compared to the aggregates.[17] Aromatic hydrocarbon solvents generally reduce the degree of aggregation of the carbanionic chain-ends compared to aliphatic solvents like cyclohexane.[18]

Several classes of initiators can be used for LAP. Soluble aromatic radical anions were one of the earliest initiators to be studied.[1] Addition of sodium to naphthalene in THF produces a stable radical anion of naphthalene. The oxidation–reduction reaction of the naphthalene with sodium is reversible upon the removal of THF. This makes it necessary to use THF as the solvent. The naphthalene radical anion can react with styrene to give the radical anion of the styrene monomer (Scheme 1-2). These monomer radical anions can then dimerize rapidly giving rise to dianions, which are the initiating species.

$$Naph + Mt \rightarrow Naph^-.Mt^+$$

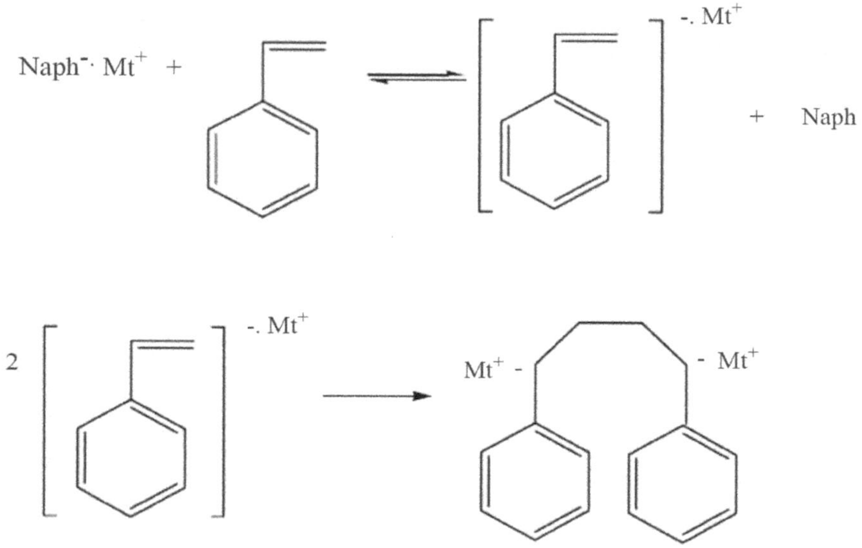

SCHEME 1-2   Synthesis of a styrene dianionic initiator using the naphthalene radical anion.

A variety of alkyllithium initiators, commercially available in hydrocarbon solvents, are widely used as initiators in LAP. The relative reactivity of these alkyllithium-intiators with respect to styrene and diene monomers is dependent on their degree of aggregation. Usually less aggregated initiators are more active initiating species. The order of reactivity of some commonly used alkyllithium initiators for the anionic polymerization of styrene are shown below.[19]

Methyllithium > *sec*-BuLi > *i*-PrLi > *i*-BuLi > *n*-BuLi > *t*-BuLi.

sec-BuLi is a commonly used initiator in the anionic polymerization of styrene and diene monomers in view of its high rates of initiation compared to propagation. Another common class of initiators are difunctional initiators which are used for the two-step synthesis of triblock copolymers. A widely employed hydrocarbon-soluble, dilithium initiator is synthesized from 1,3-*bis*-(1-phenylethylene)benzene.[20] The addition reaction of *sec*-BuLi to this compound to produce the dilithium initiator is shown in Scheme 1-3. This initiator is very active and produces polymers with narrow MWDs and controlled MWs.

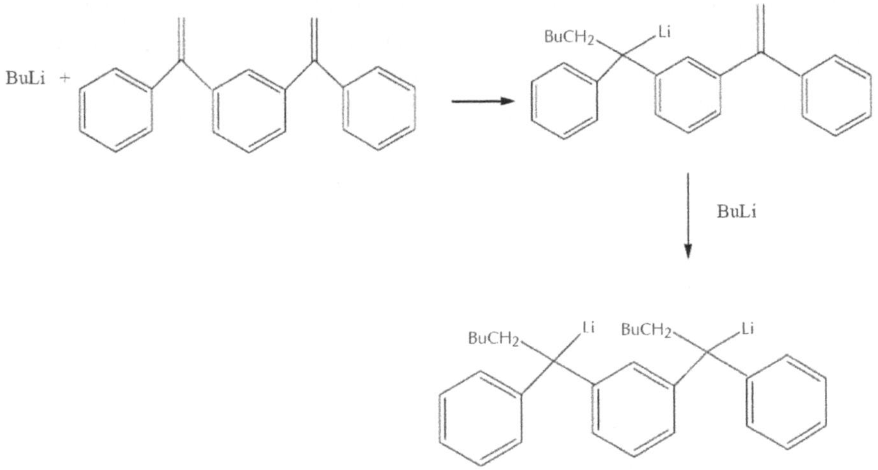

SCHEME 1-3   Synthesis of dilithium initiator from 1,3-*bis*(1-phenylethylene)-benzene.

Poly(styryl) lithium chains are associated as dimers in hydrocarbon solvents. The kinetic order of the propagation reactions is independent of the nature of the hydrocarbon solvent (aliphatic or aromatic), which is different from that observed for the initiation reactions.[17] Propagation kinetics of styrenes and dienes in hydrocarbon solvents exhibit a first order dependence on the monomer concentration. However, there is still a lack of clarity regarding the state of aggregation of the diene chain-ends in hydrocarbon solvents.[21,22]

One of the unique features of polydienes synthesized anionically in hydrocarbon solvents with lithium as the counterion is that they exhibit a high 1,4-microstructure. When isoprene is polymerized in benzene at room temperature with lithium as counterion, 70% 1,4-*cis*-; 24% 1,4-*trans*-; and 6% 3,4-microstructures are observed.[23] Use of an aliphatic solvent, like cyclohexane, results in an increase in the amount of *cis*-1,4-microstructure. However the total amount of 1,4-microstructure (*cis* + *trans*) always remains constant. Polymerization in hydrocarbon solvents at room temperature with lithium as counterion usually gives a high *cis*-1,4-microstructure. The microstructure of polydienes is known to be very sensitive to the nature of the solvent used. In presence of polar solvents with lithium as counterion, high 1,4-microstructure is no longer obtained; instead, high amounts of 1,2-polybutadiene and 3,4-polyisoprene were obtained.[17] As the size of the counterion increases, there is a greater tendency for formation of polydienes with a high 1,2-microstructure in nonpolar and polar solvents. In polar solvents, the polydienyl chain-ends are less aggregated. This apart, a polar solvent solvates a smaller counterion better. So, when lithium is present as the counterion, solvent-separated, ion-pairs exist which results in the delocalization of negative charge from the alpha carbon atom. This results in more charge on the gamma carbon atom and less on the alpha carbon atom. On the other hand, in hydrocarbon solvents the negative charge is more localized on the alpha carbon atom giving rise to high 1,4-microstructures. In conclusion, the stereochemistry of polydienes are strongly dependent on the nature of the solvent, counterion, and the presence of polar additives.[17]

1.2.1 SYNTHETIC APPROACHES TO FUNCTIONALIZATION OF LIVING ANIONIC CHAIN-ENDS

Rapid progress has been made towards the synthesis of chain-end and in-chain functionalized polymers using LAP. Anionic polymerization techniques, especially alkyllithium-initiated polymerizations of styrene and diene monomers, for the synthesis of well-defined in-chain and chain-end functionalized polymers has been an active area of research.[24–39] A large number of chain-end and in-chain functionalized polymers have been synthesized using LAP. Reaction of living chain-ends with a suitable electrophile results in end functionalized polymers.[40–49] However, adequate characterization of the degree of functionality is often lacking in the literature.

One of the methods that has been extensively used for the synthesis of well-defined chain-end functionalized polymers is by terminating the living polymer chain of styrene or diene, using a protected functional group on an alkylhalide molecule, via a nucleophilic displacement reaction (Scheme 1-4). Several chain-end functionalized polymers have been successfully prepared using this general method,[50a–d] for example, amino, mercapto, carboxy, and hydroxyl groups. Further, halogen-terminated polymers[50e] have also been prepared by reacting the living chain-ends with α,ω-dichloroalkanes in a controlled manner. Anionic polymers containing terminal monsaccharide groups[50f] have also been synthesized by carrying out the polymerization in THF at −78 °C, followed by termination of the living anionic chain-ends using benzyl chloride derivatives containing acetal protected monosaccharides, such as glucofuranose and fructopyranose. 4-Vinyl phenyl terminated polystyrenes and polyisoprenes have also been synthesized by terminating the living polymeric chain-ends using 4-(ω-haloalkyl)styrenes.[50g] Anhydride terminated polystyrene, poly(isoprene)s, poly(methyl methacrylate) (PMMA), and poly(vinylpyridine)s have been synthesized by terminating the living polymer chain-ends with di-tert-butyl maleate.[50h] Post polymerization pyrolysis reactions resulted in deprotection of the functional group at the chain-end to produce anhydride-terminated polymers. In case of polystyrene, polymerization was done in THF at −78 °C, to avoid competition between 1,2- and 1,4- addition to the enolate. DeSimone and co-workers[51a,b] developed a method based on the efficient reaction between living styrene and diene carbanionic chain-ends with silyl halides containing a perfluoralkyl group. This reaction occurs in the absence of competing side reaction and provides an excellent control over MW and MWD. In this method, the first step involves reaction of chlorosilane with an alkene, having a functional group X (perfluoroalkyl), using a platinum-catalyzed hydrosilylation reaction followed by the reaction of functionalized chlorosilane with the living chain-ends, resulting in chain-end functionalized polymers.

$$(CH_3)SiHCl \ + \quad \diagup\!\!\diagdown\!\!\diagup\!\!\diagdown{}^X \quad \xrightarrow{\text{Pt Catalyst}} \quad (CH_3)_2SiCl(CH_2)_3X$$

$$P\text{-}Li + (CH_3)_2SiCl(CH_2)_3X \quad \longrightarrow \quad P\text{-}Si(CH_3)_2(CH_2)_3X$$

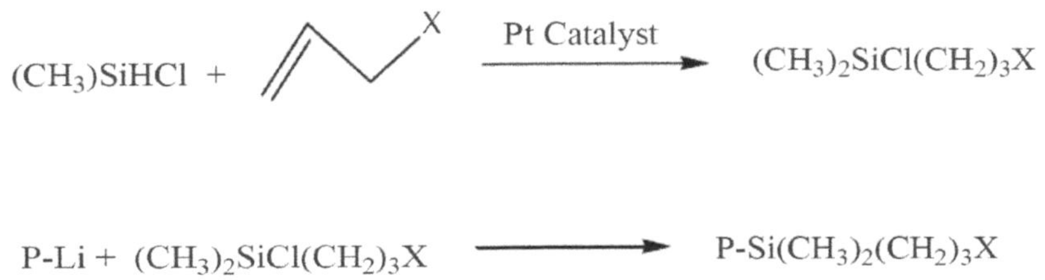

SCHEME 1-4 Use of functionalized chlorosilane for chain-end functionalization.

Another commonly used method for the synthesis of chain-end functionalized polymers is by the use of substituted 1,1'-diphenylethylene (DPE). Reactions of living polymer chain-ends occurs with functionalized DPE derivatives quantitatively and with an efficient rate of crossover.[52–57] One unique feature of this method is that it does not terminate the polymerization, instead it produces another carbanionic species which can initiate the polymerization of an added monomer unit (Scheme 1-5).

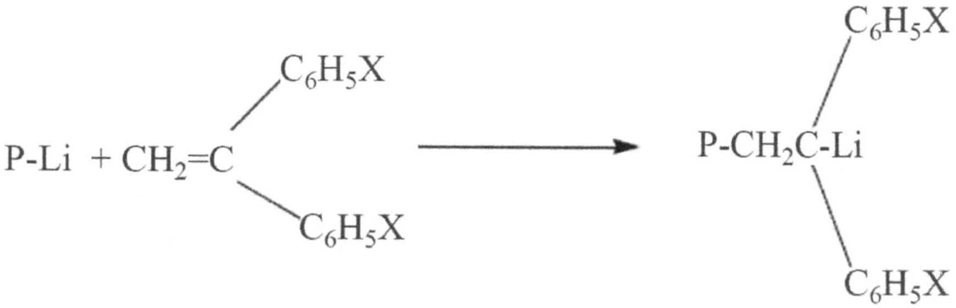

$$\text{P-Li} + CH_2{=}C\overset{\displaystyle C_6H_5X}{\underset{\displaystyle C_6H_5X}{\Big\langle}} \longrightarrow \text{P-CH}_2\overset{\displaystyle C_6H_5X}{\underset{\displaystyle C_6H_5X}{\overset{|}{\underset{|}{C}}}}\text{-Li}$$

SCHEME 1-5 Use of DPE for chain-end functionalization.

Recently, Quirk and co-workers[58a–h] reported a general method for the synthesis of chain-end functionalized polymers based on the combination of LAP and hydrosilylation chemistry (Scheme 1-6). In this method, the living polystyryllithium chain-ends are first terminated using chlorodimethylsilane, to produce silane-terminated polystyrene chains. The resulting ω-silyl hydride-functionalized polymer is then reacted with several substituted alkenes via post-polymerization hydrosilylation reaction to produce well-defined chain-end functionalized anionic polymers. Reactive functional groups are compatible with this method since the living chain-ends do not come in to contact with functional groups. This method has been successfully applied for the synthesis of amine, cyano, ethyl ether, acetate, and perfluoroalkyl chain-end functionalized polymers. The easy availability of a wide variety of substituted alkenes increases the versatility of this method.

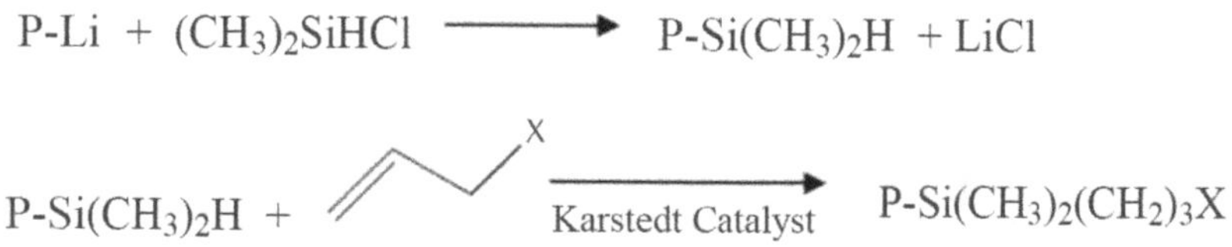

$$\text{P-Li} + (CH_3)_2\text{SiHCl} \longrightarrow \text{P-Si}(CH_3)_2\text{H} + \text{LiCl}$$

$$\text{P-Si}(CH_3)_2\text{H} + \underset{\text{Karstedt Catalyst}}{\overset{X}{\diagup\!\diagdown\!\diagup\!\diagdown}} \longrightarrow \text{P-Si}(CH_3)_2(CH_2)_3X$$

SCHEME 1-6 Chain-end functionalization using post reaction hydrosilylation chemistry.

Touris, Mays, and Hadjichristidis[59] reported the synthesis of new alkyl lithium initiators containing silyl-protected acetylene moieties

(5-triethylsilyl-4-pentynyllithium) (Scheme 1-7). The initiators were successfully employed for initiating the anionic polymerization of styrene, isoprene, and butadiene. The acetylene group was used as a versatile "click-able" site. Combination of anionic polymerization with click chemistry allowed the formation of a variety of complex architectures, such as cyclic and multiblock copolymers. An α,ω-heterotelechelic block copolymer of polystyrene and polyisoprene was synthesized using a sequential monomer addition approach with 5-triethylsilyl-4-pentynyllithium as an anionic initiator. The living polymer chain-end was reacted with 1,4-dibromobutane and sodium azide to obtain an azide functionality at the ω-chain end. Click reaction could then be carried out between the acetylene group at the α-chain end and the azide group at the ω-chain end to synthesize both block copolymer as well as cyclic copolymers at low dilutions.

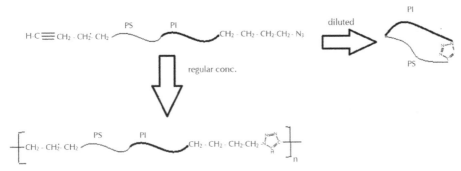

SCHEME 1-7   Use of post-polymerization alkyne-azide "click-chemistry".

Quirk and et al.[60] used a combination of anionic polymerization and thiol-ene click chemistry to produce ω-branch end-functionalized comb polystyrenes. ω-(Vinylbenzyl) polystyrene macromonomer was synthesized by *sec*-butyllithium-initiated polymerization of styrene followed by termination with 4-vinylbenzyl chloride. For the synthesis of α-4-pentenyl-ω-(*p*-vinylbenzyl) polystyrene macromonomer, an unsaturated initiator, 4-pentenyllithium, was used followed by termination with 4-vinylbenzyl chloride. Living anionic copolymerization of these two macromonomers produced well-defined comb polymers having pendant vinyl groups. The pendant vinyl groups were then subjected to thiol-ene click reaction to introduce a variety of functional groups onto the polymer backbone (Scheme 1-8).

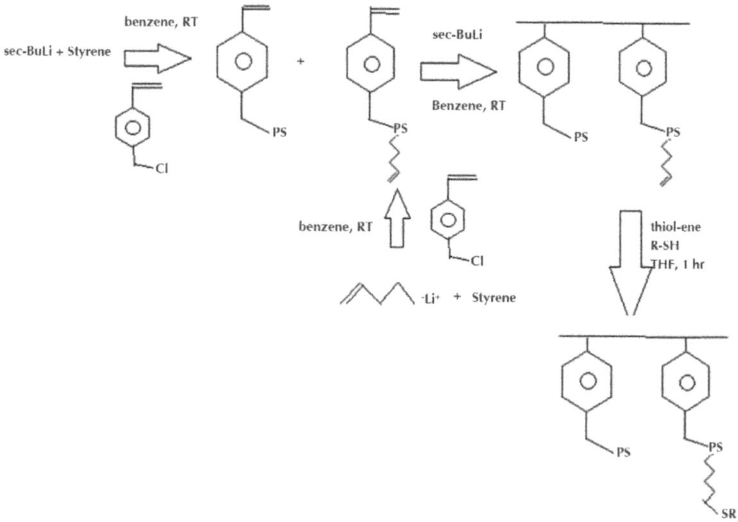

SCHEME 1-8   Synthesis of comb-like polystyrenes, using post-polymerization thiol-ene reactions.

1.2.2   IN-CHAIN FUNCTIONALIZATION VIA ANIONIC POLYMERIZATION

In-chain functionalized polymers are polymers with a pendant functional group along the polymer chain. The most obvious way to synthesize such polymers is by use of functionalized monomers. However, care must be taken to avoid reaction between the functional groups and the growing chain-ends. One way to accomplish this is to protect the functional groups. These groups can then be removed easily by post polymerization.[15] In a situation where the functional groups are nonreactive, the monomers can be directly polymerized without a protection step. A variety of functional monomers based on styrene have been polymerized by LAP.[61–90] Nakahama and Hirao[15] synthesized several styrenic in-chain functional polymers using a protection–deprotection strategy (Scheme 1-9).

LAP of styrene monomers containing many interesting in-chain substituents have been reported such as, cyclobutane (high temperatue crosslinkability),[91] adamantates (high temperature stability),[92] and diphenylamine (chromophore).[93] These groups are shown to be stable to the active anionic chain-ends under specific reaction conditions. Another strategy for anionic polymerization of functionalized monomers, was developed by Hirao and co-workers.[94–96] A variety of electron withdrawing groups, such as,

N-alkyl- and N-arylimines,[97] N,N-dialkylamides,[98] 2-oxazoline,[99] alkyl,[100] aryl esters,[101] and nitrile[102] were introduced.

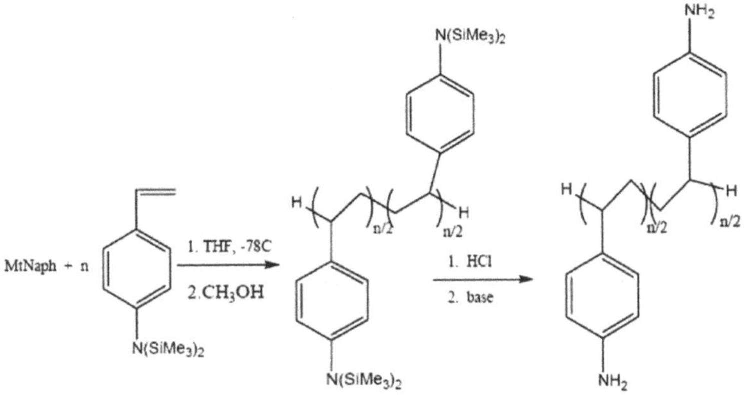

SCHEME 1-9 Synthesis of in-chain functionalized polymers, using protection–deprotection strategy.

Quirk and Chowdhury[58e] reported a general method for the synthesis of well-defined in-chain functionalized anionic polymers. This method combines anionic polymerization with hydrosilylation chemistry for the synthesis of controlled in-chain functional polymers (Scheme 1-10a and b). Living poly-styryllithium chain-ends in a hydrocarbon solvent was reacted with dichloromethylsilane at room temperature. A slight excess of polystyryllithium was used and these chains were terminated using ethylene oxide to produce a fraction of hydroxyl terminated polymer chains. The silyl hydride-functionalized polymer was separated from the hydroxyl terminated polymer by using column chromatography. In-chain functionalization of the silyl hydride groups with allyl cyanide was accomplished via hydrosilylation to produce well-defined polymers containing a functional group in the polymer chain.

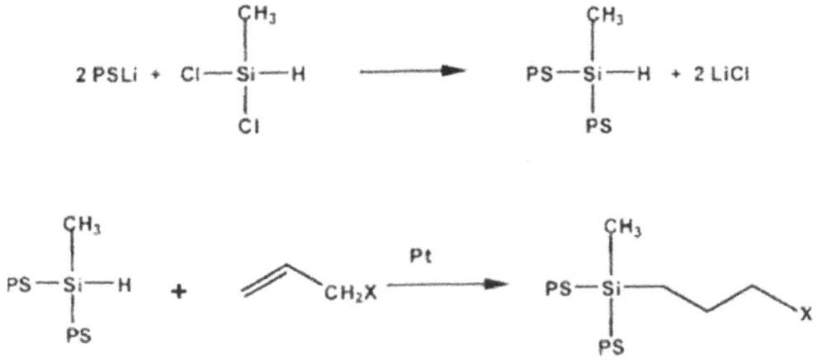

SCHEME 1-10 Synthesis of in-chain functionalized polymers.

Anionic polymerization of a series of 2-trialkoxysilyl-1,3-butadienes have been carried out in THF at −78 °C by Takenaka et al.[103] resulting in silylalkoxy functional polydienes. It was observed that the polymerization of 2-triisopropylsilyl-1,3-butadiene, proceeded exclusively to produce a cis-1,4-microstructure in THF solvent at −78 °C, in the presence of various counterions, such as, Li⁺, Na⁺, or K⁺ (Scheme 1-11).

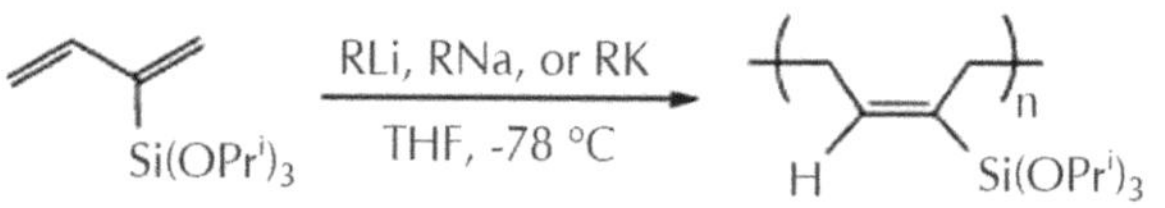

SCHEME 1-11   Controlled anionic polymerization of 2-trialkoxysilyl-1,3-butadienes.

Another interesting example is that of anionic polymerization of 1,3-butadiene, containing a bulky adamantyl group in the 2-position.[104] A high cis-1,4 (>96%) was observed on polymerization with s-BuLi in a hydrocarbon solvent. Surprisingly, a high cis-1,4 (>88%) was also observed on polymerization in THF at −78 °C. These results show that the presence of a bulky group in the 2-position results in high amounts of cis-1,4 addition.

1.3   LIVING/CONTROLLED POLYMERIZATION OF POLAR MONOMERS

1.3.1   LIVING ANIONIC POLYMERIZATION (LAP) OF METHYL METHACRYLATE (MMA)

Conventional anionic initiators such as metal alkyls when used for alkyl (meth)acrylate monomers generally yield polymers with broad MWD and in low conversions.[105] During polymerization the ester group of the monomer participates in the solvation of the counterion, thereby, becoming more susceptible to nucleophilic attack. Nucleophilic attack of the polar ester group can take place either during initiation or during propagation. Many such secondary reactions have been proposed in the literature.[105b,c] These are shown in Scheme 1-12.

Controlled anionic polymerization of acrylates pose even greater challenges than methacrylates because of higher reactivity of the carbonyl group towards nucleophiles and the presence of acidic hydrogen at the α-position to the carbonyl group.[106]

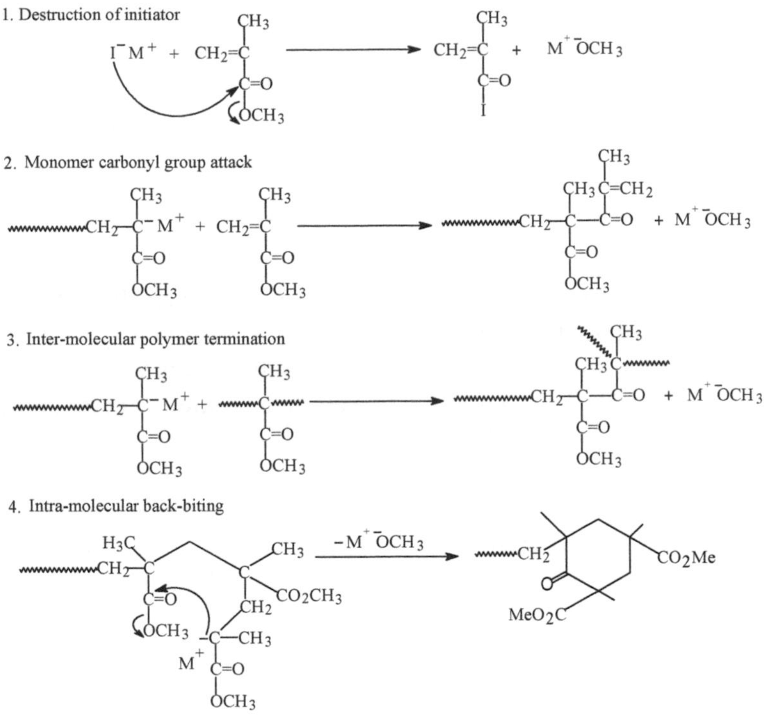

SCHEME 1-12   Secondary reactions in MMA polymerization.

The propagating centers in the anionic polymerization of alkyl (meth) acrylates are ester enolate anions which are stabilized through aggregation. The existence of ester enolates in equilibrium with aggregated form also poses a problem in controlling the living nature polymerization of alkyl (meth)acrylates.[107] Such aggregates in alkyl (meth)acrylate polymerization also influence the MWD of the polymer.[107c] A slow exchange between aggregated and nonaggregated ion pairs leading to the broadening of MWD was observed.[107c] Hence, controlling side reactions as well as aggregation–equilibrium dynamics assumes significant importance in achieving an ideal LAP of alkyl (meth)acrylates.

One of the earliest studies in the anionic polymerization of methacrylates was due to Anderson et al.[108] Diphenylhexyllithium was used as the initiator at −78 °C (Scheme 1-13). At these temperatures the bulky initiator does not react with the ester groups on the monomers and the backbiting reaction is absent. MW control and low MWDs ($M_w/M_n \leq 1.2$) were obtained.

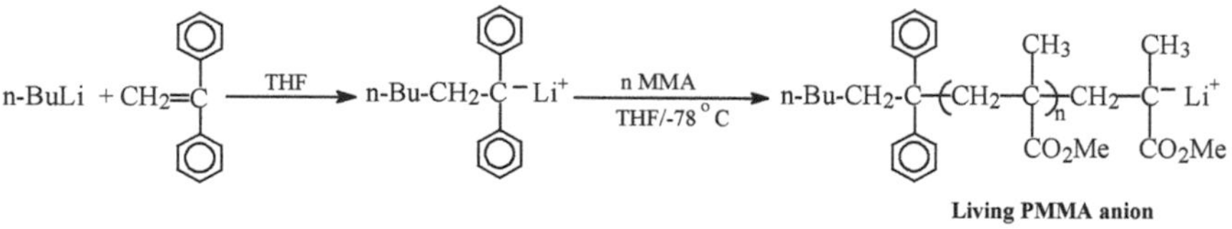

SCHEME 1-13   Anionic polymerization of MMA using 1,1′-diphenylhexyllithium as initiator.

Anionic polymerization of alkyl (meth)acrylates were best performed using bulky monofunctional initiator such as 1,1′-diphenylhexyl, 1,1′-diphenylmethyl, trityl, α-methylstyryl salts, or metalated esters with bulky counterion ($Cs^+ > K^+ > Na^+ > Li^+$) in polar solvent (e.g., dimethyl ether and THF) at low temperatures (<−65 °C).

With the objective of improving the control on chain length and MWD several ligands capable of coordinating with the enolate ion pairs have been explored.[2b] Teyssie et al.[109] classified the coordination of ligands with enolate ion pairs into (1) σ-type coordination with ligands such as crown ethers,[110] cryptands,[111a,b] and tertiary amines;[112a–d] (2) μ-type coordination with aluminum alkyls,[113a–d] alkali metal salts of alkoxide,[114a–c] halides,[115a,b] and perchlorate;[116a,b] and (3) σ,μ-type coordination with alkoxyalkoxides,[117a–d] aminoalkoxides,[118] and silanolates.[119] Ballard et al.[111c] found that bulky dialkyl aluminum phenolate additives improve the anionic polymerization of acrylic monomers in terms of better control (Scheme 1-14).

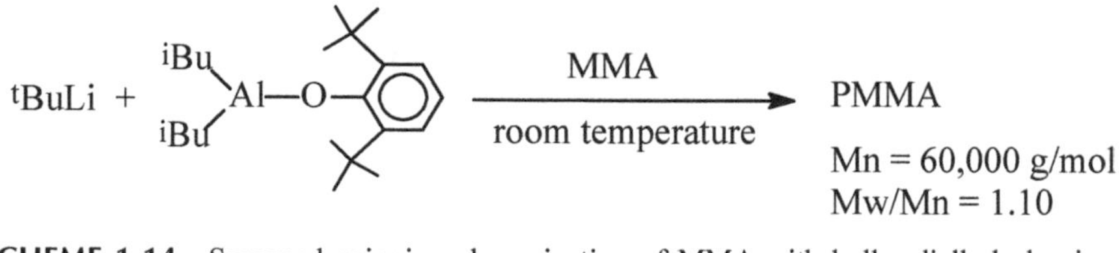

SCHEME 1-14   Screened anionic polymerization of MMA with bulky dialkyl aluminum phenolate as additive.

Metal-free initiators elicited interest, especially, on account of their ability to polymerize primary acrylates at room temperature. Metal-free carbon nucleophiles and resonance stabilized tertiary carbanions with tetrabutylammonium ion as countercations, act as initiators for the controlled polymerization of alkyl acrylates. Reetz and co-workers[120,121] first used metal-free carbanions as initiators (**1a**, Scheme 1-15) for the controlled anionic polymerization of *n*-butyl acrylate to obtain relatively narrow MWD polymers at room temperature. However, side reactions such as end group cyclization and Hoffmann elimination were observed.[122]

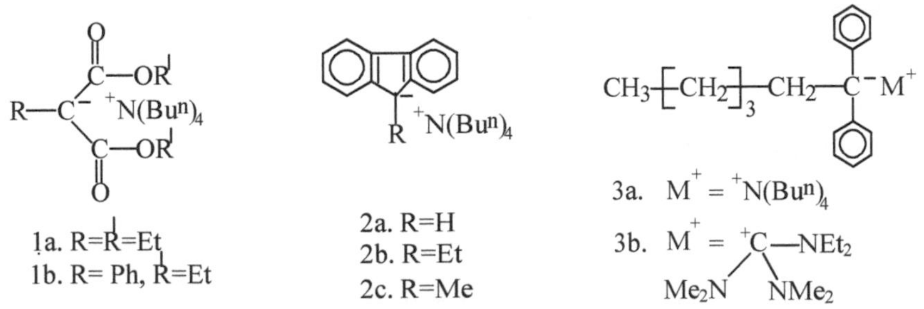

la. R=R'=Et
1b. R= Ph, R'=Et

2a. R=H
2b. R=Et
2c. R=Me

3a. $M^+ = {}^+N(Bu^n)_4$

3b. $M^+ = $ (Me$_2$N, NMe$_2$, $^+C-NEt_2$)

SCHEME 1-15   Metal-free carbanions used for alkyl (methyl)acrylate polymerization.

Quirk and Bidinger[123] used Bu_4N^+salt of 9-methylfluorene (**2c**, Scheme 1-15) to initiate polymerization of MMA in THF at ambient temperature. At very low initiator concentration, they obtained PMMA with a broad MWD ($M_w/M_n = 2.16$) in low yield (24%). At higher initiator concentration (**2b**, Scheme 1-4), Reetz et al.[124] observed a slow initiation at room temperature. The reason for a slow initiation was attributed to both aggregation and steric shielding of bulky nonmetal cation. Baskaran et al.[125] performed anionic polymerization of methyl acrylate, *n*-butyl acrylate, *t*-butyl acrylate, and MMA using various carbanions with tetrabutylammonium counterion (**1b**, **2a**, and **2b**, Scheme 1-15) in THF at 30 °C. The initiator efficiency was low and polymers obtained were characterized by broad MWDs.

Baskaran et al.[125] also studied the effect of metal and nonmetal counterions on the anionic polymerization of MMA in THF. They performed anionic polymerization of MMA in the presence of tetrabutylammonium, tetramethyldiethylguanidinium (initiator **3a**, **3b**, Scheme 1-15), and lithium counterion using 1,1'-diphenylhexyl anion as initiator at −40 °C. They reported that the polymerization in the presence of nonmetal counterion is very fast and conversion is quantitative within 2 min; however, the obtained PMMAs had broad/bimodal distribution with low initiator efficiency.

Webster[126] found that the size of the counterion is more important than the fact that it is nonmetallic. Potassium dimethyl malonate/18-crown-6 polymerizes MMA at 25–60 °C to give quantitative yields of PMMA, with MWD 1.5–1.9. Excess malonate or methanol lowered the MW of the PMMA but did not shut down the polymerization. They also postulated that a hydrogenbonded version of the Quirk intermediate (in GTP) is stabilizing the enolate ends (Scheme 1-16).

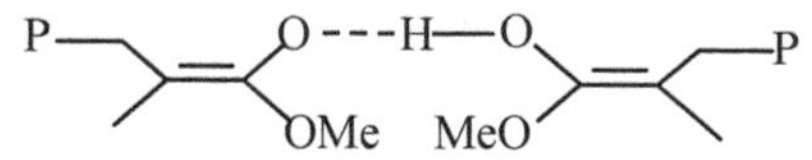

SCHEME 1-16   Hydrogen-bonded version of the stabilizing enolate ends.

Pietzonka and Seebach[127] used P$_4$ base (Scheme 1-17) as an initiator for the anionic polymerization of MMA.

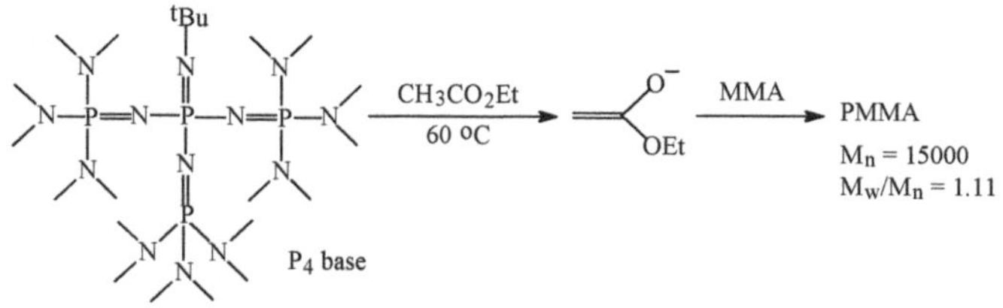

SCHEME 1-17   Anionic polymerization of MMA with P$_4$ base as an initiator.

Good control of MW and MWDs are obtained at temperatures up to 60 °C. The experimental MWs are higher than those expected by the amount of P$_4$ base used. Muller and Baskaran[128] confirmed these results using P$_5$ counterion at 20 °C (Scheme 1-18).

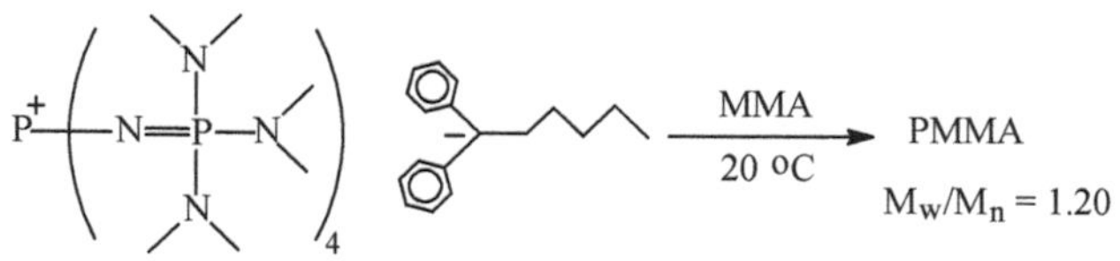

SCHEME 1-18   Anionic polymerization of MMA with P$_5$ counterion.

Muller et al.[129] used the related *bis*(triphenylposphoranylidene) ammonium ion (PNP) as a counterion (Scheme 1-19) for polymerization of MMA at 0 °C.

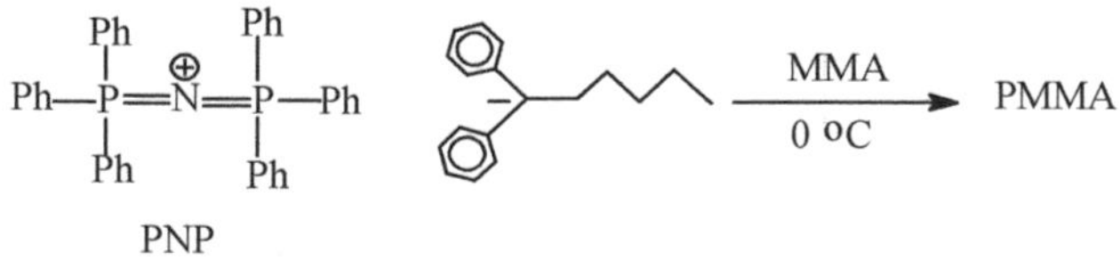

SCHEME 1-19   Anionic polymerization of MMA with PNP counterion.

Zagata and Hogen-Esch[130] as well as Baskaran and Muller[131] used a tetraphenylphosphonium counterion for controlled anionic polymerization of MMA at low temperatures (+20 to −80 °C).

The historical development and strategies adapted by several research groups for the living and controlled anionic polymerization of alkyl (meth) acrylates is discussed in detail in recent reviews.[2a–c]

1.3.2   GROUP TRANSFER POLYMERIZATION (GTP) OF METHYL METHACRYLATE (MMA)

GTP is a controlled conjugate addition of organosilicon compounds such as silyl ketene acetals to acrylate monomers at ambient or elevated temperatures in the presence of either a nucleophilic or electrophilic catalyst.[4a] One of the most important features of GTP is the formation of isolable, well-characterized silyl ketene acetal ended polymers (Scheme 1-20) whose reactive end groups can be converted into other functional groups. Addition of a new monomer at this point starts chain growth again to produce a block polymer[4b] or quenching with a proton source gives the silicon-free polymer. Since the first report[4] of GTP of methyl methacrylate (MMA) in 1983, over one hundred different methacrylates have been polymerized by this method to form polymers with controlled MWs and narrow MWDs.

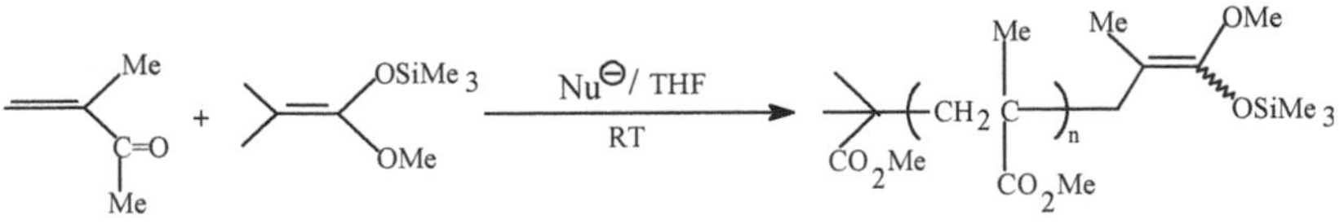

MMA (4)   MTS (5)   Silyl ketene acetal-ended PMMA (6)

SCHEME 1-20   Nucleophile catalyzed GTP of MMA

To obtain polymers with narrow MWDs in a living polymerization the rate of initiation must be faster or closer to the rate of propagation ($k_i \geq k_p$). This can be accomplished if the structure of the initiator is similar to that of the growing chain-end. The preferred initiator for GTP is silyl ketene acetal, namely, 1-methoxy-2-methyl-1-propenoxy trimethylsilane (MTS).

For nucleophile assisted GTP two mechanisms have been proposed, namely the associative pathway (Scheme 1-21) and the dissociative pathway (Scheme 1-22). In the associative mechanism the silyl ketene acetal group is activated by complexation with catalyst for addition to monomer. The silyl group transfers to incoming monomer and remains on the same polymer chain during the polymerization step. Silyl end group exchange is presumed to occur by some unknown process. The equilibrium rate for catalyst complex formation must be fast to insure MW control and low MWD.

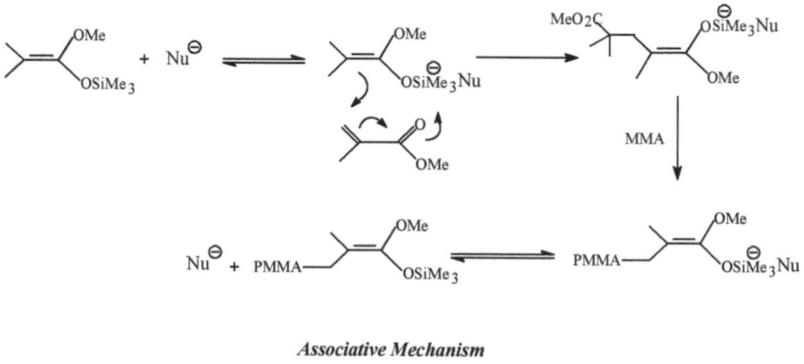

Associative Mechanism

SCHEME 1-21   Associative mechanism in GTP.

In the dissociative route the nucleophilic catalyst complexes with the silyl ketene acetal end groups and generates a reactive enolate end that adds monomer in a reversible cleavage step.

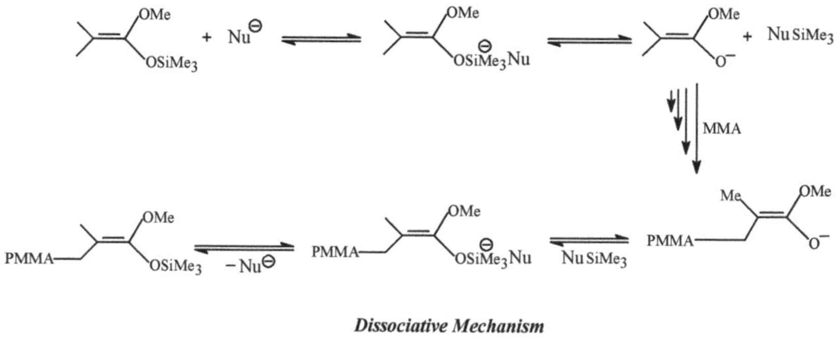

Dissociative Mechanism

SCHEME 1-22   Dissociative mechanism in GTP.

The enolate end groups are then capped by R_3SiNu to regenerate silyl ketene acetal ends. Since low MWD and controlled MW polymer is obtained at low catalyst concentrations, the equilibrium generating enolate ends must be much faster than the rate of polymerization.

In associative mechanism increasing the amount of catalyst should merely increase the rate of reaction. However, in the dissociative mechanism increasing the amount of catalyst may generate more enolate than that can be stabilized through complexation with the existing silyl ketene acetal end groups can be combined before the addition of monomer. Studies by Quirk and Kim[132] have shown that significant amounts (~ 40%) of silyl group exchange are observed during polymerization. A mechanism involving an enolate anionic propagating species reversibly complexed with the silyl ketene acetal chain-end has been proposed as shown in Scheme 1-23. Also, this ester enolate anion reacts with monomer to produce polymer. This mechanism supports the intermediacy of enolate anions as active center in GTP.

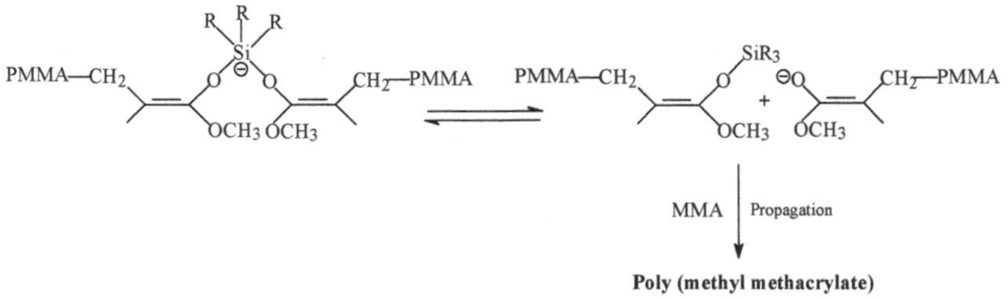

SCHEME 1-23   Intermolecular equilibrium of dissociated enolate anion with silyl ketene acetal.

Except for the low temperature exchange studies, no other evidence supports the associative mechanism. Based on the lack of exchange of added silyl fluoride with silyl ketene acetal ends it looks as though fluoride and bifluoride catalysts operate by irreversible generation of ester enolate chain-ends. On the other hand carboxylate catalysts appear to operate by a reversible generation of ester enolate ends as evidenced by rapid exchange of silyl acetate with silyl ketene acetal ends.

Nucleophile assisted GTP inevitably involves participation of enolate anion as intermediate during propagation. The most active catalysts are fluorides and bifluorides. At above ambient temperatures carboxylates and bicarboxylates are preferred.[133] However, the relative efficiency of a catalyst strongly depends on the corresponding acidity pK_a value of the carbon acids from which it is derived. The efficiency increases with increasing nucleophilicity of the catalyst. In general, bifluoride is more reactive than oxyanions which are more reactive than bioxyanions.

Backbiting reaction (cyclization) is the major termination pathway[134] in GTP. Banderman and Sitz[135] found that the polydispersity of PMMA obtained in the nucleophile catalyzed GTP increases with increasing concentration of catalyst. The termination reaction in GTP becomes insignificant when the catalyst concentration is kept low. If one assumes a dissociative mechanism, the large uncreative counterion may work better by slowing down the rate of backbiting termination and formation of ketenes[136] from ester enolate ends. In addition, in the dissociative process the large counterions would favor complex formation of enolate polymer ends with silyl ketene acetal ends. The backbiting termination reaction is more pronounced for stronger nucleophilic catalysts than for weaker ones. Thus, the concentration and nature of catalyst strongly determine the "living character" of GTP.

MWs in the range of 20,000 are easily obtained for GTP. However, preparation of higher MW polymer is somewhat more difficult. During polymerization, especially, at higher temperatures, the resulting polymer will contain upto 30% dead ends, as a consequence of backbiting reaction of enolate chain-ends and/or reaction with protic impurities.[137] Brittain and Dicker[134] found that at the trimer stage (DP = 3) the rate of backbiting is ten times higher than at DP > 3. Thus, one should start GTP with more than three equivalents of monomer to by-pass the trimer stage quickly.

Acrylates polymerize two orders of magnitude faster than methacrylates by anion catalysed GTP. However, the polymerization dies after reaching a MW of about 10,000. During the anion catalyzed polymerization of acrylates the silyl ketene acetal end groups migrate to internal positions. These ketene acetals are too hindered to act as initiators for branch formation.[138] Also, due to the presence of α-hydrogen in these monomers, side reaction occurs. The living polymerization of acrylates by GTP does proceed under Lewis acid catalysis.[139] $ZnCl_2$ or $ZnBr_2$ are effective but require concentrations of catalyst at a level of 10% based on monomer. R_2AlCl works at lower levels. However, $HgCl_2$ activated by TMS iodide is the best Lewis acid system and gives living acrylate polymers at low catalyst level.[140] Lewis acids are believed to activate monomers by coordination with carbonyl oxygen of acrylates.

In GTP the backbiting termination reaction turns into a chain transfer process. The backbiting reaction of chain-ends produces a cyclohexanone derivative and a methoxide ion. In classical anionic polymerization this terminates the chain. In GTP the methoxide can react with the latent silyl ketene acetal ends to regenerate enolate ends for further polymerization (Scheme 1-24).

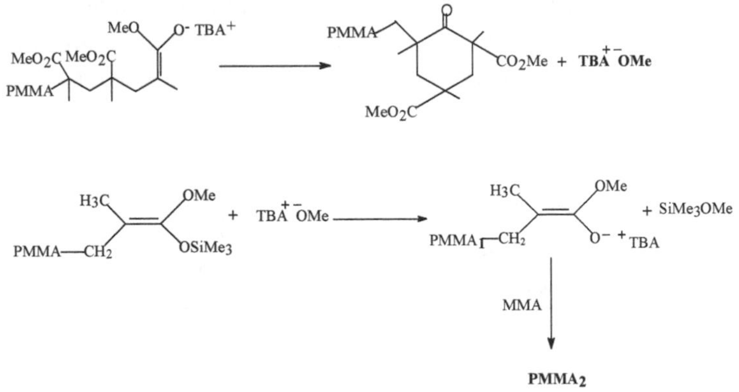

SCHEME 1-24   Termination and chain transfer reaction in MMA polymerization.

For acrylate polymerization by GTP[4c] the cyclohexanone that results from backbiting is an α-ketoester with active hydrogen that reacts with the methoxide, thus, preventing it from regenerating the enolate ion (Scheme 1-25).

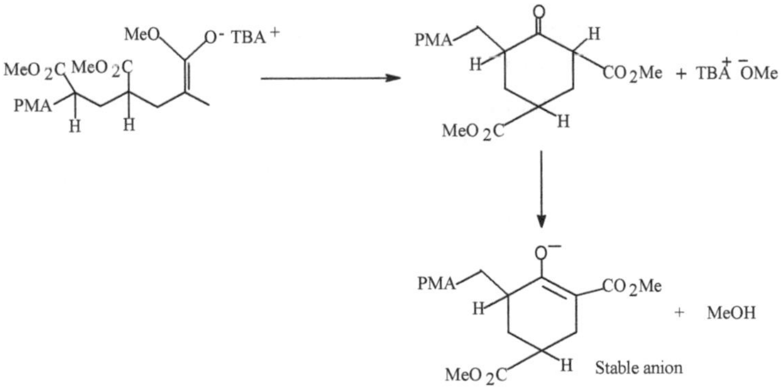

SCHEME 1-25   Termination reaction in acrylate polymerization via chain-end cyclization.

In view of the "living" nature of GTP, the method is amenable for the synthesis of well-defined random, block copolymers including amphiphilic block copolymers containing poly methyacrylic acid (PMAA) segments[141–144] by using protected monomers (Scheme 1-26), graft[145] and star-branched polymers,[146] networks,[147] macromonomers,[148] functional polymers[149] including end-functionalized polymers,[150] and telechelics.[151] Hyperbranched methacrylates were also synthesized[152] by self-condensing group transfer polymerization (SCGTP). Also, GTP can be advantageously carried out in the absence of a solvent to prepare linear polymers and randomly cross-linked networks.[153] Very recently GTP has also been performed in ionic liquids medium.[154]

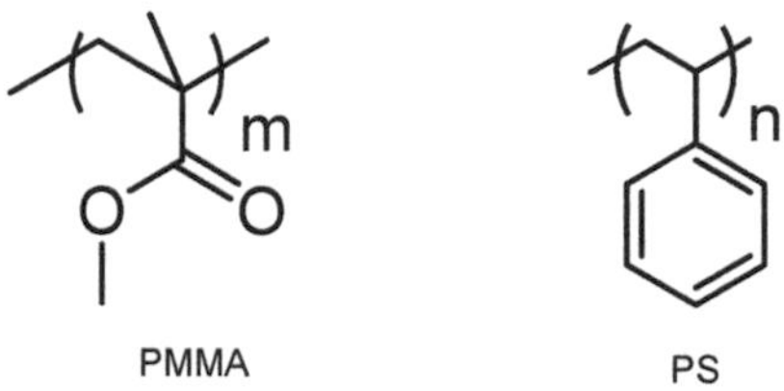

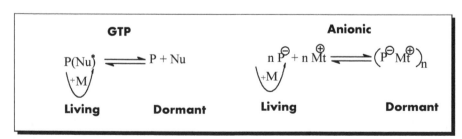

PMMA   PS

SCHEME 1-26   Chemical structure of PMMA and PS.

Known anionic initiators (ester enolates) for MMA can act as catalysts for GTP. Other factors such as, the need for large unreactive counterions, induction periods, use of living enhancing agents, and so forth support the dissociative process in GTP.[4] LAP works better at very low temperature (−78 °C) whereas GTP works at ambient or elevated temperatures. The "living nature" of GTP depends on the nature and concentration of catalyst, possibly because of the participation of enolate anions in the propagation. Several kinetic and mechanistic studies that are currently available convincingly suggest that the nucleophile catalyzed GTP of alkyl (meth)acrylate is merely a hypervalent silicon mediated anionic polymerization. Scheme 1-27 shows the dynamic equilibrium between active species and dormant species in GTP as well as in LAP. Several experimental evidences suggest that GTP is a sub-set of anionic polymerization, with similar active centers (ester enolates) and differ only with respect to the relative concentration of active centers (Table 1-1).

SCHEME 1-27   Dynamic equilibrium between active and dormant species.

TABLE 1.2 Comparison of features of GTP and IAP of MMA in THF

Entry	Parameter	GTP	Anionic
1	Chain life times (s)	$10–10^6$	$10–10^6$
2	k_p/k_t (mol/L)	250	8
3	Termination reaction	Backbiting	Backbiting
4	Chain transfer reaction	With carbon acids of $16 <$ $pK_a < 25$	With carbon acids of $16 <$ $pK_a < 25$
5	Activation energy, E_a (kJ/mol)	16.9 (TASHF$_2$)	19.5 (Cs$^+$)
6	Syndiotacticity of PMMA	(~56% (TASHF$_2$); ~54% (N$^+$Bu$_4$); ~35% (Cs$^+$)	(~61% (Li$^+$); ~56% (N$^+$Bu$_4$); ~57% (P$^+$Ph$_4$); ~34–38% (Cs$^+$)
7	Frequency exponent, log A	6.8 (TASHF$_2$)	7.3 (Cs$^+$)
8	Reactivity ratios	MMA/tBMA: 4.59:0.16 MMA/nBMA: 1.76:0.67 MMA/DMA: 1.66:0.48 MMA/EA: 0.01:14.15 EMA/MA: 0.01:14.41	MMA/tBMA: 35:0.43 MMA/nBMA: 1.04:0.81 MMA/DMA: 1.5:0.72 MMA/EA: 0.11:9.67 EMA/MA: 0.20:3.51

1.4 FUNCTIONAL POLY(METHYL METHACRYLATE)S (PMMAS) VIA GROUP TRANSFER POLYMERIZATION (GTP)

End-functional polymers and macromonomer synthesis can be achieved by two different strategies, either by deactivation of the living chain-end with a suitable electrophile or by initiation of the living polymerization with a functional initiator. Depending on the functional group and the choice of polymerization methods, protection of the end group may or may not be necessary. The use of functional initiators in living polymerization methods like GTP and anionic technique ensures that each polymer chain contains one functional group. However, initiators of this type are limited, especially, for anionic polymerizations. GTP, by virtue of the unique protected chain-end structure, is more tolerant towards functional groups. Functional anionic initiators often exhibit limited solubility in hydrocarbon solvents. On the contrary, GTP initiators are readily soluble most common solvents used for polymerization.

A drawback of the termination route for the preparation of end-functional polymers is that any polymer chain that has already been terminated will not react with the electrophile, thereby, limiting quantitative functionalization. Since living chain-ends have finite kinetic life times, quantitative end functionalization by termination of a chain-end is a formidable challenge.

Polymers with terminal hydroxyl or carboxyl groups are readily prepared[138] by using functional initiators [(2-methyl-1-[2-(trimethylsiloxy) ethoxy]-1-propenyl)oxy] trimethylsilane and 1,1'-bis(trimethylsiloxy)-2-methylpropene-1 followed by hydrolysis of the resulting polymer in refluxing methanolic tetrabutyl ammonium fluoride (TBAF) (Scheme 1-28). The high degree of monofunctionality of the resulting polymer was determined using HPLC.[155]

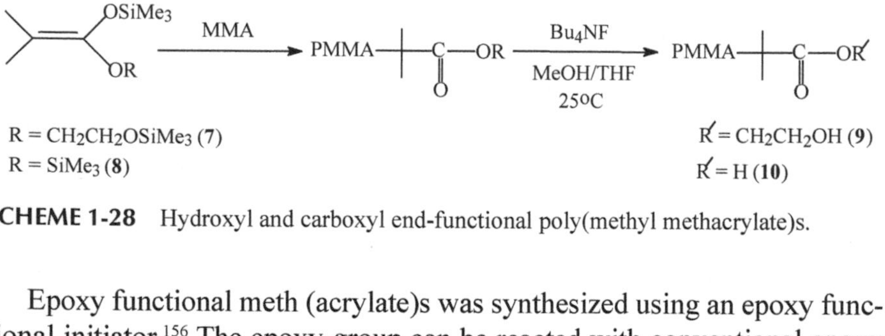

R = CH₂CH₂OSiMe₃ (7)
R = SiMe₃ (8)

$R' = CH_2CH_2OH$ (9)
$R' = H$ (10)

SCHEME 1-28   Hydroxyl and carboxyl end-functional poly(methyl methacrylate)s.

Epoxy functional meth (acrylate)s was synthesized using an epoxy functional initiator.[156] The epoxy-group can be reacted with conventional epoxy resins to make branched or triblock polymers.

A variety of phosphorous-terminated polymers[138] were synthesized by using phosphorous containing silyl ketene acetals. These initiators promoted GTP of MMA in the presence of tris-dimethyl amino sulfonium bifluoride (TASHF₂) to give functional PMMA with low polydispersity. Usually high levels of catalyst (4–11% on initiator) were required for these polymerizations due to coordination of catalyst with phosphonate group. Polymers containing a terminal silyl phosphonate group are easily converted into a terminal phosphoric acid group by hydrolyzing with dilute HCl or methanolic TBAF. Shen et al.[157] prepared phosphonium containing silyl ketene acetals which was used to initiate GTP of acrylic monomers and forms polymers with terminal phosphonium cations. The phosphonium cation was post reacted to make functional polymers. Polydispersities of the polymers were broader because the reactivity for a nucleophilic attack was reduced in the initiators substituted by the triphenylphosphonium group with electron withdrawing property. The macromolecular phosphonium salts could be converted with sodium ethanolate into the corresponding macromolecular ylides and by a subsequent Wittig reaction with an aldehyde into a macromonomer.

A variety of initiators bearing functional groups such as vinyl, allyl, styrenyl, and so forth that are not reactive under GTP conditions have been used to prepare acrylic macromonomers. Functional initiators initiated GTP of MMA in THF using TASHF$_2$ or tetrabutyl ammonium cyanide (TBACN) with quantitative conversions.[158,159] MWD's were broader (1.2–1.7) than that expected from a truly living system.

Attempts were made to prepare[158] vinyl group substitution at the alkoxy group of silyl ketene acetals, but their activity in GTP has not been examined in detail. A variety of allyl group containing initiators has also been used.[158] The rate of initiation of MMA with silyl poly(enolate)s was found to be faster than with silyl ketene acetals. A summary of end-functional PMMA's prepared by GTP using functional initiators is given in Table 1-2.

TABLE 1-2   End Functional Poly(methyl methacrylate) Using Functional Initiators

Entry	Initiator	Catalyst	M$_n$ × 10^{-3} (Theory)	M$_n$ × 10^{-3} (SEC)	M$_w$/M$_n$	Functionality
1	H$_3$C—C(CH$_3$)— O(CH$_2$)$_2$OSi(CH$_3$)$_3$ / OSi(CH$_3$)$_3$	TASHF$_2$	—	21.0	1.03	–OH
2	H$_3$C—C(CH$_3$)= OSi(CH$_3$)$_3$ / OSi(CH$_3$)$_3$	TASHF$_2$	—	21.0	1.03	–COOH
3	H$_3$C—C(CH$_3$)= OSi(CH$_3$)$_3$ / O—O	CsHF$_2$	—	—	—	(epoxide/cyclopropane structure)
4	((CH$_3$)$_3$SiO)$_2$P(=O)—CH= OSi(CH$_3$)$_3$ / OCH$_3$; R^1=CH$_3$, R^1=H	TASHF$_2$	7.57	7.05	1.03	–PO$_3$H$_2$
5	Cl$^{\ominus}$ Ph$_3$P$^{\oplus}$—CH= OSi(CH$_3$)$_3$ / OCH$_3$; R^1=CH$_3$, R^1=H	ZnBr$_2$	3.04	3.38	1.78	–PO$_3$H$_2$
6	H$_3$C—C(CH$_3$)= OSi(CH$_3$)$_3$ / OR; R=C$_2$H$_5$ or CH$_3$	TPSHF$_2$	15.0	11.2	1.45	Allyl

TABLE 1-2 *(Continued)*

Entry	Initiator	Catalyst	$M_n \times 10^{-3}$ (Theory)	$M_n \times 10^{-3}$ (SEC)	M_w/M_n	Function-ality
7		TBACN	15.0	16.76	—	Allyl
8		TASF-$_2$SiMe$_3$	10.0	10.7	1.10	Allyl
9		TBABB	6.1	7.06	1.80	Allyl
		TBAmCB	15.1	19.0	1.34	
10		TBABB	10.1	9.76	1.46	Allyl
11		TBABB	1.45	2.17	1.29	Allyl
12		TASHF$_2$	3.1	3.91	1.17	Allyl
13		TBABB	5.15	20.1	1.12	Allyl
		TBAmCB	5.12	5.91	1.15	

Macromonomer is a polymer or oligomer having a polymerizable functional group (double bond, heterocyclic group, etc.) at one chain-end or at both chain-ends, not only for vinyl polymerization, but also for polycondensation, polyaddition, or ring-opening polymerizations. Preparation of macromonomers involve initiators containing polymerizable functional groups which are inert to GTP conditions.

Classes of functional GTP initiators that have been employed for the preparation of macromonomers are shown in Table 1-3.

Asami et al.[160] used vinylphenyl ketenetrimethylsilyl acetal (VPKSA) as the initiator for the synthesis of a macromonomer with a styryl group using the catalyst TASF$_2$SiMe$_3$ (25 °C, THF). A polymer with M_n 3600–11400 g/mol (M_w/M_n = 1.09–1.10) was obtained.

Asami[161] also prepared oxazoline ring-terminated macromonomers by GTP of MMA employing trimethylsilyl group-containing oxazoline initiators. Trimethylsilyl-2-methyl-2-oxazoline or trimethylsilyl-2-ethyl-2-oxazoline and tris (dimethylamino) sulfonium difluorotrimethylsilicate as a catalyst was used at room temperature or at −78 °C. The oxazoline-terminated macromonomer obtained had a broad MWD and low initiating efficiency. However, the high functionality of the end oxazoline group was supported by the formation of the graft copolymer after cationic copolymerization with 2-methyl-2-oxazoline. This polymer can be used not only as a macro-monomer, but also as a reactive polymer for the synthesis of functional poly-mers by exploiting the reactivity of the end oxazoline group towards the carboxyl groups.

TABLE 1-3 Functional Initiators for the Synthesis of PMMA Macromonomers via Initiator Method in GTP

Entry	Initiators	Catalyst	M_n (Theory)	M_n (SEC)	M_w/M_n
1	VPKSA	TASF$_2$SiMe$_3$ TBACN	3000 7500	3600 6900	1.09 1.55
2	Trimethylsilyl-2-methyl-2-oxazoline	TASF$_2$SiMe$_3$	—	—	—
3	MTB	TPSHF$_2$	15000	12400	1.69
4	MTPD	TPSHF$_2$	15000	13620	1.55
5	MTBD	TBACN	3750	4210	1.25
6		NBu$_4$OAc	1990	3520	1.09

1-methoxy-2-methyl-1-trimethylsiloxy-1,3-butadiene (MTB) cannot be radically homopolymerized in bulk presumably due to the stability of the conjugated π-electron system towards a radical attack. Because of substituents $-OCH_3$ and $-OSiMe_3$, MTB has more polar character and cannot stabilize the radical in a free radical polymerization. It is also impossible to polymerize macromonomers of MMA[162] from MTB by Azobisisobutyronitrile (AIBN). Again, olefinic double bond at the end of each macromonomer appear to be less reactive in an addition step to a 1-cyano-1-methylethyl radical. Bandermann also observed no reactivity of these macromonomers in a free-radical copolymerization with styrene. They obtained similar results[162] with initiator 1-methoxy-2-methyl-1-trimethylsiloxy-1,4-penta-diene (MTPD).

In contrast to the initiators MTB and MTPD, the initiators, 1-methoxy-2-methyl-4-methylene-1-trimethylsiloxy-1,5-hexadiene MTBD and VPKSA (**23,** Table 1-3), underwent polymerization using AIBN as initiator. PMMA macromonomers derived from MTBD and VPKSA also underwent polymerization[162] with both 1,4- and 1,2-addition.

Bandermann[162] prepared graft copolymers from styrene and four PMMA macromonomers derived from initiator VPKSA using AIBN at 60 °C in benzene for 120 h. They separated all possible products, homopolystyrene, macromonomer and poly(macromonomer), and poly(styrene-g-MMA), the graft copolymer, according to the procedure published by Kuhn.[163]

GTP of MMA in THF with initiator at entry **6**, Table 1-3 afforded the α-silylketene acetal of ω-[2,5-bis(trimethylsilyloxy) phenyl] poly(methyl methacrylate). The polymerization was carried out under solvent reflux conditions.[164] Tetrabutylammonium acetate was used as catalyst in order to avoid cleavage of the phenolic TMS protective groups by the fluoride or bifluoride catalysts (Scheme 1-29). The living chain-end of was quenched with methanol to give the TMS-protected macromonomer with 76% yield ($M_n = 3520$–16100 g/mol; $M_w/M_n = 1.09$–1.49).

Proof of functionalization was established by UV spectroscopy. Desilylation with fluoride and workup in air gave partially oxidized polymer with $M_n = 3780$ and $M_w/M_n = 1.25$. Slow or discontinuous monomer addition resulted in broadening of the MWD. This implies that some deactivation reaction is occurring in the monomer-deficient system and hence, is not living in the true sense of the term.[164] This was shown by an experiment in which the monomer feed was paused for 7 min, which caused the reaction temperature to drop from 65 to 46 °C. Size Exclusion Chromatography (SEC) analysis of the isolated polymer revealed a bimodal MWD indicating that more than 80% of the active chain-ends had died. Resuming the monomer feed, the temperature rose to 54 °C. Five minutes after the end of monomer addition the reaction was quenched with benzoyl fluoride to give a benzoyl end-group. Additionally, the presence of fluoride liberated from the benzoyl fluoride caused a desilylation-esterification reaction of the hydroquinone moiety to give a polymer with a 2,5-bis(benzoyloxy) phenyl end-group.[164]

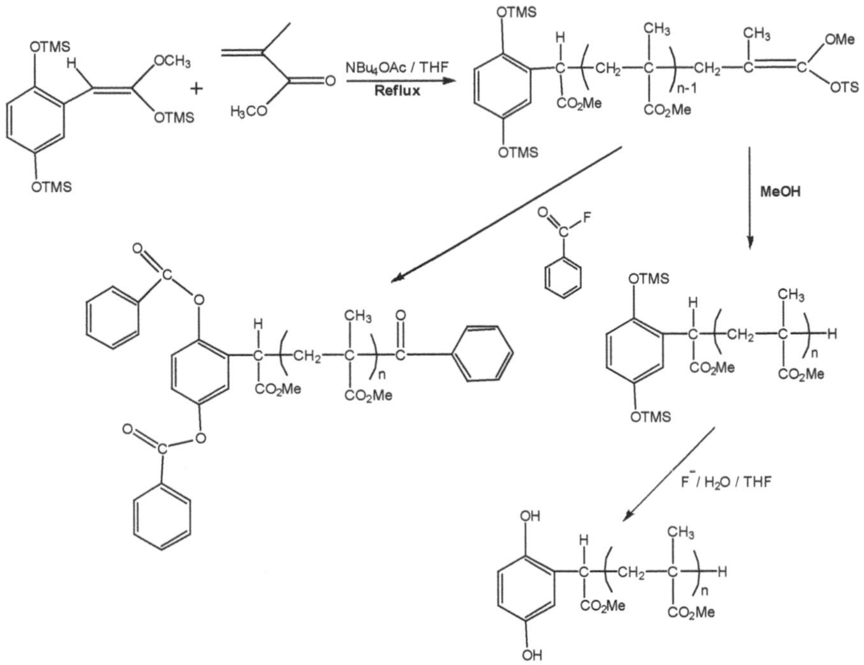

SCHEME 1-29   Poly(methyl methacrylate)s macromonomers with phenolic functionality.

The reactivity of the silyl ketene acetal end group of polymers prepared by GTP provides an opportunity for chain-end functionalization by end-capping reactions or coupling of polymer chains by reaction with polyfunctional terminating agents. Sogah et al.[138] reported the termination of silyl ketene acetal-ended PMMA with benzaldehyde using TASHF$_2$ catalyst at room temperature and after desilylation using TBAF/MeOH, gave a PMMA with a terminal benzhydryl alcohol group (Scheme 1-31). MMA was polymerized in THF using 1-(2-(trimethylsiloxy) ethoxy)-1-(trimethylsiloxy)-2-methyl-1-propene as initiator and TASHF$_2$ as catalyst at room temperature. After 2 h, the polymer was terminated with benzaldehyde using additional TASHF$_2$ catalyst and reacted overnight. After desilylation with TBAF/MeOH at reflux temperature for 1.5 h, a quantitative yield of hydroxy-PMMA was obtained with $M_n = 3200$ g/mol ($M_w/M_n = 1.18$).

Webster et al.[165] also reported the termination of GTP living chain-end with bromine to produce PMMA-Br. Grigor and Eastmond[166] prepared bromine terminated PMMA and utilized it for the formation of block copolymers. GTP of MMA was achieved by using MTS as initiator and potassium bifluoride in acetonitrile and polymerization was terminated with excess bromine to yield quantitative PMMA-Br with $M_n = 12{,}790$ g/mol and $M_w/M_n = 1.2$ (Scheme 1-30).

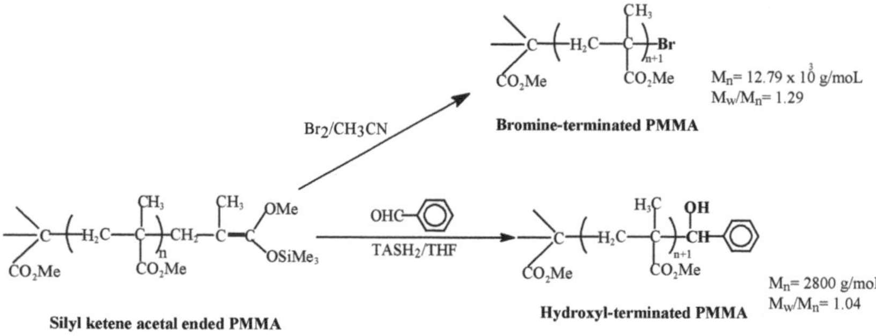

SCHEME 1-30   Bromo and hydroxyl end-functional PMMA via electrophilic termination.

The PMMA-Br was used to initiate styrene polymerization by free radical photo polymerization to form block copolymers (ABA and AB type) using dimanganese decacarbonyl and irradiated with light ($\lambda = 436$ nm) for 30 min. Under the reaction conditions employed only a small fraction of the PMMA-Br would have reacted to form radicals.

GTP of acrylates using a bifluoride catalyst generally leads to polymers with a broader MWD than that is observed in GTP of the corresponding methacrylate or GTP of acrylates using Lewis acid catalysts. Thus, polymerization of butyl acrylate in THF at 0 °C gave[170] poly(butyl acrylate) with $M_n = 27{,}200$ and $D = 2.16$ (theoretical $M_n = 26{,}100$). To obtain further insight into the causes of the MW broadening, living ethyl acrylate oligomer (degree of polymerization, DP = 4) prepared by TASF-catalyzed GTP was treated with p-nitrobenzyl bromide at −78 °C. Chromatographic purification gave 60% yield of the benzylated product containing both internal and terminal p-nitrobenzyl groups in the ratio of 9:1 (Scheme 1-31). These results suggest that the trimethylsilyl group is capable of isomerizing to an internal position of the poly(acrylate) chain. However, ^{13}C NMR studies of poly(ethyl acrylate) gave no evidence of branching, suggesting that the internal ketene silyl acetal, while capable of reacting with a benzyl halide, is too sterically hindered to initiate a branch point. No evidence is available to indicate whether a O- or C-silyl intermediate is formed.

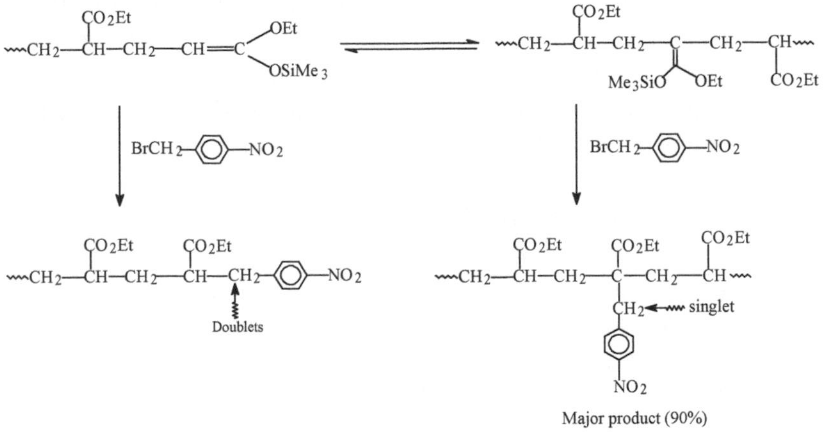

SCHEME 1-31   p-Nitrobenzyl end-functional poly(ethyl acrylate)s.

Quirk and Ren[167] developed a functionalization method for GTP using sterically hindered monomers, analogous to the substituted DPE chemistry, which has been utilized for LAP. The monomers used for functionalization were methyl-2-phenylpropenoate (MPHA), ethyl-2-phenyl-2-butenoate (EPB), ethyl-2-methyl-2-butenoate (EMB), and methyl-E-3-(2-dimethyl aminophenyl)-2-phenyl acrylate (AMPA) (Scheme 1-32). Various functional groups such as $-NR_2$, $-OMe$, $-X$ (Br, I, and Cl), $-COOMe$, and so forth could be introduced into poly(alkyl methacrylates) via substituents on the aromatic ring.

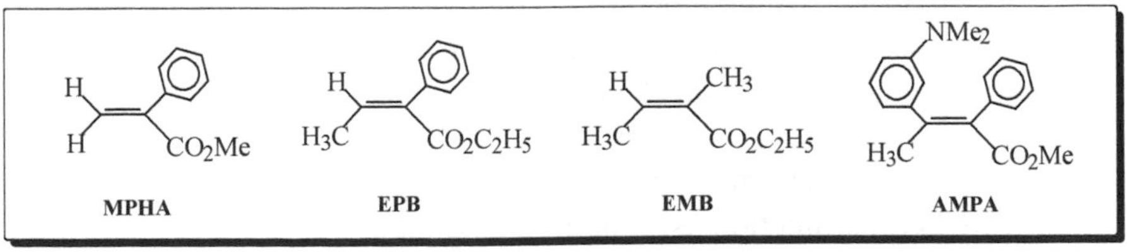

SCHEME 1-32   Sterically hindered monomers for functionalization of GTP.

The functionalization reactions of MPHA with living PMMA prepared by GTP were investigated under a variety of reaction conditions and stoichiometry. The functionalized polymer was characterized by SEC, *Vapor pressure osmometry* (VPO), NMR, and UV spectroscopy. All the results were consistent with only monoaddition of MPA to the living silyl ketene acetal chain-end at room temperature, even when two molar equivalents of MPHA were present. No evidence for oliomerization was obtained. However, when the addition of MPHA was carried out at −78 °C with two molar equivalents of MPHA, two monomer units were added to the chain-end. This is consistent with the predictions based on the reported ceiling temperature of −40 °C for MPHA.

The polymerizations were carried using silyl ketene acetal-ended PMMA with monomers (EPB, EMB, and MPB) at room temperature[167] for 0.5 h using TASHF$_2$ as catalyst. In the case of EPB, high functionalization efficiency of 95% (by $F_n = M_n$ (VPO)/M_n (NMR), was observed when one molar equivalent of EPB is used. Surprisingly, the functionalization efficiency decreased to 0.6 when functionalizations was carried out with two molar equivalents of EPB. Nucleophilic catalysts promote chain termination as well as propagation, and, at sufficiently low monomer concentrations chain termination rate is faster than propagation. In order to obtain high functionality and to minimize chain termination catalyst levels should be kept as low as possible. Therefore, a catalyst concentrating of 0.1 mol% based on initiator was used. Temperatures between 0 and 60 °C did not have significant effect on functionalization efficiency (0.92 to 0.87). However, it is reasonable to conclude that functionalization reactions will be favored at lower temperature since the probability of chain termination is low.

When two molar equivalents of EPB[167] was added to the living silyl ketene acetal-ended PMMA in the presence of 0.1 mol% of TASHF$_2$ catalyst it was presumed that a possible nonliving process which would lower the functionality could occur due to EPB functioning as a chain transfer agents. This result suggests that the functionalization reaction is reversible since the functionality decreases upon addition of a second equivalent of functionalizing monomer. A possible mechanism[167] to explain these results is shown in Scheme 1-33.

Similarly, it was reported[167] that methyl-2-phenyl-2-butenoate (MPB) as well as EMB could function as a chain transfer agent.

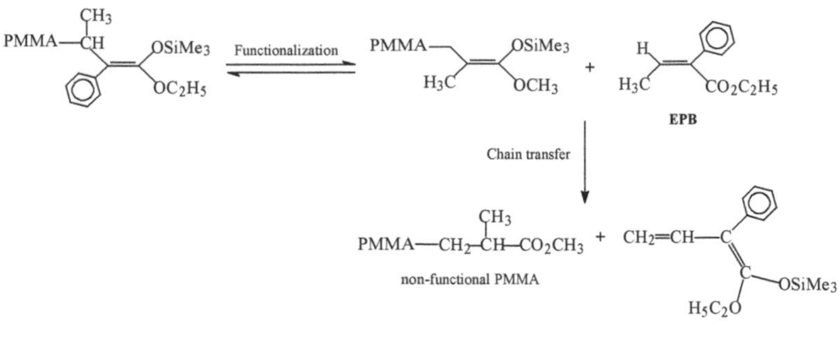

SCHEME 1-33 Chain transfer reaction of EPB in functionalization of PMMA.

The stoichiometric addition reaction[167] of 1,4-butanediol diphenylacrylate (2-phenylpropenoic acid, 1,4-butanediyl ester (DBPA) (1.2 mmol) with living trimethylsilylketene acetal-ended PMMA (M_n = 2100, M_w/M_n = 1.07) (Scheme 1-34) was carried out in THF. The MW of the product (M_n = 3900, M_w/M_n =1.06) corresponds closely to the MW expected for dimerization by simple monoaddition of one living chain-end to each acrylate unit in DBPA with a coupling efficiency of 94%.

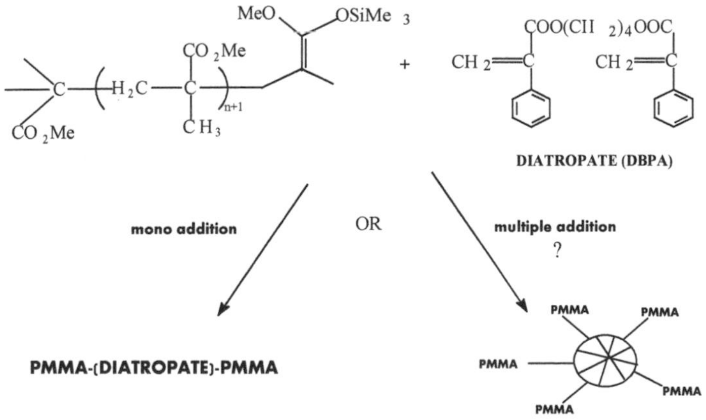

SCHEME 1-34 Functionalization of GTP living chain-end with diatropate.

Polymers containing amino groups are important because of the chemical versatility of the amino function. The reaction of AMPA with living trimethylsilyl ketene acetal-ended PMMA[167] (M_n = 2100, M_w/M_n = 1.07) in THF for an hour was investigated (Scheme 1-35). The polymer solution turned slightly yellow on addition of AMPA and the color faded when the reaction was quenched with methanol.

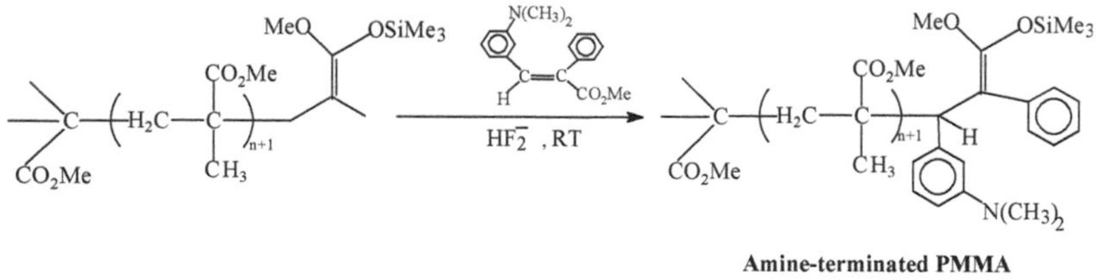

Amine-terminated PMMA

SCHEME 1-35   Amine-terminated poly(methyl methacrylate)s by GTP.

The efficiency of this functionalization reaction[167,168] (F_n = 0.90–0.93) was also determined titrimetrically as well as by SEC, VPO, and ^{1}H NMR. The results were consistent with monoaddition of the nonpolymerizable monomer to the living GTP chain-end.

Asami et al.[160] reported PMMA macromonomer with styryl terminal functional groups by termination method. The living PMMA chain-end, prepared by the polymerization of MMA with the MTS in THF using $HF_2^{\ominus}$ catalyst at room temperature, was reacted with vinylbenzyl bromide (or tosylate) using $TASF_2SiMe_3$ as catalyst (Scheme 1-36). The styryl terminal PMMA macromonomer obtained were shown to be highly monodisperse (M_w/M_n = 1.07 to 1.06 with M_n = 3250 to 4380 g/mol). When one equivalent of catalyst was used the functionality (F_n) determined by UV spectroscopy was found to be only in the range of 25–83%.

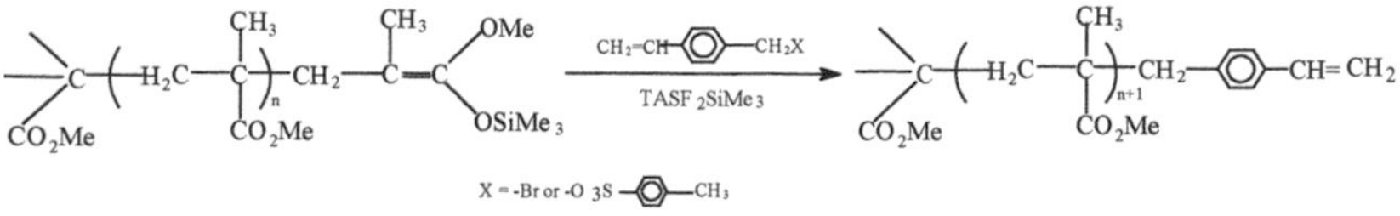

SCHEME 1-36   Macromonomer synthesis via GTP by electrophilic termination.

A summary of terminating agents that has been employed in GTP is shown in Table 1-4.

ω-Trimethylsiloxy PMMA was prepared by GTP of MMA in THF at 20 °C using Initiator (TETMP) as a functionalized initiator and tetra-n-butyl ammonium bibenzoate (TBABB) as the catalyst. Several procedures were used to convert the resulting ω-trimethyl siloxy PMMA to ω-methacryloyl-PMMA. ω-Trimethylsiloxy PMMA was either directly reacted with methacryloyl fluoride/TBAF[169] or hydrolyzed and subsequent reaction of the resulting ω-hydroxy PMMA with methacryloyl chloride/trimethylamine or methacrylic acid/dicyclohexylcarbodiimide (Scheme 1-37).

TABLE 1-4 Functional Terminating Agents for GTP

Entry	Terminators	Catalyst/Solvent	M_n (Theory)	M_n (SEC)	M_w/M_n	Functional group	F_n
1	Benzaldehyde	TASHF$_2$/THF	3300	2800	1.04	–OH	—
2	Br$_2$	—/CH$_3$CN	6100	12790	1.29	–Br	—
3	p-Vinylbenzyl bromide	TASF$_2$SiMe$_3$/THF	3000	3250	1.07	Styryl	0.83
4	p-Vinylbenzyl to sylate	TASF$_2$SiMe$_3$/THF	3000	4380	1.06	Styryl	0.73
5	MPHA	TASHF$_2$/THF	2100	3600	1.05	–C$_6$H$_5$	1.05
6	EPB	TASHF$_2$/THF	2100	1900	1.03	–C$_6$H$_5$	0.94
7	AMPA	TASHF$_2$/THF	2100	2200	1.06	–C$_6$H$_5$NMe$_2$	0.93

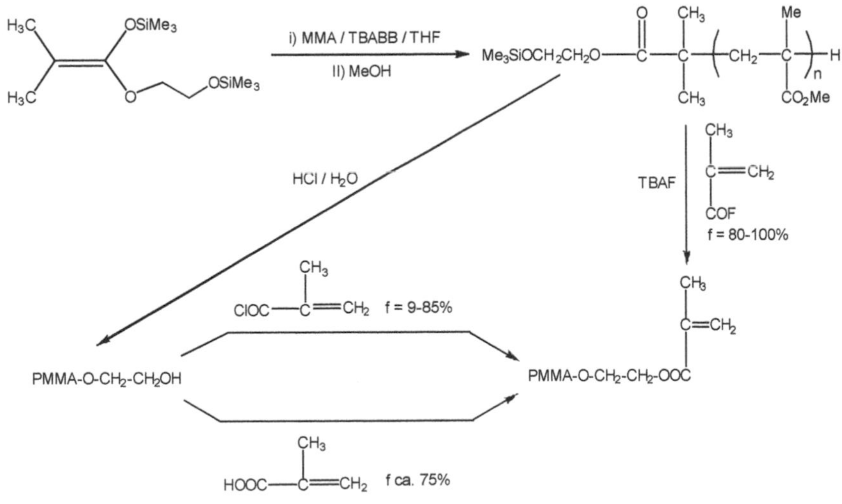

SCHEME 1-37 Synthesis of methacryloyl-terminated PMMA macromonomer by GTP.

The functionality of the macromonomers as determined by ^{1}H NMR was poor, ca. 75% for methacrylic acid/DCC and 80–100% for methacryloyl fluoride.

Cohen[169] reported that macromonomers of M_n = 3000–40,000 with upto 95% terminal vinyl functionality, prepared by extending the acylation chemistry to PMMA silyl ethers. Silyl ether was chosen for reaction with methacryloyl fluoride because the silyl ketene acetal-end of PMMA failed to give the desired product. Polycondensations (Cl⁻ catalyzed) using aryl silyl ether and acid chloride generally occur at 200–300 °C.

The macromonomer technique is the only method available to prepare well-defined graft polymers containing nearly monodisperse branches of controlled MW. A novel all-PMMA comb polymer was prepared by a combination of GTP and anionic polymerization (Scheme 1-38).169b

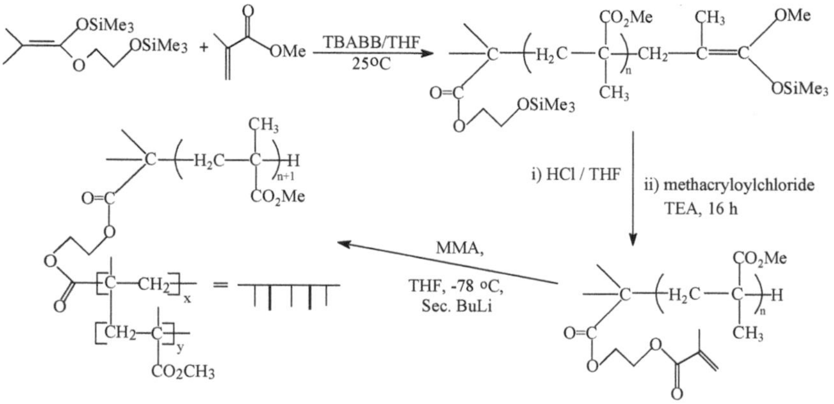

SCHEME 1-38 Comb-shaped poly(methyl methacrylate)s by a combination of GTP and anionic polymerization.

The resulting comb had a M_n = 2,48,6000 with an average of 14 grafted PMMA chains per molecule. The grafted PMMA chain had a M_n = 6300 and M_w/M_n = 1.11. Additional examples of comb-grafted polymers via GTP have been reported by Heitz and Webster,[170a] Hertler et al.,[170b] and Jenkins et al.[170c]

Nucleophile assisted GTP inevitably involves participation of enolate anion as intermediate during propagation. The concentration and nature of catalyst strongly determine the "living character" of GTP, possibly because of the participation of enolate anions in the propagation. Gnaneshwar and Sivaram undertook a detailed study of the reaction of GTP chain-ends with benzaldehyde, N-trimethyl silyl benzaldimine, and 5,6-dihydro-2H-pyran-2-one (Scheme 1-39).[171] The analogous reactions, involving small molecules, are typically conducted with strong Lewis acid catalysts and chlorinated organic solvents. These conditions are not conducive for GTP. Hence, model reactions were examined to establish the efficiency of reactions in THF using weakly nucleophilic or electrophilic catalysts. The learnings from this study were extended to the reaction of GTP chain-ends with benzaldehyde, N-trimethyl silyl benzaldimine, and 5,6-dihydro-2H-pyran-2-one, respectively. The efficiency of functionalization reactions was examined using ^{1}H NMR, SEC, and MALDI-ToF-MS.

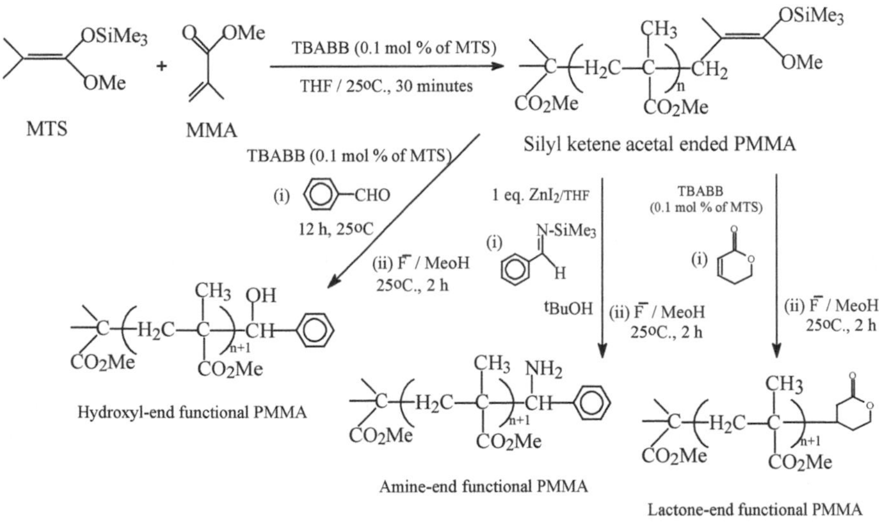

SCHEME 1-39   Electrophilic termination reactions of silyl ketene acetal ended PMMA.

Narrow MW hydroxyl, amine, and lactone-end functional PMMAs were prepared ($M_w/M_n = 1.06$–1.19) via GTP by electrophilic termination of silyl ketene acetal ended PMMA with benzaldehyde, N-trimethylsilyl benzaldimine, and 5,6-dihydro-2H-pyran-2-one, respectively. The number average degree of functionalization (F_n) as determined by NMR/SEC was in the range 0.70–0.85. Lewis acid (ZnI_2, 1 equivalent) was used for terminating silyl ketene acetal ended PMMA with N-trimethylsilyl benzaldimine whereas TBABB (0.1 mol%) catalyst was used in the case of benzaldehyde and 5,6-dihydro-2H-pyran-2-one at room temperature.

The hydroxyl functional PMMA's were examined using MALDI-ToF-MS. The expected mass peaks corresponding to the $[M+Li]^+$ molecular ions for hydroxyl end-functional PMMA $\{[M+Li]^+ = 100.12(DP) + H (1.0079) + 107.1324 (C_7H_7O) + Li^+ (6.941)\}$ are 4104.92, 4205.04, and so forth which are in agreement with the observed series viz., 4106.04, 4207.11, and so forth (with $\Delta = 2$) (**Figure 1-1**). The observation of a single series of peaks in Figure 1-1 indicates that there was no significant chain-end cyclization when low concentration of TBABB (0.1 mol%) was used.

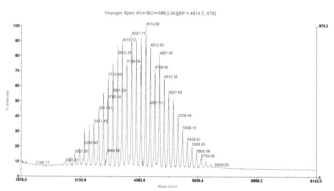

FIGURE 1-1 MALDI-TOF spectrum of hydroxyl end-functional PMMA prepared by GTP (using 0.1 mol% TBABB catalyst; entry 2, Table 1-5). $M_n = 100.12(20) + H(1.0079) + 107.1324(C_7H_7O) + K^+(39.098)$. (Matrix: Dithranol and CF_3COOK for enhancement of ion formation ($\Delta = 6$ Da).)

Figure 1-2 shows MALDI-ToF spectrum of hydroxyl end-functional PMMA in which functionalization reaction was carried out at 0 °C using 1 mol% TBABB catalyst. The mass peaks corresponding to $n = 30$ are expected to be 3118.6813, 3218.8013, 3318.9213, and so forth. The sample (**entry 9**, Table 1-5) shows a major series (Figure 1-2) 3104.56, 3204.62, 3304.26, and so forth with $\Delta = 14$ Da and a minor series of peaks of lower intensity which can be attributed to an oligomer series with cyclic end groups $[M + Li]^+ = 100.12$ (MMA) × n (DP) + H (1.0079) – OCH_3 (31) + 6.9 (Li) at 2977.98, 3077.06, and so forth with $\Delta = 3$ Da.

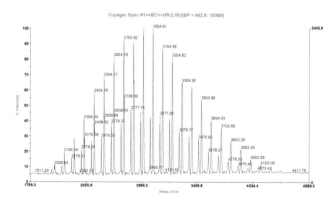

FIGURE 1-2 MALDI-TOF spectrum of hydroxyl end-functional PMMA prepared by GTP (using 1 mol% TBABB catalyst; entry 9, Table 1-5). $M_n = 100.12(20) + H(1.0079) + 107.1324(C_7H_7O) + Li^+(6.941)$. (Matrix: Dihydroxybenzoic acid and LiCl for enhancement of ion formation ($\Delta = 14$ Da).)

TABLE 1-5 α- and ω-End Functionalized Polymers Prepared Using LAP

Entry	Monomer	Functional Group	End-Capping Group	$M_n \times 10^{-4}$ (calculated)	$M_n \times 10^{-4}$ (found)	M_w/M_n	Solvent/ Temperature	Initiator
1	Styrene, Isoprene	$-NH_2$	[Si–N–Si structure]	1.8	1.6	1.16	THF/-78 °C	Li-Napth
2	Styrene, diene	$-(CH_2)_nOH$ ($n = 0, 1, 2$)	$-O-SiMe_2Bu^t$	7.4	8.0	1.04	THF/-78 °C	Li-Napth
3	Styrene, diene	$-COOH$	$-C(OCH_2)_3$	1.5	1.3	1.06	THF/-78 °C	Li-Napth
4	Styrene, diene, MMA, vinylpyridine	Succinic anhydride	[structure **1**]	2.15	2.23	1.07	THF/-78 °C	Sec-BuLi
5	Styrene, isoprene	Monosaccharides (e.g., glucofuranose)	[structure]	40	36	1.1	THF/-78 °C	n-BuLi
6	Styrene, isoprene	α,ω-Halides	1,3-dichloroalkane	0.2	0.19	1.1	THF/-78 °C	Sec-BuLi + DPE (1,1'-diphenyl-ethylene)
7	Styrene, isoprene	4-Vinylphenyl	$X-(CH_2)_3$ [vinylphenyl structure]	0.2	0.21	1.05	THF/-78 °C	Sec-BuLi

TABLE 1-5 *(Continued)*

Entry	Monomer	Functional Group	End-Capping Group	$M_n \times 10^{-4}$ (calculated)	$M_n \times 10^{-4}$ (found)	M_w/M_n	Solvent/ Temperature	Initiator
8	Styrene	Perfluoroalkyl	$(CH_3)_2SiCl(CH_2)_3X$ (X = perfluoroalkyl)	0.113	0.114	1.04	Cyclo-hexane/room temperature	*Sec*-BuLi
9	Styrene	α-,ω-Alkyne, azide	1,4-Dibromobutane + sodium azide	1.8	1.89	1.05	Toluene/room temperature	5-Triethylsilyl-4-pentynyllithium (protected alkyne)
10	Styrene	Silyl hydride	Chlorodimethylsilane	0.3	0.31	1.03	Toluene/room temperature	*Sec*-BuLi
11	Styrene	Cyanide	Chlorodimethylsilane + allyl cyanide	0.31	0.32	1.04	Toluene/room temperature	*Sec*-BuLi
12	Styrene	Acetate	Chlorodimethylsilane + allyl acetate	0.31	0.32	1.05	Toluene/room temperature	*Sec*-BuLi
13	Styrene	Ethyl ether	Chlorodimethylsilane + allyl ethyl ether	0.31	0.32	1.03	Toluene/room temperature	*Sec*-BuLi

Thus, end functional polymers with reduced loss of chain-ends via cyclization and high chain-end functionality can be obtained using a low concentration of poorly nucleophilic catalyst (0.1 mol% TBABB) and drop wise addition of MMA, while keeping monomer concentrations high.

1.5 FUNCTIONAL POLY(METHYL METHACRYLATE)S (PMMA) VIA LIVING ANIONIC POLYMERIZATION (LAP)

LAP is particularly suitable for the synthesis of functional polymers with well-defined structures.[1c,2d,14,17] Chain-end functionalization can be carried out in two ways, namely, (1) post-polymerization reactions of living chains with suitable electrophilic reagents and (2) using a functionalized initiator for polymerization.[11,12,25,33,40,43,50b,172,173] Functional initiation has advantages over electrophilic termination since it is easier to achieve high degree of end functionalization. However, functional groups on initiators are usually not stable to organolithium reagents. It is, therefore, necessary to protect the functional groups prior to initiation. A vast body of work has been reported on the synthesis of functional polymers from nonpolar monomers in hydrocarbon solvents.[2e,14,17] However, significantly less work is reported on the synthesis of end-functionalized methacrylates in polar solvents. Unlike hydrocarbon monomers, the propagating chain-ends of polar monomers such as alkyl (meth)acrylates are less stable, exist in equilibrium with the enolate form and are difficult to functionalize through termination using electrophilic reagents. The difficulty arises from the poor reactivity of the enolate anions relative to carbanions.[174–176] Hence, the use of protected anionic-initiator is often the preferred method for the synthesis of functional poly(alkyl methacrylates).[108,177,178]

Adducts of DPE with alkyllithium bearing a protected functionality provides an excellent initiator system for polymerization of methacrylates. Functional anionic initiators have been synthesized by the reaction of DPE with protected hydroxyl substitution and DPE with n-BuLi and alkyllithium bearing protected hydroxyl substitution, respectively.[179] tert-Butyldimethylsilyloxy group was used to protect the hydroxy group present in DPE or alkyllithium since it is known that a hindered silyloxy group provides better stability during anionic polymerization.[180] An initiator bearing protected aromatic hydroxyl group was prepared by reacting 1-[p-(tert-butyldimethylsilyloxy) phenyl-1'-phenylethylene with n-BuLi in THF. Other initiators, differing in the number of alkyl groups, were prepared by reacting 6-(tert-butyldimethylsilyloxy) hexyllithium and a commercially available

3-(*tert*-butyldimethylsilyloxy) propyllithium with DPE, respectively, in THF (Figure 1-3, Scheme 1-40). The efficacy of these functional initiators for the anionic polymerization of MMA at −78 °C in THF was examined in the presence of LiClO$_4$.[116a] LiClO$_4$ was used in 10:1 molar ratio with respect to the initiator to moderate the propagation kinetics of aggregated and nonaggregated enolate anions.[2b,116b]

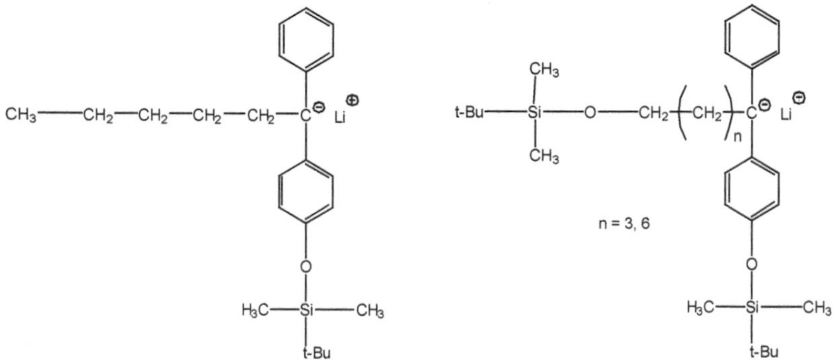

FIGURE 1-3 Anionic initiators containing hydroxyl group protected with tert-butyl dimethyl silane.

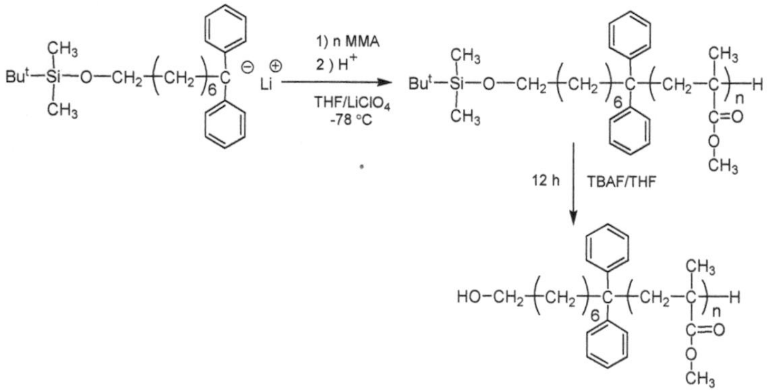

SCHEME 1-40 Synthesis of monohydroxyl functional poly(methyl methacrylate)s.

Hydroxyl end functional PMMA's was obtained in quantitative conversion (~100%) with predictable number average MW ($M_{n,cal}$) and narrow MWD's. The initiator efficiency, f, determined by the ratio of calculated $M_{n,cal}$ to the $M_{n,SEC}$ obtained using SEC was close to unity ($0.8 \leq f \leq 1$) indicating the absence of side reactions with the silylprotected hydroxyl groups of the initiators during polymerization. The silyl groups of the functional PMMAs could be deprotected using TBAF (1 M) in THF at room temperature. The functionality of the chain-ends was confirmed by MALDI-TOF-MS.

The terminal hydroxyl groups of the PMMAs were used as initiator for the ring opening anionic polymerization of EO (Scheme 1-41). To accomplish this, the terminal group of the hydroxyl functional PMMAs was converted to the corresponding alkoxide of potassium by reacting with triphenylmethyl potassium in THF at 30 °C. Polymerization of EO was carried out over 36 h at room temperature in THF. Well defined poly(methyl methacrylate)-*b*-poly(ethylene oxides)s could be synthesized. This is the only unequivocal synthesis of such amphiphilic block copolymers available in the literature.

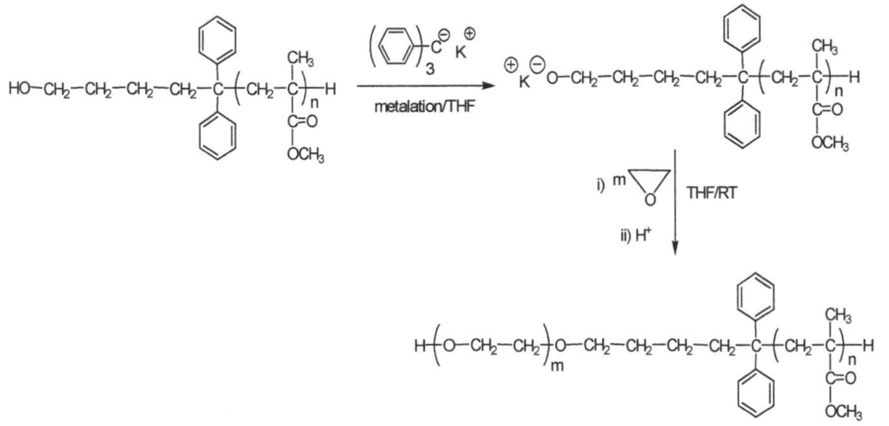

SCHEME 1-41   Synthesis of PMMA-b-EO diblock copolymer in the presence of potassium counter ion.

Coupling of living chains of silyl-protected hydroxy-PMMA chains using ethylene glycol dimethacrylate as the crosslinking agent was reported to give PMMA star polymers with protected-hydroxy group at the end of each arm of the star.[180] The protected hydroxy end-functions were deprotected by treatment with TBAF to form PMMA stars with free hydroxy groups at the chain-ends. The number of arms per star could be controlled between 6 and 10 by adjusting the ratio of initiator to the crosslinking agent. The living enolate chain-ends of PMMA can also be reacted with p-vinyl benzylchloride to give end styryl functional PMMA macromonomers.[181]

Polymers having polymerizable functional groups at each repeat unit are of significant interest as they can be used as reactive polymers in many applications, such as, cross-linked polymers, controlled post grafting, immobilization of bioactives, and thermal cross-linking agent in the production of high refractive-index transparent organic glass.[182-184] However, selective vinyl polymerization of divinyl monomers is a synthetic challenge and only in a few cases such a selectivity has been achieved.[185-187]

Sivaram et al.[188] have recently reported the synthesis styryl pendant PMMA by selective LAP of 4-vinylbenzyl methacrylate using 1,1-diphenylhexyllithium (DPHLi) and tritylpotassium (TritylK) as initiators at −78 °C (Scheme 1-42). At lower temperature, the reactivity of stabilized anions such as DPHLi or trityl anions can be modulated to selectively polymerize the monomer at its methacrylate functionality leaving the reactive styryl group as a pendant group on the polymer.

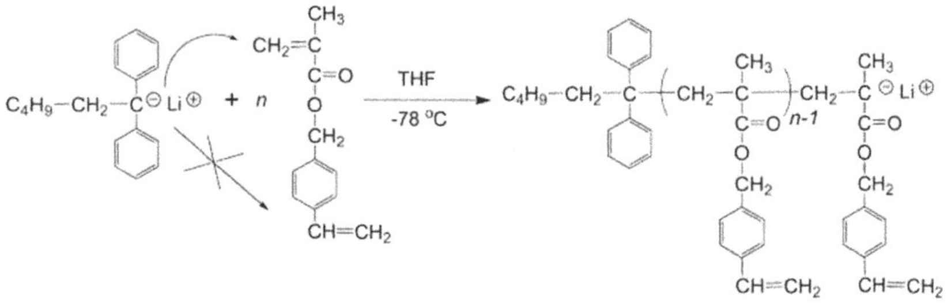

SCHEME 1-42   Anionic polymerization 4-vinylbenzyl methacrylate using DPHLi as initiator.

The selective LAP was performed using DPHLi and trityl potassium as initiators in THF at −78 °C in the presence of excess $LiClO_4$ to control the equilibrium dynamics of the propagating enolate anions. After termination, poly(4-vinylbenzyl methacrylate) was obtained with a conversion of nearly 100%. The high selectivity observed for polymerization of the acrylic double bond is attributed to a relatively higher activation energy associated with the initiation of the vinyl double bond using stabilized DPHLi or trityl potassium initiators. Thus, the nucleophilicity of the stabilized carbanions/enolate anions can be suitably modulated for selective LAP.

1.6   SUMMARY AND PERSPECTIVES

Functional polymers can be prepared either by initiation of LAP of a vinyl, diene, or acrylic monomer using a suitable functional initiator or by reaction of the "living" chain-ends with an electrophile bearing a functional group. Both methods have been extensively studied. However, careful consideration of many parameters is necessary before one can make a choice of the appropriate synthetic method for introducing a given functional group.

Use of functional initiator is obviously the method of choice for most monomers as long as the initiator does not react with a carbanion or enolate. Under carefully designed experimental conditions it is possible to obtain polymers with good control over MW and MWD as well as high degree of efficiency in functionalization. However, this method is restricted to those functional groups which can be protected prior to polymerization and deprotected after polymerization. The synthesis of functional initiators is, often, not trivial and to make them in high chemical purities, needed for LAP, poses a formidable challenge.

Termination of a living chain-end by a functional electrophile appears seemingly simple, especially, since conventional initiators can be used. However, the reactive carbanion or enolate is not so reactive because of factors such as aggregation equilibria and effect of counter cations. Solvents and reaction temperatures also play a critical role. Whereas, nucleophilic displacement reactions with electrophiles are favored at higher temperatures, reactive chain-ends are only stable at lower temperatures. To arrive at the right conditions of reactions for effective functionalization is often time consuming and involves a trial and error approach.

GTP offers a useful method for the synthesis of functional polymers, since the silyl protected enolates are stable at room temperatures and above. These are discrete, isolable species unlike enolates or carbanions. However, nucleophile assisted desilylation chemistry is still not well understood in low dielectric solvents such as THF. Consequently, the potential of this method has been far less exploited.

Notwithstanding the method used, quantitative detection of end functional groups in polymers also poses a challenge. NMR, a widely employed method, is only applicable to low MW polymers. MALDI-ToF is sensitive with only polar polymers. One of the significant gaps in the literature on this subject is often the lack of unequivocal demonstration of the efficiency of end functionalization.

Nevertheless, in the intervening 60 years following the first discovery of living nature of polymerizations, our understanding of structure of the chain-ends and mechanism of living or controlled polymerization has undergone substantial refinement. This body of cumulative knowledge should enable "a priori" the choice of best method for the synthesis of any end functional polymer by living and/or controlled polymerization methods.

ACKNOWLEDGEMENTS

I wish to express my deep sense of appreciation to former students, Dr. Mahua Ganguly Dhara and Dr. R. Gnaneshwar for their contributions to the area of functional polymers. A special thanks to Dr. D. Baskaran for his sustained contributions and support to this research programme. I wish to thank CSIR-NCL for the facilities and CSIR Bhatanagar Fellowship for financial support.

REFERENCES

1. (a). Szwarc, M. *Nature* **1956**, *178*, 1168. (b). Szwarc, M.; Levy, M.; Milkovich, R. J. *Am. Chem. Soc.* **1956**, *78*, 2656. (c). Szwarc, M. *Carbanions, Living Polymers and Electron Transfer Processes;* Interscience: New York, 1968. (d). Szwarc, M.; Van Beylen, M. *Ionic Polymerization and Living Polymers*; Chapman Hall: New York, 1993.

2. (a). Jerome, R.; Teyssie, P. H.; Vuillemin, B.; Zundel, T.; Zun, C. *J. Polym, Sci. Polym. Chem.* **1999**, *37*, 1. (b). Baskaran, D. *Prog. Polym. Sci.* **2003**, *28*, 521. (c). Baskaran, D.; Muller A. H. E. In *Controlled and Living Polymerizations: From Mechanism to Applications;* Muller, A. H. E., Matyjaszewski, K., Eds.; Wiley-VCH: Weinheim, 2009; Chapter 1. (c). Dhara, M. G.; Sivaram, S. *Living Anionic Polymerization of Methyl Methacrylate: Block Copolymers, Star Polymers and Macromonomers*; VDM Verlag Dr. Muller Aktiengesellschaft & Co KG: Saarbrucken, 2010. (d). Baskaran, D.; Muller, A. H. E. *Prog. Polym. Sci.* **2007**, *32*, 173. (e). Morton, M. *Anionic Polymerization: Principles and Practice;* Academic Press: New York, 1983.

3. (a). Kwon, Y.; Faust, R. *Adv. Polym. Sci.* **2004**, *167*, 107. (b). Kennedy, J. P.; Holden, G.; Legge, N. R.; Quirk, R. P.; Schroeder, H. E. Eds.; *Thermoplastic Elastomers,* 2nd ed.; Hanser Publishers: New York, 1999; Chapter 13.

4. (a). Webster, O. W.; Hertler, W. R.; Sogah, D. Y.; Farnham, W. B.; Rajan Babu, T. V. *J Am. Chem. Soc.* **1983**, *105*, 5706. (b). Hertler, W. R. *Silicon in Polymer Synthesis;* Kricheldorf, H. R. Ed.; Springer: Heidelberg, 1996; p 69. (c). Webster, O. W. *Adv. Polym. Sci.* **2004**, *167*, 1.

5. Matyjaszewski, K.; Xia, J. *Chem. Rev.* **2001**, *101*, 2921.

6. Hawker, C. J.; Bosman, A. W.; Harth, E. *Chem. Rev.* **2001**, *101*, 3661.

7. (a). Chiefari, J.; Chong, Y. K. B.; Ercole, F.; Krstina, J.; Jeffery, J.; Letp, T.; Mayadunne, R. T. A.; Meijs, G. F.; Moad, C. L.; Moad, G.; Rizzardo, E.; Thang, S. H. *Macromolecules* **1998**, *31*, 5559. (b). Mayadunne, R. T. A.; Rizzardo, E.; Chiefari, J.; Chong, Y. K.; Moad, G.; Thang, S. H. *Macromolecule* **1999**, *32*, 6977.

8. Reetz, M. T.; Hutte, S.; Goddard, R. J. *Phy. Org. Chem.,* **1995**, *8*, 231.

9. Matyjaszewski, K. Radical Polymeriation. In *Controlled and Living Polymerizations: From Mechanisms to Applications;* Muller, A. H. E., Matyjaszewski, K., Eds.; Wiley-VCH: Weinheim, 2009.

10. Matyjaszewski, K.; Moeller, M. Eds., *Polymer Science: A Comprehensive Reference;* Elsevier: Amsterdam, 2012.

11. (a). Goethals, E. J. *Telechelic Polymers: Synthesis and Applications*; CRC: Boca Raton, 1989. (b). Gnanou, Y. *J. Macromol. Sci.-Rev. Macromol. Chem. Phys.* **1996**, *C36* (1), 77. (c). Frechet, J. M. J. *Science* **1994**, *263*, 1710.
12. Patil, A. O.; Schulz, D. N.; Novak, B. M. *Functional Polymers, ACS Symp. Ser.* **1998**, 704.
13. Kennedy, J. P.; Ivan, B. *Designed Polymers by Carbocationic Macromolecular Engineering: Theory and Practice;* Hanser Publishers: Munich, 1992.
14. Hadjidichristidis, N.; Pitsikalis, M.; Pispas, S.; Iatrou, H. *Chem. Rev.* **2001**, *101*, 3747.
15. Nakahama, S.; Hirao, A. *Prog. Polym. Sci.* **1990**, *15*, 299.
16. Winstein, W.; Clippinger, E.; Fainberg, A. H.; Robinson, G. C. *J. Am. Chem. Soc.* **1954**, *76*, 2597.
17. Hsieh, H. L.; Quirk, R. P. *Anionic Polymerization: Principles and Practical Applications;* Marcel Dekker: New York, 1996.
18. Bywater, S.; Worsfold, D. J. *J. Organometal. Chem.* **1967**, *10*, 1.
19. Bywater, S. In *Encyclopedia of Polymer Science and Engineering,* 2nd ed.; Kroshwitz, J. I., Ed.; Wiley-Interscience: New York, 1985; Vol. 2, p 1.
20. Quirk, R. P.; Ma, J.-J. *Polym. Int.* **1991**, *24*, 197.
21. Finnegan, R. A.; Kutta, H. W. *J. Org. Chem.* **1965**, *30*, 4138.
22. Glaze, W. H.; Lin, J.; Felton, E. G. *J. Org. Chem.* **1965**, *30*, 1258.
23. Morton, M.; Rupert, J. R. In *Initiation of Polymerization;* Bailey, F. E. Jr., Ed.; ACS Symposium Series No. 212; American Chemical Society: Washington, DC, 1983; p. 284.
24. Jagur-Grodzinski, J. *React. Funct. Polym.* **2001**, *49*, 1.
25. Schulz, D. N.; Sanda, J. C.; Willoughby, B. C. In *Anionic Polymerization: Kinetics Mechanisms and Synthesis;* McGrath, J. E., Ed., ACS Symposium Series No. 166; American Chemical Society: Washington, DC, 1981; p 427.
26. Harris, F. W.; Spinelli, H. J.; Eds. *Reactive Oligomers,* ACS Symposium Series No. 282; American Chemical Society: Washington, DC, 1985.
27. Rempp, P. F.; Franta, E. *Adv. Polym. Sci.* **1984**, *58*, 1.
28. Brosse, J.-C.; Derouet, D.; Epaillard, F.; Soutif, J.-C.; Legeay, G.; Dusek, K. *Adv. Polym. Sci.* **1987**, *81*, 167.
29. Akelah, A. *J. Mat. Sci.* **1986**, *21*, 2977.
30. Paulus, G.; Jerome, R.; Teyssie, P. *Brit. Polym. J.* **1987**, *19*, 361.
31. Bergbreiter, D. E.; Martin, C. R.; Eds., *Functional Polymer;* Plenum Press: New York, 1989.
32. Quirk, R. P.; Yin, J.; Guo, S.-H.; Hu, X.-W.; Summers, G.; Kim, J.; Zhu, L.; Schock, L. E. *Makromol. Chem. Maacromol. Symp.* **1990**, *32*, 47.
33. Quirk, R. P. In *Comprehensive Polymer Science; First Supplement;* Agarwal, S. L., Russo, S., Eds.; Pergamon Press: Elmsford, NY, 1992; p 83.
34. Quirk, R. P.; Yin, J.; Guo, S.-H.; Hu, X.-W.; Summers, G.; Kim, J.; Zhu, L. F.; Ma, J.-J.; Takizawa, T.; Lynch, T. *Rubber Chem. Tech.* **1991**, *64*, 648.
35. Gnanou, Y. *Ind. J. Technol.* **1993**, *31*, 317.
36. Goethals, E. J. In *Ring-Opening Polymerization. Mechanisms, Catalysis, Structure Utility;* Brunelle, D. J., Ed.; Hanser: New York, 1993; p 295.
37. Quirk, R. P.; Kim, J. In *Ring-Opening Polymerization. Mechanisms, Catalysis, Structure Utility;* Brunelle, D. J., Ed.; Hanser: New York, 1993; p 263.
38. Mays, J. W.; Hadjichristidis, N. *Polym. Bull.* **1989**, *22*, 471.

39. Richards, D. H.; Eastmond, G. C.; Stewart, M. J. In *Telechelic Polymers: Synthesis and Applications;* Goethals, E. J., Ed.; CRC Press: Boca Raton, FL, 1989; p 33.

40. Fontainille, M. In *Comprehensive Polymer Science, Chain Polymerization I;* Eastmond, G. C., Ledwith, A., Russo, S., Sigwalt, P., Eds.; Pergamon Press: Elmsford, NY, 1989; Vol. 3, p 425.

41. Fetters, L. J. *J. Polym. Sci., Macromol. Rev.* **1967,** *2,* 71.

42. Morton, M.; Fetters, L. J. *J. Polym. Sci. Part C.* **1969,** *26,* 1.

43. Bywater, S. *Prog. Polym. Sci.* **1974,** *4,* 27.

44. Wakefield, B. J. In *Organolithium Methods;* Academic Press: San Diego, CA, 1988.

45. Brandsma, L.; Verkuijsse, H. D. In *Preparative Polar Organometallic Chemistry;* Springer-Verlag: Berlin, 1987; Vol. 1.

46. Bates, R. B.; Ogle, C. A. *Carbanion Chemistry;* Springer-Verlag: Berlin, 1983.

47. Wakefield, B. J. In *Comprehensive Organometallic Chemistry;* Wilkinson, G. Ed.; Pergamon: Oxford, 1982; Chapter 44.

48. Richards, D. H.; Eastmond, G. C.; Stewart, M. J. In *Telechelic Polymers: Synthesis and Applications;* Goethals, E. J., Ed.; CRC Press: Boca Raton, FL, 1989; p 33.

49. Wakefield, B. J. *The Chemistry of Organolithium Compounds;* Pergamon Press: Elmsford, NY, 1974.

50. (a). Ueda, K.; Hirao, A.; Nakahama, S. *Macromolecules* **1990,** *23,* 939. (b). Hirao, A.; Nagahama, H.; Ishizone, T.; Nakahama, S. *Macromolecules* **1993,** *26,* 2145. (c). Tohyama, M.; Hirao, A.; Nakahama, S.; Takenaka, K. *Macromol. Chem. Phys.* **1996,** *197,* 3135. (d). Hirao, A.; Hayashi, M.; Nakahama, S. *Macromolecules* **1996,** *29,* 3353. (e). Hirao, A.; Tohoyama, M.; Nakahama, S. *Macromolecules* **1997,** *30,* 3484. (f). Hayashi, M.; Loykulnant, S.; Hirao, A.; Nakahama, S. *Macromolecules* **1998,** *31,* 2057. (g). Hirao, A.; Hayashi, M.; Nakahama, S. *Macromolecules* **1996,** *29,* 3353. (h). Cerno-hous, J. J.; Macosko, C. W.; Hoye, T. R. *Macromolecules* **1997,** *30,* 5213.

51. (a). Hunt, M. O.; Belu, A. M.; Linton, R. W.; DeSimone, J. M. *Macromolecules* **1993,** *26,* 4854. (b). Belu, A. M.; Hunt, M. O.; DeSimone, J. M.; Linton, R. W.; *Macromolecules* **1994,** *27,* 1905.

52. Szwarc, M. *Adv. Polym. Sci.* **1983,** *49,* 1.

53. Laita, Z.; Szwarc, M. *Macromolecules* **1969,** *2,* 412.

54. Busson, R.; van Beylen, M. *Macromolecules* **1977,** *10,* 1320.

55. Ziegler, K.; Gellert, H. G. *Justus Lieb. Ann. Chem.* **1950,** *567,* 179.

56. Köbrich, G.; Stober, I. *Chem. Ber.* **1970,** *103,* 2744.

57. Wittig, G.; Scholpokf, U. *Chem. Ber.* **1954,** *87,* 1318.

58. (a). Quirk, R. P.; Kim, H. *Macromolecules* **2005,** *38,* 7895. (b). Quirk, R. P.; Hoon, K.; Chowdhury, S. R. *Polym. Prepr. (Am. Chem. Soc., Div. Polym. Chem.)* **2005,** *46* (2), 583. (c). Quirk, R. P.; Chowdhury, S. R. *Polym. Prepr. (Am. Chem. Soc., Div. Polym. Chem.)* **2006,** *47* (1), 102. (d). Quirk, R. P.; Chowdhury, S. R. *Polym. Prepr. (Am. Chem. Soc., Div. Polym. Chem.)* **2006,** *47* (2), 182. (e). Quirk, R. P.; Chowdhury, S. R. *Macromolecules* **2009,** *42,* 494. (f). Reidar, L.; Plaza-García, S.; Alegría, A.; Colmenero, J.; Janoski, J.; Chowdhury, S. R.; Quirk, R. P. *J. Non-Cryst. Solids* **2010,** *356,* 676. (g). Quirk, R. P.; Janoski, J.; Chowdhury,. *Macromolecules* **2009,** *42,* 494. (h). Lund, R.; Plaza-García, S.; Alegría, A.; Colmenero, J.; R. S Janoski, J.; Chowdhury, S. R.; Quirk, R. P. *Macromolecules* **2009,** *42,* 8875.

59. Touris, A.; Mays, J. W.; Hadjichristidis, N. *Macromolecules* **2011,** *44,* 1886.

60. Liu, B.; Quirk, R. P.; Wesdemiotis, C.; Yol, A. M.; Foster, M. D. *Macromolecules* **2012,** *45,* 9233.

61. Mays, J. W.; Hadjichristidis, N. *Polym. Bull.* **1989,** *22,* 471.
62. Quirk, R. P.; Sarkis, M. T.; Meier, D. J. In *Advances in Elastomers and Rubber Elasticity;* Lal, J., Mark, J. E., Eds., Plenum Press: New York, 1986; p 143.
63. Chen, J. C.; Fetters, L. J. *Polym. Eng. Sci.* **1987,** *27,* 1300.
64. Conlon, D. A.; Crivello, J. V.; Lee, J. L.; O'Brien, M. J. *Macromolecules* **1989,** *22,* 509.
65. Yuki, H.; Okamoto, Y.; Kuwae, Y.; Hatada, K. *J. Polym. Sci.* **1969,** *7,* 1933.
66. Geerts, J.; Van Beylen, M.; Smets, G. *J. Polym. Sci.* **1969,** *7,* 2859.
67. Se, K.; Kijima, M.; Fujimoto, T. *Polym. J.* **1988,** *20,* 791.
68. Shima, M.; Ogawa, E.; Konishi, K. *Makromol. Chem.* **1976,** *177,* 241.
69. Ogawa, E.; Shima, M. *Polym. J.* **1986,** *20,* 791.
70. Konigsberg, I.; Jagur-Grondinski, J. *J. Polym. Sci. Polym. Chem. Ed.* **1983,** *21,* 2535.
71. Konigsberg, I.; Jagur-Grondinski, J. *J. Polym. Sci. Polym. Chem. Ed.* **1983,** *21,* 2649.
72. Narita, T.; Hagiwara, T.; Hamana, H.; Irie, H.; Sugiyama, H. *Polym. J.* **1987,** *19,* 985.
73. Chaumont, P.; Beinert, G.; Herz, J.-E.; Rempp, P. *Makromol. Chem.* **1982,** *183,* 1181.
74. Hatada, K.; Okamoto, Y.; Kitayama, T.; Sasaki, S. *Polym. Bull.* **1983,** *9,* 228.
75. Okay, O.; Funke, W. *Macromolecules* **1990,** *23,* 2623.
76. Nagasaki, Y.; Ito, H.; Tsuruta, T. *Makromol. Chem.* **1986,** *187,* 23.
77. Kajiwara, K.; Suzuki, H.; Inagaki, H.; Maeda, M.; Tsuruta, T. *Makromol. Chem.* **1986,** *187,* 2257.
78. Nagasaki, Y.; Tamura, Y.; Tsuruta, T. *Makromol. Chem., Rapid Commun.* **1988,** *9,* 31.
79. Nakagawa, H.; Matsushita, Y.; Tsuge, S. *Polymer* **1987,** *28,* 1512.
80. Nasirova, R. M.; Murav'eva, L. S.; Mushina, E. A.; Krentsel, B. A. *Russ. Chem. Rev.* **1979,** *48,* 692.
81. Worsfold, D. J. *Macromolecules* **1970,** *3,* 514.
82. Young, R. N.; Fetters, L. J. *Macromolecules* **1978,** *11,* 899.
83. Ishizone, T.; Hirao, A.; Nakahama, S. *Macromolecules* **1991,** *24,* 625.
84. Ishizone, T.; Wakabayashi, S.; Hirao, A.; Nakahama, S. *Macromolecules* **1991,** *24,* 5015.
85. Whicher, S. J.; Brash, J. L. *J. Appl. Polym. Sci.* **1985,** *30,* 2297.
86. Whicher, S. J.; Brash, J. L. *J. Polym. Sci. Polym. Chem. Ed.* **1981,** *19,* 1995.
87. Kase, T.; Imahori, M.; Kazama, T.; Isono, Y.; Fujimoto, T. *Macromolecules* **1991,** *24,* 1714.
88. Ishizone, T.; Hirao, A.; Nakahama, S. *Macromolecules* **1989,** *22,* 2895.
89. Yamakazi, N.; Nakahama, S.; Hirao, A.; Goto, J.; Shiraishi, Y.; Martinez, F.; Phung, H. M. *J. Macromol. Sci., Pure Appl. Chem.* **1981,** *16,* 1129.
90. Hirao, A.; Shiraishi, Y.; Martinez, F.; Phung, H. M.; Nakahama. S.; Yamakazi, N. *Makromol. Chem.* **1983,** *184,* 961.
91. Sakellariou, G.; Baskaran, D.; Hadjichristidis, N.; Mays, W. J. *Macromolecules* **2006,** *39,* 3525.
92. Kobayashi, S.; Matsuzawa, T.; Matsuoka, S.; Tajima, H.; Ishizone, T. *Macromolecules* **2006,** *39,* 5979.
93. Higashihara, T.; Ueda, M. *Macromolecules* **2009,** *42,* 8794.
94. Hirao, A.; Loykulnant, S.; Ishizone, T. *Prog. Polym. Sci.* **2002,** *27,* 1399.
95. Ishizone, T.; Sugiyama, K.; Hirao, A. In *Polymer Science: A Comprehensive Reference*; Matyjaszewski, K., Möller, M., Eds.; Elsevier BV: Amsterdam, 2012; Vol. 3, p 591.
96. Ishizone, T.; Hirao, A. In *Anionic Polymerization: Recent Advances. In Synthesis of Polymers: New Structures and Methods*; Schlüter, D., Hawker, C. J., Sakamoto, J., Eds.; Wiley-VCH: Singapore, 2012; Vol. 1, p 81.

97. Hirao, A.; Nakahama, S. *Macromolecules* **1987,** *20,* 2968.
98. Hirao, A.; Nakahama, S. *Polymer* **1986,** *27,* 309.
99. Ishizone, T.; Wakabayashi, S.; Hirao, A.; Nakahama, S. *Macromolecules* **1991,** *24,* 5015.
100. Ishizone, T.; Hirao, A.; Nakahama, S. *Macromolecules* **1989,** *22,* 2895.
101. Ishizone, T.; Kato, H.; Yamazaki, D.; Hirao, A.; Nakahama, S. *Macromol. Chem.* **2000,** *201,* 1077
102. Ishizone, T.; Tsuchiya, J.; Hirao, A.; Nakahama, S. *Macromolecules* **1992,** *25,* 4840.
103. Takenaka, K.; Hirao, A.; Hattori, T.; Nakahama, S. *Macromolecules* **1987,** *20,* 2034.
104. Kobayashi, S.; Kataoka, H.; Ishizone, T. *Macromolecules* **2009,** *42,* 5017.
105. (a). Schreiber, H. *Makromol. Chem.* **1960,** *36,* 86. (b). Goode, W. E.; Owens, F. H.; Myers, W. *J. Polym. Sci.* **1960,** *47,* 75. (c). Graham, R. K.; Dunkelberger, D. L.; Goode, W. E. *J. Am. Chem. Soc.* **1960,** *82,* 400.
106. (a). Glusker, D. L.; Gallucio, R. A.; Evans, R. A. *J. Am. Chem. Soc.* **1964,** *86,* 187. (b). Kawabata, N.; Tsuruta, T. *Makromol. Chem.* **1965,** *80,* 231. (c). Feit, B. A. *Eur. Polym. J.* **1967,** *3,* 523.
107. (a). Jeuck, H.; Muller, A. H. E. *Makromol. Chem. Rapid Commun.* **1982,** *3,* 121. (b). Warzelhan, V.; Hocker, H.; Schulz, G. V. *Makromol. Chem.* **1980,** *181,* 149. (c). Kunkel, D.; Muller, A. H. E.; Lochmannn, L.; Janata, M. *Makromol. Chem. Macromol. Symp.* **1992,** *60,* 315.
108. Anderson, B. C.; Andrews, G. D.; Arthur, P. Jr.; Jacobson, H. W.; Melby, L. R.; Playtis, A. J.; Sharkey, W. H. *Macromolecules* **1984,** *14,* 1599.
109. Wang, J. S.; Jerome, R.; Teyssie, P. *J. Phys. Org. Chem.* **1995,** *8,* 208.
110. Varshney, S. K.; Jerome, R.; Bayard, P.; Jacobs, C.; Fayt, R.; Teyssie, P. *Macromolecules* **1992,** *25,* 4457.
111. (a). Wang, J.-S.; Jerome, R.; Bayard, P.; Baylac, L.; Patin, M.; Teyssie, P. *Macromolecules* **1994,** *27,* 4615. (b). Johann, C.; Muller, A. H. E. *Makromol. Chem. Rapid Commun.* **1981,** *2,* 687. (c). Ballard, D. G. H.; Bowles, R. J.; Haddleton, D. M.; Richards, S. N.; Sellens, R.; Twose, D. L. *Macromolecules* **1992,** *25,* 5907.
112. (a). Baskaran, D.; Chakrapani, S.; Sivaram, S. *Macromolecules* **1995,** *28,* 7315. (b). Baskaran, D, Muller, A. H. E.; Sivaram, S. *Macromol. Chem. Phys.* **2000,** *201,* 1901. (c). Marchal, J.; Gnanou, Y.; Fontanille, M. *Makromol. Chem. Macromol. Symp.* **1996,** *107,* 27. (d). Gia, H.-B.; McGrath, J. E. In *Recent Advances in Anionic Polymerization;* Hogen-Esch, T., Smid, J., Eds.; Elsevier: New York, 1987; p 173.
113. (a). Kitayama, T.; Shinozaki, T.; Sakamoto, T.; Yamamoto, M.; Hatada, K. *Macromol. Chem. Suppl.* **1989,** *15,* 167. (b). Schlaad, H.; Schmitt, B.; Muller, A. H. E.; Jungling, S.; Weiss, H. *Macromolecules* **1998,** *31,* 573. (c). Schlaad, H.; Schmitt, B.; Muller, A. H. E. *Angew. Chem. Int. Ed. Engl.* **1998,** *37,* 1389.
114. (a). Lochmann, L.; Kolarik, J.; Doskocilova, D.; Vozka, S.; Trekoval, J. *J. Polym. Sci., Polym. Chem. Ed.* **1979,** *17,* 1727. (b). Lochmann, L.; Muller, A. H. E. *Makromol. Chem.* **1990,** *191,* 1657. (c). Vlcek, P.; Lochmann, L.; Otoupalova, J. *Macromol. Chem. Rapid Commun.* **1992,** *13,* 163.
115. (a). Varshney, S. K.; Hautekeer, J.-P.; Fayt, R.; Jerome, R.; Teyssie, P. *Macromolecules* **1990,** *23,* 2618. (b). Varshney, S. K.; Bayard, P.; Jacobs, C.; Jerome, R.; Fayt, R.; Teyssie, P. *Macromolecules* **1992,** *25,* 5578.
116. (a). Baskaran, D.; Sivaram, S. *Macromolecules* **1997,** *30,* 1550. (b). Baskaran, D.; Muller, A. H. E.; Sivaram, S. *Macromolecules* **1999,** *32,* 1357.

117. (a). Baskaran, D. *Macromol. Chem. Phys.* **2000**, *201*, 890. (b). Wang, J.-S.; Jerome, R.; Bayard, P.; Patin, M.; Teyssie, P. *Macromolecules* **1994**, *27*, 4635. (c). Wang, J. S.; Jerome, R.; Bayard, P.; Teyssie, P. *Macromolecules* **1994**, *27*, 4908. (d). Maurer, A.; Marcarian, X.; Muller, A. H. E.; Navarro, C.; Vuillemin, B. *Polym Prepr. (Am Chem Soc, Div Polym Chem)* **1997**, *38* (1), 467.
118. Marchal, J.; Fontanille, M.; Gnanou, Y. *Proceedings of IP 97 (Paris)* 1997.
119. Zundel, T.; Zune, C.; Teyssie, P.; Jerome, R. *Macromolecules* **1998**, *31*, 4089.
120. Reetz, M. T.; Ostarek, R. *Chem. Commun.* **1988**, 213.
121. Reetz, M. T.; Knauf, T.; Minet, U.; Bingel, C. *Angew. Chem. Int. Ed. Engl.* **1988**, *100*, 1422.
122. Reetz, M. T.; Hutte, S.; Goddard, R. *J. Phys. Org. Chem.* **1995**, *8*, 231.
123. Quirk, R. P.; Bidinger, G. P. *Polym. Bull.* **1989**, *22*, 63.
124. Reetz, M. T.; Herzog, H. M.; Konen, W. *Macromol. Rapid Commun.* **1996**, *17*, 383.
125. Baskaran, D.; Chakrapani, S.; Sivaram, S.; Hogen-Esch, T. E.; Muller, A. H. E. *Macromolecules* **1999**, *32*, 2865.
126. Webster, O. W. *J. Macromol. Sci., Pure Appl. Chem.* **1994**, *A31*, 927.
127. Pietzonka, T.; Seebach, D. *Angew. Chem. Int. Ed. Engl.* **1993**, *32*, 716.
128. Baskaran, D.; Muller, A. H. E. *Macromol. Rapid. Commun.* **2000**, *21*, 390.
129. Konigsmann, H.; Jungling, S.; Muller, A. H. E. *Macromol. Rapid Comm.* **2000**, *21*, 758.
130. Zagata, A. P.; Hogen-Esch, T. E. *Macromolecules* **1996**, *29*, 3038.
131. Baskaran, D.; Muller, A. H. E. *Macromolecules* **1997**, *30*, 1869.
132. Quirk, R. P.; Kim, J.-S. *J. Phy. Org. Chem.* **1995**, *8*, 242.
133. Dicker, I. B.; Cohen, G. M.; Farnham, W. B.; Hertler, W. R.; Laganis, E. D.; Sogah, D. Y. *Macromolecules* **1990**, *23*, 4034.
134. Brittain, W. J.; Dicker, I. B. *Macromolecules* **1989**, *22*, 1054.
135. Sitz, H.-D.; Bandermann, F. In *Recent Advances in Mechanistic and Synthetic Aspects of Polymerization*; Fontanille, M., Guyot, M., Eds.; Reidel: Dordrecht, 1987; p 41.
136. Seebach, D. *Angew. Chem. Int. Ed. Engl.* **1988**, *100*, 1624.
137. Simms, J. A. *Rubber Chem. Technol.* **1991**, *64*, 139.
138. Sogah, D. Y.; Hertler, W. R.; Webster, O. W.; Cohen, G. M. *Macromolecules* **1987**, *20*, 1473.
139. Hertler, W. R.; Sogah, D. Y.; Webster, O. W.; Trost, B. M. *Macromolecules* **1984**, *17*, 1415.
140. Zhuang, R.; Muller, A. H. E. *Macromolecules* **1995**, *28*, 8043.
141. Doherty, M. A.; Muller A. H. E. *Makromol. Chem.* **1989**, *190*, 527.
142. Mykytiuk, J.; Armes, S. P.; Billingham, N. C. *Polym. Bull.* **1992**, *29*, 139.
143. (a). Rannard, S. P.; Billingham, N. C.; Armes, S. P.; Mykytiuk, J. *Eur. Polym. J.* **1993**, *29*, 407. (b). Lowe, A. B.; Billingham, N. C.; Armes, S. P. *Chem. Commun.* **1997**, 1035.
144. Mori, H.; Muller, A. H. E. *Prog. Polym. Sci.* **2003**, *28*, 1403.
145. (a). Radke, W.; Muller, A. H. E. *Makromol. Chem. Macromol. Symp.* **1992**, *54/55*, 583. (b). Asami, R.; Takaki, M.; Moriyama, Y. *Polym. Bull.* **1986**, *16*, 125.
146. (a). Webster, O. W. *Macrmol. Chem. Macromol. Symp.* **1990**, *33*, 133. (b). Spinelli, H. J. *US Patent* 4, 627, 913 to DuPont. (c). Haddleton, D. M.; Crossman, M. C. *Macromol. Chem.* **1997**, *198*, 871. (d). Georgious, T. K.; Vamvakaki, M.; Patrickios, C. S. *Biomacromolecules* **2004**, *5*, 2221.
147. Simmons, M. R.; Yamasaki, E. N.; Patrickios, C. S. *Macromolecules* **2000**, *33*, 3176.
148. Spinelli, H. J. *Adv. Org. Coat. Sci. Technol.* **1990**, *12*, 34.

149. (a). Webster, O. W.; Sogah, D. Y. In *Recent Advances in Mechanistic and Synthetic Aspects of Polymerization*; Fontanille, M., Guyot, M., Eds.; Reidel: Dordrecht, 1987; p 3. (b). Sannigrahi, B. Ph.D. thesis, 1997, Pune University, Pune, India.
150. Cohen, G.; Mreich, H. J. *US Patent* 4, 983, 679, DuPont.
151. Sogah, D. Y.; Webster, O. W. *J. Polym. Sci: Polym. Lett. Ed.* **1983**, *21*, 927.
152. Simon, P. F. W.; Muller, A. H. E. *Macromolecules* **2004**, *37*, 7548.
153. Yamasaki, E. N.; Patrickios, C. S. *Eur. Polym. J.* **2003**, *39*, 609.
154. Vijayaraghavan, R.; MacFarlance, D. R. *Chem. Commun.* **2005**, 1149.
155. Andrews, G. D.; Vatvars, A. *Macromolecules* **1981**, *14*, 1603.
156. Simms, J. A.; Spinelli, H. J. *J. Coat. Technol.* **1987**, *57*, 125.
157. Shen, W.-P.; Chu, W.-D.; Yang, M.-F.; Wang, L. *Macromol. Chem.* **1989**, *190*, 3061.
158. Hertler, W. R.; Rajan Babu, T. V. *Polym. Prepr. (Am. Chem. Soc., Div. Polym. Chem.)* **1988**, *29* (2), 71.
159. Witkowski, R.; Bandermann, F. *Macromol. Chem.* **1989**, *190*, 2173.
160. Asami, R.; Kondo, Y.; Takak, M. *Polym. Prepr. (Am. Chem. Soc., Div. Polym. Chem.)* **1986**, *27* (1), 186.
161. Asami, R. Jpn. Kokai Tokkyo Koho JP 63/86703, Toagosei Chemical Industry.
162. Witkowski, R.; Bandermann, F. *Macromol. Chem.* **1989**, *190*, 2173.
163. Kuhn, R. *Makromol. Chem.* **1976**, *177*, 1525.
164. Webster, O. W.; Heitz, T. *Makromol. Chem.* **1991**, *192*, 2463.
165. Webster, O. W.; Sogah, D. Y.; Hertler, W. R.; Farnham, W. B.; Rajan Babu, T. V. *J. Macromol. Sci. Chem.* **1984**, *A21*, 943.
166. Grigor, J.; Eastmond, G. C. *Macromol. Chem. Rapid Commun.* **1986**, *7*, 375.
167. (a). Ren, J.; Quirk, R. P. *Polymer International* **1993**, *32*, 205. (b). Ren, J.; Quirk, R. P. *Makromol. Symp.* **1994**, *88*, 17.
168. (a). Spinelli, H. J. *Adv. Org. Coat. Sci. Technol. Ser.* **1990**, *12*, 34. (b). Simms, J. A. *J. Coat. Tech.* **1987**, *59*, 125.
169. (a). Cohen, G. M. *Polym. Prepr.* **1998**, *29* (2), 46. (b). DeSimone, J. M.; Hellstern, A. M.; Siochi, E. J.; Smith, S. D.; Ward, T. C.; Krukonis, V. J.; McGrath, J. E. *Makromol. Chem. Macromol. Symp.* **1990**, *32*, 21.
170. (a). Heitz, T.; Webster, O. W. *Macromol. Chem.* **1991**, *192*, 2463. (b). Hertler, W.; Sogah, D. Y.; Boetlcher, F. P. *Macromolecules* **1990**, *23*, 1264. (c). Jenkins, A. D.; Tsartolia, E.; Walton, D. R. M.; Hoska-Jekins, J.; Kratchovil, P.; Stejskal, J.; *Macromol. Chem.* **1990**, *191*, 2511.
171. Gnaneshwar, R.; Sivaram, S. *J. Polym. Sci., Polym. Chem. Ed.* **2007**, *45*, 2514.
172. Hayashi, M.; Nakahama, S.; Hirao, A. *Macromolecules* **1999**, *32*, 1325.
173. Fetters, L. J.; Firer, E. R. *Polymer* **1977**, *18*, 306.
174. Fontaine, F.; Ledent, J.; Sobey, R.; Francoise, E.; Jerome, R.; Teyssie, P. *Macromolecules* **1993**, *26*, 1480.
175. Müller, A. H. E. In *Recent Advances in Anionic Polymerization*; Hogen-Esch, T., Smid, J., Eds.; Elsevier: New York, 1987; p 205.
176. Wang, J. S.; Jérôme, R.; Teyssié, P. *J. Polym. Sci., Polym. Chem. Ed.* **1992**, *30*, 2251.
177. Ohata, M.; Isono, Y. *Polymer* **1993**, *34*, 1546.
178. Mason, J. P.; Hattori, T.; Hogen-Esch, T. E. Polym. *Prepr. (Am. Chem. Soc., Div. Polym. Chem.)* **1989**, *30* (1), 259.
179. Dhara, M. G.; Baskaran, D.; Sivaram, S. *J. Polym. Sci., Polym. Chem. Ed.* **2008**, *46*, 2132.
180. Dhara, M. G.; Sivaram, S.; Baskaran, D. *Polym. Bull.* **2009**, *63*, 185.

181. Dhara, M. G.; Sivaram, S. *J. Macromol. Sci., Pure Appl. Chem.* **2009,** *46,* 983.
182. Li, Z.; Day, M.; Ding, J.; Faid, K. *Macromolecules* **2005,** *38,* 2620.
183. Percec, V.; Auman, B. C. *Macromol. Chem.* **1984,** *185,* 2319.
184. Percec, V.; Auman, B. C. *Macromol. Chem.* **1984,** *185,* 1867.
185. Yu, Y. S.; Dubois, P.; Jerome, R.; Teyssie, P. *Macromolecules* **1996,** *29,* 1753.
186. Ishizone, T.; Uehara, G.; Hirao, A.; Nakahama, S.; Tsuda, K. *Macromolecules* **1998,** *31,* 3764.
187. Ruckenstein, E.; Zhang, H. *Macromolecules* **1999,** *32,* 6082.
188. Murali Mohan, M.; Raghunadh, V.; Sivaram, S.; Baskaran, D. *Macromolecules* **2012,** *45,* 3387.

CHAPTER 2

ATOM TRANSFER RADICAL POLYMERIZATION: A KEY TOOL TOWARDS THE DESIGN AND SYNTHESIS OF FUNCTIONAL POLYMERS

VENKATANARASIMHAN SWARNALATHA,
MUKUNDAMURTHY KANNAN and
RAGHAVACHARI DHAMODHARAN

CONTENTS

2.1   INTRODUCTION

Functional polymers are important among the several primary representatives of modern day materials. Polymers are similar to small organic and inorganic molecules in certain ways and yet possess novel and unusual characteristics such as molecular entanglement, viscosity, viscoelasticity, glass transition, and few other specific properties. Polymers are primarily synthesized by the chain growth or step growth kinetics. The most common method of synthesizing commercial polymers in very large volume or bulk is the conventional free radical polymerization (CFRP), which is one form of chain growth polymerization. The others include cationic, anionic, and organometallic based polymerizations. The major concern associated with CFRP is that the polymers synthesized exhibit broad molecular weight distribution (MWD) and therefore distribution of properties. The physical properties of polymers such as mechanical strength, melt viscosity, and diffusion coefficient in solution depend critically on the molecular weight and MWD. In addition, they invariably cannot be used for the preparation of polymers with varying molecular architecture. For example, chain extension using other monomers to prepare block copolymers requires an active chain end. This led to the development of new approaches called "controlled/living radical polymerizations" (CRP) that were able to minimize the limitations of CFRP. The IUPAC nomenclature for CRP is reversible-deactivation radical polymerization (RDRP). In RDRP, the lifetime of the growing chains is relatively higher (>1 h) than that in CFRP (~1 s).[1] The RDRP relies on a peculiar kinetic behavior called persistent radical effect (PRE)[2,3] whereas in CFRP, a steady state concentration of radicals (~10^{-8} M) is achieved due to similar rates of initiation and termination steps. In CFRP, at the end of polymerization, all the chain ends are dead (not radical in nature) while in RDRP, the chain ends in the polymers can be reactivated. These active chain ends enable further functionalization of the polymer chains. In CFRP, bimolecular termination occurs between two growing chains with a free radical chain end, leading to the formation of dead polymer chains. On the other hand, in RDRP, the termination rate decreases with time due to PRE. The consequence of these kinetics steps is that in CFRP the molecular weight of the polymer increases dramatically at very low monomer conversion and remains the same almost until the end of the conversion of all the monomers. In the case of RDRP, in view of the control achieved over the chain propagation step (and with minimum termination) the chain continues to grow with monomer conversion resulting in a linear variation of molecular weight with monomer conversion. In simpler words, polymers with controllable molecular weight, desired architecture, and narrow MWD can be successfully synthesized using RDRP.

Several RDRP methods were developed in mid-1990s and early-2000s, which led to a revolution in the synthesis of advanced functional materials. The important RDRP methods include nitroxide mediated polymerization (NMP),[4] atom transfer radical polymerization (ATRP),[5,6] single-electron transfer living radical polymerization (SET-LRP),[7] reversible addition-fragmentation chain transfer polymerization (RAFT),[8] and single electron transfer–reversible addition-fragmentation chain transfer polymerization (SET-RAFT).[9] Among these, ATRP is the most extensively studied, due to its robust nature, versatility and being simple to perform. The ATRP technique has been exploited to synthesize a wide range of polymers with pre-defined molecular weights and narrow MWD. The essential components of an ATRP reaction are monomer, initiator, and a catalytic system involving a metal salt (capable of switching from the lower oxidation state to the higher oxidation state by electron transfer) and a ligand. The underlying conception of a typical ATRP reaction involves establishing a reversible equilibrium between a low concentration of active propagating chains and a large concentration of dormant chains (containing transition metal complexes) via an inner sphere electron transfer mechanism. The fundamentals of ATRP and the role of all the components are not discussed here and the readers are directed to read more comprehensive reviews on the basics of ATRP.[10–14] The proposed mechanism for a transition metal catalyzed ATRP technique is presented in Scheme 2-1.

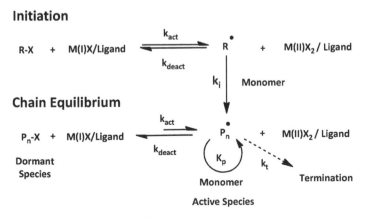

Initiation

$R\text{-}X$ + $M(I)X/Ligand$ $\underset{k_{deact}}{\overset{k_{act}}{\rightleftharpoons}}$ $R^{\bullet}$ + $M(II)X_2/Ligand$

k_i | Monomer

Chain Equilibrium

$P_n\text{-}X$ + $M(I)X/Ligand$ $\underset{k_{deact}}{\overset{k_{act}}{\rightleftharpoons}}$ $P_n^{\bullet}$ + $M(II)X_2/Ligand$

Dormant Species

K_p | k_t

Monomer   Termination

Active Species

SCHEME 2-1   Mechanism of an ATRP reaction.

Besides the normal ATRP, a number of variations of the ATRP method has been reported. These include AGET-ATRP, ICAR-ATRP, ARGET-ATRP, e-ATRP, and SARA-ATRP. In activators generated by electron transfer-ATRP (AGET-ATRP), sufficient concentration of a reducing agent (e.g., ascorbic acid and tin(II) ethylhexanoate) is additionally added in order to consume the trace quantity of oxygen present,[15,16] so that the freeze-pump-thaw cycle performed before the polymerization is avoided. ATRP makes use of a large concentration of metal catalysts and therefore the final polymer obtained is typically contaminated by metals. This contamination restricts the implementation of ATRP in industries or in other commercial processes.[17] These constraints were minimized by the ATRP techniques viz. initiators for continuous activator regeneration ATRP (ICAR-ATRP),[18] activators regenerated by electron transfer ATRP (ARGET-ATRP),[19] electrochemically mediated ATRP (eATRP),[20,21] and photo ATRP.[22] A newly added ATRP technique in these low catalyst ATRP systems is supplemental activators and reducing agents ATRP (SARA-ATRP), in which zero valent metals such as copper, zinc, iron, and magnesium are employed as supplemental activators and reducing agents.[23] Very recently, a modification of SARA-ATRP was demonstrated to polymerize cationic monomers successfully,[24] in aqueous solutions. The recent advances in RDRP have allowed the preparation of precursor macromolecules of well-defined structure and architecture. The mechanisms of the modified ATRP processes are illustrated in Schemes 2-2–2-4.

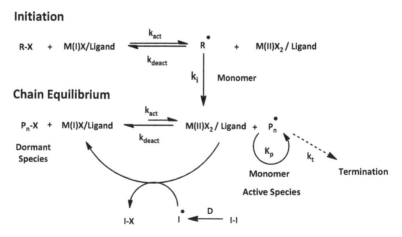

SCHEME 2-2 Mechanism of ICAR-ATRP.

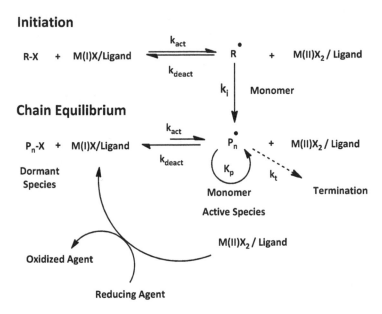

SCHEME 2-3   Mechanism of ARGET-ATRP.

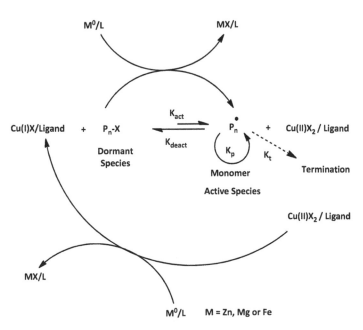

SCHEME 2-4   Mechanism of SARA-ATRP.

All the RDRP methods including ATRP have been used to prepare a wide variety of new functional polymers with controlled macromolecular architecture as illustrated in Figure 2-1.

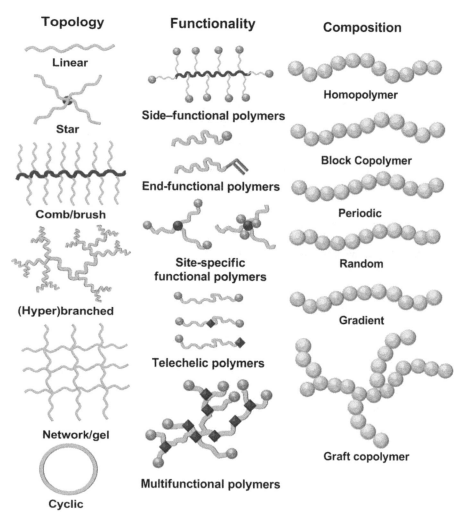

FIGURE 2-1   Functional polymers (with controlled topology, composition, and functionality) synthesized using ATRP. Adapted from reference 25.

A number of research reviews with different viewpoints have been published on the synthesis of polymers using ATRP. This chapter summarizes the various cutting edge research activities pursued in the field of using ATRP towards developing functional polymers with divergent properties and applications. Different kinds of functional materials have been divided and presented in the sub sections: (1) polymer brushes, (2) stimuli–responsive polymers, (3) microspheres, (4) microcapsules/nanocapsules, (5) degradable polymers, (6) self-healing polymers, and (7) energy harvesting materials and sensors.

2.2 POLYMER BRUSHES

Polymer brushes garnered the limelight in the world of materials science, few decades ago and still remain to be one of the most intriguing branches of polymer chemistry. Polymer brushes can be described as ensembles of thin layers of polymer chains tethered to a solid surface or an interface with one end anchored to the material–polymer interface with much of the polymer backbone present in the extended form. When polymer chains are terminally attached to any surface, closely, they tend to stretch away from the surface due to overlapping of the chains. The extent of stretching is determined by the balance between the repulsive interactions arising out of overlapping and the loss of conformational entropy (elastic free energy) by stretching. The structural and dynamic aspects of the polymer brushes, which provide the models for several interfacial systems has been studied, widely, for several years. The tethering of polymer in the brush form on a solid substrate enables the modification of the surface properties of the particular substrate and is an appealing way to introduce targeted functionalities and physicochemical properties. Polymer brushes have also emerged as fascinating materials because of their wide-spread applications in biomedical devices,[26–31] energy generation and storage,[32–34] sensing,[35-40] and environmental remediation.[41,42]

2.2.1 POLYMER BRUSHES VIA GRAFTING FROM AND GRAFTING TO

Polymer brushes can be achieved through two different experimental approaches viz. *grafting to* and *grafting from*, each of which possess unique characteristics. In *grafting to* technique, end-functionalized, pre-synthesized, polymer chains are physically (physisorption) or covalently attached (chemisorption) to the reactive sites on a surface to form a monolayer. In contrast, *grafting from* technique involves two steps. In the initial step, an initiator group is immobilized on the surface of the substrate, and in the following step, the polymer chains are grown from the initiator sites in situ.

Between these methodologies, *grafting from* is believed to be more suitable to provide high grafting densities because of the easy diffusion of the monomers towards the reactive sites. In the *grafting to* methodology, the already attached polymer chains sterically hinderthe new-coming polymer chains and may mask the reactive centers on the surface, ensuing low grafting densities. Therefore *grafting from* has been usually preferred to synthesize dense and ordered polymer brushes with tailored properties. But, few studies have reported achieving high grafting densities via *grafting to* method.[43,44] The merit of *grafting to* strategy is the prior knowledge of the molecular weight and polydispersity of the pre-synthesized polymers, which is preferred from the industrial perspective.[45]

Polymer brushes of different kinds such as copolymer brushes,[46] binary brushes,[47] hyperbranched brushes,[48] cross-linked brushes,[49] bimodal brushes,[50] and Janus polymer brushes[51] can be generated using ATRP technique. The upcoming paragraphs discuss the prominent trends in surface initiated ATRP (SI-ATRP) to develop functional materials for high end applications.

2.2.2 SI-ATRP FOR INORGANIC/POLYMER AND ORGANIC/POLYMER INTERFACES

A wide range of methodologies have been proposed to develop polymer brushes in the active research period spanning two decades, but achieving well-defined polymer brushes with precise control over the structure, composition, and grafting densities is still challenging. Among the several existing methods to synthesize dense polymer brushes via *grafting from* method viz. anionic polymerization, cationic polymerization, ring opening polymerization, and surface-initiated controlled radical polymerization (SI-CRP) has been frequently adopted as an effective tool to grow covalently bonded polymer chains with high grafting densities.[52] Owing to the attractive features such as versatility and precise control of structural parameters, composition, and dispersity, SI-CRP has become an indispensable route to synthesize polymer brushes. The most common SI-CRP techniques are surface-initiated atom transfer radical polymerization (SI-ATRP), surface-initiated nitroxide-mediated polymerization (SI-NMP), surface-initiated reversible addition fragmentation chain transfer polymerization (SI-RAFT), and surface initiated photo-iniferter mediated polymerization (SI-PIMP). Klok et al. has published a detailed review on the topic of development of polymer brushes via all of the above mentioned SI-CRP techniques.[52] SI-ATRP has been vastly practiced to attain a wide spectra of polymer/inorganic and polymer/organic hybrid materials with controlled structural aspects and improved properties and an updated review on the same has been published more recently.[53] It is also possible that the important structural parameters viz. grafting density, shell thickness, crosslinking density, and so forth can be well tuned by varying the reaction conditions while adapting SI-ATRP. As SI-ATRP provides several salient features to the synthesized materials to meet the qualities for high end applications, it has indeed become one of the primary choices for the modification of almost all kinds of surfaces. The implementation of SI-ATRP method through *grafting from* mode for growing polymer brushes from flat surfaces, nanostructured materials, and natural substrates and their successive applications have been discussed in the following paragraphs.

2.2.2.1 POLYMER BRUSHES GROWN FROM PLANAR SUBSTRATES

Silicon wafer is the most commonly used planar substrate for growing brushes due to the possibility to synthesize Si-C bonds (strong) through covalent bond formation and the availability of planar Si wafers. The required functional groups are immobilized on to this substrate surface by one or multi steps. Often, UV induced hydrosilylation coupling of p-/o-chloromethylstyrene is employed to provide a monolayer of stable ATRP initiator on the silicon surface by creating stable Si-C bonds. Polymer chains of the monomers such as 2-hydroxyethyl methacrylate (HEMA),[54] N,N-(dimethylamino)ethyl methacrylate (DMAEMA),[54] HEMA-*b*-DMAEMA,[54] DMAEMA-*b*-HEMA,[54] sodium 4-styrenesulfonate (SSS),[54] poly(ethylene glycol) monomethacrylate (PEGMA),[54] tertiary butylacrylate (*t*BA),[55] pentafluorostyrene,[56] glycidyl methacrylate (GMA),[57] HEMA,[58] glycidyl methacrylate-*cb*-N-isopropylacrylamide (GMA-*cb*-NIPAM),[59] PEGMA, and HEMA[60] have been covalently attached to silicon surface using the aforementioned UV induced hydrosilylation procedure. Apart from the UV coupling, direct halogenations via N-bromosuccinimide (NBS) or PCl_5 have been further exploited for the polymerization of HEMA and DMAEMA.[61] Occasionally, polymer chains are also tethered onto planar metal substrates such as nickel,[62] copper,[62] gold,[61,63] and iron surfaces.[64] SI-ATRP on metal surfaces to grow polymer brushes is more appealing over conventional polymer coatings in the context that the SI-ATRP leads to stable adhesion through covalent bonding whereas the conventional polymer coating through weak forces can be peeled off easily by stresses.[62]

2.2.2.2. POLYMER BRUSHES PRODUCED FROM NANOSTRUCTURED SURFACES

Surface derivatization chemistry of nanostructured materials has evolved tremendously over the years with the ultimate intention of producing advanced functional materials. SI-ATRP has been an effective means to create well-defined polymer/nanocomposites for a wide spectrum of monomers with good dimension and composition, among other surface modification protocols.[65] The ability of the nanomaterials to uniformly disperse in solvents increases after the surface modification with polymers due to the steric stabilization imparted by the polymeric shells in their peripheries. Polymer/nanocomposites generally display thermal, optical, conducting, and mechanical properties that are superior in relation to the parent polymers or mere nanomaterials. The nano-surfaces with several types of properties and applications such as TiO_2 (photoactive), ZnO, CdSe, and CdS (quantum dots), carbon nanotubes, carbon nanofibres, graphene (conductive), gold (metal nanoparticles), Fe_3O_4, and Fe_2O_3 (magnetically sensitive), SiO_2 and Al_2O_3 have been modified via Si-ATRP and an inventory involving few of these examples is given in Table 2-1.

TABLE 2-1 Summary of Literature on Growing Polymer Brush from Nanostructured Surfaces

Inorganic Substrate	Polymers	Major Remarks/Applications	References
TiO_2 nanoparticles (TiO_2 NP)	Poly(methyl methacrylate) (PMMA)	Effective stabilization of metal oxide nanoparticles	[66]
	PMMA	Increased the ionic conductivity of PVDF in lithium ion batteries	[67]
	PNIPAM	Thermoresponsive and self-flocculating and used in pollutant degradation	[68]
	Poly (4-chloromethyl styrene-*g*-4-vinylpyridine) (PC) (PCMSt-*g*-P4VP)	Improves the optical properties and serve as UV filters	[69]
	Poly(ethyleneoxide-*b*-dimethylaminoethyl methacrylate-*b*-styrene)	Yields efficient and safe UV screening by eliminating photocatalytic activity of titania	[70]
	HEMA	Used for further fabrication of hybrid ultrafiltration membrane with polysulphone	[71]

TABLE 2-1 *(Continued)*

Inorganic Substrate	Polymers	Major Remarks/Applications	References
	Polystyrene (PS)	Elated thermal stability and T_g	[72]
	Poly(3-sulfopropyl methacrylate) (PSPMA)	Self catalytic UV induced ATRP	[73]
	Poly(oligoethyleneglycolm ethacrylate) (POEGMA)		
	PDMAEMA		
	PNIPAM		
Alumina NP Alumina (Porous)	Poly(oxyethylene methacrylate) (POEM)	Used in the formation of organized meosporous TiO_2 electrodes for solid state dye sensitized solar cells (DSSC)	[74]
	Poly(acrylic acid) (PAA)	Delivers surface functionality to anodic aluminium oxide (AAO) membrane	[75]
Quantum Dots ZnO nanowires	PGMA	Might be useful for self assembly in semiconductor devices	[76]
ZnO NP	PMMA	Possesses excellent dispersability in organic solvents	[77]
	Poly[poly(ethylene glycol) methyl ether monomethacrylate] (PPEGMEMA)	Stronger near band edge emission	[78]
	PMMA Poly(carbazole ethyl methacrylate) (PCEM)	Improved thermal and optical behaviours, respectively	[79]
ZnO nanorods	PHEMA	Introduction of functionality	[80]
CdSe	Poly(N-vinyl carbazole) (PNVK)	Improved photovoltaic properties and device performance in solar cells in blending with poly(3-hexylthiophene)	[81]
CdS NP	Poly(acrylate) (PA)	Evenly dispersed quantum dots (QDs) with a blue shift	[82]
Carbon Based Nanomaterials Multi-wall carbon nanotubes (MWNT)	Poly(3-sulfopropylamino methacrylate)	Works as scaffold to synthesize other quantum dots/magnetite nanoparticles	[83]

TABLE 2-1   *(Continued)*

Inorganic Substrate	Polymers	Major Remarks/Applications	References
	Poly(styrene sulfonic acid) (PSSA)	Used in membrane separation and sensor electrodes with chitosan	[84]
	PS	Soluble in THF and CHCl$_3$.	[85]
	PS-*b*-PNIPAM	Amphiphilic in CHCl$_3$/water interface	
	PGMA	Used as a catalyst for Suzuki reaction after loaded with in situ synthesized Pd nanoparticles	[86]
	Poly(azidoethyl methacrylate) (PAzEMA)	Facilitates LBL-click chemistry due to the clickable azido groups on the nanosurface	[87]
Single-wall carbon nanotubes (SWNT)	PNIPAM	Have tunable antibacterial activity below with respect to the LCST of PNIPAM	[88]
	Poly(vinylquinoline)	Optoelectronic properties of CNTs are influenced by electron accepting and photoactive quinoline group	[89]
Carbon nanofibre (CNF)	P*t*BA	Fillers for polyamide-12 matrix for boosting the thermomechanical properties	[90]
Graphitic CNF	PnBA	Very high surface density of polymer chains and possible applications as sensors	[91]
	Poly(isobutyl methacrylate) (P*i*BMA)		
	P*t*BA		
Graphene Oxide	PS	Inflated solubility or dispersability	[92]
	PMMA		
	PBA		
	P*t*BA	Exhibits bistable electrical conductivity switching and nonvolatile electronic memory while integrated into poly(3-hexylthiophene) thin films	[93]
	PMMA	Reinforcement filler to enhance thermal and mechanical character	[94]

TABLE 2-1 *(Continued)*

Inorganic Substrate	Polymers	Major Remarks/Applications	References
	PNIPAM	Thermosensitive and controlled system to uptake and release the drug, ibuprofen	[95]
	PDMAEMA	Transition from LCST type thermosensitivity to UCST type thermosensitivity after quarternization of the polymer	[96]
	PSSS	Catalyst for synthesizing isoamyl benzoate	[97]
	PNIPAM Poly(3-sulfopropyl methacrylate) (PSPMA) POEGMA PDMAEMA	A soft lithography protocol to yield anionic, cationic and neutral polymers. POEGMA brushes at reduced GO specifically recognizes FITC conjugated streptavidin	[98]
Functional-ized Graphene sheets	PAA	pH tailored dispersion/ aggregation states	[99]
Gold NPs	PNIPAM-*b*-POEGMA	Thermally modulated catalytic activity with two thermosensitive points	[100]
	PNIPAM	Thermoresponsive behavior above and below the LCST	[101]
	PHEMA	Colorimetric detection for immunoassay	[102]
	Poly(methyl methacrylate-*b*-hexyl acrylate-*b*-methyl methacrylate) (PMMA-*b*-PHA-*b*-PMMA)	Preparation of organically dispersible nanoaggregates	[103]
	PBA and crosslinker ethylene glycol dimethacrylate (EGDMA)	Dual polymer layers (crosslinked PEGDMA shell and linear tethered PBA brushes) protect the linear polymer brushes dissociating from the surface of the NPs at high temperatures	[104]
	POEGMA PHEMA	Usage of functionalized DNA as ATRP initiator and naked eye DNA sensing	[105]

TABLE 2-1 *(Continued)*

Inorganic Substrate	Polymers	Major Remarks/Applications	References
	Poly(N-(2-hydroxyethyl) acrylamide) (PHEAA)	Anti-fouling coating with excellent resistance to protein adsorption and bacteria adhesion	[106]
Iron oxide Fe_3O_4 NP	PMMA	Good dispersibility in organic solvents	[107, 108]
	PMMA Poly(benzyl methacrylate) (PBnMA) PS PS-*b*-PHEMA	Stability in organic solvents and very high grafting density for PBnMA	[109]
	Poly(hydroxyl ethyl acrylate-*b*-N-isopropyl acrylamide) (PHEA-*b*-PNIPAM)	The inner polymer shell could improve drug encapsulation capacity and the outer layer provides thermoresponse	[110]
	PHEMA	Polymer obtained after the grafting of poly(ε-caprolactone) (PCL) is used in encapsulating the drug chlorambucil and the release rate relies on the chain length of PCL blocks	[111]
	P4VP	Lower pH enhanced the particle stabilization due to the positive charge repulsion of the protonated PVP brushes	[112]
	Poly(poly (ethyleneglycol) methyl ether methacrylate *stat*-azobenzene acrylate (PPEGMEMA)-*stat*-PABA)	Photocontrollable release of prednisolone, a corticosteroid drug with magnetic character	[113]
	Poly(2-(methacryloyloxy) ethyl phosphorylcholine) (PMPC)	Can be applied as magnetic resonance imaging (MRI) agents due to having biocompatibility and high relaxivity ratio	[114]
	Poly{9-(4-vinylbenzyl)-9H-carbazole-*b*-poly (ethyleneglycol) methyl ether methacrylate} (PVBK-*b*-PPEGMEMA)	Double functionality of and magnetism and fluorescence. MRI agent with negative contrast (T_2 type).	[115]

TABLE 2-1 *(Continued)*

Inorganic Substrate	Polymers	Major Remarks/Applications	References
	Poly(poly (ethylene-glycol) methyl ether methacrylate-*stat*-2-vinyl-4,4-dimethylazlactone) (PPEGMEMA-*stat*-PVDM)	Good dispersability in polar solvents and thymine peptide nucleic acid monomer can be grafted	[116]
	PDMAEMA	Quartenized polymer is antibacterial to towards *E. coli*	[117]
	PEGMEMA	MRI contrasting agent	[118]
Fe$_2$O$_3$ NP	PS	Solvent free ATRP is reported for achieving well-defined core–shell nanoparticles	[119]
	PDMAEMA	Gene delivery agent and selective separation of the transfected cells	[120]
Silica NP	Poly{methoxydi (ethylene glycol) meth-acrylate}(PDEGMMA) Poly{methoxytri(ethylcnc glycol) methacrylate} (PTEGMMA)	Introduction of thermosensitivity	[121]
	PBnMA	Brushes with high grafting density stabilizing NPs in organic solvents	[122]
	PS	Comb-like polymer brushes through using PVBC based macroinitiator	[123]
	PS PS-*b*-PMMA	Highly ordered isoporous membranes can be fabricated	[124]
	POEM PSSA	Well dispersed in Alcohol due to the hydrophilic polymers	[125]
	Poly(2-(diisopropylamino) ethyl methacrylate) (PDPA)	Modification of roughness and the hydrophilicity/hydro-phobicity of various substrate surfaces using pH responsive SiNPs of distinct sizes	[126]
	PDMAEMA	Dual stimuli (pH and tempera-ture) responsivity	[127]

TABLE 2-1 *(Continued)*

Inorganic Substrate	Polymers	Major Remarks/Applications	References
	PMPC	Effect of ionic strengths on dimensions and interactions of silica bound and free biocompatible polyampholyte was probed	[128]
	PHEMA	Wormlike structure on the silica NP surface from the comb-coil polymer brushes of poly(DL-lactide)-nBA) (PLA-PnBA) formed from PHEMA brushes via ATRP and ROP	[129]
	PDMAEMA	pH responsiveness	[130]
	PNIPAM	Double phase transitions depending on temperature	[131]

The naked eye detection of DNA by SI-ATRP modified gold NPs and T_2 type MRI contrast by SI-ATRP modified Fe_3O_4 NPs are shown in the following Figures 2-2 and 2-3, respectively.

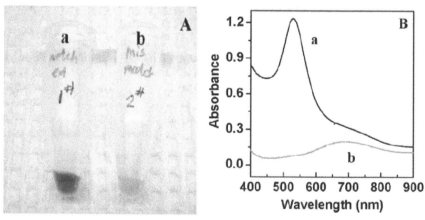

FIGURE 2-2 DNA sensing by DNA-polymer-hybrid-coated core–shell Au nanoparticles. (Adapted from reference 105. (From Lou, X.; Wang, C.; He, L. Core–shell Au Nanoparticle Formation with DNA-Polymer Hybrid Coatings Using Aqueous ATRP. Biomacromolecules 2007, 8, 1385–1390. © Copyright 2007 American Chemical Society. Used with permission.)

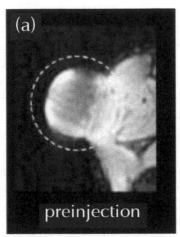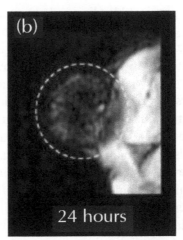

FIGURE 2-3 MRI contrasting using PEGMA coated Fe_3O_4 NPs. Adapted from reference 118. (From Ohno, K.; Mori, C.; Akashi, T.; Yoshida, S.; Tago, Y.; Tsujii, Y.; Tabata, Y. Fabrication of Contrast Agents for Magnetic Resonance Imaging from Polymer-Brush-Afforded Iron Oxide Magnetic Nanoparticles Prepared by Surface-Initiated Living Radical Polymerization. Biomacromolecules 2013, 14, 3453–3462. © Copyright 2012 American Chemical Society. Used with permission.)

Sometimes metal/meal oxide nanoparticles are coated with silica nanoparticles to form core/shell materials prior to the SI-ATRP. From simple hydrosilylation method the core NPs can be covered by functionalized SiO_2 NPs, followed by SI-ATRP. One such report is silica coated Fe_3O_4 NP where the silica nanoparticles are then used as substrate for the ATRP of PEGMEMA and GMA.[132] Carbon disulphide, a NIR fluorescent chromophore could then be attached with the pre-grown polymer brushes and the fluorescence intensity can be regulated by changing the feed ratio. The NPs obtained possess multiple functional properties such as NIR fluorescence and magnetism, which in turn can be effectively used as contrasting agents in biomedicine.

2.2.2.3 POLYMER BRUSHES TETHERED FROM NATURAL SUBSTRATES

SI-ATRP can be performed on natural substrates like natural polymers and clay minerals. Natural polymers, in spite of having the advantages of being biodegradable and biocompatible, often do not exhibit versatile properties that are required for several functional and structural applications, unlike common synthetic polymers. A great deal of research has long been dedicated to develop new composite materials based on natural polymers for high performance applications. Besides the commonly adopted practices viz. physical blending and chemical modification, ATRP technique can aid the grafting of polymer chains on natural macromolecules. Generation of pure graft copolymers via CFRP is quite difficult due to the formation of homopolymers, which will be obviated in ATRP. Surface initiated ATRP (SI-ATRP) is an easy and effectual strategy to modify these natural polymers, by growing new polymer side chains from the substrate. This modification through SI-ATRP can potentially make the polymers derived from renewable resources such as polysaccharides, polypeptides, and polynucleic acids suitable for advanced applications. This work particularly has opened a new avenue in green chemistry as the environmental concerns posed by common synthetic polymers can also be minimized by producing more green plastics based on natural polymers. The immobilization of the ATRP initiator (e.g., 2-bromoisobutyryl bromide) on the substrate and the subsequent polymerization of a new monomer on the modified polymer substrate will afford polymeric materials with tailored properties for high end applications. In a very recent research work, it has been substantiated that *grafting from* is superior to *grafting to* technique in terms of obtaining better polymer content in the surface, while employing cellulose as substrate.[45]

The most frequently experimented natural polymer substrate to fabricate composites is cellulose, where the natural fibers of cellulose function as reinforcements for the resulting materials. The properties chiefly imported to the substrate polymers are flame retardancy, reduction in hydrophilicity, antibacterial activity, and abrasion resistance. Various polymers have been grafted on the surface of cellulose via SI-ATRP technique and some of them are filter paper-g-poly(methyl acrylate),[133] filter paper-g-PDMAEMA,[134] cellulose acetate-g-PMMA,[135] cellulose acetate-g-poly(L-lactide),[136] cellulose-g-PMMA,[137] cellulose-g-PS,[137] cellulose-g-PMMA,[138] cellulose-g-poly[2-(methacryloyloxy) ethyl]-trimethylammoniumchloride,[139] wood pulp cellulose-g-poly (ethyl acrylate),[140] bacterial cellulose-g-PMMA,[141] bacterial cellulose-g-PBA,[141] and bacterial cellulose-g-PMMA-co-PBA.[141] Zwitterionic brushes bearing carboxybetaine were constructed from cellulose membrane via SI-ARGET-ATRP method to improve blood compatibility.[142] An interesting case is the ARGET ATRP preparation of cellulose-g-PnBA-co-PMMA), which is reported to be a third-generation thermoplastic elastomer.[143] In this case, grafted copolymer side chains act as the rubbery matrix while minor proportion of cellulose behaves as the rigid domains. The transparency and the shape recovery of the material are shown in Figure 2-4.

Few other instances for the surface modification of natural polysaccharides through SI-ATRP include natural rubber-g-PMMA,[144] chitosan-g-poly (acrylamide),[145] chitosan-g-poly (methacrylic acid),[146] dextran-g-PNIPAM,[147] silk-*g*-poly (dimethyl methacryloyloxyethyl phosphate-*b*-PDMAEMA,[148] poly(4-aminoantipyrine)-*g*-guar gum,[149] and so forth. Wang et al. has prepared disulfide-linked cationic PDMAEMA side chains on nonionic dextran, by a two-step protocol:[150] (1) the introduction of reduction sensitive disulphide linked initiator sites containing cystamine onto dextran backbone and (2) the ATRP of PDMAEMA from the initiator sites. The presence of disulphide linkages aids reduction sensitivity to the polymers and the PDMAEMA side chains could therefore easily be cleaved from the dextran via reduction. Such kind of redox responsive polymers are broadly used in intracellular drug and gene delivery[151] and the synthesized polymers with well-defined, biocleavable, and comb-shaped vectors were also found to have promising applications in gene therapy.

FIGURE 2-4 Photos of transparent thermoplastic elastomeric sample for tensile test. After bending (A) and post-recovery (B). Adapted from reference 143. (From Jiang, F.; Wang, Z.; Qiao, Y.; Wang, Z.; Tang, C. A Novel Architecture Toward Third-Generation Thermoplastic Elastomers by a Grafting Strategy. Macromolecules 2013,46, 4772–4780. © Copyright 2013 American Chemical Society. Used with permission.)

In addition to these polymers, peptides are also well–known to be employed as substrates for SI-ATRP. One such peptide monomer with the well-known β-sheet-forming peptide sequence of alanine-glycine-alanine-glycine was used as macroinitiator for the polymerization of MMA and the ABA triblock copolymers produced were shown to retain the β-sheet secondary structure.[152] Similarly, de Graaf et al. demonstrated the synthesis of ABC block copolymers by attaching two different polymer chains from the two termini (C-terminus and N-terminus) of a native hybrid peptide via ATRP.[153] Initially, POEGMA was polymerized from the C-terminus of the peptide followed by the polymerization of PNIPAM from the N-terminus of the peptide. The peptide linkages in the polymer blocks could be cleaved by collagenase (a metalloprotease) and in this process the micellar corona containing hydrophilic POEGMA blocks were shed by the action of the enzyme thereby making the synthesized peptide block copolymers, a suitable material for enzyme-triggered drug delivery.

From the time the seminal work carried out by Giannelis and Vaia in 1990s, polymer-clay nanocomposites have been studied extensively.[154,155] SI-ATRP can be performed on several natural clays to synthesize various clay mineral-polymer nanocomposites where the inner clay particles are enwrapped by the protective polymer shell. Some examples of the SI-ATRP modification of natural clay involving attapulgite, halloysite, and montmorillonite are compiled in the following Table 2-2.

TABLE 2-2   Surface-Initiated ATRP from clay surfaces

Clay Mineral	Polymers	Major Remarks/Applications	References
Attapulgite (ATP)	Polyacrylamide (PAM)	Adsorbent for mercury ions and dyes	[156]
	Poly(styrene-*b*-divinylbenzene-*g*-acrylonitrile) (PS-*b*-PDVB-*g*-PAN)	Adsorbent for lead ions and phenol after the transformation of acrylonitrile groups to acrylamide oxime	[157]
	PS PS-*b*-PMMA	Good dispersion of ATP nanorods in organic solvents	[158]
Halloysite	PS PAN	Dissolution of clay leads to polymer nanotubes/Nanowires. Non-woven porous fabric can be prepared by casting and thermal crosslinking.	[159]
	P4VP	Further immobilization with methyltrioxorhenium(VII) selectively and effectively catalyses the epoxidation of soybean oil	[160]
	PMMA-*b*-PNIPAM	Reverse ATRP technique has been adopted	[161]
	PMMA	Selective modification of inner lumen of clay nanotubes due to strong interactions of catechol with alumina rather than silica on the outer surface	[162]
Montmorillonite	PS	Further block copolymerization with THF achieved through cationic ring opening polymerization (CROP) gives rise to a copolymer with both hard (PS) and soft (PTHF) sections which can be used as thermoplastic	[163]

TABLE 2-2 *(Continued)*

Clay Mineral	Polymers	Major Remarks/Applications	References
	PHEMA	Intercalation and exfoliation, but majorly intercalation, gained thermal stability and declined T_g	[164]
	PS	A seven-fold increase in PS growth rate on MMT is observed compared to that of bulk polymerization kinetic rate, at the grafting density of 1 chains/nm². However bulk kinetics were restored at lower grafting density	[165]
	PMMA	A3 star polymer with increased T_g and thermal stability	[166]
	Poly(2-ethylhexyl acrylate) (PEHA) (PEHA-*b*-PMMA)	Better thermal stability. Two definite glass transitions with two distinct (soft PHEA and hard PMMA) domains	[167]

2.3 STIMULI–RESPONSIVE POLYMERS BY ATRP

Stimuli–responsive polymers are smart synthetic polymers that exhibit a response such as change in size, shape, assembly-disassembly, corresponding to a change in the environment. These exclusive properties of stimuli–responsive polymers have attracted the huge interest of many research groups, in view of their applications as smart materials in several areas such as drug delivery,[168–171] sensors[172] and tissue engineering,[173,174] and gene delivery and imaging.[175] Stimuli–responsive polymers tend to undergo reversible conformational and morphological transitions in response to external triggers such as temperature, pH, light, pI, or mechanical stress, which in turn leads to changes in physicochemical properties of the materials.[176]

Mother Nature has provided several inspirations (crucial example being the signal transmission along cell membranes via the biomacromolecules present in the human body) for developing new kind of smart materials, which are capable of effectively as well as rapidly responding to stimulus/stimuli in an inevitable manner. Smart polymers in short, can be defined as materials which receive process and respond to a stimulus, reversibly. Such smart polymers contain functionalities either in their backbones or in pendants that are sensitive towards the changes in their immediate environment, eventually resulting in dramatic transformations in the properties of the materials at macroscopic level. In general, the response should be specific both to the site and the stimuli. The rate of response should be easily adjustable so that the material can oppose stimulus when encountered as an inherent feature of the system under consideration in case of in-vivo applications. There are a plenty of exhaustive review articles published based on stimuli–responsive polymers.[177–190]

Polymers with diverse composition and architecture, including simple homopolymers, statistical/block copolymers, graft copolymers, molecular brushes, and polymers attached to various surfaces are capable of responding to the external stimuli and are therefore eminently suited for smart material applications. This section highlights the cutting edge trends in the field of the smart polymeric materials synthesized by using ATRP.

2.3.1   THERMORESPONSIVE POLYMERS BY ATRP

Among the thermoresponsive polymers, more attention has been offered on the synthesis of thermoresponsive linear copolymers rather than on graft thermoresponsive copolymers.[191] The lower critical solution temperature (LCST) of a thermoresponsive polymer can be modulated by varying the comonomer feed ratio. The most frequently reported thermosensitive polymers are either based on NIPAM or 2-(2-methoxyethoxy)ethyl methacrylate (MEO$_2$MA) monomers, as they are more relevant in biological applications with their LCSTs at around human body temperature (32 and 26 °C, respectively). PEO based polymers are synthesized and used as temperature sensitive polymers as well.

A very interesting application of thermosensitive polymers reported involves SI-ATRP assisted grafting of PNIPAM on monolithic silica rods, with the aim of constructing temperature responsive chromatographic column matrices.[192] The PNIPAM modified monolithic silica rods were used for separating five different hydrophobic steroids, which showed effective separation with short analysis time (10 min). The thermal phase transitions associated with the PNIPAM units also enabled temperature dependent elution of the hydrophobic steroids, as is visible in HPLC curves in Figure 2-5 and the transition occurring with respect to temperature is also represented in Figure 2-5. The precedent work was the synthesis of thermosensitive hydrogel based glass capillary lumens[193] and silica beads,[194] which were also capable of separating steroids and peptides via ATRP.

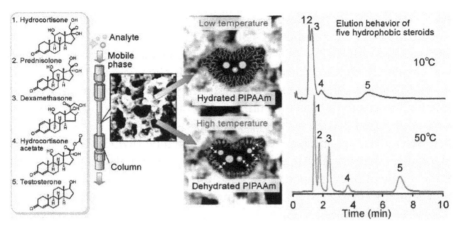

FIGURE 2-5 Structures of five steroids to be separated, temperature assisted phase transitions of PNIPAM, and HPLC traces of the analytes with respect to two different temperatures. Adapted from reference 192. (From Nagase, K.; Kobayashi, J.; Kikuchi, A.; Akiyama, Y.; Kanazawa, H.; Okano, T. Thermoresponsive Polymer Brush on Monolithic-Silica-Rod for the High-Speed Separation of Bioactive Compounds. Langmuir 2011, 27, 10830–10839. © Copyright 2011 American Chemical Society. Used with permission.)

Another example for the thermally responsive stationary phases for all aqueous HPLC is porous polymer monoliths grafted with PMEO$_2$MA-*co*-POEGMA, which were obtained by the two-step ATRP method. Under isocratic conditions, the synthesized porous polymer monoliths were found to effectively separate the test mixture in pure aqueous mobile phase.[195]

Development of stimuli–responsive inorganic–organic composite involving PNIPAM on the periphery of porous silicon was reported using SI-ATRP and the material has applicability in temperature sensing and stimuli controlled drug delivery.[196] Modulation of the temperature in the range of its LCST was shown to affect the rate of delivery of the anticancer drug camptothecin, by the synthesized composite material. Anionic block copolymers consisting of sodium 2-acrylamido-2-methylpropanesulfonate (AMPS) and NIPAM could also prepared by ATRP.[197]

Biocompatible, biodegradable, four-arm star copolymers containing a hydrophobic PCL block, a hydrophilic POEOMA block and a thermoresponsive PMEO$_2$MA block were synthesized by a combination of controlled ROP and ATRP. Being amphiphilic copolymers, the triblock copolymers could self-assemble in aqueous solution at room temperature and were able to undergo sol–gel transitions between room temperature (22 °C) and human body temperature (37 °C).[198]

Allyl-functionalized MEO$_2$MA-based thermosensitive copolymers were prepared via ATRP.[199] The presence of unreacted double bonds in the developed copolymers affords a versatile pathway for further chemical modifications and preparation of polymer conjugates. For instance, a thiol functionalized BODIPY (4,4-difluoro-4-bora-3a,4adiaza-s-indacene) dye was covalently attached to each pendant allylic functional group, through in-situ hydrolysis/thiol-ene click chemistry to yield fluorescent copolymers with thermosensitive character. Furthermore, thermoresponsive co-networks were synthesized starting with the copolymers as precursors for the UV-initiated free radical polymerization of NIPAM and the P(MEO$_2$MA-l-NIPAM) copolymers were revealed to exhibit sharp volume thermal transitions (VTT) in the region of body temperature.

Thermally responsive star polymers consisting of MEO$_2$MA and OEGMA were synthesized via ATRP.[200] The thermoresponsive star polymers were then used for modifying commercially used tissue culture grade polystyrene surface, by depositing a thin film of functionalized star polymers via UV induced cross-linking at 365 nm. The surface modified substrates are apt for cell adhesion above the LCST since the attached cells shrank from the surface of the substrate when the temperature was brought down below 4 °C. Thermosensitive triblock copolymer hydrogels composed of P(MEO$_2$MA-b-OEOMA-b-HEMA) were also synthesized, which again contained MEO$_2$MA blocks as the temperature responsive segments.[201] Well-defined, double hydrophilic temperature sensitive graft copolymers of poly[poly(ethylene glycol) methyl ether acrylate] backbone and poly[poly(ethylene glycol) ethyl ether methacrylate] (PPEGMEA-g-PPEGEEMA) side chains were synthesized by the combination of SET-LRP and ATRP.[191]

Schizophrenic diblock copolymers consisting of poly(propylene oxide) (PPO) and DMAEMA which could self-assemble to either micelles or reverse micelles in water were synthesized by ATRP.[202] Schizophrenic polymers can be potently used in the branch of surfactants and biomedicine. The effect of coronal chain branching on the micellization of hydrophilic–hydrophobic block copolymers had been addressed using OEGMA and PPO based semi-branched copolymers, which were analogous to the renowned Pluronic based non-ionic surfactants.[203]

Other stimuli–responsive polymers synthesized via ATRP were PNIPAM,[204] P(MEO$_2$MA-co-OEGMA),[205] n-octadecyl-poly(ethylene glycol)-b-poly(N-isopropylacrylamide) (C$_{18}$-PEG-b-PNIPAM),[206] poly(2-(2-bromopropionyloxy)ethyl methacrylate-g-(N,N-dimethylacrylamide-$stat$-butyl acrylate) (PBPEM-g-PDMA-$stat$-PBA), and poly[(2-(2-bromoisobutyryloxy)ethyl methacrylate)-g-(2-(dimethylamino)ethyl methacrylate)-$stat$-methyl methacrylate)] (PBIBM-g-PDMAEMA-$stat$-PMMA).[207] Thermally sensitive porous membranes were also fabricated based on PNIPAM polymers such as PNIPAM grafted PET membranes[208,209] and PNIPAM grafted AAO membranes.[210]

Another fascinating work reported on PNIPAM based thermoresponsive polymers via ATRP deals with the preparation of thermally modulated biointerfaces to be applied for regenerative medicine and tissue engineering.[211] Thermoresponsive hydrophobic copolymer brushes of PNIPAM-*co*-PBMA) were grown on glass substrates via SI-ATRP, to be used further for cell adhesion. These thermosensitive polymer brushes exhibited cell adhesion at 37 °C and detachment at 20 or 10 °C, for four types of human cells namely human umbilical vein endothelial cells (HUVECs), normal human dermal fibroblasts (NHDFs), human aortic smooth muscle cells (SMCs), and human skeletal muscle myoblast cells (HSMMs). HUVECs and NHDFs exhibited their effective detachment temperature at 20 and 10 °C, respectively (Figure 2-6). Thus, the designed smart copolymer brushes were capable of separating cells effectively by manipulating the cells' intrinsic temperature sensitivity for detachment.

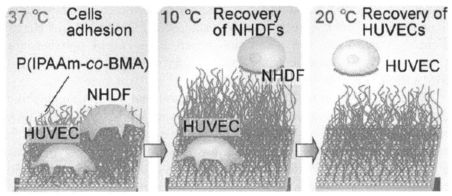

FIGURE 2-6 Cell adhesion and detachment of HUVECs and NHDFs from modified culture medium by varying time and temperature. (Adapted from reference 211. (From Nagase, K.; Hatakeyama, Y.; Shimizu, T.; Matsuura, K.; Yamato, M.; Takeda, N.; Okano, T. Hydrophobized Thermoresponsive Copolymer Brushes for Cell Separation by Multistep Temperature Change. Biomacromolecules 2013, 14, 3423–3433. © Copyright 2013 American Chemical Society. Used with permission.)

2.3.2 pH RESPONSIVE POLYMERS BY ATRP

Acrylic acid and DMAEMA are the most recurrently employed monomers for achieving pH responsive polymers via ATRP. Functionalized gold surface with pH-responsive tethered brushes made up of PAA containing a disulfide bond (PAA–S–S–PAA) were developed via *grafting to* method.[212] In the first step, ATRP of 1-ethoxyethyl acrylate (EEA) with a disulfide-containing initiator was carried out in the beginning and in the subsequent step, deprotection of PEEA was done by heating to give in the pH sensitive PAA chains. Random-type amphiphilic pH-responsive hybrid copolymers, with pH-responsive hydrophilic acrylic acid and hydrophobic and acrylate-polyhedral oligomeric silsesquioxane (POSS) were synthesized in two steps.[213] The pH dependent self assembly character of the synthesized hybrid copolymers to form nanoaggregates was also demonstrated. C_{60} fullerene based pH responsive polymers, that is, PMAA-b-C_{60}[214] and PAA-b-C_{60}[215] were also reported and their aggregation properties with respect to pH were also investigated. Similarly, pH responsive silica hybrids were prepared by SI-ARGET-ATRP of DMAEMA.[216] PCL-b-PAA copolymer was prepared by selective hydrolysis of poly(ε-caprolactone)-b-poly(methoxymethyl acrylate) (PCL-b-PMOMA) block copolymer, synthesized by combining ROP and ATRP.[217]

The primary intention behind the preparation of pH responsive polymeric materials is their smartness as delivery vehicles in drug delivery and gene delivery. There are numerous work dedicated towards developing pH responsive polymers targeting for drug/small molecule encapsulation and their controlled release such as ethyl cellulose-g-PDMAEMA for rifampicin,[218] linear, 4-arm and 8-arm star-shaped aminoPGMA polymers for Congo red,[219] polycationic nanoparticles carrying PEG, DEAEMA, and tBMA for fluorescein,[220] by applying ATRP.

By a combination of ROP and ATRP, 8-arm and 12-arm amphiphilic dendrimer-like pH sensitive block copolymers were synthesized.[221] The PAA carrying dendrimer-like block copolymers starting from second generation dendrimer-like PCL were synthesized in two steps. The PCL polymer was grown using and ROP and then the hydroxyl-end groups quantitatively converted into bromoester groups for the ATRP of tBA. The tBA chains in the last step resulted in PAA chains. The drug encapsulation efficiency and the pH dependent release were evaluated by using quercetin, an anti-cancer drug. The in-vitro cytotoxicity studies also revealed the possible usage of the material as an excellent drug delivery system. Likewise, pH responsive PEO based dendrimers like scaffold with inner PAA were also synthesized using ATRP.[222] Another instance for the dendritic pH responsive polymers is the synthesis of amphiphilic pH-responsive dendritic star-block poly(L-lactide)-b-poly(2-(N,N-diethylamino)ethyl methacrylate)-b-poly(ethylene oxide) (DPLLA-b-PDMAEMA-b-PEO) terpolymers, which were synthesized through a combination of ROP, ATRP, and click chemistry.[223] The amphiphilic terpolymers could self-assemble into micelles with PLLA segments as core, in water. The size and conformation of the micelles could be adjusted upon changing the pH values of the solutions.

pH sensitive poly(L-glutamate)-g-oligo(2-aminoethyl methacrylate hydrochloride) (PLG-g-OAMA) was prepared by successive ROP and ATRP.[224] The amphiphilic poly(amino acid) could self-assemble into vesicles in PBS and the hollow core of the vesicles was made use for drug delivery. The in-vitro release of doxorubicin, a common chemotherapeutic, from the vesicle in PBS could be controlled by the solution pH. DNA binding studies also accentuate the possibility of the polymers to be gene carriers as well. pH responsive poly(2-(methacryloyloxy) ethyl trimethyl ammonium-g-N-hydroxyethyl acrylamide) (PTM-g-PHEAA) based nanogels were prepared by inverse microemulsion ATRP and SI-ATRP, in two steps.[225] Biodegradable polyHEAA nanogels cross-linked with acid-liable ethylidenebis(oxy-2,1-ethanediyl) ester (EOE) were also synthesized, which exhibited controlled release of encapsulated rhodamine 6G at acidic conditions. The synthesized nanogels had excellent antifouling property and stability, biodegradability, low toxicity, and pH-responsive intracellular drug release which might be employed in targeted drug delivery systems.

Well-defined double hydrophilic graft copolymers containing PPEGMEA backbone and P2VP side chains were synthesized by combining SET-LRP and ATRP which aggregated to form micelles with P2VPcore at pH above 5.0.[226]

A series of pH-responsive methoxyPEG-b-PDMAEMA diblock copolymers synthesized via ATRP were able to self-assemble into core–shell micelles depending on the solution pH.[227] Additionally, core cross-linked (CCL) micelles were also with pH-responsive PDMAEMA cores and biocompatible PEG coronas using N,N'-methylene bisacrylamide (MBA) as the cross-linker. The reversible pH-dependent swelling–shrinking phenomenon of diblock copolymers and CCL micelles has been shown in Figure 2-7. These structurally stable CCL micelles were then tested for drug delivery by investigating the pH controlled release profile of model hydrophobic drugs, dipyridamole, ibuprofen, and famotidine.

Poly(glycerol monomethacrylate)-b-poly(2-(diethylamino)ethyl methacrylate) [PGlyMA-b-PDMAEMA] and poly(glycerol monomethacrylate)-b-poly(2-(diisopropylamino)ethyl methacrylate) [PGlyMA-b-PDPA] with outer blocks comprised pH-responsive PDMAEMA or PDPA units were synthesized via ATRP.[228] Nanoporous platforms that are pH responsive were synthesized by growing PMAA brushes on SiN$_4$ films for preparing nanoscale valves.[229] The transport behavior across the nanopores with respect to pH was followed by the addition of a fluorescent marker and it was found that the diffusion was faster at pH 4 in comparison with at 8 due to the neutral state of the PMAA at the former pH.

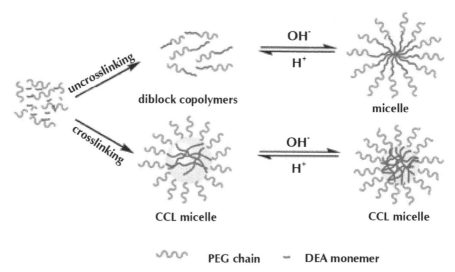

FIGURE 2-7 Micellization behavior of the mPEG-*b*-PDMAEMA in aqueous solution (top) and the CCL micelles (bottom), upon changing the pH of the media. Adapted from reference 227. (From Zeng, J.; Du, P.; Liu, P. One-Pot Self-Assembly Directed Fabrication of Biocompatible Core Cross-Linked Polymeric Micelles as a Drug Delivery System. RSC Adv. 2013, 3, 19492–19500. © Copyright 2013 Royal Society of Chemistry. Used with permission.)

Zwitterionic diblock copolymer poly[2-(dimethylamino)ethyl methacry-late-*b*-(4-vinylbenzoic acid)](PDMAEMA-*b*-PVBA) synthesized by ATRP could self-assemble to micelles both in acidic and basic solution (PVBA formed the micelle cores in acidic pH and the scenario was reversed at basic pH).[230] The zwitterionic triblock copolymer poly(ethylene oxide)-*b*-poly[2-(dimethylamino)ethyl methacrylate-*b*-(2-succinyloxyethyl methacrylate)] was synthesized by initially preparing PEO-*b*-PDMAEMA-*b*-PHEMA copolymer and converting the hydroxy groups of the HEMA blocks by esterification.[231] These triblock copolymers could form micelles in both acidic and basic pH. The driving forces for the micellization were hydrogen bonding at low pH, inter polyelectrolyte complexation at intermediate pH, and hydrophobic interactions at high pH, respectively.

Zwitterionic pH responsive diblock copolymer using the monomers, 2-(methacryloyloxy)ethyl phosphorylcholine, and 2-(diisopropylamino) ethyl methacrylate), were synthesized from ATRP.[232] As the formed PMPC-*b*-PDPA vesicles were colloidally stable at physiological pH and dissociated at pH < 6, the synthesized biocompatible polymer could possibly be engaged for intracellular delivery of water-soluble drugs and proteins. The presence of biomimetic phosphorylcholine group, which was analogous to conventional liposomes, also raised the possibility of the usage of the polymers as drug delivering vehicles.

Du and Armes have reported the synthesis of pH-responsive triblock copolymer poly[(ethylene oxide)-*b*-2-(diethylamino)ethylmethacrylate-*stat*-3-(trimethoxysilyl)propyl methacrylate], [PEO-*b*-P(DMAEMA-*stat*-TMSPMA)] by ATRP where the hydrolytically self-cross-linkable copolymer could constitute vesicles in THF/water.[233] The synthesized triblock copolymers were able to reduce Au(III) ionsto Au (0), thus forming gold nanoparticles which were confined within the walls of the vesicles as shown in Figure 2-8. These vesicle-supported metal nanoparticles may possibly be used in catalysis.

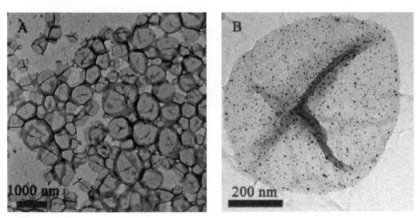

FIGURE 2-8   TEM images of (A) vesicles prepared using the PEO-*b*-P(DMAEMA-*stat*-TMSPMA) copolymer in 1:2 v/v THF/water and (B) vesicles with gold nanoparticles which have located solely within the vesicle walls. (Adapted from reference 233. (From Du, J. and Armes, S. P. pH-Responsive Vesicles Based on a Hydrolytically Self-Cross-Linkable Copolymer. J. Am. Chem. Soc. 2005, 127, 12800–12801. © Copyright 2005 American Chemical Society. Used with permission.)

2.3.3   IONIC STRENGTH RESPONSIVE POLYMERS BY ATRP

By grafting of PDMAEMA on the silicon nanowire arrays (NWA) by SI-ATRP, the wettability of the surfaces could be controlled by varying pH and concentration of NaCl.[234] The effect of pH and salt concentration altogether on the wettability of the NWAs is presented in Figure 2-9. The ionic strength tunable attachment of proteins and bacteria on the modified SiNWAs was demonstrated by using horseradish peroxidase, lyzozyme, and *Escherichia coli* as models.

Electrolyte sensitive quaternized PMAEMA brushes synthesized via ATRP were found to collapse in the presence of a high concentration of monovalent salt.[235] But these cylindrical polymer brushes were observed to form an intermediate state with helical conformation before collapsing, in the presence of di/trivalent salts.[236] The interaction of these cationic brushes with anionic surfactant sodium dodecyl sulfate was examined and found that an ionic complex was formed leading to spherical shape from worm-like shape.[237]

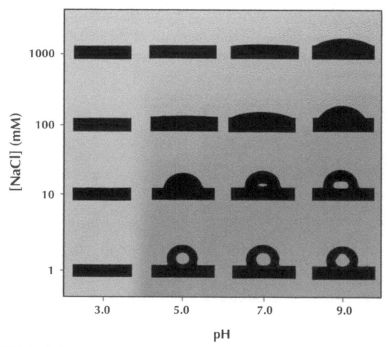

FIGURE 2-9   Influence of pH and NaCl on the water contact angle of PDMAEMA-modified SiNWAs. (Reproduced from Wang, L.; Wang, H.; Yuan, L.; Yang, W.; Wu, Z.; Chen, H. Step-Wise Control of Protein Adsorption and Bacterial Attachment on a Nanowire Array Surface: Tuning Surface Wettability by Salt Concentration. J. Mater. Chem. 2011, 21, 13920–13925. With permission from The Royal Society of Chemistry).

Ionic strength sensitive membrane was prepared by SI-ATRP of sulfobetaine methacrylate (SBMA) on regenerated cellulose (RC).[238] The analyses revealed the dependence of permeability of the RC-g-PSBMA membrane on electrolyte (NaCl) concentration. This tunable permeation behavior of the synthesized membrane was exploited for the separation of proteins from impurities by changing the NaCl concentration, by taking bovine serum albumin (BSA) as model protein and polystyrene nanoparticles as impurities.

2.3.4 DOUBLE/MULTI RESPONSIVE POLYMERS BY ATRP

By the judicial choice of the monomers, dual responsive block copolymers which respond towards any two stimuli, can be prepared. Double responsive polymers are usually produced by synthesizing diblock copolymers consisting of two different monomers, which could respond to changes in any two different stimuli. The most common dual responsive polymers reported are polymers which respond to temperature and pH. The majority of the smart polymers exhibiting pH and temperature response had been based on PDMAEMA, PMO$_2$EMA, PNIPAM, and PAA polymers. Several dual responsive polymers have been synthesized via ATRP such as PNIPAM-b-P4VP,[239] PLLA-b-PDMAEMA,[240] MEO$_2$MA-grad-PAA, MEO$_2$MA-b-PAA,[241] PEO-b-[PGMA-g-(PDMAEMA) (PMEO$_2$MA)],[242] P(MEO$_2$MA-stat-MAA), P(MEO$_2$MA-stat-DMAEMA), P(MEO$_2$MA-stat-MAA-stat-DMAEMA),[243] PDMAEMA-b-PAA,[244]   CS(-g-PDMAEMA)-g-PNIPAM,[245]   PMEO$_2$MA-b-PDMAEMA-b-PEG-b-PDMAEMA-b-PMEO$_2$MA,[246]   PDMAEMA-b-PDEGMEMA,[247] and PEG-b-P(NIPAM-g-DMAEMA).[248] Double-hydrophilic diblock copolymers, poly[(ethyl-ethylene phosphate)-b-2-(dimethylamino) ethyl methacrylate] (PEEP-b-PDMAEMA), have been synthesized via successive ROP and ATRP, which exhibit reproducible temperature and pH responses. The dual sensitive copolymers were also shown to condense DNA, which possibly could be applied for gene delivery.[249]

C$_{60}$ fullerene based water-soluble, pH and temperature responsive block copolymer of (PMAA-b-PDMAEMA-b-C$_{60}$) was synthesized by ATRP technique.[250] This ampholytic block copolymer with end-capped C$_{60}$ displayed dual responsiveness by pH and temperature dependent solubility. At room and elevated temperatures and at low and high pH, the polymers were soluble in water, whereas at intermediate pH range (5.4–8.8), phase separation occurred. The self-assembly behavior of the synthesized copolymer was unlike from that of standard PMAA-b-PDMAEMA block copolymer.

Dual-responsive (pH and temperature) intelligent cellulose surfaces were also obtained through synthesizing of PNIPAM-b-P4VP brushes via SI-ATRP.[251] These modified cellulose surfaces exhibited a reversible response to both pH and temperature, as witnessed by changes occurred in wettability of the surfaces. Dual responsive nylon membranes were produced from SI-ATRP of NIPAM and DMAEMA and the permeation of the aqueous solution through the surface modified nylon membranes was shown to experience abrupt change within the temperature range of 30–35°C and pH range of 6–8.[252]

pH tunable thermoresponsive polymers were synthesized by conducting the ATRP to prepare PHEMA-alkyne and the successive click chemistry with 2-azidoethylamine, 2-azido-N,N-dimethylethylamine and 2-azido-N,N-diethylethylamine.[253] Unlike other dual responsive polymers which normally possess two different functionalities which respond to the stimuli separately, here in this case, a single functional group responded to temperature and pH. The dual responsive nature was confirmed by the drastic changes in the LCSTs with variation in solution pH.

Gold nanoparticles coated with protein-polymer conjugates exhibiting pH dependent thermosensitive character were synthesized by grafting P(MEO$_2$MA-co-OEGMA) onto BSA capped gold nanoparticles via SI-ATRP.[254] BSA provided pH sensitivity and the grafted polymer brushes displayed thermal response. The dependence of LCST on the solution pH and the reversible aggregation are shown in Figure 2-10. In the case of pH > pI of BSA (Figure 2-10A and C), the nanoparticles are negatively charged and in the case of Figure 2-10B and D, no net charge is present on the nanoparticles. At temperature lower than LCST, the brushes (Figure 2-10A and B) were stable due to steric stabilization, but got collapsed when the temperature exceeded the LCST (Figure 2-10C and D). Between Figure 2-10C and D, Figure 2-10C charged NPs did not aggregate due to the electrostatic repulsions arised from BSA, while in Figure 2-10D, neutral NPs formed aggregates due to the lack of these electrostatic repulsions and increased hydrophobic interactions.

Multiresponsive composite nanoparticles with response towards pH and light were reported by a combination of RAFT and ATRP.[255] In the initial step, crosslinked nanoparticles were prepared containing MMA, 4-VBC, and DVB. In the following step, by ATRP, 1-pyrenylmethyl methacrylate (PPyMA) and DMAEMA were polymerized and incorporated in the composite nanoparticles. The PPyMA units imparted photosensitiveness whereas the pH responsivity was stemmed from PDMAEMA. Using poly(ionic liquid) (PIL) as core and PNIPAM as shell, nanoparticles which were ionic strength and pH responsive were prepared by combining dispersion polymerization and ATRP.[256]

Other than these afore-described stimuli sensitive polymers, magnetically responsive polymers can also be synthesized via conducting SI-ATRP on magnetically active nuclei such as magnetite, hematite, maghemite, and so forth. Examples for the same have been given in the polymer brushes section.

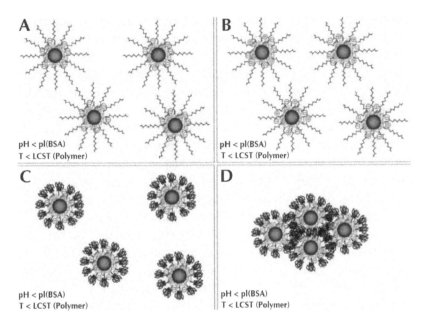

FIGURE 2-10 pH-dependent thermoresponsive behavior of Au@BSA@P(MEO$_2$MA$_{85}$-co-OEGMA$_{15}$). (From Strozyk, M. S.; Chanana, M.; Pastoriza-Santos, I.; Pérez-Juste, J.; Liz-Marzán, L. M. Protein/Polymer-Based Dual-Responsive Gold Nanoparticles with pH-Dependent Thermal Sensitivity. Adv. Funct. Mater. 2012, 22, 1436–1444. Copyright © 2012 WILEY-VCH Verlag GmbH & Co. KGaA, Weinheim. Used with permission.)

2.4   BIODEGRADABLE POLYMERS VIA ATRP

In recent years, development of biodegradable polymers with novel properties have attracted attention in view of their applications as materials in tissue engineering, controlled drug delivery, regenerative medicine, and gene therapy.[257] The backbones of these polymers consist of weak and labile chemical bonds that are readily cleavable in the presence of chemicals or light. The functional groups such as sulfide (SS), ester (COO), amide (CONH), and so forth, present in the backbone of a biodegradable polymer can be introduced through the polycondensation reactions and these groups are easily broken under certain conditions. Biodegradable polymers can be derived either from biological sources or through well-known chemical synthesis. All though biologically derived biodegradable polymers are expected to be biocompatible, they may generate immune response in the body (especially if they are from a foreign body).[258] Moreover, the chemical modifications of biologically derived biodegradable polymers are rather tricky and may alter the properties of the native polymers.[258] Therefore, synthetic polymers that are degradable are more beneficial in terms of designing and tailoring, without the loss of desirous bulk properties, in comparison with biologically derived biodegradable polymers.

A large number of synthetic degradable polymers have been synthesized by adopting ATRP method. The degradable groups can be inserted into polymers by the use of initiators or monomers with appropriate functional groups via an ATRP.[259] The synthetic scheme can be so tailored that the degradation can be photochemical,[260] electrochemical,[261] or simple hydrolysis.[260]

Tsarevsky and Matyjaszewski synthesized degradable linear polymers of MMA, tBMA, and BnMA by using bis[2-(2-bromoisobutyryloxy)ethyl] disulphide as the ATRP initiator.[262] The disulphide bond (SS) present in the polymers was then reductively cleaved by tributylphosphine and the molecular weights of the resulting polymers were determined to be halved, thus validating the degradability of the materials. The disulphide (SS) linkage is well-known to undergo scission under reducing conditions and light and is a primary example for a biodegradable functional group. The "SS" functionality can be in corporate in the polymer either through a proper functional initiator or a monomer or by the reaction of halogen end capped polymers with sulphur containing nucleophiles. In this way, ATRP provides an additional advantage through the availability of end-terminated halide groups that can be functionalized further. The sulphide/thiol (SS/SH) couple have been given special interest in the context to their relevance to several phenomena occurring in cells and their key role in protein folding and protein stabilization.[258] This unique property of the cleavability of the disulphide bonds makes disulphide containing polymers highly efficient in drug delivery[151] and MRI.[263] Star polymers of PEG, PDMAEMA, and disulphide cross-linker (SS functionalized EGDMA) with degradable cationic core were synthesized by the "arm first" method.[264] The degradation of the star polymers to individual polymers by glutathione redox process was confirmed by dynamic light scattering (DLS) experiments as shown in Figure 2-11. The function of SS bonds in polymer degradation had been explained by performing control degradation studies on the same polymer but without SS links. The synthesized star polymers offered an effectual path for conducting gene and siRNA delivery with no adverse cellular response and biocompatibility.

Reductively degradable amphiphilic block copolymers of polyesters and POEOMA, with cleavable sulphide bonds in the backbone, were synthesized by combining polycondensation and ATRP.[265] The polymers are self-assembled in water forming core/shell micelles. The controlled release of model dye from the micelles by reductively cleaving the SS bonds was shown, which could be extended for drug delivery. Biodegradable nanogels of

PEO-*b*-POEOMA and disulphide functionalized cross-linker were prepared using inverse miniemulsion strategy.[266] The living nature of the degraded polymer chains were proved by extending the POEOMA chains further via the ATRP of styrene.

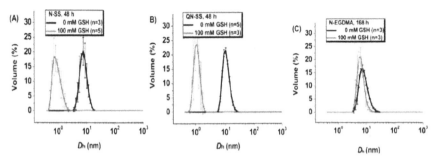

FIGURE 2-11 Size distribution of N-SS, QN-SS, and N-EGDMA star polymers after glutathione (GSH) treatment in 48 h, determined by DLS analysis {N-DMAEMA; QN-QDMAEMA (quaternized)}. Adapted from reference 264. (From Cho, H. Y.; Srinivasan, A.; Hong, J.; Hsu, E.; Liu, S.; Shrivats, A.; Kwak, D.; Bohaty, A. K.; Paik, H.-J.; Hollinger, J. O.; Matyjaszewski, K. Synthesis of Biocompatible PEGBased Star Polymers with Cationic and Degradable Core for siRNA Delivery. Biomacromolecules 2011, 12, 3478–3486. © Copyright 2011 American Chemical Society. Used with permission.)

More recently, degradable high internal phase emulsions (polyHIPE) were prepared by the AGET ATRP of 2-ethylhexyl methacrylate (EHMA) and bis(2-methacryloyloxyethyl) disulfide (DSDMA), a degradable cross-linker, for the first time.[267] The HIPEs are in general, highly viscous and possess large volume of internal/droplet phase.[268] Due to the porous nature and low densities, polyHIPEs are used extensively in various areas.[267] The Young's modulus of the polyHIPEs obtained using SS-BPMODA/CuBr$_2$ [SS-N,N-bis(2-pyridylmethyl) octadecylamine] was found to be greater than that obtained by using SS-bpy/CuBr$_2$ as a catalyst system, due to the uniform network structure present in the former case. Such polyHIPES with biocompatible backbones and degradable crosslink agent are materials of great interest, as porous scaffolds, in tissue engineering.

Biodegradable model networks (MNs) were proposed where ozonisable tBA MNs were developed composed of an α,ω-azido-poly(tBA) macromonomer cross-linked with tri- and tetra-acetylenecross-linkers.[269] The degradability of model network polymers thus obtained in the presence of light was ascertained in a later report by the same group.[270] Comb shaped degradable PGMA derivative vectors consisting of ethanolamine functionalized PGMA (*c*-PGEA) were prepared by ATRP and ring opening reactions.[271] The hydrolytically degradable *c*-PGEA vectors possess low cytotoxicity and good buffering capacity and hence are potential materials for gene therapy. The hydrolysable and degradable PGEA vectors were copolymerized with PEGEEMA in order to enhance the gene transfection efficiency. Similar kind of comb shaped vectors (SS-PHPD) having poly(N-3-hydroxypropyl)aspartamide (PHPA) backbones and disulphide functionalized PDMAEMA side chains were synthesized via ATRP.[272] The synthesized copolymers were readily cleaved into their individual polymer chains under reducible conditions. The low cytotoxicity, good transfection efficiency, and good pDNA condensation capacity exhibited by the SS-PHPD vectors can be made use of in-vivo gene therapy.

Bian et al. has reported the synthesis and self-assembly of biodegradable poly(ethylethylene phosphate)-*b*-poly[2-(succinyloxy)ethyl methacrylate] (PEEP-*b*-PSEMA) diblock copolymers via a combination of ROP, ATRP, and a polymer reaction (Scheme 2-5).[273] The diblock copolymers prepared thus were self-assembled into micelles of various sizes in accordance with pH, which were used for drug delivery. Naproxen (NAP), an anti-inflammatory model drug, was encapsulated in the core of the micelle and the in-vitro release of NAP from the micelles was demonstrated. The drug release relied on pH and was accelerated by phosphodiesterase I, an enzyme used to catalyze the degradation of polyphosphoesters. The schematic illustration of the work is presented in Scheme 2-5. The cytotoxicity assays also suggested the biocompatibility of the synthesized double hydrophilic block copolymers to HeLa cells making them suitable for drug delivery system (DDS).

Dextran/synthetic glycopolymer biohybrids were synthesized by ATRP using dextran, lactobionamidoethyl methacrylate (LAMA) glycomonomer and di(ethylene glycol) methyl ether methacrylate (DEGMA).[274] The synthesized biodegradable, biocompatible, biomimetic, and thermoresponsive polymers were shown to have potential for specific biomolecular recognition with ricinus communis agglutinin 120 and for possible drug delivery. Using AGET ATRP, biodegradable POEOMA-*co*-PHEMA nanogels were prepared and the nanogels were shown to release the encapsulated drug, RITC-Dx (rhodamine B isothiocyanate-dextran) in a controlled fashion, on hydrolysis.[275] Further acrylation of the synthesized polymer nanogels and a Michael-type addition reaction with thiolated hyaluronic acid led to nanostructured and biodegradable hyaluronic acid based hydrogels, which have enormous potential as candidates for biomedicinal applications.

Biodegradable PHEMA hydrogels were synthesized by using a hydrolytically and enzymatically degradable crosslinker and macroinitiator, based on PCL.[276] The hydrogels lost 30% of their mass by enzymatic degradation in 16 weeks. These PHEMA hydrogels can be utilized as scaffolds for tissue engineering and for drug delivery as they are degradable and exhibit no adverse effects.

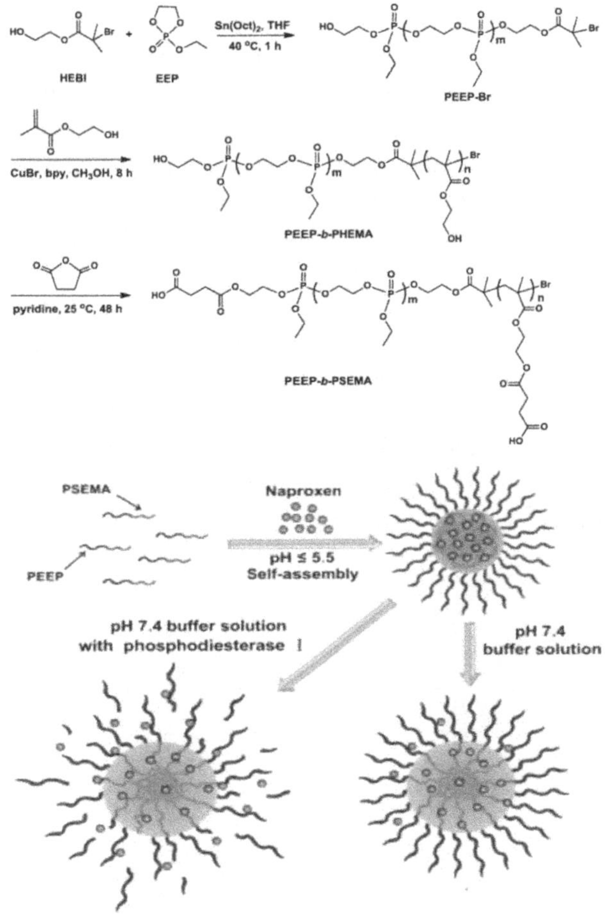

SCHEME 2-5 Synthesis of pH responsive polymer brushes and their applications in drug delivery. (From Bian, J.; Zhang, M.; He, J.; Ni, P. Preparation and Self-Assembly of Double Hydrophilic Poly(ethylethylene phosphate)-Block-Poly[2-(succinyloxy)ethyl methacrylate] Diblock Copolymers for Drug Delivery. React. Funct. Polym. 2013, 73, 579–587. Copyright © 2012 Elsevier Ltd. Used with permission.)

2.5 MICROSPHERES

Polymeric microspheres have gained substantial attention in the field of materials science due to their great potential as functional scaffolds.[277] The development of highly crosslinked and monodisperse microspheres was first carried out by Li and Stöver via precipitation polymerization.[278] Precipitation polymerization is an ideal route for synthesizing polymeric microspheres as it is very easy to operate and does not involve any surfactant or stabilizer.

Linear and hyperbranched glycopolymer chains were grafted on the surface of microspheres consisting of PDVB core via a combination of ATRP and self-condensing vinyl copolymerization (SCVCP).[279] The synthesized glycopolymers adsorbed wheat germ agglutinin (WGA) efficiently, which is a glucosamine-specific lectin. Especially the hyperbranched polymer showed maximum WGA adsorption and the affinity to lectin was found to increase with extent of branching.

ATRP can also be applied as an effective tool for the direct generation of polymeric spheres. A new facile methodology called atom transfer radical precipitation polymerization (ATRPP) was suggested by Zhang et al. in 2011 for the first time, for obtaining monodisperse, surface functionalized, and highly crosslinked polymeric microspheres using functional comonomers of 4-VP, acrylamide and HEMA with EGDMA.[280] ATRPP technique is based on the introduction of the ATRP mechanism into a conventional precipitation polymerization. The SEM images of few of the polymeric microspheres synthesized via ATRPP are shown in Figure 2-12, which proves that

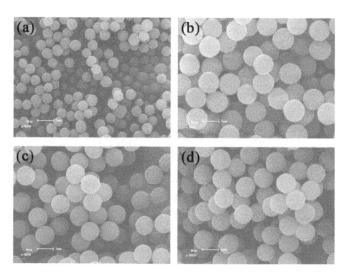

FIGURE 2-12 SEM images of (a) P(HEMA-*co*-EGDMA), (b) P(AM-*co*-EGDMA), (c) PNIPAM brushes-grafted P(4-VP-*co*-EGDMA), and (d) PHEMA-grafted P(4-VP-*co*-EGDMA) microspheres. The scale bar represents 2 μm. (Source: Jiang, J.; Zhang, Y.; Guo, X.; Zhang, H. Narrow or Monodisperse, Highly Cross-Linked, and "Living" Polymer Microspheres by Atom Transfer Radical Precipitation Polymerization. Macromolecules 2011, 44, 5893–5904. © 2007, American Chemical Society. Adapted with permission.)

monodisperse particles can be obtained by the ATRPP process. The polydispersity indices were found to be lower than 1.01 and the sizes of the microspheres could be modulated by altering the reaction parameters. The microspheres thus obtained could also facilitate further surface functionalization, if needed. The efficiency of the method using different monomers has also been substantiated by the same group, to produce special types of polymers like molecularly imprinted polymers.[281,282]

2.6 NANOCAPSULES/MICROCAPSULES

In the past few years, substantial amount of effort is being placed on developing new synthetic routes towards hollow polymeric nanostructures/microstructures or "nanocapsules/microcapsules". Nanocapsules (NC) typically have the size ranging from 1 to 1000 nm[283] and the nanodimension of the polymers is anticipated to introduce unique and distinct physicochemical/mechanical properties. These materials have gained a significant place in science and technology as a result of their extensive applications as delivery vehicles in pharmaceutics,[284–287] as catalysts,[288,289] as contrasting agent in diagnostics,[290] as sensor,[291] as nano/microreactors,[292–294] among others. There are numerous ways by which functional polymeric NC can be generated like emulsion polymerization,[295] suspension polymerization,[296] sacrificial core–shell template approach[297,298] and self-assembly of copolymers.[299] In the core–shell technique, polymer brushes are initially grown from the surface of the colloidal templates and then the colloids in the inner core are chemically etched to get structurally uniform nanocapsules (Figure 2-13). The diameter of the NC obtained can be altered by choosing the template core of proper size. A variety of precisely defined polymeric nano/microcapsules can be prepared via SI-ATRP technique from the colloidal templates due to the reliability of ATRP. A very recent and comprehensive review on SI-ATRP for the preparation of nano/microcapsules is available.[300]

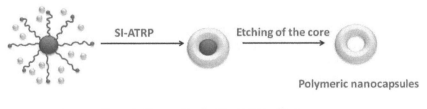

FIGURE 2-13 Preparation of polymeric nanocapsules via core–shell templating technique.

The most favored sacrificial colloidal template for the development of nano/microcapsules is silica particles. The predominant usage of silica in templating chemistry might be due to the ease involved in the synthesis and surface functionalization. The selective removal of the uniform colloidal silica by HF treatment also simplifies the preparation of nano/microcapsules. Magnetically active thermoresponsive polymeric nanocapsules of PNIPAM and MBA were developed via SI-ATRP from the surfaces of $Fe_3O_4@SiO_2$ nanoparticles followed by selective silica etching.[301] The superparamagnetism and temperature dependent phase transition present in the synthesized materials make them fit for encapsulation and targeted delivery applications. Cross-linked polymeric nanocapsules of PMA(temperature responsive),[302] PCMSt and PHEA[303] were prepared by the same strategy from the surface of silica NPs. Dual responsive cross-linked nanocapsules containing PtBA (acidolyzed) and PNIPAM were synthesized by two sequential ATRP from silica core.[304] Another example of dual responsive capsules was synthesized by polymerizing the monomers MEO_2MA and OEGMA via ATRP, from polydopamine coated silica core.[305] The dual responsiveness was demonstrated by the loading/unloading of the dyes Rh6G and methyl orange with changes in pH and temperature.

Hollow PNIPAM NCs and Ag/PNIPAM hybrid NCs were synthesized by combining SI-ATRP and click chemistry.[306] In the first step, NIPAM and 3-azidopropylacrylamide were polymerized from the surface of SiO_2 NPs. Then the azide groups in the polymer shells were cross-linked with 1,1,1-tris(4-(2-propynyloxy)phenyl)ethane by click chemistry. The consecutive removal of the silica core led to hollow NCs. Silica NPs with cross-linked polymer shell could also serve as nanoreactors for the in-situ synthesis of Ag NPs. The thermoresponsive PNIPAM and Ag/PNIPAM NCs synthesized could be applied in controlled release, catalysis, and optoelectronic devices. Hollow polyelectrolyte nanospheres were also obtained using silica NPs as core and crosslinked polyionic liquids consisting of poly [1-(4-vinylbenzyl)-3-butyl imidazolium] cation and PDVB as shell.[307]

Another approach to prepare polymeric nanocapsules via ATRP is miniemulsion polymerization. Polymeric nanocapsules were prepared by interfacially confined AGET-ATRP of PBMA with divinyl cross-linker (EGDMA/DSDMA) in a miniemulsion system, by using PEO-*b*-PBMA-Cl as a macroinitiator and stabilizer.[308] Using DSDMA crosslinker, degradable polymeric nanocapsules were obtained containing disulphide bonds that could be degraded under reducible conditions. In the same miniemulsion method, a dual reactive surfactant (DRS), which is an amphiphilic block copolymer containing an ATRP functional group for initiating polymerization and an azido functional group to undergo click chemistry could also be used.[309] By using α-azido-ω-2-chloroisobutyrate-poly(oligo(ethylene oxide) monomethylether methacrylate)-b-poly(n-butyl methacrylate) (N$_3$-POEOMA-PBMA-Cl) as a DRS, the monofunctional reactive surfactant, poly(ethylene oxide)-b-poly(n-butyl methacrylate) (PEO-PBMA-Cl) had been copolymerized with *n*BMA and EGDMA, DSDMA and DMAEP as cross-linkers in miniemulsion, forming nanocapsules with cross-linked shells. The introduction of azido group in the polymer expedites attaching small molecules such as dyes, further on the polymer shell. By using cleavable cross-linking agents, degradable nanocapsules can be produced. This work has been pictorially presented in Scheme 2-6.

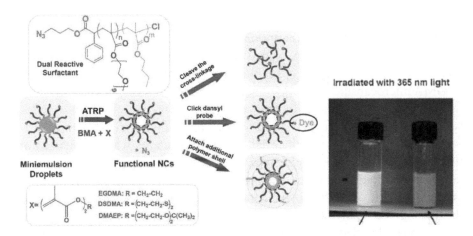

SCHEME 2-6   Representation of the NCs with and without dansyl probe modification via click chemistry. Adapted from reference 309. (From Li, W.; Yoon, J. A.; Matyjaszewski, K. Dual-Reactive Surfactant Used for Synthesis of Functional Nanocapsules in Miniemulsion. J. Am. Chem. Soc. 2010, 132, 7823–7825. © Copyright 2010 American Chemical Society. Used with permission.)

Double walled microcapsules were prepared via Pickering emulsion template interfacial transfer radical polymerization (PETI-ATRP).[310] The surface of the cationic LUDOX CL (silica) NPs was electrostatically coated with anionic poly(sodium styrene sulfonate-*co*-2-(2-bromoisobutyryloxy) ethyl methacrylate). The modified silica NPs with ATRP initiator could stabilize o/w Pickering emulsions and polymerize water soluble cross-linking monomers. The sequential PETI-ATRP and in-situ polymerization of the encapsulated monomers resulted in double walled MCs. The suggested method was shown to be versatile to several monomers.

2.7   SELF-HEALING POLYMERS

Biological wound healing process has been a real inspiration to the opening of self-healing polymers. Currently, several research activities are ongoing in this field. Self-healing polymers have the capacity to restore their shape and properties following mechanical damage in the form of cracks or scratches. The main working principle behind the self-healing polymers is based on the crack induced discharge of healing agents from microcapsules present in the polymers. On the other hand, making polymers heal through reversible chemical covalent bonds or non-covalent interactions (H-bonding, etc.) is another plausible alternative.[311]

Microcapsulation has been an important strategy to derive self-healing polymers which do not need any external reagents for repairing the damages. Preparation of self-healing thermoplastics comprising of PMMA composites incorporated with GMA-loaded microcapsules had been reported by combining ATRP and microcapsulation method.[312] Crack triggered breakage of GMA microcapsules in PMMA matrix and the consequent ATRP allowed the molecules to self-heal. In the same way, multilayered microcapsules were prepared and used as ATRP micro-reactors for constructing self-healing thermoplastics, in which latent ATRP reactants were introduced at the microcapsules in multiple steps.[313] When a polymer containing these microcapsules suffers a crack, the monomer is released and polymerization is initiated. The re-bonding in the damaged segments in the polymer leads to the healing of the polymer. In a much later work reported by the same group, it was concluded that preparation of self-healing polymers via microcapsulation involving RAFT polymerization is better than that by ATRP.[314] The sensitivity of metal ions involved in ATRP to air made ATRP less versatile and less feasible than RAFT.

Polymers bearing disulphide linkages are known to introduce self-healing properties due to the reversibility of the SS/SH redox reaction.[315] PnBA star polymers grafted from cross-linked poly(ethylene glycol diacrylate) core was prepared by ATRP and the chain was further extended using bis(2-methacryloyloxyethyl disulfide) to get self-healing polymers. The ability of the polymers to heal is demonstrated in the photographs shown in Figure 2-14.

Kavitha and Singha have reported the synthesis of thermoreversible self-healing triblock copolymers via ATRP.[316] In this work, poly(furfuryl methacrylate)-b-poly(2-ethylhexyl acrylate)-b-poly(furfuryl methacrylate) [PFMA-b-PEHA-b-PFMA] with a reactive furfuryl group (diene) was initially prepared via ATRP and Diels–Alder reaction between the reactive furfuryl group and bismaleimide (dienophile) resulted in thermoreversible self-healing polymers. By combining click chemistry involving alkoxy-amines and azides of star-like oligomers prepared via ATRP, poly(nBA) networks with property to heal under UV irradiation were prepared.[317]

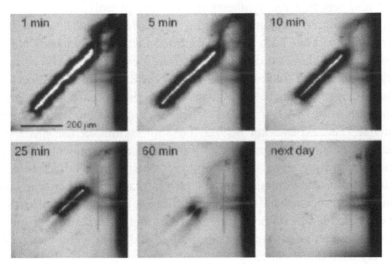

FIGURE 2-14 Optical microscope images of the self-healing polymers with disulphide linkage. Adapted from reference 315. (From JYoon, J. A.; Kamada, J.; Koynov, K.; Mohin, J.; Nicolaÿ, R.; Zhang, Y.; Balazs, A. C.; Kowalewski, T.; Matyjaszewski, K. Self-Healing Polymer Films Based on Thiol-Disulfide Exchange Reactions and Self-Healing Kinetics Measured Using Atomic Force Microscopy. Macromolecules 2012, 45, 142–149.. © Copyright 2012 American Chemical Society. Used with permission.)

Core–shell self-healing polymers with cross-linked PS core and polyacrylate-amidecorona were developed via ATRP.[311] The self-healing property was furnished by reversible hydrogen bonding of secondary amides present in the PA-amide brush. Stiffness was imparted by PS cores whereas self-healing ability and ductility were provided by PA-amide shell. The self-healing property was dependent on the extent of chain lengths of the corona.

2.8 ENERGY HARVESTING MATERIALS AND SENSORS

2.8.1 FLUORESCENT AND CONDUCTING POLYMERS

Fluorescent polymers are macromolecules, which display fluorescence and the fluorescence can either be intrinsic or can be introduced by attaching a fluorophore. Fluorescent polymers have been a topic of huge importance due to their expansive applications in sensors,[318,319] detection of nucleic acids,[320] imaging,[321] light harvesting,[322] and in several optoelectronic devices. Fluorescent polymers can be productively synthesized either by using a fluorescent monomer or a fluorescent initiator. Several research groups have worked and reported the synthesis of fluorescent polymers by ATRP by using (1) fluorescent monomer[323–328] and (2) fluorescent initiator.[329–333]

Surprisingly, it has been recently reported that fluorescent polymers can be synthesized without introducing any fluorescent agent.[334] The work dealt with the synthesis of fluorescent polymers via RAFT and ATRP methods. The fluorescence signals became stronger with the increase in the polymerization interval of MA (Figure 2-15), when using (bromomethyl) benzene as initiator. The explanation behind the photoluminescence was the p–p interactions occurring among the polymerization components. When the polymerization time increased, the polymer chains grew longer and coiled nanostructures were formed. Since the p–p interactions of benzene and the carbonyl groups occurred at the center of these coiled nanostructures, longer polymer chains led to stable p–p interactions and in turn stronger luminescence. As these polymers were biocompatible and fluorescent, they could be effectively used in cell labeling.

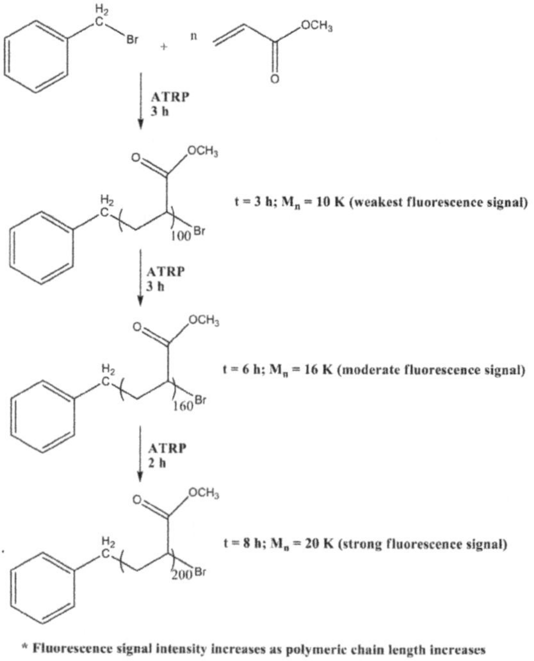

* Fluorescence signal intensity increases as polymeric chain length increases

FIGURE 2-15 Fluorescence microscope images of ATRP of methyl acrylate in a capillary at various polymerization intervals using (bromomethyl) benzene initiator. Adapted from reference 334.

Fluorescent polymers with tunable double color emission were synthesized via ATRP by using 1-(4-vinylstyryl)naphthalene (NVS) and 2-(2,3-dihydro-2-(4-methoxyphenyl)-1,3-dioxo-1H-phenalen-7-ylamino)-ethyl-2-chloroacetate (NAPH) as monomer and initiator, respectively.[335] The resultant polymers exhibited a composite emission spectrum corresponding to a blue emission for the monomer and a green emission for the initiator. By manipulating the polymer chain length, the peak intensity of the blue and green bands could be tuned.

3-, 5-, and 8-arm star polymers of DMAEMA and block copolymers of MMA and DMAEMA were synthesized via ATRP.[336] Fluorescent character was introduced to the polymers by the attachment of a fluorescent tag called 2-(8-methacryloyloy-3,6-dioxaoctyl)thioxantheno[2,1,9-dej]isoquinoline-1,3-dione. All the synthesized multi-arm star polymers exhibited fluorescence at broad pH range whereas the maximum emission was observed at pH 4. Particularly, hostosol tagged 3-arm star polymer was exemplified to have better mucoadhesion over bioadhesion in rat tissue.

Polythiophene (PT) and its derivatives are prominent materials in the literature for developing conjugated polymers. Poly(3-hexylthiophene) (P3HT) is of very frequent use as an electron donating species to fabricate organic bulk heterojunction (BHJ) solar cells because of its high conductivity. This bulk heterojunction is created by phase separated and highly ordered morphology through self-assembly, in a blend system. The utility of rod–coil block copolymers in the fabrication of organic photovoltaic devices is well-known due to the fact that they can self-assemble and form well-defined morphologies of high order.[337] The resulting morphologies have huge influence on the physical properties of the photovoltaics and in turn their performance.

Several conjugated rod–coil block copolymers synthesized via ATRP based on P3HT, polyfluorene (PF) or poly (p-phenylene vinylene) (PPV) are: PPV-*b*-PS,[338] P[{(1,4-fullerene)-*alt*-[1,4-dimethylene-2,5-bis (cyclo-hexylmethyl ether)phenylene]}-*b*-(3-hexylthiophene)] (PFDP-*b*-P3HT),[339] P[(3-hexylthiophene)-*b*-2,4-diphenyl-6-(4-vinylphenyl)quinoline] (P3HT-*b*-PSDPQ),[340] P(fluorene-b-2-(9-carbazolyl)ethyl methacrylate),[341] PF-b-PDMAEMA,[342] PF-b-PMAA,[343] and polyfluorene-b-poly(γ-methacryloxyp ropyltrimethoxysilane) (PF-b-PMPS).[344]

By blending rod–coil P3HT-*b*-PDMAEMA block copolymer prepared via ATRP, with CdSe QDs, BHJ solar cells with bicontinuous electron-donor/electron-acceptor networks were made.[345] In this case, the P3HT channels were electron donors whereas the CdSe QDs channels played the role of electron-acceptors. P3HT-*b*-PS initially synthesized by Grignard metathesis polymerization and ATRP was used for preparing composite films with [6,6]-phenyl-C61-butyric acid methyl ester.[346] Photovolatic cells were fabricated from the composite and the effect of the annealing process on the morphology of the composite film was studied.

Using oxidative coupling polymerization and ATRP, 2,5-poly(3-[1-ethyl-2-(2-[poly(3-(N-carbazolyl)propyl acrylate)]propionate)] thiophene) (PT-*g*-PCPA), a light emitting and hole transporting graft copolymer was developed.[347] The investigations confirmed the energy transfer process from PCPA side chains to the PT backbone in solid and solution states. The PCPA side chains also trapped the PT backbone in a solution-like conformation which avoided the aggregation of the PT backbone. Devices developed from this copolymer also confirmed that they could be used as light emitting and hole transporting materials. Thermally sensitive PNIPAM brushes were also grown on PT backbone via ATRP to yield material with unique optical, electronic properties.[348]

Upon combining ATRP and a sol–gel process, double layer photoelectrodes containing TiO_2 nanospheres for DSSC were developed.[349] PVC-g-POEM, an amphiphilic copolymer was synthesized by ATRP and in subsequent step, polymer electrolytes were prepared by complexing the PVC-g-POEM polymer with a metal salt or an ionic liquid. As the electrolyte resided in the hydrophilic POEM domains, there was an increase in the d-spacing between the PVC domains. The microphase separated morphology showed that the POEM channels were well interconnected, which led to a very high ionic conductivity to be suitable for DSSC.[350] Mesoporous TiO_2 electrodes were fabricated by combination of POEM grafted alumina with titanium isopropoxide.[74] An inorganic-polymer hybrid material using CdSe nanocrystals and PVK with high electron mobility was prepared by ATRP. The grafting of PVK onto the surface of CdSe nanocrystals decreased the band gap of PVK and caused a red shift in the emission.[81]

Polymeric light emitting material made up of 8-hydroxyqunioline end-capped polystyrene was prepared through ATRP. Then the 8-hydroxyquinoline end group present in the polymer was chelated with triethylaluminum to form electroluminescent complex of high luminescent efficiency. Besides aluminium, the ligand was able to chelate with Zn, Mg, and other metal ions. The single layer LED with the wavelength of 570 nm was prepared by common spin-coating method.[351]

2.8.2 SENSORS

By combining surface plasmon resonance (SPR) and molecular imprinting, an SPR chip with ametryn-imprinted sensing film was fabricated by using MAA as the functional monomer and EGDMA as the crosslinker, for the detection of ametryn, a human-toxic pesticide.[352] SPR analyses revealed that the imprinted sensing film had more affinity towards its own template molecule ametryn, among several other structurally analogous compounds and the imprinted film exhibited larger response while compared to the non-imprinted one. MIP grafted magnetic ZnO nanorods were prepared via the ARGET ATRP by using MAA as functional monomer, 3-(trimethoxysilyl) propyl methacrylate coated γ-Fe_2O_3 as magnetic monomer, EGDMA as crosslinker, and sulfamethazine as template molecule. These nanorods acted as fluorescence based sensors for the selective recognition of the antibiotic, sulfamethazine.[353]

Mesoporous poly(methyl methacrylate-co-(4-vinylphenylboronic acid) nanospheres synthesized by surface-confined ATRP (SC ATRP) were shown to be sensing glucose.[354] By applying SI-ATRP, cationic polymer brushes of the monomer DMAEMA were grown on the surface of cellulose paper and the modified cellulose surface was used as a bioanalytical tool.[355] The linear cationic polymer was able to specifically detect DNA hybridization in pico-molar level, when combined with PicoGreen. A 3D antibody-immobilized gel layer gold chip was developed by copolymerizing acrylamide and an acryloyl antibody with polymerizable group, through ATRP.[356] The fabricated SPR sensor chips were found to respond to a target antigen by undergoing a large shift in the resonance angle. The resonance angle shift was larger in the present case than the directly antibody immobilized chips, which had been synthesized by adopting the amino coupling procedure. Based on the SPR method, PHEMA biotinylated gold substrates were designed to detect protein toxins[357] and membrane bound proteins.[358] Indium tin oxide (ITO) grafted with PPEGMA brushes were used for immobilizing anti-CEA and could be used as electrochemical immunosensor in protein diagnostics and assay.[359]

Fluorescent silicon wafer based sensor with surface PGMA brushes and pyrene derivatives (spacer molecules) was fabricated by SI-ATRP. The solid sensor was proved to be highly selective to nitrite, a known toxin, in the presence of other anions.[360] Recently Nandi et al. reported the synthesis of pH and temperature responsive fluorescent polymer, PT-g-P(MeO$_2$MA-co-DMAEMA) by ATRP.[361] The fluorescence intensity of the dual responsive polymer was quenched in the presence of nitroaromatic compounds such as picric acid, dinitrophenol, and so forth in both the solid and solution states.

By preparing PS-PEG-PS via ATRP and achieving composite material with carbon black, gas sensors were developed, which were demonstrated for sensing solvent vapors.[362] The volume expansion created by the gas molecules in the amorphous regions of the polymer cut down the conduction of the carbon black leading to rapid increase in electrical resistivity. Folic acid imprinted quartz crystal microbalance (QCM) sensors for selectively sensing folic acid were developed by polymerizing and forming a molecularly imprinted polymer film on the surface of gold crystal, using 1,3,5-trisacrylamide-2,4,6-triazine and EGDMA through AGET-ATRP, with folic acid as template.[363]

2.9 CONCLUSIONS

The developments in the atom transfer radical polymerization method has contributed, in parallel, to the growth of applications in a number of sophisticated areas in applied polymer science such as polymer brushes, stimuli–responsive polymers, polymeric microspheres, microcapsules/nanocapsules, degradable polymers, self-healing polymers, and materials for optoelectronics (sensors, light and energy harvesting materials). In sophistication, ATRP is matched by other competitive methods such as SET-LRP and RAFT. There are few limitations of ATRP in comparison with RAFT, the competing and alternate of reversible deactivation radical polymerization method but these details do not fall under the scope of this article.

REFERENCES

1. Braunecker, W. A.; Matyjaszewski, K. Controlled/Living Radical Polymerization: Features, Developments, and Perspectives. *Prog. Polym. Sci.* **2007,** *32*, 93–146.
2. Goto, A.; Fukuda, T. Kinetics of Living Radical Polymerization. *Prog. Polym. Sci.* **2004,** *29*, 329–385.
3. Fischer, H. The Persistent Radical Effect: A Principle for Selective Radical Reactions and Living Radical Polymerizations. *Chem. Rev.* **2001,** *101*, 3581–3610.
4. Georges, M. K.; Veregin, R. P. N.; Kazmaier, P. M.; Hamer, G. K. Narrow Molecular Weight Resins by a Free-Radical Polymerization Process. *Macromolecules* **1993,** *26*, 2987–2988.
5. Wang, J. S.; Matyjaszewski, K. Controlled "Living"Radical Polymerization. Atom-transfer Radical Polymerization in the Presence of Transition-Metal Complexes. *J. Am. Chem. Soc.* **1995,** *117*, 5614–5615.
6. Kato, M.; Kamigaito, M.; Sawamoto, M.; Higashimura, T. Polymerization of Methyl Methacrylate with the Carbon Tetrachloride/Dichlorotris-(triphenylphosphine)

ruthenium(II)/methylaluminum bis(2,6-di-tert-butylphenoxide) Initiating System: Possibility of Living Radical Polymerization. *Macromolecules* **1995**, *28*, 1721–1723.

7. Percec, V.; Popov, A. V.; Ramirez-Castillo, E.; Monteiro, M.; Barboiu, B.; Weichold, O.; Asandei, A. D.; Mitchell, C. M. Aqueous Room Temperature Metal-Catalyzed Living Radical Polymerization of Vinyl Chloride. *J. Am. Chem. Soc.* **2002**, *124*, 4940–4941.

8. Chiefari, J.; Chong, Y. K. B.; Ercole, F.; Krstina, J.; Jeffery, J.; Le, T. P. T.; Mayadunne, R. T. A.; Meijs, G. F.; Moad, C. L.; Moad, G.; Rizzardo, E.; Thang, S. H. Living Free-Radical Polymerization by Reversible Addition-Fragmentation Chain Transfer: The RAFT Process. *Macromolecules* **1998**, 31, 5559–5562.

9. Subramanian, S. H.; Babu, R. P.; Dhamodharan, R. Ambient Temperature Polymerization of Styrene by Single Electron Transfer Initiation, Followed by Reversible Addition Fragmentation Chain Transfer Control. *Macromolecules* **2008**, *41*, 262–265.

10. Matyjaszewski, K.; Xia, J. Atom Transfer Radical Polymerization. *Chem. Rev.* **2001**, *101*, 2921–2990.

11. Kamigaito, M.; Ando, T.; Sawamoto, M. Metal-Catalyzed Living Radical Polymerization. *Chem. Rev.* **2001**, *101*, 3689–3745.

12. Coessens, V.; Pintauer, T.; Matyjaszewski, K. Functional Polymers by Atom Transfer Radical Polymerization. *Prog. Polym. Sci.* **2001**, *26*, 337–377.

13. Matyjaszewski, K. Atom Transfer Radical Polymerization: From Mechanisms to Applications. *Isr. J. Chem.* **2012**, *52*, 206–220.

14. Matyjaszewski, K. Atom Transfer Radical Polymerization (ATRP): Current Status and Future Perspectives. *Macromolecules* **2012**, *45*, 4015–4039.

15. Jakubowski, W.; Matyjaszewski, K. Activator Generated by Electron Transfer for Atom Transfer Radical Polymerization. *Macromolecules* **2005**, *38*, 4139–4146.

16. Oh, J. K.; Min, K.; Matyjaszewski, K. Preparation of Poly(oligo(ethylene glycol) Monomethyl Ether Methacrylate) by Homogeneous Aqueous AGET ATRP. *Macromolecules* **2006**, *39*, 3161–3167.

17. Lou, Q.; Shipp, D. A. Recent Developments in Atom Transfer Radical Polymerization (ATRP): Methods to Reduce Metal Catalyst Concentrations. *Chem. Phys. Chem.* **2012**, *13*, 3257–3261.

18. Matyjaszewski, K.; Jakubowski, W.; Min, K.; Tang, W.; Huang, J. Q.; Braunecker, W. A.; Tsarevsky, N. V. Diminishing Catalyst Concentration in Atom Transfer Radical Polymerization with Reducing Agents. *Proc. Natl. Acad. Sci. USA* **2006**, *103*, 15309–15314.

19. Jakubowski, W.; Matyjaszewski, K. Activators Regenerated by Electron Transfer for Atom-Transfer Radical Polymerization of (Meth)acrylates and Related Block Copolymers. *Angew. Chem. Int. Ed.* **2006**, *45*, 4482–4486.

20. Magenau, A. J. D.; Strandwitz, N. C.; Gennaro, A.; Matyjaszewski, K. Electrochemically Mediated Atom Transfer Radical Polymerization. *Science* **2011**, *332*, 81–84.

21. Bortolamei, N.; Isse, A. A.; Magenau, A. J. D.; Gennaro, A.; Matyjaszewski, K. Controlled Aqueous Atom Transfer Radical Polymerization with Electrochemical Generation of the Active Catalyst. *Angew. Chem. Int. Ed.* **2011**, *50*, 11391–11394.

22. Konkolewicz, D.; Schroder, K.; Buback, J.; Bernhard, S.; Matyjaszewski, K. Visible Light and Sunlight Photoinduced ATRP with ppm of Cu Catalyst. *ACS Macro Lett.* **2012**, *1*, 1219–1223.

23. Zhang, Y.; Wang, Y.; Matyjaszewski, K. ATRP of Methyl Acrylate with Metallic Zinc, Magnesium, and Iron as Reducing Agents and Supplemental Activators. *Macromolecules* **2011**, *44*, 683–685.

24. Visnevskij, C.; Makuska, R. ICAR ATRP with ppm Cu Catalyst in Water. *Macromolecules* **2013**, *46*, 4764–4771.

25. Matyjaszewski, K.; Tsarevsky, N. V. Nanostructured Functional Materials Prepared by Atom Transfer Radical Polymerization. *Nat. Chem.* **2009**, *1*, 276–288.

26. Wei, H.; Pahang, J. A.; Pun, S. H. Optimization of Brush-Like Cationic Copolymers for Nonviral Gene Delivery. *Biomacromolecules* **2013**, *14*, 275–284.

27. Zolotarskaya, O. Y.; Yuan, Q.; Wynne, K. J.; Yang, H. Synthesis and Characterization of Clickable Cytocompatible Poly(ethylene glycol)-Grafted Polyoxetane Brush Polymers. *Macromolecules* **2013**, *46*, 63–71.

28. Shi, J.; Schellinger, J. G.; Johnson, R. N.; Choi, J. L.; Chou, B.; Anghel, E. L.; Pun, S. H. Influence of Histidine Incorporation on Buffer Capacity and Gene Transfection Efficiency of HPMA-co-oligolysine Brush Polymers. *Biomacromolecules* **2013**, *14*, 1961–1970.

29. Hu, H.; Xiu, K. M.; Xu, S. L.; Yang, W. T.; Xu, F. J. Functionalized Layered Double Hydroxide Nanoparticles Conjugated with Disulfide-Linked Polycation Brushes for Advanced Gene Delivery. *Bioconjugate Chem.* **2013**, *24*, 968–978.

30. Song, J.; Zhou, J.; Duan, H. Self-Assembled Plasmonic Vesicles of SERS-Encoded Amphiphilic Gold Nanoparticles for Cancer Cell Targeting and Traceable Intracellular Drug Delivery. *J. Am. Chem. Soc.* **2012**, *134*, 13458–13469.

31. Shi, J.; Choi, J. L.; Chou, B.; Johnson, R. N.; Schellinger, J. G.; Pun, S. H. Effect of Polyplex Morphology on Cellular Uptake, Intracellular Trafficking, and Transgene Expression. *ACS Nano* **2013**, *7*, 10612–10620.

32. Wang, Y.-H.; Hung, M.-K.; Lin, C.-H.; Lin, H.-C.; Lee, J.-T. Patterned Nitroxide Polymer Brushes for Thin-Film Cathodes in Organic Radical Batteries. *Chem. Commun.* **2011**, *47*, 1249–1251.

33. Lin, H.-C.; Li, C.-C.; Lee, J.-T. Nitroxide Polymer Brushes Grafted onto Silica Nanoparticles as Cathodes for Organic Radical Batteries. *J. Power Sources* **2011**, *196*, 8098–8103.

34. Hung, M.-K.; Wang, Y.-H.; Lin, C.-H.; Lin, H.-C.; Lee, J.-T. Synthesis and Electrochemical Behaviour of Nitroxide Polymer Brush Thin-Film Electrodes for Organic Radical Batteries. *J. Mater. Chem.* **2012**, *22*, 1570–1577.

35. Li, Y.; Zhang, J.; Fang, L.; Wang, T.; Zhu, S.; Li, Y.; Wang, Z.; Zhang, L.; Cui, L.; Yang, B. Fabrication of Silicon/Polymer Composite Nanopost Arrays and Their Sensing Applications. *Small* **2011**, *7*, 2769–2774.

36. Brault, N. D.; Sundaram, H. S.; Huang, C.-J.; Li, Y.; Yu, Q.; Jiang, S. Two-Layer Architecture using Atom Transfer Radical Polymerization for Enhanced Sensing and Detection in Complex Media. *Biomacromolecules* **2012**, *13*, 4049–4056.

37. Huang, C.-J.; Brault, N. D.; Li, Y.; Yu, Q.; Ji, S. Controlled Hierarchical Architecture in Surface-Initiated Zwitterionic Polymer Brushes with Structurally Regulated Functionalities. *Adv. Mater.* **2012**, *24*, 1834–1837.

38. Wan, X.; Wang, D.; Liu, S. Fluorescent pH-Sensing Organic/Inorganic Hybrid Mesoporous Silica Nanoparticles with Tunable Redox-Responsive Release Capability. *Langmuir* **2010**, *26*, 15574–15579.

39. Tokarev, I.; Tokareva, I.; Minko, S. Optical Nanosensor Platform Operating in Near-Physiological pH Range via Polymer-Brush-Mediated Plasmon Coupling. *ACS Appl. Mater. Interfaces* **2011**, *3*, 143–146.

40. Gupta, S.; Agrawal, M.; Conrad, M.; Hutter, N. A.; Olk, P.; Simon, F.; Eng, L. M.; Stamm, M.; Jordan, R. Poly(2-(dimethylamino)ethyl methacrylate) Brushes with

Incorporated Nanoparticles as a SERS Active Sensing Layer. *Adv. Funct. Mater.* **2010,** *20*, 1756–1761.

41. Daniels, C. R.; Tauzin, L. J.; Foster, E.; Advincula, R. C.; Landes, C. F. On the pH-Responsive, Charge-Selective, Polymer-Brush-Mediated Transport Probed by Traditional and Scanning Fluorescence Correlation Spectroscopy. *J. Phys. Chem. B* **2013,** *117*, 4284–4290.

42. Bayramoglu, G.; Arica, M. Y. Preparation of Comb-Type Magnetic Beads by Surface-Initiated ATRP: Modification with Nitrilotriacetate Groups for Removal of Basic Dyes. *Ind. Eng. Chem. Res.* **2012,** *51*, 10629–10640.

43. Nebhani, L.; Schmiedl, D.; Barner, L.; Barner-Kowollik, C. Quantification of Grafting Densities Achieved via Modular "Grafting-To" Approaches onto Divinylbenzene Microspheres. *Adv. Funct. Mater.* **2010,** *20*, 2010–2020.

44. Nebhani, L.; Sinnwell, S.; Inglis, A. J.; Stenzel, M. H.; Barner-Kowollik, C.; Barner, L. Efficient Surface Modification of Divinylbenzene Microspheres via a Combination of RAFT and Hetero Diels–Alder Chemistry. *Macromol. Rapid Commun.* **2008,** *29*, 1431–1437.

45. Hansson, S.; Trouillet, V.; Tischer, T.; Goldmann, A. S.; Carlmark, A.; Barner-Kowollik, C.; Malmström, E. Grafting Efficiency of Synthetic Polymers onto Biomaterials: A Comparative Study of Grafting-From versus Grafting-To. *Biomacromolecules* **2013,** *14*, 64–74.

46. Liu, H.; O'Mahony, C. T.; Audouin, F.; Ventura, C.; Morris, M.; Heise, A. Random Poly(methyl methacrylate-*co*-styrene) Brushes by ATRP to Create Neutral Surfaces for Block Copolymer Self-Assembly. *Macromol. Chem. Phys.* **2012,** *213*, 108–115.

47. Ye, P.; Dong, H.; Zhong, M.; Matyjaszewski, K. Synthesis of Binary Polymer Brushes via Two-Step Reverse Atom Transfer Radical Polymerization. *Macromolecules* **2011,** *44*, 2253–2260.

48. Gao, C.; Zheng, X. Facile Synthesis and Self-Assembly of Multihetero-Arm Hyperbranched Polymer Brushes. *Soft Matter* **2009,** *5*, 4788–4796.

49. Mazurowski, M.; Gallei, M.; Li, J.; Didzoleit, H.; Stühn, B.; Rehahn, M. Redox-Responsive Polymer Brushes Grafted From Polystyrene Nanoparticles by Means of Surface Initiated Atom Transfer Radical Polymerization. *Macromolecules* **2012,** *45*, 8970–8981.

50. Wever, D. A. Z.; Raffa, P.; Picchioni, F.; Broekhuis, A. A. Acrylamide Homopolymers and Acrylamide-N-isopropylacrylamide Block Copolymers by Atomic Transfer Radical Polymerization in Water. *Macromolecules* **2012,** *45*, 4040–4045.

51. Ahmad, H.; Saito, N.; Kagawa, Y.; Okubo, M. Preparation of Micrometer-sized, Monodisperse "Janus" Composite Polymer Particles having Temperature-Sensitive Polymer Brushes at Half of the Surface by Seeded Atom Transfer Radical Polymerization. *Langmuir* **2008,** *24*, 688–691.

52. Barbey, R.; Lavanant, L.; Paripovic, D.; Schüwer, N.; Sugnaux, C.; Tugulu, S.; Klok, H.-A. Polymer Brushes via Surface-Initiated Controlled Radical Polymerization: Synthesis, Characterization, Properties, and Applications. *Chem. Rev.* **2009,** *109*, 5437–5527.

53. Hui, C. M.; Pietrasik, J.; Schmitt, M.; Mahoney, C.; Choi, J.; Bockstaller, M. R.; Matyjaszewski, K. Surface-Initiated Polymerization as an Enabling Tool for Multifunctional (Nano-)engineered Hybrid Materials. *Chem. Mater.* **2014,** *26*, 745–762.

54. Xu, F. J.; Cai, Q. J.; Kang, E. T.; Neoh, K. G. Surface-Initiated Atom Transfer Radical Polymerization from Halogen-Terminated Si(111) (Si–X, X=Cl, Br) Surfaces for the Preparation of Well-Defined Polymer-Si Hybrids. *Langmuir* **2005,** *21*, 3221–3225.

55. Raghuraman, G. K.; Dhamodharan, R.; Prucker, O.; Rühe, J. A Robust Method for the Immobilization of Polymer Molecules on SiO$_2$ Surfaces. *Macromolecules* **2008**, *41*, 873–878.

56. Xu, F. J.; Yuan, Z. L.; Kang, E. T.; Neoh, K. G. Branched Fluoropolymer-Si Hybrids via Surface-Initiated ATRP of Pentafluorostyrene on Hydrogen-Terminated Si(100) Surfaces. *Langmuir* **2004**, *20*, 8200–8208.

57. Xu, F. J.; Cai, Q. J.; Li, Y. L.; Kang, E. T.; Neoh, K. G. Covalent Immobilization of Glucose Oxidase on Well-Defined Poly(glycidyl methacrylate)-Si(111) Hybrids from Surface-Initiated Atom-Transfer Radical Polymerization. *Biomacromolecules* **2005**, *6*, 1012–1020.

58. Xu, F. J.; Zhong, S. P.; Yung, L. Y. L.; Tong, Y. W.; Kang, E. T.; Neoh, K. G. Collagen-Coupled Poly(2-hydroxyethyl methacrylate)-Si(111) Hybrid Surfaces for CellImmobilization. *Tissue Eng.* **2005**, *11*, 1736–1748.

59. Xu, F. J.; Zhong, S. P.; Yung, L. Y. L.; Tong, Y. W.; Kang, E.-T.; Neoh, K. G. Thermoresponsive Comb-Shaped Copolymer-Si(1 0 0) Hybrids for Accelerated Temperature-Dependent Cell Detachment. *Biomaterials* **2006**, *27*, 1236–1245.

60. Peng, J. W.; Liu, Z. F.; Mo, R. H.; Zhong, Y. W.; Qin, J.; Deng, W. X. Star Hydrophilic Polymer Brushed Nanoparticles Covalently Tethered on Si(100) Substrates via Surface-Initiated Atom Transfer Radical Polymerization (ATRP). *Adv. Mat. Res.* **2013**, *662*, 44–54.

61. Xu, L. Q.; Wan, D.; Gong, H. F.; Neoh, K.-G.; Kang, E.-T.; Fu, G. D. One-Pot Preparation of Ferrocene-Functionalized Polymer Brushes on Gold Substrates by Combined Surface-Initiated Atom Transfer Radical Polymerization and "Click Chemistry". *Langmuir* **2010**, *26*, 15376–15382.

62. Chen, R.; Zhu, S.; Maclaughlin, S. Grafting Acrylic Polymers from Flat Nickel and Copper Surfaces by Surface-Initiated Atom Transfer Radical Polymerization. *Langmuir* **2008**, *24*, 6889–6896.

63. Lilge, I.; Schönherr, H. Covalently Cross-Linked Poly(acrylamide) Brushes on Gold with Tunable Mechanical Properties via Surface-Initiated Atom Transfer Radical Polymerization. *Eur. Polym. J.* **2013**, *49*, 1943–1951.

64. Matrab, T.; Chehimi, M. M.; Perruchot, C.; Adenier, A.; Guillez, A.; Save, M.; Charleux, B.; Cabet-Deliry, E.; Pinson J. Novel Approach for Metallic Surface-Initiated Atom Transfer Radical Polymerization using Electrografted Initiators Based on Aryl Diazonium Salts. *Langmuir* **2005**, *21*, 4686–4694.

65. Pyun, J., Kowalewski, T.; Matyjaszewski, K. Synthesis of Polymer Brushes using Atom Transfer Radical Polymerization. *Macromol. Rapid Commun.* **2003**, *24*, 1043–1059.

66. Raghuraman, G. K.; Rühe, J.; Dhamodharan, R. Grafting of PMMA Brushes on Titania Nanoparticulate Surface via Surface-Initiated Conventional Radical and "Controlled" Radical Polymerization (ATRP). *J. Nanopart. Res.* **2008**, *10*, 415–427.

67. Cui, W.-W.; Tang, D.-Y.; Gong, Z.-L. Electrospun Poly(vinylidene Fluoride)/Poly(methyl Methacrylate) Grafted TiO$_2$ Composite Nanofibrous Membrane as Polymer Electrolyte for Lithium-Ion Batteries. *J. Power Sources* **2013**, *223*, 206–213.

68. Gong, Z.-L.; Tang, D.-Y.; Guo, Y.-D. The Fabrication and Self-Flocculation Effect of Hybrid TiO$_2$ Nanoparticles Grafted with Poly(N-isopropylacrylamide) at Ambient Temperature via Surface-Initiated Atom Transfer Radical Polymerization. *J. Mater. Chem.* **2012**, *22*, 16872–16879.

69. Jaymand, M. Synthesis and Characterization of Well-Defined Poly(4-chloromethyl styrene-g-4-vinylpyridine)/TiO$_2$ Nanocomposite via ATRP Technique. *J. Polym. Res.* **2011**, *18*, 1617–1624.

70. Xiao, J.; Chen, W.; Wang, F.; Du, J. Polymer/TiO$_2$ Hybrid Nanoparticles with Highly Effective UV Screening But Eliminated Photocatalytic Activity. *Macromolecules* **2013,** *46,* 375–383.

71. Zhang, G.; Lu, S.; Zhang, L.; Meng, Q.; Shen, C.; Zhang, J. Novel Polysulfone Hybrid Ultrafiltration Membrane Prepared with TiO$_2$-g-HEMA and Its Antifouling Characteristics. *J. Membr. Sci.* **2013,** *436,* 163–173.

72. Wang, W.; Cao, H.; Zhu, G.; Wang, P. A Facile Strategy to Modify TiO$_2$ Nanoparticles via Surface-Initiated ATRP of Styrene. *J. Polym. Sci., Part A: Polym. Chem.* **2010,** *48,* 1782–1790.

73. Yan, J.; Li, B.; Zhou, F.; Liu, W. Ultraviolet Light-Induced Surface-Initiated Atom-Transfer Radical Polymerization. *ACS Macro Lett.* **2013,** *2,* 592–596.

74. Park, J. T.; Chi, W. S.; Roh, D. K.; Ahn, S. H.; Kim, J. H. Hybrid Templated Synthesis of Crack-Free, Organized Mesoporous TiO$_2$ Electrodes for High Efficiency Solid-State Dye-Sensitized Solar Cells. *Adv. Funct. Mater.* **2013,** *23,* 26–33.

75. Wang, W.-C.; Wang, J.; Liao, Y.; Zhang, L.; Cao, B.; Song, G.; She, X. Surface Initiated ATRP of Acrylic Acid on Dopamine-Functionalized AAO Membranes. *J. Appl. Polym. Sci.* **2010,** *117,* 534–541.

76. Zhang, B.; Hu, N.; Wang, Y.; Wang, Z.; Wang, Y.; Kong, E. S.; Zhang, Y. Poly(glycidyl Methacrylates)-Grafted Zinc Oxide Nano-Wire by Surface-Initiated Atom Transfer Radical Polymerization. *Nano-Micro Lett.* **2010,** *2,* 285–289.

77. Abbasian, M.; Aali, N. K.; Shoja, S. E. Synthesis of Poly(methyl Methacrylate)/Zinc Oxide Nanocomposite with Core–Shell Morphology by Atom Transfer Radical Polymerization. *J. Macromol. Sci., Pure Appl. Chem.* **2013,** *50,* 966–975.

78. Peng, X.; Chen, Y.; Li, F.; Zhou, W.; Hu, Y. Preparation and Optical Properties of ZnO@ PPEGMA Nanoparticles. *Appl. Surf. Sci.* **2009,** *255,* 7158–7163.

79. Sato, M.; Kawata, A.; Morito, S.; Sato, Y.; Yamaguchi, I. Preparation and Properties of Polymer/Zinc Oxide Nanocomposites using Functionalized Zinc Oxide Quantum Dots. *Eur. Polym. J.* **2008,** *44,* 3430–3438.

80. Hou, X.; Wang, L.; Hao, J. C. Design, Synthesis, and Characterization of One-Dimensional ZnO/Polymer Nanohybrids. *Mater. Lett.* **2013,** *107,* 162–165.

81. Wang, T.-L.; Yang, C.-H.; Shieh, Y.-T.; Yeh, A.-C. Synthesis of CdSe–poly(N-vinylcarbazole) Nanocomposite by Atom Transfer Radical Polymerization for Potential Optoelectronic Applications. *Macromol. Rapid Commun.* **2009,** *30,* 1679–1683.

82. Esteves, A. C. C.; Bombalski, L.; Trindade, T.; Matyjaszewski, K.; Barros-Timmons, A. Polymer Grafting from CdS Quantum Dots via AGET ATRP in Miniemulsion. *Small* **2007,** *3,* 1230–1236.

83. Llarena, I.; Romero, G.; Ziolo, R. F.; Moya1, S. E. Carbon Nanotube Surface Modification with Polyelectrolyte Brushes Endowed with Quantum Dots and Metal Oxide Nanoparticles through In Situ Synthesis. *Nanotechnology* **2010,** *21,* 055605 (p 8).

84. Liu, Y.-L.; Chen, W.-H.; Chang, Y.-H. Preparation and Properties of Chitosan/Carbon Nanotube Nanocomposites using Poly(styrene sulfonic acid)-Modified CNTs. *Carbohydr. Polym.* **2009,** *76,* 232–238.

85. Liu, Y.-L.; Chen, W.-H. Modification of Multiwall Carbon Nanotubes with Initiators and Macroinitiators of Atom Transfer Radical Polymerization. *Macromolecules* **2007,** *40,* 8881–8886.

86. Chergui, S. M.; Ledebt, A.; Mammeri, F.; Herbst, F.; Carbonnier, B.; Romdhane, H. B.; Delamar, M.; Chehimi, M. M. Hairy Carbon Nanotube@nano-Pd Heterostructures:

Design, Characterization, and Application in Suzuki C–C Coupling Reaction. *Langmuir* **2010**, *26*, 16115–16121.

87. Zhang, Y.; He, H.; Gao, C.; Wu, J. Covalent Layer-by-Layer Functionalization of Multi-walled Carbon Nanotubes by Click Chemistry. *Langmuir* **2009**, *25*, 5814–5824.

88. Pangilinan, K. D.; Santos, C. M.; Estillore, N. C.; Rodrigues, D. F.; Advincula, R. C. Temperature-Responsiveness and Antimicrobial Properties of CNT-PNIPAM Hybrid Brush Films. *Macromol. Chem. Phys.* **2013**, *214*, 464–469.

89. Chochos, C. L.; Stefopoulos, A. A.; Campidelli, S.; Prato, M.; Gregoriou, V. G.; Kallitsis, J. K. Immobilization of Oligoquinoline Chains on Single-Wall Carbon Nanotubes and Their Optical Behavior. *Macromolecules* **2008**, *41*, 1825–1830.

90. Ghislandi, M.; Prado, L. A. S. de A.; Schulte, K.; Barros-Timmons, A. Effect of Filler Functionalization on Thermo-Mechanical Properties of Polyamide-12/Carbon Nanofibers Composites: A Study of Filler–Matrix Molecular Interactions. *J. Mater. Sci.* **2013**, *48*, 8427–8437.

91. Li, L.; Lukehart, C. M. Synthesis of Hydrophobic and Hydrophilic Graphitic Carbon Nanofiber Polymer Brushes. *Chem. Mater.* **2006**, *18*, 94–99.

92. Lee, S. H.; Dreyer, D. R.; An, J.; Velamakanni, A.; Piner, R. D.; Park, S.; Zhu, Y.; Kim, S. O.; Bielawski, C. W.; Ruoff, R. S. Polymer Brushes via Controlled, Surface-Initiated Atom Transfer Radical Polymerization (ATRP) from Graphene Oxide. *Macromol. Rapid Commun.* **2010**, *31*, 281–288.

93. Li, G. L.; Liu, G.; Li, M.; Wan, D.; Neoh, K. G.; Kang, E. T. Organo- and Water-Dispersible Graphene Oxide-Polymer Nanosheets for Organic Electronic Memory and Gold Nanocomposites. *J. Phys. Chem. C* **2010**, *114*, 12742–12748.

94. Gonçalves, G.; Marques, P. A. A. P.; Barros-Timmons, A.; Bdkin, I.; Singh, M. K.; Emami, N.; Grácio, J. Graphene Oxide Modified with PMMA via ATRP as a Reinforcement Filler. *J. Mater. Chem.* **2010**, *20*, 9927–9934.

95. Zhu, S.; Li, J.; Chen, Y.; Chen, Z.; Chen, C.; Li, Y.; Cui, Z.; Zhang, D. Grafting of Graphene Oxide with Stimuli-Responsive Polymers by Using ATRP for Drug Release. *J. Nanopart. Res.* **2012**, *14*, 1132.

96. Yuan, W.; Wang, J.; Shen, T.; Ren, J. Surface Modification of Grapheme Oxide with Thermoresponsive Polymers via Atom Transfer Radical Polymerization: Transition from LCST to UCST. *Mater. Lett.* **2013**, *107*, 243–246.

97. Sun, X.; Wang, W.; Wu, T.; Qiu, H.; Wang, X.; Gao, J. Grafting of Graphene Oxide with Poly(sodium 4-styrenesulfonate) by Atom Transfer Radical Polymerization. *Mater. Chem. Phys.* **2013**, *138*, 434–439.

98. Gao, T.; Wang, X.; Yu, B.; Wei, Q.; Xia, Y.; Zhou, F. Noncovalent Microcontact Printing for Grafting Patterned Polymer Brushes on Graphene Films. *Langmuir* **2013**, *29*, 1054–1060.

99. Bak, J. M.; Lee, H. pH-Tunable Aqueous Dispersion of Graphene Nanocomposites Functionalized with Poly(acrylic acid) Brushes. *Polymer* **2012**, *53*, 4955–4960.

100. Li, D.; Cui, Y.; Wang, K.; He, Q.; Yan, X.; Li, J. Thermosensitive Nanostructures Comprising Gold Nanoparticles Grafted with Block Copolymers. *Adv. Funct. Mater.* **2007**, *17*, 3134–3140.

101. Chakraborty, S.; Bishnoi, S. W.; Pérez-Luna, V. H. Gold Nanoparticles with Poly(N-isopropylacrylamide) Formed via Surface Initiated Atom Transfer Free Radical Polymerization Exhibit Unusually Slow Aggregation Kinetics. *J. Phys. Chem. C* **2010**, *114*, 5947–5955.

102. Shi, H.; Yuan, L.; Wu, Y.; Liu, S. Colorimetric Immunosensing via Protein Functional-ized Gold Nanoparticle Probe Combined with Atom Transfer Radical Polymerization. *Biosens. Bioelectron.* **2011**, *26*, 3788–3793.

103. Datta, H.; Bhowmick, A. K.; Singha, N. K. Methacrylate/Acrylate ABA Triblock Copolymers by Atom Transfer Radical Polymerization; Their Properties and Applica-tion as a Mediator for Organically Dispersible Gold Nanoparticles. *Polymer* **2009**, *50*, 3259–3268.

104. Dong, H.; Zhu, M.; Yoon, J. A.; Gao, H.; Jin, R.; Matyjaszewski, K. One-Pot Synthesis of Robust Core/Shell Gold Nanoparticles. *J. Am. Chem. Soc.* **2008**, *130*, 12852–12853.

105. Lou, X.; Wang, C.; He, L. Core–shell Au Nanoparticle Formation with DNA-Polymer Hybrid Coatings Using Aqueous ATRP. *Biomacromolecules* **2007**, *8*, 1385–1390.

106. Zhao, C.; Zheng, J. Synthesis and Characterization of Poly(N-hydroxyethylacrylamide) for Long-Term Antifouling Ability. *Biomacromolecules* **2011**, *12*, 4071–4079.

107. Raghuraman, G. K.; Dhamodharan R. Surface-Initiated Atom Transfer Radical Polym-erization of Methyl Methacrylate from Magnetite Nanoparticles at Ambient Tempera-ture. *J. Nanosci. Nanotechnol.* **2006**, *6*, 2018–2024.

108. Babu, K.; Dhamodharan, R. Synthesis of Polymer Grafted Magnetite Nanoparticle with the Highest Grafting Density via Controlled Radical Polymerization. *Nanoscale Res. Lett.* **2009**, *4*, 1090–1102.

109. Babu, K.; Dhamodharan, R. Grafting of Poly(methyl methacrylate) Brushes from Magnetite Nanoparticles Using a Phosphonic Acid Based Initiator by Ambient Temper-ature Atom Transfer Radical Polymerization (ATATRP). *Nanoscale Res. Lett.* **2008**, *3*, 109–117.

110. Mu, B.; Wang, T.; Wu, Z.; Shi, H.; Xue, D.; Liu, P. Fabrication of Functional Block Copolymer Grafted Superparamagnetic Nanoparticles for Targeted and Controlled Drug Delivery. *Colloids Surf., A* **2011**, 375, 163–168.

111. Yuan, W.; Yuan, J.; Zhou, L.; Wu, S.; Hong, X. Fe_3O_4@poly(2-hydroxyethyl methacrylate)-graft-poly(3-caprolactone) Magnetic Nanoparticles with Branched Brush Polymeric Shell. *Polymer* **2010**, *51*, 2540–2547.

112. Rutnakornpituk, B.; Wichai, U.; Vilaivan, T.; Rutnakornpituk, M. Surface-Initiated Atom Transfer Radical Polymerization of Poly(4-vinylpyridine) From Magnetite Nanoparticle. *J. Nanopart. Res.* **2011**, *13*, 6847–6857.

113. Theamdee, P.; Traiphol, R.; Rutnakornpituk, B.; Wichai, U.; Rutnakornpituk, M. Surface Modification of Magnetite Nanoparticle with Azobenzene-Containing Water Dispersible Polymer. *J. Nanopart. Res.* **2011**, *13*, 4463–4477.

114. Sun, X.-Y.; Yu, S.-S.; Wan, J.-Q.; Chen, K.-Z. Facile Graft of Poly(2-methacryloyloxy-ethyl phosphorylcholine) onto Fe_3O_4 Nanoparticles by ATRP: Synthesis, Properties, and Biocompatibility. *J. Biomed. Mater. Res. Part A* **2013**, *101A*, 607–612.

115. Liu, J.; He, W.; Zhang, L.; Zhang, Z.; Zhu, J.; Yuan, Chen, L. H.; Cheng, Z.; Zhu, X. Bifunctional Nanoparticles with Fluorescence and Magnetism via Surface-Initiated AGET ATRP Mediated by an Iron Catalyst. *Langmuir* **2011**, *27*, 12684–12692.

116. Prai-in, Y.; Tankanya, K.; Rutnakornpituk, B.; Wichai, U.; Montembault, V.; Pascual, S.; Fontaine, L.; Rutnakornpituk, M. Azlactone Functionalization of Magnetic Nanopar-ticles Using ATRP and Their Bioconjugation. *Polymer* **2012**, *53*, 113–120.

117. Dong, H.; Huang, J.; Koepsel, R. R.; Ye, P.; Russell, A. J.; Matyjaszewski, K. Recy-clable Antibacterial Magnetic Nanoparticles Grafted with Quaternized Poly(2-(dimethylamino)ethyl methacrylate) Brushes. *Biomacromolecules* **2011**, *12*, 1305–1311.

118. Ohno, K.; Mori, C.; Akashi, T.; Yoshida, S.; Tago, Y.; Tsujii, Y.; Tabata, Y. Fabrication of Contrast Agents for Magnetic Resonance Imaging from Polymer-Brush-Afforded Iron Oxide Magnetic Nanoparticles Prepared by Surface-Initiated Living Radical Polymerization. *Biomacromolecules* **2013**, *14*, 3453–3462.

119. Wang, Y.; Teng, X.; Wang, J.-S.; Yang, H. Solvent-Free Atom Transfer Radical Polymerization in the Synthesis of Fe_2O_3@polystyrene Core–Shell Nanoparticles, *Nano Lett.* **2003**, *3*, 789–793.

120. Majewski, A. P.; Schallon, A.; Jérôme, V.; Freitag, R.; Müller, A. H. E.; Schmalz, H. Dual-Responsive Magnetic Core–Shell Nanoparticles for Nonviral Gene Delivery and Cell Separation. *Biomacromolecules* **2012**, *13*, 857–866.

121. Li, D.; Jones, G. L.; Dunlap, J. R.; Hua, F.; Zhao, B. Thermosensitive Hairy Hybrid Nanoparticles Synthesized by Surface-Initiated Atom Transfer Radical Polymerization. *Langmuir* **2006**, *22*, 3344–3351.

122. Munirasu, S.; Karunakaran, R. G.; Rühe, J.; Dhamodharan, R. Synthesis and Morphological Study of Thick Benzyl Methacrylate-Styrene Diblock Copolymer Brushes. *Langmuir* **2011**, *27*, 13284–13292.

123. Liu, P.; Liu, W.; Xue, Q. Preparation of Comb-Like Styrene Grafted Silica Nanoparticles. *J. Macromol. Sci., Pure Appl. Chem.* **2004**, *41*, 1001–1010.

124. Nyström, D.; Antoni, P.; Malsmström, E.; Johansson, M.; Whittaker, M.; Hult, A. Highly-Ordered Hybrid Organic–Inorganic Isoporous Membranes from Polymer Modified Nanoparticles. *Macromol. Rapid Commun.* **2005**, *26*, 524–528.

125. Park, J. T.; Seo, J. A.; Ahn, S. H.; Kim, J. H.; Kang, S. W. Surface Modification of Silica Nanoparticles with Hydrophilic Polymers. *J. Ind. Eng. Chem.* **2010**, *16*, 517–522.

126. Kotsuchibashi, Y.; Wang, Y.; Kim, Y.-J.; Ebara, M.; Aoyagi, T.; Narain, R. Simple Coating with pH-Responsive Polymer-Functionalized Silica Nanoparticles of Mixed Sizes for Controlled Surface Properties. *ACS Appl. Mater. Interfaces* **2013**, *5*, 10004–10010.

127. Zhou, L.; Yuan, W.; Yuan, J.; Hong, X. Preparation of Double-Responsive SiO_2-*g*-PDMAEMA Nanoparticles via ATRP. *Mater. Lett.* **2008**, *62*, 1372–1375.

128. Matsuda, Y.; Kobayashi, M.; Annaka, M.; Ishihara, K.; Takahara, A. Dimensions of a Free Linear Polymer and Polymer Immobilized on Silica Nanoparticles of a Zwitterionic Polymer in Aqueous Solutions with Various Ionic Strengths. *Langmuir* **2008**, *24*, 8772–8778.

129. Zhao, H.; Kang, X.; Liu, L. Comb-Coil Polymer Brushes on the Surface of Silica Nanoparticles. *Macromolecules* **2005**, *38*, 10619–10622.

130. Kotsuchibashi, Y.; Faghihnejad, A.; Zeng, H.; Narain, R. Construction of "Smart" Surfaces with Polymer Functionalized Silica Nanoparticles. *Polym. Chem.* **2013**, *4*, 1038–1047.

131. Wu, T.; Zhang, Y.; Wang, X.; Liu, S. Fabrication of Hybrid Silica Nanoparticles Densely Grafted with Thermoresponsive Poly(N-isopropylacrylamide) Brushes of Controlled Thickness via Surface-Initiated Atom Transfer Radical Polymerization. *Chem. Mater.* **2008**, *20*, 101–109.

132. He, W.; Cheng, L.; Zhang, L.; Liu, Z.; Cheng, Z.; Zhu, X. Facile Fabrication of Biocompatible and Tunable Multifunctional Nanomaterials via Iron-Mediated Atom Transfer Radical Polymerization with Activators Generated by Electron Transfer. *ACS Appl. Mater. Interfaces* **2013**, *5*, 9663–9669.

133. Carlmark, A., Malmstrom, E. E. Atom Transfer Radical Polymerization from Cellulose Fibers at Ambient Temperature. *J. Am. Chem. Soc.* **2002**, *124*, 900–901.

134. Lee, S. B.; Koepsel, R. R.; Morley, S. W.; Matyjaszewski, K. Permanent, Nonleaching Antibacterial Surfaces 1. Synthesis by Atom Transfer Radical Polymerization. *Biomacromolecules* **2004**, *5*, 877–882.

135. Yamanaka, H.; Teramoto, Y.; Nishio, Y. Orientation and Birefringence Compensation of Trunk and Graft Chains in Drawn Films of Cellulose Acetate-Graft-PMMA Synthesized by ATRP. *Macromolecules* **2013**, *46*, 3074–3083.

136. Unohara, T.; Teramoto, Y.; Nishio, Y. Molecular Orientation and Optical Anisotropy in Drawn Films of Cellulose Diacetate-Graft-PLLA: Comparative Investigation with Poly(vinyl acetate-co-vinyl alcohol)-Graft-PLLA. *Cellulose* **2011**, *18*, 539–553.

137. Meng, T.; Gao, X.; Zhang, J.; Yuan, J.; Zhang, Y.; He, J. Graft Copolymers Prepared by Atom Transfer Radical Polymerization (ATRP) from Cellulose. *Polymer* **2009**, *50*, 447–454.

138. Zhong, J.-F.; Chai, X.-S.; Fu, S.-Y. Homogeneous Grafting Poly(methyl methacrylate) on Cellulose by Atom Transfer Radical Polymerization. *Carbohydr. Polym.* **2012**, *87*, 1869–1873.

139. Glaied, O.; Dube, M.; Chabot, B.; Daneault, C. Synthesis of Cationic Polymer-Grafted Cellulose by Aqueous ATRP. *J. Colloid Interface Sci.* **2009**, *333*, 145–151.

140. Zampano, G.; Bertoldo, M.; Bronco, S. Poly(ethyl acrylate) Surface-Initiated ATRP Grafting from Wood Pulp Cellulose Fibers. *Carbohydr. Polym.* **2009**, *75*, 22–31.

141. Lacerda, P. S. S.; Barros-Timmons, A. M. M. V.; Freire, C. S. R.; Silvestre, A. J. D.; Neto, C. P. Nanostructured Composites Obtained by ATRP Sleeving of Bacterial Cellulose Nanofibers with Acrylate Polymers. *Biomacromolecules* **2013**, *14*, 2063–2073.

142. Wang, M.; Yuan, J.; Huang, X.; Cai, X.; Li, L.; Shen, J. Grafting of Carboxybetaine Brush onto Cellulose Membranes via Surface-Initiated ARGET-ATRP for Improving Blood Compatibility. *Colloids Surf., B* **2013**, *103*, 52–58.

143. Jiang, F.; Wang, Z.; Qiao, Y.; Wang, Z.; Tang, C. A Novel Architecture Toward Third-Generation Thermoplastic Elastomers by a Grafting Strategy. *Macromolecules* **2013**, *46*, 4772–4780.

144. Vayachuta, L.; Phinyocheep, P.; Derouet, D.; Pascual, S. Synthesis of NR-g-PMMA by "Grafting From" Method Using ATRP Process. *J. Appl. Polym. Sci.* **2011**, *121*, 508–520.

145. Li, N.; Bai, R.; Liu, C. Enhanced and Selective Adsorption of Mercury Ions on Chitosan Beads Grafted with Polyacrylamide via Surface-Initiated Atom Transfer Radical Polymerization. *Langmuir* **2005**, *21*, 11780–11787.

146. Huang, L.; Yuan, S.; Lv, L.; Tan, G.; Liang, B.; Pehkonen, S. O. Poly(methacrylic acid)-Grafted Chitosan Microspheres via Surface Initiated ATRP for Enhanced Removal of Cd(II) Ions From Aqueous Solution. *J. Colloid Interface Sci.* **2013**, *405*, 171–182.

147. Patrizi, M. L.; Piantanida, G.; Coluzza, C.; Masci, G. ATRP Synthesis and Association Properties of Temperature Responsive Dextran Copolymers Grafted with Poly(N-isopropylacrylamide). *Eur. Polym. J.* **2009**, *45*, 2779–2787.

148. Xing, T.; Hu, W.; Li, S.; Chen, G. Preparation, Structure and Properties of Multi-Functional Silk via ATRP Method. *Appl. Surf. Sci.* **2012**, *258*, 3208–3213.

149. Mishra, V.; Kumar, R. Grafting of 4-Aminoantipyrine from Guar Gum Substrates Using Graft Atom Transfer Radical Polymerization (ATRP) Process. *Carbohydr. Polym.* **2011**, *86*, 296–303.

150. Wang, Z.-H.; Zhu, Y.; Chai, M.-Y.; Yang, W.-T.; Xu, F.-J. Biocleavable Comb-Shaped Gene Carriers from Dextran Backbones with Bioreducible ATRP Initiation Sites. *Biomaterials* **2012**, *33*, 1873–1883.

151. Meng, F.; Hennink, W. E.; Zhong, Z. Reduction-Sensitive Polymers and Bioconjugates for Biomedical Applications. *Biomaterials* **2009,** *30,* 2180–2198.

152. Ayres, L.; Adams, P. H. H. M.; Löwik, D. W. P. M.; van Hest, J. C. M. β-Sheet Side Chain Polymers Synthesized by Atom-Transfer Radical Polymerization. *Biomacromolecules* **2005,** *6,* 825–831.

153. de Graaf, A. J.; Mastrobattista, E.; Vermonden, T.; van Nostrum, C. F.; Rijkers, D. T. S.; Liskamp, R. M. J.; Hennink, W. E. Thermosensitive Peptide-Hybrid ABC Block Copolymers Obtained by ATRP: Synthesis, Self-Assembly, and Enzymatic Degradation. *Macromolecules* **2012,** *45,* 842–851.

154. Giannelis, E. P. Polymer Layered Silicate Nanocomposites. *Adv. Mater.* **1996,** *8,* 29–35.

155. Vaia, R. A.; Jandt, K. D.; Kramer, E.; Giannelis, E. P. Microstructural Evolution of Melt Intercalated Polymer-Organically Modified Layered Silicates Nanocomposites. *Chem. Mater.* **1996,** *8,* 2628–2635.

156. Liu, P.; Guo, J. Polyacrylamide Grafted Attapulgite (PAM-ATP) via Surface-Initiated Atom Transfer Radical Polymerization (SI-ATRP) for Removal of Hg(II) Ion and Dyes. *Colloids Surf., A* **2006,** *282–283,* 498–503.

157. Jin, X.; Li, Y.; Yu, C.; Ma, Y.; Yang, L.; Hu, H. Synthesis of Novel Inorganic–Organic Hybrid Materials for Simultaneous Adsorption of Metal Ions and Organic Molecules in Aqueous Solution. *J. Hazard. Mater.* **2011,** *198,* 247–256.

158. Zhang, H.; Ye, W.; Zhou, F. Preparation of Monodispersed and Lipophilic Attapulgite and Polystyrene Nanorods via Surface-Initiated Atom Transfer Radical Polymerization. *J. Appl. Polym. Sci.* **2011,** *122,* 2876–2883.

159. Li, C.; Liu, J.; Qu, X.; Guo, B.; Yang, Z. Polymer-Modified Halloysite Composite Nanotubes. *J. Appl. Polym. Sci.* **2008,** *110,* 3638–3646.

160. Jiang, J.; Zhang, Y.; Cao, D.; Jiang, P. Controlled Immobilization of Methyltrioxorhenium(VII) Based on SI-ATRP of 4-Vinyl Pyridine from Halloysite Nanotubes for Epoxidation of Soybean Oil. *Chem. Eng. J.* **2013,** *215–216,* 222–226.

161. Wang, L.-P.; Wang, Y.-P.; Pei, X.-W.; Peng, B. Synthesis of Poly(methyl methacrylate)-*b*-poly(N-isopropylacrylamide) (PMMA-*b*-PNIPAM) Amphiphilic Diblock Copolymer Brushes on Halloysite Substrate via Reverse ATRP. *React. Funct. Polym.* **2008,** *68,* 649–655.

162. Yah, W. O.; Xu, H.; Soejima, H.; Ma, W.; Lvov, Y.; Takahara, A. Biomimetic Dopamine Derivative for Selective Polymer Modification of Halloysite Nanotube Lumen. *J. Am. Chem. Soc.* **2012,** *134,* 12134–12137.

163. Yenice, Z.; Tasdelen, M. A.; Oral, A.; Guler, C.; Yagci, Y. Poly(styrene-b-tetrahydrofuran)/Clay Nanocomposites by Mechanistic Transformation. *J. Polym. Sci., Part A: Polym. Chem.* **2009,** *47,* 2190–2197.

164. Oral, A.; Shahwan, T.; Güler, Ç. Synthesis of Poly-2-hydroxyethyl Methacrylate–Montmorillonite Nanocomposite via In Situ Atom Transfer Radical Polymerization. *J. Mater. Res.* **2008,** *23,* 3316–3322.

165. Behling, R. E.; Williams, B. A.; Staade, B. L.; Wolf, L. M.; Cochran, E. W. Influence of Graft Density on Kinetics of Surface-Initiated ATRP of Polystyrene from Montmorillonite. *Macromolecules* **2009,** *42,* 1867–1872.

166. Aydin, M.; Tasdelen, M. A.; Uyar, T.; Yagci, Y. In Situ Synthesis of A_3-Type Star Polymer/Clay Nanocomposites by Atom Transfer Radical Polymerization. *J. Polym. Sci., Part A: Polym. Chem.* **2013,** *51,* 5257–5262.

167. Haloi, D. J.; Ata, S.; Singha, N. K. Synthesis and Characterization of All Acrylic Block Copolymer/Clay Nanocomposites Prepared via Surface Initiated Atom Transfer Radical Polymerization (SI-ATRP). *Ind. Eng. Chem. Res.* **2012**, *51*, 9760–9768.
168. Lee, S.-M.; Nguyen, S. T. Smart Nanoscale Drug Delivery Platforms from Stimuli-Responsive Polymers and Liposomes. *Macromolecules* **2013**, *46*, 9169–9180.
169. Surnar, B.; Jayakannan, M. Stimuli-Responsive Poly(caprolactone) Vesicles for Dual Drug Delivery Under the Gastrointestinal Tract. *Biomacromolecules* **2013**, *14*, 4377–4387.
170. Cheng, R.; Meng, F.; Deng, C.; Klok, H.-A.; Zhong, Z. Dual and Multi-Stimuli Responsive Polymeric Nanoparticles for Programmed Site-Specific Drug Delivery. *Biomaterials* **2013**, *34*, 3647–3657.
171. Yang, P.; Li, D.; Jin, S.; Ding, J.; Guo, J.; Shi, W.; Wang, C. Stimuli-Responsive Biodegradable Poly(methacrylic acid) Based Nanocapsules for Ultrasound Traced and Triggered Drug Delivery System. *Biomaterials* **2014**, *35*, 2079–2088.
172. Ulijn, R. V.; Bibi, N.; Jayawarna, V.; Thornton, P. D.; Todd, S. J.; Mart, R. J.; Smith, A. M.; Gough, J. E. Bioresponsive Hydrogels. *Mater. Today* **2007**, *10*, 40–48.
173. Kelley, E. G.; Albert, J. N. L.; Sullivan, M. O.; Epps, T. H. Stimuli-Responsive Copolymer Solution and Surface Assemblies for Biomedical Applications. *Chem. Soc. Rev.* **2013**, *42*, 7057–7071.
174. Hoffman, A. S. Stimuli-Responsive Polymers: Biomedical Applications and Challenges for Clinical Translation. *Adv. Drug Deliver. Rev.* **2013**, *65*, 10–16.
175. Shim, M. S.; Kwon, Y. J. Stimuli-Responsive Polymers and Nanomaterials for Gene Delivery and Imaging Applications. *Adv. Drug Deliver. Rev.* **2012**, *64*, 1046–1059.
176. Lee, H.; Pietrasik, J.; Sheiko, S. S.; Matyjaszewski, K. Stimuli-Responsive Molecular Brushes. *Prog. Polym. Sci.* **2010**, *35*, 24–44.
177. Bajpai, A. K.; Shukla, S. K.; Bhanu, S.; Kankane, S. Responsive Polymers in Controlled Drug Delivery. *Prog. Polym. Sci.* **2008**, *33*, 1088–1118.
178. Chen, T.; Ferris, R.; Zhang, J.; Ducker, R.; Zauscher, S. Stimulus-Responsive Polymer Brushes on Surfaces: Transduction Mechanisms and Applications. *Prog. Polym. Sci.* **2010**, *35*, 94–112.
179. Roy, D.; Cambre, J. N.; Sumerlin, B. S. Future Perspectives and Recent Advances in Stimuli-Responsive Materials. *Prog. Polym. Sci.* **2010**, *35*, 278–301.
180. Kumar, A.; Srivastava, A.; Galaev, I. Y.; Mattiasson, B. Smart Polymers: Physical Forms and Bioengineering Applications. *Prog. Polym. Sci.* **2007**, *32*, 1205–1237.
181. McCormick, C. L.; Kirkland, S. E.; York, A. W. Synthetic Routes to Stimuli Responsive Micelles, Vesicles, and Surfaces via Controlled/Living Radical Polymerization. *Polym. Rev.* **2006**, *46*, 421–443.
182. Stuart, M. A. C.; Huck, W. T. S.; Genzer, J.; Müller, M.; Ober, C.; Stamm, M.; Sukhorukov, G. B.; Szleifer, I.; Tsukruk, V. V.; Urban, M.; Winnik, F.; Zauscher, S.; Luzinov, I.; Minko, S. Emerging Applications of Stimuli-Responsive Polymer Materials. *Nat. Mater.* **2010**, *9*, 101–113.
183. Yamato, M.; Akiyama, Y.; Kobayashi, J.; Yang, J.; Kikuchi, A.; Okano, T. Temperature-Responsive Cell Culture Surfaces for Regenerative Medicine with Cell Sheet Engineering. *Prog. Polym. Sci.* **2007**, *32*, 1123–1133.
184. Schmaljohann, D. Thermo- and pH-Responsive Polymers in Drug Delivery. *Adv. Drug. Deliver. Rev.* **2006**, *5*, 1655–1670.
185. Nelson, A. Stimuli-Responsive Polymers: Engineering Interactions. *Nat. Mater.* **2008**, *7*, 523–525.

186. Pelah, A.; Bharde, A.; Jovin, T. M. Protein Manipulation by Stimuli Responsive Polymers Encapsulated in Erythrocyte Ghosts. *Soft Matter* **2009,** 1006–1010.
187. Jeong, B.; Gutowska, A. Lessons from Nature: Stimuli-Responsive Polymers and Their Biomedical Applications. *Trends Biotechnol.* **2002,** *20,* 305–311.
188. Urban, M. W. Stratification, Stimuli-Responsiveness, Self-Healing, and Signaling in Polymer Networks. *Prog. Polym. Sci.* **2009,** *34,* 679–687.
189. Alarcon, C. D. H.; Pennadam, S.; Alexander, C. Stimuli Responsive Polymers for Biomedical Applications. *Chem. Soc. Rev.* **2005,** *34,* 276–285.
190. Chaterji, S.; Kwon, I. K.; Park, K. Smart Polymeric Gels: Redefining the Limits of Biomedical Devices. *Prog. Polym. Sci.* **2007,** *32,* 1083–1122.
191. Zhai, S.; Wang, B.; Feng, C.; Li, Y.; Yang, D.; Hu, J.; Lu, G.; Huang, X. Thermoresponsive PPEGMEA-g-PPEGEEMA Well-Defined Double Hydrophilic Graft Copolymer Synthesized by Successive SET-LRP and ATRP. *J. Polym. Sci., Part A: Polym. Chem.* **2010,** *48,* 647–655.
192. Nagase, K.; Kobayashi, J.; Kikuchi, A.; Akiyama, Y.; Kanazawa, H.; Okano, T. Thermoresponsive Polymer Brush on Monolithic-Silica-Rod for the High-Speed Separation of Bioactive Compounds. *Langmuir* **2011,** *27,* 10830–10839.
193. Idota, N.; Kikuchi, A.; Kobayashi, J.; Akiyama, Y.; Sakai, K.; Okano, T. Thermal Modulated Interaction of Aqueous Steroids using Polymer-Grafted Capillaries. *Langmuir* **2006,** *22,* 425–430.
194. Nagase, K.; Kobayashi, J.; Kikuchi, A.; Akiyama, Y.; Kanazawa, H.; Okano, T. Interfacial Property Modulation of Thermoresponsive Polymer Brush Surfaces and Their Interaction with Biomolecules. *Langmuir* **2007,** *23,* 9409–9415.
195. Li, N.; Qi, L,.; Shen, Y.; Li, Y.; Chen, Y. Thermoresponsive Oligo(ethylene glycol)-Based Polymer Brushes on Polymer Monoliths for All-Aqueous Chromatography. *ACS Appl. Mater. Interfaces* **2013,** *5,* 12441–12448.
196. Vasani, R. B.; McInnes, S. J. P.; Cole, M. A.; Md Jani, A. M.; Ellis, A. V.; Voelcker, N. H. Stimulus-Responsiveness and Drug Release from Porous Silicon Films ATRP-Grafted with Poly(N-isopropylacrylamide). *Langmuir* **2011,** *27,* 7843–7853.
197. Masci, G.; Diociaiuti, M.; Crescenzi, V. ATRP Synthesis and Association Properties of Thermoresponsive Anionic Block Copolymers. *J. Polym. Sci., Part A: Polym. Chem.* **2008,** *46,* 4830–4842.
198. Zhu, W.; Nese, A.; Matyjaszewski, K. Thermoresponsive Star Triblock Copolymers by Combination of ROP and ATRP: From Micelles to Hydrogels. *J. Polym. Sci., Part A: Polym. Chem.* **2011,** *49,* 1942–1952.
199. Liras, M.; García-García, J. M.; Quijada-Garrido, I.; Gallardo, A.; París, R. Thermo-Responsive Allyl-Functionalized 2-(2-methoxyethoxy)ethylmethacrylate-Based Polymers as Versatile Precursors for Smart Polymer Conjugates and Conetworks. *Macromolecules* **2011,** *44,* 3739–3745.
200. Park, S.; Cho, H. Y.; Yoon, J. A.; Kwak, Y.; Srinivasan, A.; Hollinger, J. O.; Paik, H.-J.; Matyjaszewski, K. Photo-Cross-Linkable Thermoresponsive Star Polymers Designed for Control of Cell-Surface Interactions. *Biomacromolecules* **2010,** *11,* 2647–2652.
201. Yoon, J. A.; Kowalewski, T.; Matyjaszewski, K. Comparison of Thermoresponsive Deswelling Kinetics of Poly(oligo(ethylene oxide) methacrylate)-Based Thermoresponsive Hydrogels Prepared by "Graft-From" ATRP. *Macromolecules* **2011,** *44,* 2261–2268.
202. Liu, S.; Billingham, N. C.; Armes, S. P. A Schizophrenic Water-Soluble Diblock Copolymer. *Angew. Chem. Int. Ed.* **2001,** *40*(12), 2328–2331.

203. Robinson, K. L.; de Paz-Banez, M. V.; Wang, X. S.; Armes, S. P. Synthesis of Well-Defined, Semibranched, Hydrophilic–Hydrophobic Block Copolymers using Atom Transfer Radical Polymerization. *Macromolecules* **2001**, *34*, 5799–5805.

204. Li, C.; Gunari, N.; Fischer, K.; Janshoff, A.; Schmidt, M. New Perspectives for the Design of Molecular Actuators: Thermally Induced Collapse of Single Macromolecules fromCylindrical Brushes to Spheres. *Angew. Chem., Int. Ed.* **2004**, *43*, 1101–1104.

205. Lutz, J.-F.; Akdemir, O. Hoth, A.; Point by Point Comparison of Two Thermosensitive Polymers Exhibiting a Similar LCST: Is the Age of Poly(NIPAM) Over? *J. Am. Chem. Soc.* **2006**, *128*, 13046–13047.

206. Quan, Z.; Zhu, K.; Knudsen, K. D.; Nyström, B.; Lund, R. Tailoring the Amphiphilicity and Self-Assembly of Thermosensitive Polymers: End-Capped PEG–PNIPAAM Block Copolymers. *Soft Matter* **2013**, *9*, 10768–10778.

207. Pietrasik, J.; Sumerlin, B. S.; Lee, R. Y.; Matyjaszewski, K. Solution Behavior of Temperature-Responsive Molecular Brushes Prepared by ATRP. *Macromol. Chem. Phys.* **2007**, *208*, 30–36.

208. Friebe, A.; Ulbricht, M. Controlled Pore Functionalization of Poly(ethylene terephthalate) Track-Etched Membranes via Surface-Initiated Atom Transfer Radical Polymerization. *Langmuir* **2007**, *23*, 10316–10322.

209. Alem, H.; Duwez, A. S.; Lussis, P.; Lipnik, P.; Jonas, A. M.; Demoustier-Champagne, S. Microstructure and Thermo-Responsive Behavior of Poly(N-isopropylacrylamide) Brushes Grafted in Nanopores of Track-Etched Membranes. *J. Membr. Sci.* **2008**, *308*, 75–86.

210. Li, P.-F.; Xie, R.; Jiang, J.-C.; Meng, T.; Yang, M.; Ju, X.-J.; Yang, L.; Chu, L.-Y. Thermo-Responsive Gating Membranes with Controllable Length and Density of Poly(N-isopropylacrylamide) Chains Grafted by ATRP Method. *J. Membr. Sci.* **2009**, *337*, 310–317.

211. Nagase, K.; Hatakeyama, Y.; Shimizu, T.; Matsuura, K.; Yamato, M.; Takeda, N.; Okano, T. Hydrophobized Thermoresponsive Copolymer Brushes for Cell Separation by Multistep Temperature Change. *Biomacromolecules* **2013**, *14*, 3423–3433.

212. Camp, W. V.; Du Prez, F. E.; Alem, H.; Demoustier-Champagne, S.; Willet, N.; Grancharov, G.; Duwez, A.-S. Poly(acrylic acid) with Disulfide Bond for the Elaboration of pH-Responsive Brush Surfaces. *Eur. Polym. J.* **2010**, *46*, 195–201.

213. Wang, Z.; Tan, B. H.; Hussain, H.; He, C. pH-Responsive Amphiphilic Hybrid Random-Type Copolymers of Poly(acrylic acid) and Poly(acrylate-POSS): Synthesis by ATRP and Self-Assembly in Aqueous Solution. *Colloid Polym. Sci.* **2013**, *291*, 1803–1815.

214. Ravi, P.; Dai, S.; Tan, C. H.; Tam, K. C. Self-Assembly of Alkali-Soluble [60]Fullerene Containing Poly(methacrylic acid) in Aqueous Solution. *Macromolecules* **2005**, *38*, 933–939.

215. Wang, C.; Ravi, P.; Tam, K. C. Morphological Transformation of [60]Fullerene-Containing Poly(acrylic acid) Induced by the Binding of Surfactant. *Langmuir* **2006**, *22*, 2927–2930.

216. Cheesman, B. T.; Willott, J. D.; Webber, G. B.; Edmondson, S.; Wanless, E. J. pH-Responsive Brush-Modified Silica Hybrids Synthesized by Surface-Initiated ARGET ATRP. *ACS Macro Lett.* **2012**, *1*, 1161–1165.

217. Li, C.; Gu, C.; Zhang, Y.; Lang, M. Synthesis and Self-Assembly of pH-Responsive Amphiphilic Poly(ε-caprolactone)-Block-Poly(acrylic acid) Copolymer. *Polym. Bull.* **2012**, *68*, 69–83.

218. Wang, D.; Tan, J.; Kang, H.; Ma, L.; Jin, X.; Liu, R.; Huang, Y. Synthesis, Self-Assembly and Drug Release Behaviors of pH-Responsive Copolymers Ethyl Cellulose-Graft-PDEAEMA through ATRP. *Carbohydr. Polym.* **2011,** *84,* 195–202.

219. Hui, G.; Yanan, M.; Xueyou, L.; Yongri, L.; Baoquan, C.; Jianbiao, M. pH-Responsive Nano-Assemblies of Amino Poly(glycerol methacrylate). *Eur. Polym. J.* **2011,** *47,* 1232–1239.

220. Forbes, D. C.; Creixell, M.; Frizzell, H.; Peppas, N. A. Polycationic Nanoparticles Synthesized using ARGET ATRP for Drug Delivery. *Eur. J. Pharm. Biopharm.* **2013,** *84,* 472–478.

221. Sedghi, R.; Oskooie, H. A.; Heravi, M. M.; Nabid, M. R.; Zarnani, A. H. Divergent Synthesis of Dendrimer-like pH-Responsive Macromolecules Through a Combination of ATRP and ROP for Controlled Release of Anti-Cancer Drug. *J. Mater. Chem. B* **2013,** *1,* 773–786.

222. Feng, X.; Taton, D.; Borsali, R.; Chaikof, E. L.; Gnanou, Y. pH Responsiveness of Dendrimer-like Poly(ethylene oxide)s. *J. Am. Chem. Soc.* **2006,** *128,* 11551–11562.

223. Yuan, W.; Zhang, J.; Wei, J.; Zhang, C.; Ren, J. Synthesis and Self-Assembly of pH-Responsive Amphiphilic Dendritic Star-Block Terpolymer by the Combination of ROP, ATRP, and Click Chemistry. *Eur. Polym. J.* **2011,** *47,* 949–958.

224. Ding, J.; Xiao, C.; He, C.; Li, M.; Li, D.; Zhuang, X.; Chen, X. Facile Preparation of a Cationic Poly(amino acid) Vesicle for Potential Drug and Gene Co-Delivery. *Nanotechnology* **2011,** *22,* 494012.

225. Zhao, C.; Patel, K.; Aichinger, L. M.; Liu, Z.; Hu, R.; Chen, H.; Li, X.; Li, L.; Zhang, G.; Chang, Y.; Zheng, J. Antifouling and Biodegradable Poly(N-hydroxyethyl acrylamide) (polyHEAA)-Based Nanogels. *RSC Adv.* **2013,** *3,* 19991–20000.

226. Zhai, S.; Song, X.; Yang, D.; Chen, W.; Hu, J.; Lu, G.; Huang, X. Synthesis of Well-Defined pH-Responsive PPEGMEA-g-P2VP Double Hydrophilic Graft Copolymer via Sequential SET-LRP and ATRP. *J. Polym. Sci., Part A: Polym. Chem.* **2011,** *49,* 4055–4064.

227. Zeng, J.; Du, P.; Liu, P. One-Pot Self-Assembly Directed Fabrication of Biocompatible Core Cross-Linked Polymeric Micelles as a Drug Delivery System. *RSC Adv.* **2013,** *3,* 19492–19500.

228. Iddon, P. D.; Armes, S. P. Synthesis of Stimulus-Responsive Block Copolymer Gelators by Atom Transfer Radical Polymerization. *Eur. Polym. J.* **2007,** *43,* 1234–1244.

229. de Groot, G. W.; Santonicola, M. G.; Sugihara, K.; Zambelli, T.; Reimhult, E.; Vörös, J.; Vancso, G. J. Switching Transport through Nanopores with pH-Responsive Polymer Brushes for Controlled Ion Permeability. *ACS Appl. Mater. Interfaces* **2013,** *5,* 1400–1407.

230. Liu, S.; Armes, S. P. Polymeric Surfactants for the New Millennium: A pH-Responsive, Zwitterionic, Schizophrenic Diblock Copolymer. *Angew. Chem. Int. Ed.* **2002,** *41,* 1413–1416.

231. Cai, Y.; Armes, S. P. A Zwitterionic abc Triblock Copolymer that Forms a "Trinity" of Micellar Aggregates in Aqueous Solution. *Macromolecules* **2004,** *37,* 7116–7222.

232. Du, J.; Tang, Y.; Lewis, A. L.; Armes, S. P. pH-Sensitive Vesicles Based on a Biocompatible Zwitterionic Diblock Copolymer. *J. Am. Chem. Soc.* **2005,** *127,* 17982–17983.

233. Du, J. and Armes, S. P. pH-Responsive Vesicles Based on a Hydrolytically Self-Cross-Linkable Copolymer. *J. Am. Chem. Soc.* **2005,** *127,* 12800–12801.

234. Wang, L.; Wang, H.; Yuan, L.; Yang, W.; Wu, Z.; Chen, H. Step-Wise Control of Protein Adsorption and Bacterial Attachment on a Nanowire Array Surface: Tuning Surface Wettability by Salt Concentration. *J. Mater. Chem.* **2011,** *21,* 13920–13925.

235. Xu, Y.; Bolisetty, S.; Drechsler, M.; Fang, B.; Yuan, J. pH and Salt Responsive Poly(N,N-dimethylaminoethyl methacrylate) Cylindrical Brushes and their Quaternized Derivatives. *Polymer* **2008,** *49,* 3957–3964.
236. Xu, Y.; Bolisetty, S.; Drechsler, M.; Fang, B.; Yuan, J. Manipulating Cylindrical Polyelectrolyte Brushes on the Nanoscale by Counterions: Collapse Transition to Helical Structures. *Soft Matter* **2009,** *5,* 379–384.
237. Xu, Y.; Bolisetty, S, Ballauff, M.; Mueller, A. H. E. Switching the Morphologies of Cylindrical Polycation Brushes by Ionic and Supramolecular Inclusion Complexes. *J. Am. Chem. Soc.* **2009,** *131,* 1640–1641.
238. Zhao, Y.-H.; Wee, K.-H.; Bai, R. A Novel Electrolyte-Responsive Membrane with Tunable Permeation Selectivity for Protein Purification. *ACS Appl. Mater. Interfaces* **2010,** *2,* 203–211.
239. Xu, Y.; Shi, L.; Ma, R.; Zhang, W.; An, Y.; Zhu, X. Synthesis and Micellization of Thermo- and pH-Responsive Block Copolymer of Poly(N-isopropylacrylamide)-Block-Poly(4-vinylpyridine). *Polymer* **2007,** *48,* 1711–1717.
240. Mao, J.; Ji, X.; Bo, S. Synthesis and pH/Temperature-Responsive Behavior of PLLA-b-PDMAEMA Block Polyelectrolytes Prepared via ROP and ATRP. *Macromol. Chem. Phys.* **2011,** *212,* 744–752.
241. Medel, S.; García, J. M.; Garrido, L.; Quijada-Garrido, I.; París, R. Thermo- and pH-Responsive Gradient and Block Copolymers Based on 2-(2-Methoxyethoxy)ethyl Methacrylate Synthesized via Atom Transfer Radical Polymerization and the Formation of Thermoresponsive Surfaces. *J. Polym. Sci., Part A: Polym. Chem.* **2011,** *49,* 690–700.
242. Li, C.; Ge, Z.; Fang, J.; Liu, S. Synthesis and Self-Assembly of Coil-Rod Double Hydrophilic Diblock Copolymer with Dually Responsive Asymmetric Centipede-Shaped Polymer Brush as the Rod Ssegment. *Macromolecules* **2009,** *42,* 2916–2924.
243. Yamamoto, S.-I.; Pietrasik, J.; Matyjaszewski, K. Temperature- and pH-Responsive Dense Copolymer Brushes Prepared by ATRP. *Macromolecules* **2008,** *41,* 7013–7020.
244. Xiong, Z.; Peng, B.; Han, X.; Peng, C.; Liu, H.; Hu, Y. Dual-Stimuli Responsive Behaviors of Diblock Polyampholyte PDMAEMA-*b*-PAA in Aqueous Solution. *J. Colloid Interface Sci.* **2011,** *356,* 557–565.
245. Bao, H.; Li, L.; Gan, L. H.; Ping, Y.; Li, J.; Ravi, P. Thermo- and pH-Responsive Association Behavior of Dual Hydrophilic Graft Chitosan Terpolymer Synthesized via ATRP and Click Chemistry. *Macromolecules* **2010,** *43,* 5679–5687.
246. Ren, T.; Lei, X.; Yuan, W. Synthesis and Self-Assembly of Double-Hydrophilic Pentablock Copolymer with pH and Temperature Responses via Sequential Atom Transfer Radical Polymerization. *Mater. Lett.* **2012,** *67,* 383–386.
247. Schmalz, A.; Schmalz, H.; Müller, A. H. E. Smart Hydrogels Based on Responsive Star-Block Copolymers. *Soft Matter* **2012,** *8,* 9436–9445.
248. Zhang, B.-Y.; He, W.-D.; Li, W.-T.; Li, L.-Y.; Zhang, K.-R.; Zhang, H. Preparation of Block-Brush PEG-*b*-P(NIPAM-*g*-DMAEMA) and its Dual Stimulus-Response. *Polymer* **2010,** *51,* 3039–3046.
249. Liu, X.; Ni, P.; He, J.; Zhang, M. Synthesis and Micellization of pH/Temperature-Responsive Double-Hydrophilic Diblock Copolymers Polyphosphoester-block-poly[2-(dimethylamino)ethyl methacrylate] Prepared via ROP and ATRP. *Macromolecules* **2010,** *43,* 4771–4781.
250. Teoh, S. K.; Ravi, P.; Dai, S.; Tam, K. C. Self-Assembly of Stimuli-Responsive Water-Soluble [60]Fullerene End-Capped Ampholytic Block Copolymer. *J. Phys. Chem. B* **2005,** *109,* 4431–4438.

251. Lindqvist, J.; Nyström, D.; Östmark, E.; Antoni, P.; Carlmark, A; Johansson, M.; Hult, A.; Malmström, E. Intelligent Dual-Responsive Cellulose Surfaces via Surface-Initiated ATRP. *Biomacromolecules* **2008,** *9,* 2139–2145.

252. Zhang, Z. B.; Zhu, X. L.; Xu, F. J.; Neoh, K. G.; Kang, E. T. Temperature- and pH-Sensitive Nylon Membranes Prepared via Consecutive Surface-Initiated Atom Transfer Radical Graft Polymerizations. *J. Membr. Sci.* **2009,** *342,* 300–306.

253. Jung, S.-H.; Song, H.-Y.; Lee, Y.; Jeong, H. M.; Lee, H. Novel Thermoresponsive Polymers Tunable by pH. *Macromolecules* **2011,** *44,* 1628–1634.

254. Strozyk, M. S.; Chanana, M.; Pastoriza-Santos, I.; Pérez-Juste, J.; Liz-Marzán, L. M. Protein/Polymer-Based Dual-Responsive Gold Nanoparticles with pH-Dependent Thermal Sensitivity. *Adv. Funct. Mater.* **2012,** *22,* 1436–1444.

255. Feng, H.; Zhao, Y.; Pelletier, M.; Dan, Y.; Zhao, Y. Synthesis of Photo- and pH-Responsive Composite Nanoparticles Using a Two-Step Controlled Radical Polymerization Method. *Polymer* **2009,** *50,* 3470–3477.

256. Men, Y.; Drechsler, M.; Yuan, J. Double-Stimuli-Responsive Spherical Polymer Brushes with a Poly(ionic liquid) Core and a Thermoresponsive Shell. *Macromol. Rapid Commun.* **2013,** *34,* 1721–1727.

257. Nair, L. S.; Laurencin, C. T. Biodegradable Polymers as Biomaterials. *Prog. Polym. Sci.* **2007,** *32,* 762–798.

258. Tian, H.; Tang, Z.; Zhuang, X.; Chen, X.; Jing, X. Biodegradable Synthetic Polymers: Preparation, Functionalization and Biomedical Application. *Prog. Polym. Sci.* **2012,** *37,* 237–280.

259. Tsarevsky, N. V.; Matyjaszewski, K. Environmentally Benign Atom Transfer Radical Polymerization: Towards "Green"Processes and Materials. *J. Polym. Sci., Part A: Polym. Chem.* **2006,** *44,* 5098–5112.

260. Chung, I. S.; Matyjaszewski, K. Synthesis of Degradable Poly(methyl methacrylate) via ATRP: Atom Transfer Radical Ring-Opening Copolymerization of 5-Methylene-2-phenyl-1,3-dioxolan-4-one and Methyl Methacrylate. *Macromolecules* **2003,** *36,* 2995–2998.

261. Tsarevsky, N. V.; Matyjaszewski, K. Reversible Redox Cleavage/Coupling of Polystyrene with Disulfide or Thiol Groups Prepared by Atom Transfer Radical Polymerization. *Macromolecules* **2002,** *35,* 9009–9014.

262. Tsarevsky, N. V.; Matyjaszewski, K. Combining Atom Transfer Radical Polymerization and Disulfide/Thiol Redox Chemistry: A Route to Well-Defined (Bio)degradable Polymeric Materials. *Macromolecules* **2005,** *38,* 3087–3092.

263. Lu, Z. R.; Wang, X. H.; Parker, D. L.; Goodrich, K. C.; Buswell, H. R. Poly(L-glutamic acid) Gd(III)-DOTA Conjugate with a Degradable Spacer for Magnetic Resonance Imaging. *Bioconjugate Chem.* **2003,** *14,* 715–719.

264. Cho, H. Y.; Srinivasan, A.; Hong, J.; Hsu, E.; Liu, S.; Shrivats, A.; Kwak, D.; Bohaty, A. K.; Paik, H.-J.; Hollinger, J. O.; Matyjaszewski, K. Synthesis of Biocompatible PEG-Based Star Polymers with Cationic and Degradable Core for siRNA Delivery. *Biomacromolecules* **2011,** *12,* 3478–3486.

265. Nelson-Mendez, A.; Aleksanian, S.; Oh, M.; Limb, H.-S.; Oh, J. K. Reductively Degradable Polyester-Based Block Copolymers Prepared by Facile Polycondensation and ATRP: Synthesis, Degradation, and Aqueous Micellization. *Soft Matter* **2011,** *7,* 7441–7452.

266. Oh, J. K.; Tang, C.; Gao, H.; Tsarevsky, N. V.; Matyjaszewski, K. Inverse Miniemulsion ATRP: A New Method for Synthesis and Functionalization of Well-Defined Water-Soluble/Cross-Linked Polymeric Particles. *J. Am. Chem. Soc.* **2006,** *128,* 5578–5584.

267. Lamson, M.; Epshtein-Assor, Y.; M. S. Silverstein; Matyjaszewski, K. Synthesis of Degradable PolyHIPEs by AGET ATRP. *Polymer* **2013**, *54*, 4480–4485.
268. Kimmins, S. D.; Cameron, N. R. Functional Porous Polymers by Emulsion Templating: Recent Advances. *Adv. Funct. Mater.* **2011**, *21*, 211–225.
269. Johnson, J. A.; Lewis, D. R.; Díaz, D. D.; Finn, M. G.; Koberstein, J. T.; Turro, N. J. Synthesis of Degradable Model Networks via ATRP and Click Chemistry. *J. Am. Chem. Soc.* **2006**, *128*, 6564–6565.
270. Johnson, J. A.; Finn, M. G.; Koberstein, J. T.; Turro, N. J. Synthesis of Photocleavable Linear Macromonomers by ATRP and Star Macromonomers by a Tandem ATRP-Click Reaction: Precursors to Photodegradable Model Networks. *Macromolecules* **2007**, *40*, 3589–3598.
271. Yang, X. C.; Chai, M. Y.; Zhu, Y.; Yang, W. T.; Xu, F. J. Facilitation of Gene Transfection with Well-Defined Degradable Comb-Shaped Poly(glycidyl methacrylate) Derivative Vectors. *Bioconjugate Chem.* **2012**, *23*, 618–626.
272. Zhu, Y.; Tang, G.-P.; Xu, F.-J. Efficient Poly(N-3-hydroxypropyl)aspartamide-Based Carriers via ATRP for Gene Delivery. *ACS Appl. Mater. Interfaces* **2013**, *5*, 1840–1848.
273. Bian, J.; Zhang, M.; He, J.; Ni, P. Preparation and Self-Assembly of Double Hydrophilic Poly(ethylethylene phosphate)-Block-Poly[2-(succinyloxy)ethyl methacrylate] Diblock Copolymers for Drug Delivery. *React. Funct. Polym.* **2013**, *73*, 579–587.
274. Li, F.; Pei, D.; Huang, Q.; Shi, T.; Zhang, G. Synthesis and Properties of Novel Biomimetic and Thermo-Responsive Dextran-Based Biohybrids. *Carbohydr. Polym.* **2014**, *99*, 728–735.
275. Bencherif, S. A.; Washburn, N. R.; Matyjaszewski, K. Synthesis by AGET ATRP of Degradable Nanogel Precursors for In SituFormation of Nanostructured Hyaluronic Acid Hydrogcl. *Biomacromolecules* **2009**, *10*, 2499–2507.
276. Atzet, S.; Curtin, S.; Trinh, P.; Bryant, S.; Ratner, B. Degradable Poly(2-hydroxyethyl methacrylate)-co-polycaprolactone Hydrogels for Tissue Engineering Scaffolds. *Biomacromolecules* **2008**, *9*, 3370–3377.
277. Barner, L. Synthesis of Microspheres as Versatile Functional Scaffolds for Materials Science Applications. *Adv. Mater.* **2009**, *21*, 2547–2553.
278. Li, K.; Stöver, H. D. H. Synthesis of Monodisperse Poly(divinylbenzene) Microspheres. *J. Polym. Sci, Part A: Polym. Chem.* **1993**, *31*, 3257–3263.
279. Pfaff, A.; Müller, A. H. E. Hyperbranched Glycopolymer Grafted Microspheres. *Macromolecules* **2011**, *44*, 1266–1272.
280. Jiang, J.; Zhang, Y.; Guo, X.; Zhang, H. Narrow or Monodisperse, Highly Cross-Linked, and "Living" Polymer Microspheres by Atom Transfer Radical Precipitation Polymerization. *Macromolecules* **2011**, *44*, 5893–5904.
281. Jiang, J.; Zhang, Y.; Guo, X.; Zhang, H. Ambient Temperature Synthesis of Narrow or Monodisperse, Highly Crosslinked, and "Living" Polymer Microspheres by Atom Transfer Radical Precipitation Polymerization. *RSC Advances* **2012**, *2*, 5651–5662.
282. Fang, L.; Chen, S.; Guo, X.; Zhang, Y.; Zhang, H. Azobenzene-Containing Molecularly Imprinted Polymer Microspheres with Photo- and Thermoresponsive Template Binding Properties in Pure Aqueous Media by Atom Transfer Radical Polymerization. *Langmuir* **2012**, *28*, 9767–9777.
283. Fu, G.-D.; Li, G. L.; Neoh, K. G.; Kang, E. T. Hollow Polymeric Nanostructures-Synthesis, Morphology and Function. *Prog. Polym. Sci.* **2011**, *36*, 127–167.

284. De Cock, L. J.; De Koker, S.; De Geest, B. G.; Grooten, J.; Vervaet, C.; Remon, J. P.; Sukhorukov, G. B.; Antipina, M. N. Polymeric Multilayer Capsules in Drug Delivery. *Angew. Chem. Int. Ed.* **2010**, *49*, 6954–6973.
285. Delcea, M.; Mohwald, H.; Skirtach, A. G. Stimuli-Responsive LbL Capsules and Nanoshells for Drug Delivery. *Adv. Drug Deliv. Rev.* **2011**, *63*, 730–747.
286. Lomas, H.; Johnston, A. P. R.; Such, G. K.; Zhu, Z. Y.; Liang, K.; van Koeverden, M. P.; Alongkornchotikul, S.; Caruso, F. Polymersome-Loaded Capsules for Controlled Release of DNA. *Small* **2011**, *7*, 2109–2119.
287. De Koker, S.; De Cock, L. J.; Rivera-Gil, P.; Parak, W. J.; Velty, R. A.; Vervaet, C.; Remon, J. P.; Grooten, J.; De Geest, B. G. Polymeric Multilayer Capsules Delivering Biotherapeutics. *Adv. Drug Deliv. Rev.* **2011**, *63*, 748–761.
288. Dong, F.; Guo, W.; Park, S. K.; Ha, C. S. Controlled Synthesis of Novel Cyanopropyl Polysilsesquioxane Hollow Spheres Loaded with Highly Dispersed Au Nanoparticles for Catalytic Applications. *Chem. Commun.* **2012**, *48*, 1108–1110.
289. Poe, S. L.; Kobaslija, M.; McQuade, D. T. Mechanism and Application of a Microcapsule Enabled Multicatalyst Reaction. *J. Am. Chem. Soc.* **2007**, *129*, 9216–9221.
290. Narayan, P.; Marchant, D.; Wheatley, M. A. Optimization of Spray Drying by Factorial Design for Production of Hollow Microspheres for Ultrasound Imaging. *J. Biomed. Mat. Res.* **2001**, *56*, 333–341.
291. Duchesne, T. A.; Brown, J. Q.; Guice, K. B.; Lvov, Y. M.; McShane, M. J. Encapsulation and Stability Properties of Nanoengineered Polyelectrolyte Capsules for Use as Fluorescent Sensors. *Sens. Mater.* **2002**, *14*, 293–308.
292. Lu, Y.; Proch, S.; Schrinner, M.; Drechsler, M.; Kempe, R.; Ballauff, M. Thermosensitive Core–Shell Microgel as Nanoreactor for Catalytic Active Metal Nanoparticles. *J. Mater. Chem.* **2009**, *19*, 3955–3961.
293. Vriezema, D. M.; Garcia, P. M. L.; Oltra, N. S.; Natzakis, N. S.; Kuiper, S. M.; Nolte, R. J. M.; Rowan, A. E.; van Hest, J. C. M. Positional Assembly of Enzymes in Polymersome Nanoreactors for Cascade Reactions. *Angew. Chem. Int. Ed.* **2007**, *46*, 7378–7382.
294. Renggli, K.; Baumann, P.; Langowska, K.; Onaca, O.; Bruns, N.; Meier, W. Selective and Responsive Nanoreactors. *Adv. Funct. Mater.* **2011**, *21*, 1241–1259.
295. Jang, J.; Ha, H. Fabrication of Hollow Polystyrene Nanospheres in Microemulsion Polymerization Using Triblock Copolymers. *Langmuir* **2002**, *18*, 5613–5618.
296. Konishi, Y.; Okubo, M.; Minami, H. Phase Separation in the Formation of Hollow Particles by Suspension Polymerization for Divinylbenzene/Toluene Droplets Dissolving Polystyrene. *Colloid. Polym. Sci.* **2003**, *281*, 123–129.
297. Li, G. L.; Xu, L. Q.; Neoh, K. G.; Kang, E. T. Hairy Hybrid Microrattles of Metal Nanocore with Functional Polymer Shell and Brushes. *Macromolecules* **2011**, *44*, 2365–2370.
298. Choi, I.; Malak, S. T.; Xu, W.; Heller, W. T.; Tsitsilianis, C.; Tsukruk, V. V. Multicompartmental Microcapsules from Star Copolymer Micelles. *Macromolecules* **2013**, *46*, 1425–1436.
299. Addison, T.; Cayre, O. J.; Biggs, S.; Armes, S. P.; York, D. Polymeric Microcapsules Assembled from a Cationic/Zwitterionic Pair of Responsive Block Copolymer Micelles. *Langmuir* **2010**, *26*, 6281–6286.
300. Liu, P. Design of Controllable-Structured Polymeric Nanocapsules via SI-ATRP from Colloidal Templates. *Curr. Org. Chem.* **2013**, *17*, 39–48.
301. Du, P.; Mu, B.; Wang, Y.; Shi, H.; Xue, D.; Liu, P. Facile Approach for Temperature-Responsive Polymeric Nanocapsules with Movable Magnetic Cores. *Mater. Lett.* **2011**, *65*, 1579–1581.

302. Mu, B.; Liu, P. Novel Temperature-Sensitive Crosslinked Polymeric Nanocapsules. *Mater. Lett.* **2010,** *64,* 1978–1980.

303. Du, P.; Liu, P. Crosslinked Polymeric Nanocapsules via Surface-Initiated Atom Transfer Radical Polymerization from SiO_2 Nano-Templates. *J. Macromol. Sci., Pure Appl. Chem.* **2010,** *47,* 1080–1083.

304. Mu, B.; Liu, P. Temperature and pH Dual Responsive Crosslinked Polymeric Nanocapsules via Surface-Initiated Atom Transfer Radical Polymerization. *React. Funct. Polym.* **2012,** *72,* 983–989.

305. Ma, Z.; Jia, X.; Hu, J.; Zhang, G.; Zhou, F.; Liu, Z.; Wang, H. Dual-Responsive Capsules with Tunable Low Critical Solution Temperatures and Their Loading and Release Behaviour. *Langmuir* **2013,** *29,* 5631–5637.

306. Wu, T.; Ge, Z.; Liu, S. Fabrication of Thermoresponsive Cross-Linked Poly(N-isopropylacrylamide) Nanocapsules and Silver Nanoparticle-Embedded Hybrid Capsules with Controlled Shell Thickness. *Chem. Mater.* **2011,** *23,* 2370–2380.

307. He, X.; Yang, W.; Yuan, L.; Pei, X.; Gao, J. Fabrication of Hollow Polyelectrolyte Nanospheres via Surface-Initiated Atom Transfer Radical Polymerization. *Mater. Lett.* **2009,** *63,* 1138–1140.

308. Li, W.; Matyjaszewski, K.; Albrecht, K.; Möller, M. Reactive Surfactants for Polymeric Nanocapsules via Interfacially Confined Miniemulsion ATRP. *Macromolecules* **2009,** *42,* 8228–8233.

309. Li, W.; Yoon, J. A.; Matyjaszewski, K. Dual-Reactive Surfactant Used for Synthesis of Functional Nanocapsules in Miniemulsion. *J. Am. Chem. Soc.* **2010,** *132,* 7823–7825.

310. Li, J.; Hitchcock, A. P.; Stöver, H. D. H. Pickering Emulsion Templated Interfacial Atom Transfer Radical Polymerization for Microencapsulation. *Langmuir* **2010,** *26,* 17926–17935.

311. Chen, Y.; Guan, Z. Self-Assembly of Core–Shell Nanoparticles for Self-Healing Materials. *Polym. Chem.* **2013,** *4,* 4885–4889.

312. Wang, H. P.; Yuan, Y. C.; Rong, M. Z.; Zhang, M. Q. Self-Healing of Thermoplastics via Living Polymerization. *Macromolecules* **2010,** *43,* 595–598.

313. Zhu, D. Y.; Rong, M. Z.; Zhang, M. Q. Preparation and Characterization of Multilayered Microcapsule-Like Microreactor for Self-Healing Polymers. *Polymer* **2013,** *54,* 4227–4236.

314. Yao, L.; Rong, M. Z.; Zhang, M. Q.; Yuan, Y. C. Self-Healing of Thermoplastics via Reversible Addition–Fragmentation Chain Transfer Polymerization. *J. Mater. Chem.* **2011,** *21,* 9060–9065.

315. Yoon, J. A.; Kamada, J.; Koynov, K.; Mohin, J.; Nicolaÿ, R.; Zhang, Y.; Balazs, A. C.; Kowalewski, T.; Matyjaszewski, K. Self-Healing Polymer Films Based on Thiol-Disulfide Exchange Reactions and Self-Healing Kinetics Measured Using Atomic Force Microscopy. *Macromolecules* **2012,** *45,* 142–149.

316. Kavitha, A. A.; Singha, N. K. Smart "All Acrylate" ABA Triblock Copolymer Bearing Reactive Functionality via Atom Transfer Radical Polymerization (ATRP): Demonstration of a "Click Reaction" in Thermoreversible Property. *Macromolecules* **2010,** *43,* 3193–3205.

317. Telitel, S.; Amamoto, Y.; Poly, J.; Morlet-Savary, F.; Soppera, O.; Lalevée, J.; Matyjaszewski, K. Introduction of Self-Healing Properties into Covalent Polymer Networks via the Photodissociation of Alkoxyamine Junctions. *Polym. Chem.* **2014,** *5,* 921–930.

318. Liu, B.; Gaylord, B. S.; Wang, S.; Bazan, G. C. Effect of Chromophore-Charge Distance on the Energy Transfer Properties of Water-Soluble Conjugated Ooligomers. *J. Am. Chem. Soc.* **2003,** *125,* 6705–6714.

319. Thomas, S. W., III.; Joly, G. D.; Swager, T. M. Chemical Sensors Based on Amplifying Fluorescent Conjugated Polymers. *Chem. Rev.* **2007,** *107,* 1339–1386.

320. Doré, K.; Dubus, S.; Ho, H.-A.; Lévesque, I.; Brunette, M.; Corbeil, G.; Boissinot, M.; Boivin, G.; Bergeron, M. G.; Boudreau, D.; Leclerc, M. Fluorescent Polymeric Transducer for the Rapid, Simple, and Specific Detection of Nucleic Acids at the Zeptomole Level. *J. Am. Chem. Soc.* **2004,** *126,* 4240–4244.

321. Kim, K.; Lee, M.; Park, H.; Kim, J.-H.; Kim, S.; Chung, H.; Choi, K.; Kim, I.-S.; Seong, B. L.; Kwon, I. C. Cell-Permeable and Biocompatible Polymeric Nanoparticles for Apoptosis Imaging. *J. Am. Chem. Soc.* **2006,** *128,* 3490–3491.

322. Chen, M.; Ghiggino, K. P.; Rizzardo, E.; Thang, S. H.; Wilson, G. J. Controlled Synthesis of Luminescent Polymers Using a Bis-dithiobenzoate RAFT Agent. *Chem. Commun.* **2008,** 1112–1114.

323. Haridharan, N.; Bhandary, R.; Ponnusamy, K.; Dhamodharan, R. Synthesis of Fluorescent, Dansyl End-Functionalized PMMA and Poly(methyl methacrylate-b-phenanthren-1-yl-methacrylate) Diblock Copolymers, at Ambient Temperature. *J. Polym. Sci., Part A: Polym. Chem.* **2012,** *50,* 1491–1502.

324. Wang, J.; Leung, L. M. Self-Assembly and Aggregation of ATRP Prepared Amphiphilic BAB Tri-Block Copolymers Contained Nonionic Ethylene Glycol and Fluorescent 9,10-Di(1-naphthalenyl)-2-vinyl-anthracene/1-vinyl-pyrene Segments. *Eur. Polym. J.* **2013,** *49,* 3722–3733.

325. Yaşayan, G.; Magnusson, J. P.; Sicilia, G.; Spain, S. G.; Allen, S.; Davies, M. C.; Alexander, C. Multi-Modal Switching in Responsive DNA Block Co-Polymer Conjugates. *Phys. Chem. Chem. Phys.* **2013,** *15,* 16263–16274.

326. Cho, H. Y.; Gao, H.; Srinivasan, A.; Hong, J.; Bencherif, S. A.; Siegwart, D. J.; Paik, H.-J.; Hollinger, J. O.; Matyjaszewski, K. Rapid Cellular Internalization of Multifunctional Star Polymers Prepared by Atom Transfer Radical Polymerization. *Biomacromolecules* **2010,** *11,* 2199–2203.

327. Lu, J.-M.; Xu, Q.-F.; Yuan, X.; Xia, X.-W.; Wang, L.-H. Synthesis of AB-Type Block Copolymers Containing Benzoxazole and Anthracene Groups by ATRP and Fluorescent Property. *J. Polym. Sci., Part A: Polym. Chem.* **2007,** *45,* 3894–3901.

328. Chen, W.-H.; Liawa, D.-J.; Wang, K.-L.; Lee, K.-R.; Lai, J.-Y. New Amphiphilic Fluorescent CBABC-Type Pentablock Copolymers Containing Pyrene Group by Two-Step Atom Transfer Radical Polymerization (ATRP) and Its Self-Assembled Aggregation. *Polymer* **2009,** *50,* 5211–5219.

329. Summers, G. J.; Maseko, R. B.; Beebeejaun, B. M. P.; Summers, C. A. Synthesis of Aromatic Oxazolyl- and Carboxyl-Functionalized Polymers: Atom Transfer Radical Polymerization of Styrene Initiated by 2-[(4-Bromomethyl)phenyl]-4,5-dihydro-4,4-dimethyloxazole. *J. Polym. Sci., Part A: Polym. Chem.* **2011,** *49,* 2601–2614.

330. Spiniello, M.; Blencowe, A.; Qiao, G. G. Synthesis and Characterization of Fluorescently Labeled Core Cross-Linked Star Polymers. *J. Polym. Sci., Part A: Polym. Chem.* **2008,** *46,* 2422–2432.

331. Zhao, K.; Cheng, Z.; Zhang, Z.; Zhu, J.; Zhu, X. Synthesis of Fluorescent Poly(methyl methacrylate) via AGET ATRP. *Polym. Bull.* **2009,** *63,* 355–364.

332. Gu, P.-Y.; Lu, C.-J.; Ye, F.-L.; Ge, J.-F.; Xu, Q.-F.; Hu, Z.-J.; Li, N.-J.; Lu, J.-M. Initiator-Lightened Polymers: Preparation of End-Functionalized Polymers by ATRP and Their

Intramolecular Charge Transfer and Aggregation-Induced Emission. *Chem. Commun.* **2012**, *48*, 10234–10236.

333. Du, F.; Tian, J.; Wang, H.; Liu, B.; Jin, B.; Bai, R. Synthesis and Luminescence of POSS-Containing Perylene Bisimide-Bridged Amphiphilic Polymers. *Macromolecules* **2012**, *45*, 3086–3093.

334. Yan, J.-J. Wang, Z.-K.; Lin, X.-S.; Hong, C.-Y.; Liang, H.-J.; Pan, C.-Y.; You Y.-Z. Polymerizing Nonfluorescent Monomers Without Incorporating Any Fluorescent Agent Produces Strong Fluorescent Polymers. *Adv. Mater.* **2012**, *24*, 5617–5624.

335. Jin, Z.; Xu, Q.; Li, N.; Lu, J.; Xia, X.; Yan, F.; Wang, L. Facile Di-Color Emission Tuning of Poly[1-(4-vinylstyryl)naphthalene] with Naphthalimide End Group via ATRP. *Eur. Polym. J.* **2008**, *44*, 1752–1757.

336. Limer, A. J.; Rullay, A. K.; Miguel, V. S.; Peinado, C.; Keely, S.; Fitzpatrick, E.; Carrington, S. D.; Brayden, D.; Haddleton, D. M. Fluorescently Tagged Star Polymers by Living Radical Polymerisation for Mucoadhesion and Bioadhesion. *React. Funct. Polym.* **2006**, *66*, 51–64.

337. Yassar, A.; Miozzo, L.; Gironda, R.; Horowitz, G. Rod–Coil and All-Conjugated Block Copolymers for Photovoltaic Applications. *Prog. Polym. Sci.* **2013**, *38*, 791–844.

338. Brochon, C.; Sary, N.; Mezzenga, R.; Ngov, C.; Richard, F.; May, M.; Hadziioanno, G. Synthesis of Poly(paraphenylene vinylene)-Polystyrene-Based Rod–Coil Block Copolymer by Atom Transfer Radical Polymerization: Toward a Self-Organized Lamellar Semiconducting Material. *J. Appl. Polym. Sci.* **2008**, *110*, 3664–3670.

339. Hiorns, R. C.; Cloutet, E.; Ibarboure, E.; Khoukh, A.; Bejbouji, H.; Vignau, L.; Cramai, H. Synthesis of Donor–Acceptor Multiblock Copolymers Incorporating Fullerene Backbone Repeat Units. *Macromolecules* **2010**, *43*, 6033–6044.

340. Economopoulos, S. P.; Chochos, C. L.; Gregoriou, V. G.; Kallitsis, J. K.; Barrau, S.; Hadziioannou, G. Novel Brush-Type Copolymers Bearing Thiophene Backbone and Side Chain Quinoline Blocks. Synthesis and Their Use as a Compatibilizer in Thiophene-Quinoline Polymer Blends. *Macromolecules* **2007**, *40*, 921–927.

341. Lu, S.; Liu, T.; Ke, L.; Ma, D.-G.; Chua, S.-J.; Huang, W. Polyfluorene-Based Light-Emitting Rod–Coil Block Copolymers. *Macromolecules* **2005**, *38*, 8494–8502.

342. Lu, S.; Fan, Q.-L.; Chua, S.-J. Synthesis of Conjugated–Ionic Block Copolymers by Controlled Radical Polymerization. *Macromolecules* **2003**, *36*, 304–310.

243. Lu, S.; Fan, Q.-L.; Liu, S.-Y.; Chua, S.-J. Synthesis of Conjugated–Acidic Block Copolymers by Atom Transfer Radical Polymerization. *Macromolecules* **2002**, *35*, 9875–9881.

344. Wang, R.; Wang, W.-Z.; Lu, S.; Liu, T. Controlled Radical Synthesis of Fluorene-Based Blue-Light-Emitting Copolymer Nanospheres with Core–Shell Structure via Self-Assembly. *Macromolecules* **2009**, *42*, 4993–5000.

345. Kim, S.; Kim, D.; Kim, H.-J.; Lee, Y.-J.; Hwang, S. S.; Baek, K.-Y. Hybrid Nanocomposite of CdSe Quantum Dots and a P3HT-*b*-PDMAEMA Block Copolymer for Photovoltaic Applications. *Mater. Sci. Forum* **2012**, *700*, 120–124.

346. Gu, Z.; Tan, Y.; Tsuchiya, K.; Shimomura, T.; Ogino, K. Synthesis and Characterization of Poly(3-hexylthiophene)-*b*-polystyrene for Photovoltaic Application. *Polymers* **2011**, *3*, 558–570.

347. Shen, J.; Masaoka, H.; Tsuchiya, K.; Ogino, K. Synthesis and Properties of a Novel Brush-Type Copolymers Bearing Thiophene Backbone and 3-(N-carbazolyl)propyl Acrylate Side Chains for Light-Emitting Applications. *Polym. J.* **2008**, *40*, 421–427.

348. Balamurugan, S. S.; Bantchev, G. B.; Yang, Y.; McCarley, R. L. Highly Water Soluble Thermally Responsive Poly(thiophene)-Based Brushes. *Angew. Chem., Int. Ed.* **2005**, *44*, 4872–4876.

349. Park, J. T.; Roh, D. K.; Chi, W. S.; Patel, R.; Kim, J. H. Fabrication of Double Layer Photoelectrodes Using Hierarchical TiO_2 Nanospheres for Dye-Sensitized Solar Cells. *J. Ind. Eng. Chem.* **2012**, *18*, 449–455.

350. Roh, D. K.; Park, J. T.; Ahn, S. H.; Ahn, H.; Ryu, D. Y.; Kim, J. H. Amphiphilic Poly(vinyl chloride)-*g*-poly(oxyethylene methacrylate) Graft Polymer Electrolytes: Interactions, Nanostructures and Applications to Dye-Sensitized Solar Cells. *Electrochim. Acta* **2010**, *55*, 4976–4981.

351. Liu, C.-M.; Qiu, J.-J.; Bao, R.; Xu, Y.; Cheng, X.-J.; Hu, F. End-Capped Polystyrene with 8-Hydroxyquinoline Group by ATRP Method. *Polymer* **2006**, *47*, 2962–2969.

352. Zhao, N.; Chen, C.; Zhou, J. Surface Plasmon Resonance Detection of Ametryn Using a Molecularly Imprinted Sensing Film Prepared by Surface-Initiated Atom Transfer Radical Polymerization. *Sens. Actuators, B* **2012**, *166–167*, 473–479.

353. Xu, L.; Pan, J.; Dai, J.; Cao, Z.; Hang, H.; Li, X.; Yan, Y. Magnetic ZnO Surface-Imprinted Polymers Prepared by ARGET ATRP and the Application for Antibiotics Selective Recognition. *RSC Advances* **2012**, *2*, 5571–5579.

354. Banerjee, S.; Paira, T. K.; Kotal, A.; Mandal, T. K. Surface-Confined Atom Transfer Radical Polymerization From Sacrificial Mesoporous Silica Nanospheres for Preparing Mesoporous Polymer/Carbon Nanospheres with Faithful Shape Replication: Functional Mesoporous Materials. *Adv. Funct. Mater.* **2012**, *22*, 4751–4762.

355. Aied, A.; Zheng, Y.; Pandit, A.; Wang, W. DNA Immobilization and Detection on Cellulose Paper Using a Surface Grown Cationic Polymer via ATRP. *ACS Appl. Mater. Interfaces* **2012**, *4*, 826–831.

356. Kuriu, Y.; Ishikawa, M.; Kawamura, A.; Uragami, T.; Miyata, T. SPR Signals of Three-Dimensional Antibody-Immobilized Gel Layers Formed on Sensor Chips by Atom Transfer Radical Polymerization. *Chem. Lett.* **2012**, *41*, 1660–1662.

357. Liu, Y.; Dong, Y.; Jauw, J.; Linman, M. J.; Cheng, Q. Highly Sensitive Detection of Protein Toxins by Surface Plasmon Resonance with Biotinylation-Based Inline Atom Transfer Radical Polymerization Amplification. *Anal. Chem.* **2010**, *82*, 3679–3685.

358. Liu, Y.; Cheng, Q. Detection of Membrane-Binding Proteins by Surface Plasmon Resonance with an All-Aqueous Amplification Scheme. *Anal. Chem.* **2012**, *84*, 3179–3186.

359. Wang, X.; Zhou, M.; Zhu, Y.; Miao, J.; Mao, C.; Shen, J. Preparation of a Novel Immunosensor for Tumor Biomarker Detection Based on ATRP Technique. *J. Mater. Chem. B* **2013**, *1*, 2132–2138.

360. Peng, Q.; Xie, M.-G.; Neoh, K.-G.; Kang, E.-T. A New Nitrite-Selective Fluorescent Sensor Fabricated From Surface-Initiated Atom-Transfer Radical Polymerization. *Chem. Lett.* **2005**, *34*, 1628–1629.

361. Das, S.; Chatterjee, D. P.; Samanta, S.; Nandi, A. K. Thermo and pH Responsive Water Soluble Polythiophene Graft Copolymer Showing Logic Operation and Nitroaromatic Sensing. *RSC Adv.* **2013**, *3*, 17540–17550.

362. Li, J. R.; Xu, J. R.; Zhang, M. Q.; Rong, M. Z.; Zheng, Q. The Role of Crystalline Phase in Triblock Copolymer PS–PEG–PS Based Gas Sensing Materials. *Polymer* **2005**, *46*, 11051–11059.

363. Madhuri, R.; Tiwari, M. P.; Kumar, D.; Mukharji, A.; Prasad, B. B. Biomimetic Piezoelectric Quartz Sensor for Folic Acid Based on a Molecular Imprinting Technology. *Adv. Mat. Lett.* **2011**, *2*, 264–267.

CHAPTER 3

SYNTHESIS OF CHAIN END FUNCTIONAL POLYMERS BY LIVING CATIONIC POLYMERIZATION METHOD

SANJIB BANERJEE, BADRI NATH JHA, RAJEEV KUMAR, BINOY MAITI, UJJAL HALDAR, and PRIYADARSI DE

CONTENTS

3.1 INTRODUCTION

Cationic polymerization is applicable to vinyl monomers having electron-rich double bond and propagation reactions that take place by electrophilic addition of the monomers to the active cationic sites. Since cationic species are very reactive, conventional cationic polymerization reactions involve chain termination, chain transfer, and other chain-breaking reactions which destroy propagating cations. Therefore, it is difficult to prepare functional polymers, also known as chain-end functional polymers (polymers with functional groups selectively placed at the chain end(s)) by traditional cationic polymerization.

Development of the controlled/living cationic polymerization[1,2] has led to the growth of functional polymers with well-defined macromolecular architectures. Undesired chain termination and chain transfer reactions are eliminated or suppressed during the living polymerizations and as a result precise functionality were obtained. In addition to well-defined chain ends, living polymerizations provide controlled molecular weights and narrow molecular weight distributions.Generally, initiation by functional initiators results functional polymers by living cationic polymerization, where functional initiator with a protected or unprotected functional group is used during the polymerization. When the functional group in the initiator is reactive under the polymerization conditions, protection of the functional group is necessary. Alternatively, end quenching of living cationic polymerizations with appropriate nucleophiles results functional polymers. This method (reported as "termination by functional terminators") is less attractive because undesired side reactions happen during the *in-situ* end functionalization of the propagating polymer ends with nucleophiles.This chapter involves the most recent developments on the synthesis of functional polymers by using living cationic polymerization techniques for the following monomers: isobutylene (IB), vinyl ethers (VE), styrene monomers, indene, *N*-vinylcarbazole, and β-pinene. Cationic polymerization derived chain-end functional polymers are of great importance due to their applications in many areas such as surface modification, motor oil additives, and so forth. Note that functional polymers possess the combination of the physical properties of the polymer and the chemical reactivities of the attached functional group(s) in the polymer chain end(s).

3.2   LIVING CATIONIC POLYMERIZATION

Controlled initiation and propagating steps are essential to achieve living cationic polymerization.[3] Apart from producing well-defined polymers with controlled molecular weight and narrow molecular weight distribution, living cationic polymerization also provide an efficient way for the preparation of end-functionalized polymers. With a few exceptions,[4,5] the initiator/Lewis acid (coinitiator) binary system is used to initiate living cationic polymerization of various vinyl monomers. Generally, after the reaction with Lewis acid the initiator produces carbocation, which will have similar structure to that from a monomer.[6] The deactivation equilibrium must be dynamic and the exchange between active (ionic) and dormant (covalent) species (Scheme 3-1) must be faster than propagation so that a very small amount of active species are present. Experimentally, living nature of the polymerization can be verified from the linear $\ln([M]_0/[M])$ versus time and number average molecular weight (M_n) versus monomer conversion plots, where $[M]_0$ and $[M]$ are the monomer concentrations at time $t = 0$ and $t = t$, respectively. The linearity of $\ln([M]_0/[M])$ versus time plot means a constant concentration of the propagating center (absence of termination) and the linearity of M_n versus conversion plot proves the absence of chain transfer during the polymerization. During the living cationic polymerization of vinyl monomers, several reports used nucleophiles, proton traps, and salts to obtain livingness. For example, the living polymerization of IB and styrene co-initiated with $TiCl_4$ or BCl_3 in the absence of nucleophilic additives but in the presence of 2,6-di-*tert*-butylpyridine (DTBP) as proton trap were reported.[7,8] The addition of nucleophilic additives had no effect on the rates of polymerization, M_n and molecular weight distributions. Now, we will discuss recent developments on the synthesis of functional polymers by using living cationic polymerization techniques for various vinyl monomers.

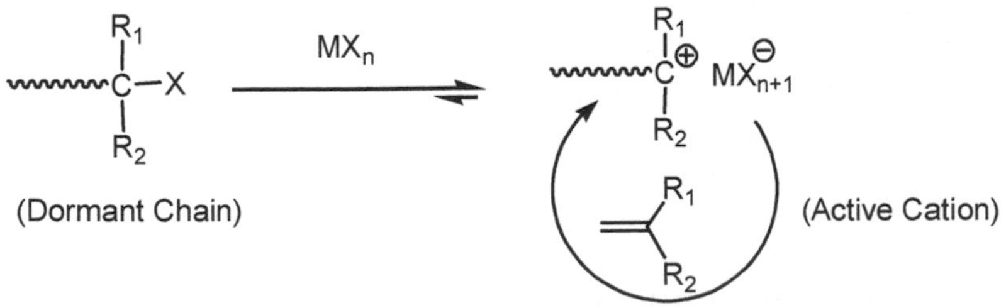

SCHEME 3-1   Dormant chain end and active propagating cation equilibrium in the living polymerization of vinyl monomers.

3.2.1   ISOBUTYLENE (IB)

The most studied monomer for cationic polymerizations is IB, which can only polymerize by a cationic mechanism. Faust and Kennedy discovered living cationic polymerization of IB using BCl_3,[9,10] and subsequently $TiCl_4$ and organoaluminum halide coinitiators have been developed. Organic esters, halides, ethers, and alcohols have been used as initiator for the living polymerization of IB. End-functionalized polyisobutylene (PIB)polymershave been prepared by using functional initiators and functional terminators. Due to their potential applications in diverse areas including surface modification, adhesion, drug delivery, polymeric catalysts, compatibilization of polymer blends, motor oil additives, and so forth the functional PIB polymers have been receiving much interest in recent years.[11]

Initially we reviewedthe production of end functional PIB via living cationic polymerization of IB, followed by end-capping of the living PIB polymer chains with appropriate nucleophiles. Storey et al. reported synthesis of N-methylpyrrole-terminated PIB by capping living PIB chains with N-methylpyrrole. Living PIB chains were synthesized by cationic polymerization of IB using 2-chloro-2,4,4-trimethylpentane/$TiCl_4$/2,6-dimethylpyridine system.[12] Simison et al. used the strategy involving quantitative β-proton abstraction with hindered bases as a valuable method to produce exo-olefin-terminated PIB in one pot.[13] De and Faust developed a novel and efficient method for the synthesis of chloroallyl chain-end-functionalized PIB (PIB-Allyl-Cl) by capping living PIB chains with 1,3-butadiene in hexanex/methyl chloride (MeCl) 60/40 (v/v) solvent mixtures at −80 °C (Scheme 3-2).[14] Monoaddition of 1,3-butadiene followed by instantaneous halide transfer from the counter anion and selective formation of the 1,4-adduct were observed at [1,3-butadiene] ≤ 0.05 mol/L. Faust et al. reported the synthesis of poly(isobutylene-b-methyl methacrylate) (PIB-b-PMMA) by a novel coupling approach combining living cationic and anionic polymerization techniques. First, they produced PIB-Allyl-Cl by the method described above. PIB-Allyl-Cl was then quantitatively converted into PIB-Allyl-Br by halogen exchange reaction using an excess of LiBr. Subsequently, PIB-Allyl-Br was coupled with PMMALi in tetrahydrofuran (THF) at −78 °C yielding PIB-b-PMMA with high coupling efficiency (>95%).[15] Cho et al. achieved synthesis of PMMA-b-PIB-b-PMMA and poly(hydroxyethyl methacrylate-b-isobutylene-b-hydroxyethyl methacrylate) (PHEMA-b-PIB-b-PHEMA) triblock copolymers by the initiation of the polymerization of methacrylates from the PIB macroanions in THF at −78 °C. The end-capping of living PIB chains with 1,4-bis(1-phenylethenyl)

benzene (DPE), followed by the methylation of the resulting diphenyl carbenium ion with dimethylzinc (Zn(CH$_3$)$_2$) produces PIB-DPE. This was then
quantitatively metalated with *n*-butyllithium (*n*-BuLi) in THF producing
the PIB macro-anion. The resulting block copolymers were used as drug
delivery matrixes for coatings on coronary stents.[16]

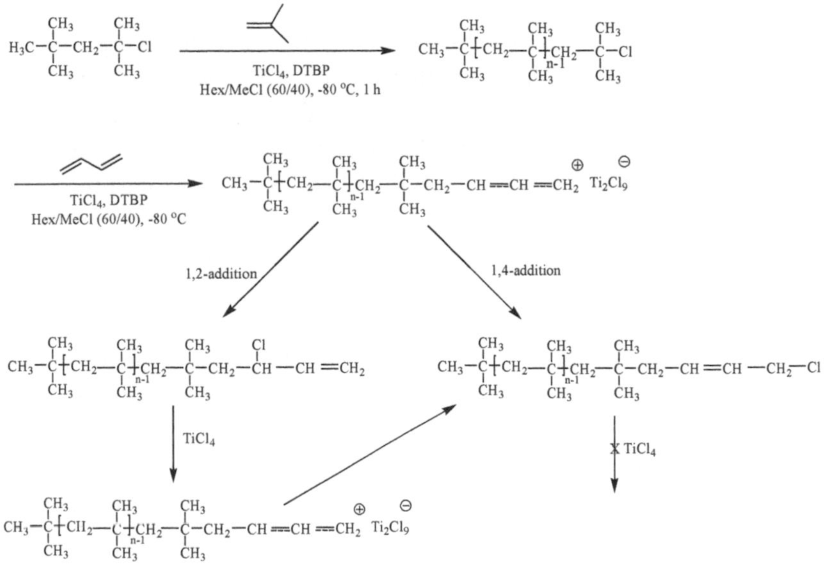

SCHEME 3-2   Capping of living PIB chain with 1,3-butadiene in hexanes/MeCl 60/40 (v/v)
solvent mixture at −80 °C.
(**Source:** Reproduced partially from De, P.; Faust, R. *Macromolecules* **2006**, *39*, 6861–6870.)

Higashihara and Faust synthesized AB diblock copolymers comprised
of PIB and poly(vinylferrocene)s (PVFc) segments by the coupling reaction between PIB-Allyl-Br and PVFcLi. Well-defined PVFc polymers were
prepared by living anionic polymerization of vinylferrocene initiated with
n-BuLi in THF at −28 °C.[17] Feng et al. synthesized P(MMA-*co*-HEMA)-*b*-
PIB-*b*-P(MMA-*co*-HEMA) triblock copolymers with different HEMA/MMA
ratios by the combination of living cationic and anionic polymerizations.
They first prepared DPE-PIB-DPE by the reaction of living difunctional PIB
with DPE, followed by the methylation of the resulting diphenyl carbenium
ion with Zn(CH$_3$)$_2$. Subsequently, PIB macro-anion was produced by quantitative metalation of the DPE ends with *n*-BuLi in THF. This macro-anion
induced polymerization of methacrylates yielding the triblock copolymers.
These new block copolymers exhibited characteristic stress–strain behavior

of TPEs.[18] Bouchekif et al. prepared poly(p-hydroxystyrene-b-isobutylene-b-p-hydroxystyrene) triblock copolymer (PHOS-b-PIB-b-PHOS) by employing 5-tert-butyl-1,3-bis (1-methoxy-1-methylethyl)benzene (DCE) as a difunctional initiator for the living polymerization of IB followed by capping with 1,1-ditolylethylene (DTE) and substitution of $TiCl_4$ with $SnBr_4$ for the polymerization of t-butoxystyrene (t-BuOS). Deprotection of the triblock copolymer in the presence of catalytic amount of HCl yielded the desired PHOS-b-PIB-b-PHOS triblock copolymer. The resulting PHOS-b-PIB-b-PHOS exhibited typical characteristic of a TPEs with tensile strength of 18 MPa and elongation of 300.[19] Faust et al. employed simple nucleophilic substitution reactions on PIB-Allyl-Cl or PIB-Allyl-Br to synthesize a wide range of end-functional PIBs including hydroxy, amino, carboxy, azide, propargyl, methoxy, and thymine end groups. They also found that the rate of substitution was faster with PIB-Allyl-Br compared to PIB-Allyl-Cl. Gel permeation chromatography (GPC) analysis of the precursor PIB-Allyl-X (X = Cl, Br) and the end functionalized PIB indicated that the polymer chain is unaffected by the substitution reactions. The methodology was extended to the synthesis of PIB-b-poly(ethylene oxide) (PEO) block copolymer by the nucleophilic substitution of PIB-Allyl-Cl with $PEO-O^-Na^+$.[20] Higashihara et al. synthesized an array of ω- or α,ω-multifunctional PIBs with 2, 4, and 8 benzyl bromide moieties using ω- or α,ω-functional PIBs with allyl halide moieties as the starting materials. They used these multifunctional PIBs for the synthesis of A_2B, A_4B, and A_8B asymmetric star polymers as well as A_2BA_2, A_4BA_4, and A_8BA_8 pompom polymers, where A and B were PMMA and PIB segments, respectively by coupling living PMMA with multifunctional PIBs in THF at $-40\ °C$. The A_nBA_n type pompom polymers showed good elastomeric properties (279–444% elongations at break and 12.6–20.7 MPa tensile strengths).[21] Faust et al. improved their original method for the synthesis of PIB-DPE by capping living PIB with DPE, followed by hydride transfer reaction with tributylsilane. PIB-DPE was then quantitatively lithiated with 1.5-fold excess n-BuLi and the resulting macroanion was used for the polymerization of alkyl methacrylates producing PMMA-b-PIB-b-PMMA, PHEMA-b-PIB-b-PHEMA, and poly(tert-butyl methacrylate) (PtBMA)-b-PIB-b-PtBMA with high blocking efficiency.[22]

Tripathy et al. reported the synthesis of new telechelic PIB macromonomers with methacrylate, acrylate, vinyl ether, and glycidyl ether end-functionality using simple nucleophilic substitution reactions of PIB-Allyl-Br (Scheme 3-3). They used a phase transfer catalyst (tetrabutylammonium bromide) to significantly increase the rate of substitution reactions.[23] The same group then synthesized PIB-based UV cured networks by photopolymerization of well-defined polyisobutylene methacrylate (PIB-MA), polyisobutylene acrylate (PIB-A), and polyisobutylene vinyl ether (PIB-VE) di- and trifunctional macromonomers. These materials are potentially useful as sealants.[24] Employing similar strategy Kennedy and Ummadesetty also synthesized bromo, hydroxyl, amine, and methacrylate functional PIBs with quantitative functionality.[25] Nagy et al. reported the synthesis of a series of amphiphilic PIB-*b*-poly(vinyl alcohol) (PIB-*b*-PVA) copolymers by living sequential cationic polymerization. They first synthesized PIB-*b*-poly(*tert*-butyl vinyl ether) (PIB-*b*-PtVE) and then removed the *tert*-butyl group by hydrolysis with HBr. They also tested the drug carrying compartment by doping the aqueous copolymer solutions with indomethacin.[26]

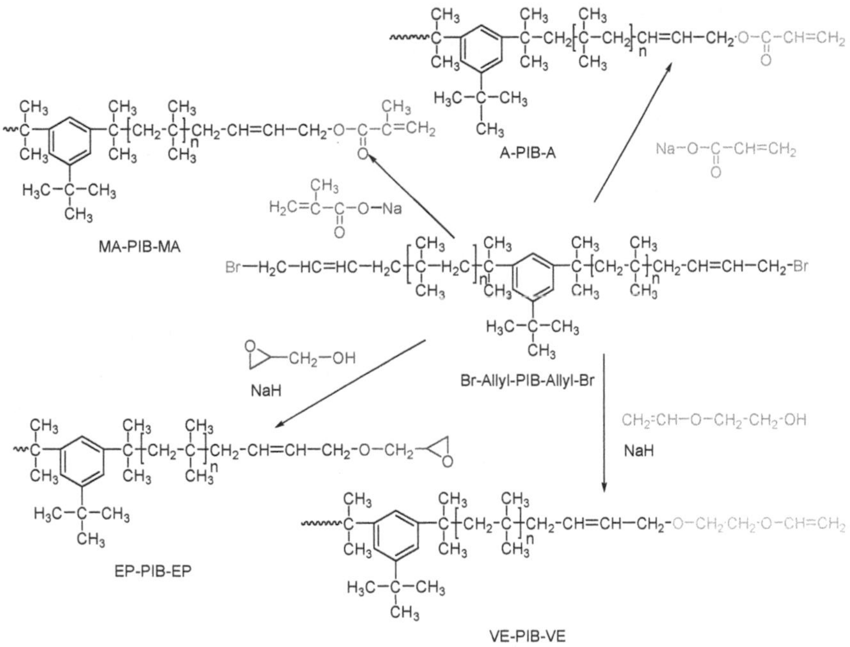

SCHEME 3-3 General methodologies for synthesis of methacrylate(MA), acrylate(A), vinyl ether (VE), and epoxy (EP) end functionalized telechelic PIB macromonomers.

Higashiharaand Faust synthesized ABA triblock copolymers comprised of PIB and poly(γ-benzyl-L-glutamate) (PBLG) segments by the polymerization of γ-benzyl-L-glutamate-*N*-carboxyanhydride (BLG-NCA) initiated with well-defined α,ω-primary amino-functional PIBs. α,ω-primary amino functional PIB was synthesized from α, ω-chloroallyl-functional PIB based on the facile Gabriel amine forming reaction.[27] Ojha et al. reported the synthesis of {-poly(L-lactide) (PLLA)-b-polyisobutylene (PIB)-}$_n$ multiblock copolymers by chain extension of PLLA-b-PIB-b-PLLA triblock copolymers with 4,4'-methylenebis(phenylisocyanate) (MDI). The triblock copolymer was prepared by anionic ring-opening polymerizations of L-lactide initiated with hydroxyallyl telechelic PIB (HO-Allyl-PIB-Allyl-OH) in toluene at 110 °C.[28] They also studied the hydrolytic degradation of a series of PLLA-PIB multiblock copolymers in phosphate buffer solution (pH ~ 7.4) at 37 °C.[29]

Epoxy telechelic PIBs (EP-PIB-EP), potentially useful to obtain flexible epoxies, were synthesized by reacting Br-Allyl-PIB-Allyl-Br with glycidol in the presence of NaH. The EP-PIB-EP macromonomers were mixed with bisphenol A diglycidyl ether (DGEBA) and triethylenetetramine as curing agent and cured at elevated temperature. These flexible networks possess superior mechanical properties.[30] Espinosa et al. reacted azido end-functionalized polyethylenes (PE-N$_3$) with functionality higher than 88% with alkyne end-functionalized or telechelic polyisobutenes using the 1,3-dipolar cycloadddition reaction to produce di- and triblock copolymers (PE-b-PIB and PE-b-PIB-b-PE).[31]

The PIB block was synthesized *via* living cationic polymerization and end-functionalized with chain transfer agent (CTA) to produce PIB-based macro-CTA.[32,33] The PIB-based macro-CTAs were used to prepare diblock copolymers consisting of PIB and either PMMA or polystyrene or poly(N-isopropylacrylamide) block segments by a site transformation approach combining living cationic and reversible addition-fragmentation transfer (RAFT) polymerizations. Recently, Bauri et al. synthesized diblock copolymers of IB with amino acid-based monomers by a combination of living cationic and RAFT polymerizations using PIB-based macro-CTA. PIB macro-CTA was prepared by the condensation reaction of PIB-OH and 4-cyano-4-(dodecylsulfanylthiocarbonyl)sulfanyl pentanoic acid (CDP). RAFT polymerization of *tert*-butyloxycarbonyl (Boc)-protected amino acid-based monomers such as Boc-L-alanine methacryloyloxyethyl ester (Boc-L-Ala-HEMA) and Boc-L-leucine methacryloyloxyethyl ester (Boc-L-Leu-HEMA) in the presence of PIB macro-CTA resulted corresponding AB diblock copolymers. These block copolymers form core–shell type micellar structure in methanol, which is a good solvent for side-chain amino acid block but bad solvent for PIB segment.[34]

Zhang et al. synthesized PIB with arylamino terminal group by alkylation of living PIB synthesized by living cationic polymerization with triphenylamine.[35] They also synthesized PIB with *sec*-arylamino terminal group by carrying out the alkylation with diphenylamine (DPA).[36] Hackethal et al. developed a technique to introduce polar moieties in a direct copolymerization approach into PIB polymers. They directly copolymerized polar styrene monomers with IB using living cationic polymerization with TiCl$_4$/TMPCl initiating system.[37] Ma et al. synthesized poly(styrene-*co*-isopropenyl acetate)-*g*-polyisobutylene graft copolymers via combination of radical polymerization with cationic polymerization. First, random copolymers of poly(styrene-*co*-isopropenyl acetate) (SIPA) were synthesized by free radical copolymerization of styrene. Then SIPA was used as the macroinitiator for the grafting cationic polymerization of IB to produce the graft copolymers.[38] Ren et al. reported the synthesis of novel functional ABA triblock copolymer thermoplastic elastomers (TPEs) bearing pendant hydroxyl groups with PIB as rubbery segments by controlled cationic polymerization. The precursor poly{(styrene-*co*-4-[2-(*tert*-butyldimethylsiloxy)ethyl] styrene)-*b*-isobutylene-*b*-(styrene-*co*-4-[2-(*tert*-butyldimethyl-siloxy)ethyl] styrene)}(P(St-*co*-TBDMES)-PIB-P(St-*co*-TBDMES)) triblock copolymer was first synthesized by living sequential cationic copolymerization of IB with styrene and 4-[2-(*tert*-butyldimethylsiloxy)ethyl]styrene (TBDMES) using 1,4-di(2-chloro-2-propyl)benzene (DCC)/ TiCl$_4$)/2,6-di-*tert*-butylpyridine as the initiating system. Then, it was hydrolyzed in the presence of *tetra*-butylammonium fluoride to yield poly{[styrene-*co*-4-(2-hydroxyethyl) styrene]-*b*-isobutylene-*b*-[styrene-*co*-4-(2-hydroxyethyl)styrene]} (P(St-*co*-HOES)-PIB-P(St-*co*-HOES)) with pendant hydroxyl groups. They also investigated P(St-*co*-HOES)-PIB-P(St-co-HOES) as the paclitaxel carrier.[39]

Functional initiators with a protected/unprotected functional group were used to polymerize IB to produce head functional PIBs. When the functional group of the functional initiator is unreactive under the polymerization conditions, protection is not needed. There are few reports of production of functional PIB using functional initiators. Puskas et al. reported a polyhedral oligomeric silsesquioxanes (POSS)-functionalized PIB by carbocationic polymerization using an epoxy-POSS/TiCl$_4$ initiating system in hexane/methyl chloride (60:40 v/v) solvent mixture at −80 °C. The ^{1}H NMR study confirmed the incorporation of one epoxy-POSS per polymer chain.[40] Puskas et al. also synthesized functional PIB using 1,2-epoxycyclohexane and bis[3,4-(epoxycyclohexyl)ethyl]-tetramethyl-disiloxane, in conjunction with TiCl$_4$.[41] Breland and Storey synthesized a PIB-based miktoarm star polymers via a combination of cationic and atom transfer radical polymerization (ATRP). First, PIB-*b*-PS copolymers and PIB homopolymers were synthesized via cationic polymerization from 3,3,5-trimethyl-5-chlorohexyl acetate (initiator), which contains a protected hydroxyl group.Subsequently, using a strong base, the acetate head group of the resulting block copolymer was cleaved to yield a hydroxyl group using a strong base and then esterified with the branching agent 2,2-bis((2-bromo-2-methyl)propionatomethyl)propionyl chloride (BPPC) to create initiating sites for ATRP of*tert*-butyl acrylate.[42] They also developed a new dual initiator to synthesize poly(isobutylene-*b*-methyl acrylate) (PIB-b-PMA) diblock copolymers via combined quasi-living cationic polymerization and ATRP.[43] Synthesis of a PIB-based miktoarm star polymers from a designed dicationic monoradical dual initiator 3-[3,5-bis(1-chloro-1-methylethyl)phenyl]-3-methylbutyl 2-bromo-2-methylpropionate (DCCBMP) was reported.[44]

There is an increasing demand of PIB based ashless dispersants for motor oil and fuel additives in the industrial settings.[45] PIB based dispersants are low molecular weight ($M_n \sim$ 500–5000 g/mol) oil soluble PIB or polybutenes (copolymers of IB with C4 olefins) with polar oligoamine end-groups.[46] Simison et al. used end-quenching of living PIB chains with hindered bases such as 2,5-dimethylpyrrole, 1,2,2,6,6-pentamethylpiperidine, or 2-*tert*-butylpyridine at −60 to −40 °C for quantitative formation of exo-olefin-terminated PIB.[13] Ummadisettyand Storey achieved one-pot/two-step quantitative synthesis of mono- and telechelicexo-olefin-terminated PIB by quenching living PIB chains with diisopropyl ether. They explored various quenching methods for production of exo-olefin-terminated PIB, such as ether, sulfide, hindered base, and methallyltrimethylsilane. Their study revealed that ether and sulfide quenching were superior to other methods with regard to maximizing exo-olefin end groups.[47] Liu et al. developed a cost-effective methodology for the synthesis of exo-olefin-terminated PIB via cationic polymerization coinitiated by $AlCl_3$ at temperatures ranging from −20 to +20 °C.[48] The same group also reported production of exo-olefin-terminated PIB via cationic polymerization of IB with H_2O/$FeCl_3$/dialkyl ether initiating system in dichloromethane at −20 to +20°C.[49] Storey et al. also developed a methodology for the synthesis of exo-olefin terminated PIB in one pot by quenching living PIB with a dialkyl (or) diaryl sulfide at −60 to −40 °C, followed by addition of a base (e.g., triethylamine), followed by warming of the reaction mixture to −20 to −10 °C and final termination with methanol.[50] Kostjuk et al. reported the synthesis exo-olefin-terminated PIB via $AlCl_3 \cdot OBu_2$-coinitiated cationic polymerization of IB in nonpolar solvents (toluene, *n*-hexane) at elevated temperatures (−20 to 30 °C).[51] Faustet al. reported the synthesis of exo-olefin-terminated PIB via $AlCl_3 \cdot OBu_2$-coinitiated cationic polymerization of IB by $AlCl_3$/ether complexes in dichloromethane/hexanes (80/20 v/v) at −40 °C.[52] They also reported production of exo-olefin-terminated PIB by cationic polymerization of IB by $GaCl_3$ or $FeCl_3$/ether complexes in hexanes in the -20 to $10°C$ temperature range.[53] For the same system, steric and electronic effects of ethers on the polymerization rates and exo-olefin contents were studied in detail.[54] Guo et al. reported synthesis of PIB with exo-olefin terminals via controlled cationic polymerization with $H_2O/FeCl_3$/iPrOH initiating system in nonpolar hydrocarbon media such as n-hexane or mixed C4 fractions at -40 to $20°C$.[55] Faust et al. reported synthesis of exo-olefin-terminated PIB via cationic polymerization using $EtAlCl_2$/bis(2-chloroethyl) ether soluble complex in hexanes.[56] Hydride transfer reaction between living PIB^+ cation and living PIB capped with 1,1-ditolylethylene ($PIB\text{-}DTE^+$) withtributylsilane in hexanes/methyl chloride 60/40 (v/v) solvent mixtures at different temperatures were studied to synthesize halogen-free PIB.[57]

3.2.2   STYRENIC MONOMERS

Since there is no strongly electron donating substituent, styrene is among the poorly reactive monomers in cationic polymerizations.[58] Living cationic polymerization of styrenic monomers is rather difficult to control compared to VEand IB. This is due to the fact that styrenic monomers, when polymerized with conventional initiators form an unstable growing carbocations that is prone to undergo chain transfer (such as β-proton elimination and intra- and/or intermolecular Friedel-Crafts alkylation), chain termination, and other undesirable side reactions.[59] As a result, the synthesis of functional polymers via living cationic polymerization of styrene based monomers(Scheme 3-4) has been rather limited as compared to VE and IB.

Styrene (St): Living cationic polymerization of St was first reported in 1990 using $CH_3SO_3H/SnCl_4$[60] and HCl adduct of $St/SnCl_4$[61] in the presence of an added quaternary ammonium halide salt (n-Bu_4NCl) in dichloromethane at $-15°C$. Lu and Larock synthesized a novel biobased nanocomposites of conjugated soybean oil (CSOY) or conjugated LoSatSoy oil (CLS) with St and divinylbenzene (DVB) by the cationic polymerization with a reactive organomodified montmorillonite (VMMT) clay as a reinforcing phase.[62] They also prepared a polymer resin by the cationic polymerization of conjugated corn oil, St and DVB, using $BF_3\text{-}OEt_2$ modified with Norway fish oil as the initiator.[63] Camerlynck et al. prepared branched polystyrenes via cationic copolymerization of St with DVB using $SnCl_4$ based system in dichloromethane solution at $0°C$.[64] Gao et al. reported synthesis of lamellar and hexagonal–coil–cylinder self-assembled structures of ABA type triblock copolymers containing mesogen-jacketed liquid

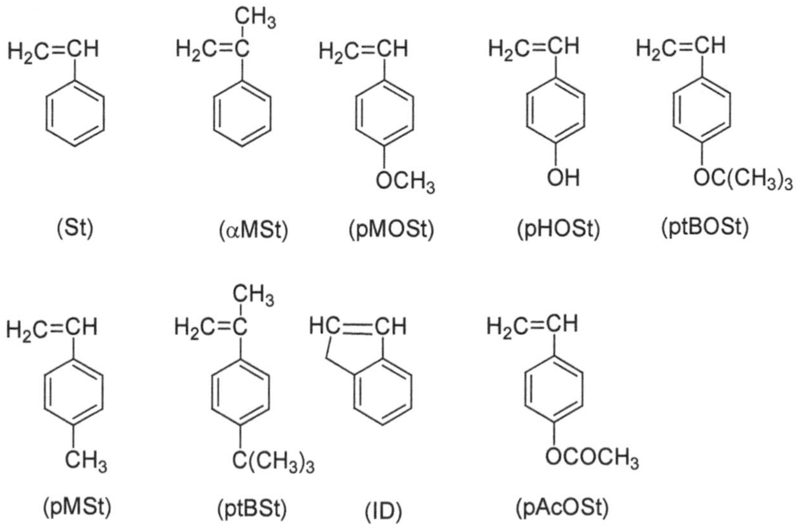

SCHEME 3-4   Chemical structure of different styrenic monomers.

crystalline polymer (MJLCP) as the rod block, and PIB as the coil middle block. For this, they first synthesizedPIB by living cationic polymerization of IB initiated by 1,4-bis(2-chloro-2-propyl)benzene (p-DCC), and then a small amount of St was introduced at the end of the PIB chains to form the difunctional PIB macroinitiator with–$CH_2CH(C_6H_5)Cl$ end groups for further ATRP. 2,5-Bis[(4-methoxyphenyl)oxycarbonyl]styrene (MPCS) was then polymerized from the difunctional PIB macroinitiators at 110 °C via ATRP to produce the ABA type triblock copolymer.[65] Landsmann et al. prepared polystyrene-polyoxometallate (PS-POM) surfactant particles using POM surfactant both as an emulsifying agent and as a source of the protons for the cationic polymerization of St.[66] Kesharwani et al. developed a very simple method for the preparation of nanoscale fibers and tubes by the cationic copolymerization of St and DVB at room temperature.[67] Hondred et al. used rare earth triflate initiators in the cationic polymerization of tung oil-based thermosetting polymers comprised of predominantly tung oil with St and DVB. St and DVB were used as a reactive diluent and to improve the initiator solubility and the homogeneity of the resulting thermoset. They also used the synthesized material for self-healing applications.[68] Wu et al. reported synthesis of long-chain branched isotactic-rich polystyrene via cationic polymerization using $AlCl_3$.[69] They also reported synthesis of triblock copolymers of PS-b-PIB-b-PS with different chain length of PS by sequential living cationic polymerizations of IB and St with DCC/FeCl$_3$/iPrOH initiating system at −80 °C.[70]

para-Alkoxy Styrenes [p-Methoxy Styrene (pMOSt), p-tert-Butoxy Styrene (ptBOSt), and p-Hydroxy Styrene (pHOSt)]: para-Alkoxy styrenes are one of the most reactive monomers in cationic polymerization among St series. Higashimura et al. first reported living cationic polymerization of pMOSt.[71] Long-lived propagating species were obtained with iodine initiator at −15 °C in a nonpolar solvent (CC1$_4$) or in a polar solvent (CH$_2$C1$_2$) containing a common ion salt ((n-C$_4$H$_9$)$_4$NI), where the ionic dissociation of the growing macro-cation was suppressed. Subsequently, Shohi et al. produced an end-functionalized poly(pMOSt) or its macromonomer using various HI adducts of VE having a functional group in conjunction with ZnI$_2$.[72] They also synthesized a tri-armed star polymer of pMOSt via cationic polymerization using a trifunctional initiator with ZnI$_2$ in toluene at 0 °C. Subsequent end capping by 2-hydroxyethyl methacrylate resulted in the production of a tri-armed poly(pMOSt) having three methacrylate groups.[73] They also prepared poly(ptBOSt) via cationic polymerization using HI/ZnCl$_2$ in toluene or in CH$_2$Cl$_2$ (with n-Bu$_4$NCl) at temperatures up to +25 °C. Subsequent hydrolysis of the polymer resulted in the production of poly(pHOSt) and various functionalized poly(ptBOSt)s. Kamigaito and co-workers reported copolymerizations of the naturally occurring β-methylstyrenes with pMOS via aqueous-controlled cationic copolymerization controlled cationic using BF$_3$•OEt$_2$. The obtained copolymer can be regarded as a well-defined linear lignin analogue with 4-hydroxy-3-methoxyphenyl groups. This is also the first example of the alternating copolymerization by a cationic polymerization pathway.[74]

R-Methyl Styrene (RMSt): It is difficult to control the cationic polymerization of RMSt due to steric hindrance, its low ceiling temperature, and the presence of five acidic protons susceptible to elimination. Ashida et al. reported an added base (ethyl acetate)-containing initiating system for living cationic polymerization of RMSt comprising of (CEVE-OAc)-EtAlCl$_2$/SnCl$_4$ in CH$_2$Cl$_2$ at −78 °C. This initiating system allowed synthesis of random copolymers of RMSt with another St derivative at 0 °C.[75] Sacristan et al. reported the production of new silicon-containing soybean-oil-based copolymers from soybean oil, styrene, divinylbenzene, and p-trimethylsilylstyrene by cationic polymerization using boron trifluoride etherate as initiator.[76] Recently, Mandal et al. reported the synthesis of a series of diblock copolymer of α-methylstyrene and isobutyl vinyl ether using a FeCl$_3$ based initiating system. They also studied the block-length-dependent vesicular aggregation of the diblock copolymers.[77]

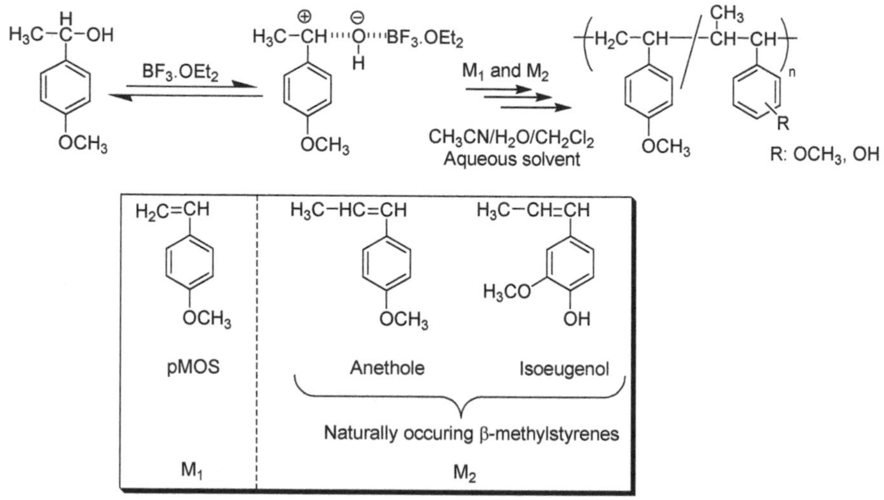

SCHEME 3-5 Schematic representation of controlled cationic polymerization of naturally occurring β-methylstyrenes.
(**Source:** Redrawn from Satoh, K.; Saitoh, S.; Kamigaito, M. J. Am. Chem. Soc.2007,129, 9586–9587.)

Other Styrene Derivatives [p-Alkyl Styrene, Indene (ID), and p-Acetoxy Styrene (pAcOSt)]: End functionalized homopolymers and/or block copolymers with a very narrow molecular weight distributions were obtained by living cationic polymerization of p-methylstyrene using HI/ZnCl$_2$ or ZnI$_2$ in toluene or CH$_2$Cl$_2$ below 0 °C.[78] The VE-HCl adduct/SnCl$_4$ in CH$_2$Cl$_2$ at −15 °C with n-Bu$_4$NCl (an added salt),[79] 2-chloro-2,4,4-trimethylpentane (*TMPCl*)/TiCl$_4$:Ti(iPrO)$_4$ with DMA or Et$_3$N (an added base),[80] pMSt-Cl/SnCl$_4$ with DTBP (a proton trap) in CH$_2$Cl$_2$ at −70 to −15 °C.[81] End functional polymer of ptBSt was produced by Kennedy et al. using TMPCl/TiCl$_4$ with DTBP (a proton trap) system in CH$_3$Cl/methyl cyclohexane at −80 °C.[82] The carbocationic polymerization of 2,4,6-trimethylstyrene initiated by the 1-chloro-1-(2,4,6-trimethylphenyl)ethane/BCl$_3$ initiating system in dichloromethane proceeds in a living fashion without side reactions in the temperature range of −70 to −20 °C. Competition experiments with 2-chloropropene capping agent at −70 °C in dichloromethane produced dichloro chain ended poly(2,4,6-trimethylstyrene).[83]

Due to the high glass transition temperature (T_g) of poly(ID) (200 °C) and its potential use as TPEs, cationic polymerization of ID has been of substantial interest. Sigwalt, Kennedy and co-workers have developed several initiating systems for living cationic polymerization of ID such as

Cumyl-OCH$_3$ or TMPCl/TiCl$_4$ or TiCl$_3$(BuO) in the presence of additives such as DMSO, DMA, Et$_3$N, DTBP.[84] The cumyl chloride/BCl$_3$ initiating system was also used. Kennedy et al. also reported the synthesis of PID-b-PIB-b-PID triblock copolymer using TMPCl/TiCl$_4$ system via sequential living cationic polymerization. The resulted polymer exhibited excellent thermoplastic elastomeric character.

Ashida et al. reported synthesis of random and block copolymers of pAcOSt with other styrene derivatives. Alkaline hydrolysis of the poly(pAcOSt) segment produces poly(HOSt) and then the copolymer exhibited a pH-responsive transition from a clear solution to an opaque one upon decreasing pH of the solution.[85,86] Recently, Shinke et al. reported synthesis of graft copolymers via living cationic polymerization of p-acetoxystyrene followed by Friedel–Crafts-type termination reaction which is generally considered to be a common side reaction, of cationically prepared living polymers.[87]

3.2.3   FUNCTIONAL POLYVINYL ETHERS

Last few decades have witnessed a great advancement in the synthesis of functionalized polyvinyl ethers via cationic polymerizations. End-functional polymers with hydroxyl, carboxyl, and primary amino terminal group were reported by living cationic polymerization of isobutyl vinyl ether using functional initiator and EtAlCl$_2$.[88,89] Well-defined macromonomers of poly(butyl vinyl ether), poly(isobutyl vinyl ether), and poly(ethyl vinyl ether) with ω-methacrylateend-group were prepared by reacting the propagating macrocation of the corresponding monomers with 2-hydroxyethyl methacrylate. Also, Well-defined thermosensitive vinyl ether polymers with various end groups of varying hydrophobicities were prepared via living cationic polymerization in the presence of an added base using functionalized initiator (cationogen) with Et$_{1.5}$AlCl$_{1.5}$ and 1,4-dioxane. The authors examined these polymers to understand the effect of the end group and molecular weight on phase transitions in an aqueous medium.[90]

Many of the poly(vinyl ether)s (PVE) with functional side chains exhibit responsive behaviors towards various external stimuli (temperature, light, and pH).[91] For example, aqueous solutions of PVEs with pendant oxyethylene side chains induce phase-separation behaviors in response to subtle change of temperature. Poly Azo-VE or its derivatives containing an azobenzene side group are known for cis–trans isomerization by light irradiation. Random copolymers of Azo-VEs containing both thermally responsive and azobenzene units showsdifferent solubility of polymers in water upon irradiation with UV or visible light at a constant temperature.

Aoshima et al. has shown thatSnCl$_4$/base initiation system is capable of fast living polymerization of polar functional monomers.[92] The authors studied the polymerization behavior of some representative vinyl ether with various polar functional groups using Et$_x$AlCl$_{3-x}$ and SnCl$_4$ in the presence of added base (**Table 3.1**). Et$_x$AlCl$_{3-x}$ being hard acid interacts with hard base like amide functional group and gets deactivated showing either slow or no polymerization. On the other hand SnCl$_4$ being soft acid interacts with soft base such as chlorine at the propagating end of polymer (HSAB theory).[93]

TABLE 3.1   Cationic Polymerization of Vinyl Ethers Using Hard (Et$_x$AlCl$_{3-x}$) and Soft (SnCl$_4$) Lewis Acid

Monomer			
Et$_x$AlCl$_{3-x}$	Slow and living polymerization	Oligomerization	No polymerization
SnCl$_4$	Fast and living polymerization	Living polymerization	Living polymerization

Source: Idea from Yonezumi, M.; Okumoto, S.; Kanaoka, S.; Aoshima, S. *J. Polym. Sci., Part A: Polym. Chem.***2008**,*46*, 6129–6141.

There is yet another method to make PVEs with polar functional group where there can be a direct living polymerization of a profunctional precursor monomer followed by a post-polymerization transformation of its pendent groups into desired functionalities.[94,95] Precursor monomer should be designed in such a way that the pendent profunctional group should not interfere with its living polymerization and that its post-polymerization reaction should be selective and quantitative under mild conditions without damaging the parent polymer architecture. The described strategy has been well addressed in the polymerization of azide-carrying monomer, 2-azidoethyl vinyl ether

(AzVE) catalyzed by SnCl$_4$ as a catalyst and the HCl adduct of an alkyl vinyl ether as an initiator.The azide pendent groups therein were quantitatively and mildly converted into amine, hydroxyl, and carboxyl by the Staudinger reduction or copper-catalyzed azide-alkyne 1,3-cycloaddition at room temperature without any acidic or basic treatment.[96] Therefore, living cationic polymerization of VE catalyzed by Lewis acid offers a versatile class of pendent-functionalized polymers of controlled architectures.[97] The 2-chloroethyl vinyl ether (CEVE) was used as versatile precursor (parent) monomer to obtain various functionalized VE monomers by simple nucleophilic substitution reactions of the pendent chloroethyl group.On the other hand, quenching the polymerization of vinyl ether with the silyl enol ethers[98] and silyl ketene acetals[99,100] have been successfully used to prepare end-functionalized poly(vinyl ethers). However, functionalization of living polymers by end quenching method remained limited up until recently because of the chemical reaction between the quenching agent and Lewis acid.

3.2.4   N-VINYLCARBAZOL

N-Vinylcarbazol is very reactive monomer for cationic polymerization. Although living cationic polymerization of N-vinylcarbazol was established,[101,102] only one report is available on functional poly(N-vinylcarbazol) whereanthracene-labeled poly(N-vinylcarbazol) with various molecular weights were synthesized by livingcationicpolymerization.[103]

3.2.5   β-PINENE

Living cationic isomerization polymerization of β-pinene was achieved with an initiating system consisting of the HCl-2-chloroethyl vinyl ether adduct/ TiCl$_3$(OiPr) initiating system in the presence of n-Bu$_4$NCl in CH$_2$Cl$_2$ at −40 and −78 °C.[104] Living polymerization was also feasible with the HCl-styrene adduct or 1-phenylethyl chloride.Random copolymers of β-pinene and IB with controlled molecular weights and narrow molecular weight distributions were synthesized by the living cationic copolymerization of the two monomers with 1-phenylethyl chloride/TiCl$_4$/Ti(OiPr)$_4$/n-Bu$_4$NCl initiating system in CH$_2$Cl$_2$ at −40 °C. Reactivity ratio studies showed that the two monomers exhibit almost equal reactivity.[105] Synthesis of end-functionalized poly(β-pinene) was achieved by using functional initiator method, where the initiators were the adducts of HCl and functionalized VE in conjunction with $TiCl_3(OiPr)$ and n-Bu_4NCl. This work was extended to obtain end-functionalized poly(p-methylstyrene)-*block*-poly(β-pinene) and methacrylate-capped macromonomers.[106]

3.3 CONCLUSIONS

In this report, we discussed advances in functional polymer synthesis using living cationic polymerization technique. Many innovative developments have been emerging, and the future of this field looks very promising to prepare IB and VE based functional polymers, which can only be prepared by cationic polymerization method. However, the scope of the living cationic polymerization should be expanded to other monomers such as β-pinene, a renewable strained bicyclic compound derived from wood and plant, to prepare end-functional poly(β-pinene). This will allow tailor-designed new materials from renewable resources.Also, end functional polymer from indene is not yet reported although several reports have been published on the living cationic polymerization of indene.[107] Real potential of various other cationic polymerization derived functional polymers has yet to be realised, but now maybe achievable with the advances in living cationic polymerization technique.

ACKNOWLEDGEMENTS

We acknowledge financial supports from the Department of Science and Technology(DST), India [Project No.: SR/S1/OC-51/2010];Council of Scientific and Industrial Research (CSIR), India [Project No.: 01(2474)/11/EMR-II]; andDefence Research & Development Organisation (DRDO), India [Project No.: ERIP/ER/1001116/M/01].

REFERENCES

1. Miyamoto, M.; Sawamoto, M.; Higashimura, T. *Macromolecules* **1984**, *17*, 265.
2. Kennedy, J. P. *J. Polym. Sci., Part A: Polym. Chem.* **1999**, *37*, 2285.
3. Matyjaszewski, K.; Sawamoto, M. In *Cationic Polymerizations:Mechanism, Synthesis, and Applications*; Matyjaszewski, K., Ed.; Marcel Dekker: New York, 1996; Chapter 4.
4. Cho, C. G.; Feit, B. A.; Webster, O. W. *Macromolecules* **1990**, *23*, 1918.
5. Lin, C.-H.; Matyjaszewski, K. *Polym. Prep.* **1990**, *31*, 599.
6. Aoshima, S.; Kanaoka, S. *Chem. Rev.* **2009**,*109*, 5245.
7. Gyor, M.; Wang, H.-C.; Faust, R. *J. Macromol. Sci., Pure Appl. Chem.***1992**, *A29*, 639.

8. Balogh, L.; Faust, R. *Polym. Bull.* **1992**, *28*, 367.
9. Faust, R.; Kennedy, J. P. *Polym. Bull.* **1986**, *15*, 317.
10. Faust, R.; Kennedy, J. P. *J. Polym. Sci., Polym. Chem. Ed.* **1987**, *25*, 1847.
11. De, P.; Faust, R. Carbocationic Polymerization. In *Synthesis of Polymers*; Schluter, A. D., Hawker, C. J., Sakamoto, J. Eds.; Wiley-VCH: Weinheim, Germany, 2012; Vol. 2, p 775–817.
12. Storey, R. F.; Stokes, C. D.; Harrison, J. J. *Macromolecules* **2005**, *38*, 4618.
13. Simison, K. L.; Stokes, C. D.; Harrison, J. J.; Storey, R. F. *Macromolecules* **2006**,*39*, 2481.
14. De, P.; Faust, R. *Macromolecules* **2006**, *39*, 6861.
15. Higashihara, T.; Feng, D.; Faust, R. *Macromolecules* **2006**, *39*, 5275.
16. Cho, J. C.; Cheng, G.; Feng, D.; Faust, R.; Richard, R.; Schwarz, M.; Chan, K.; Boden, M. *Biomacromolecules* **2006**, *7*, 2997.
17. Higashihara, T.; Faust, R. *Macromolecules* **2007**, *40*, 7453.
18. Feng, D.; Chandekar, A.; Whitten, J. E.; Faust, R. *J. Macromol. Sci., Part A: Pure Appl. Chem.* **2007**, *44*, 1141.
19. Bouchekif, H.; Som, A.; Sipos, L.; Faust, R. *J. Macromol. Sci., Part A: Pure Appl. Chem.* **2007**, *44*, 359.
20. Ojha, U.; Rajkhowa, R.; Agnihotra, S. R.; Faust, R. *Macromolecules* **2008**, *41*, 3832.
21. Higashihara, T.; Faust, R.; Inoue, K.; Hirao, A. *Macromolecules* **2008**, *41*, 5616.
22. Feng, D.; Higashihara, T.; Faust, R. *Polymer* **2008**, *49*, 386.
23. Tripathy, R.; Ojha, U.; Faust, R. *Macromolecules* **2009**, *42*, 3958.
24. Tripathy, R.; Crivello, J. V.; Faust, R. *J. Polym. Sci., Part A: Polym. Chem.* **2013**,*51*, 305.
25. Ummadesetty, S.; Kennedy, J. P. *J. Polym. Sci., Part A: Polym. Chem.* **2008**, *46*, 4236.
26. Nagy, M.; Szollosi, L.; Keki, S.; Faust, R.; Zsuga, M. *J. Macromol. Sci., Part A: Pure Appl. Chem.* **2009**, *46*, 331.
27. Higashihara, T.; Faust, R. *React. Funct. Polym.* **2009**, *69*, 429.
28. Ojha, U.; Kulkarni, P.; Singh, J.; Faust, R. *J. Polym. Sci., Part A: Polym. Chem.* **2009**, *47*, 3490.
29. Ojha, U.; Kulkarni, P.; Cozzens, D.; Faust, R. *J. Polym. Sci., Part A: Polym. Chem.* **2010**, *48*, 3767.
30. Tripathy, R.; Ojha, U.; Faust, R. *Macromolecules* **2011**, *44*, 6800.
31. Espinosa, E.; Charleux, B.; D'Agosto, F.; Boisson, C.; Tripathy, R.; Faust, R.; Soulie-Ziakovic, C. *Macromolecules* **2013**, *46*, 3417.
32. Magenau, A. J. D.; Martinez-Castro, N.; Storey, R. F. *Macromolecules* **2009**, *42*, 2353.
33. Magenau, A. J. D.; Martinez-Castro, N.; Savin, D. A.; Storey, R. F. *Macromolecules* **2009**, *42*, 8044.
34. Bauri, K.; De, P.; Shah, P. N.; Li, R.; Faust, R. *Macromolecules* **2013**, *46*, 5861.
35. Zhang, C.-L.; Wu, Y.-X.; Xu, X.; Li, Y.; Feng, L.; Wu, G.-Y. *J. Polym. Sci., Part A: Polym. Chem.* **2008**, *46*, 936.
36. Zhang, C.-L.; Wu, Y.-X.; Meng, X.-Y.; Huang, Q.; Wu, G.-Y.; Xu, R.-W. *Chin. J. Polym. Sci.* **2009**, *27*, 551.
37. Hackethal, K.; Dohler, D.; Tanner, S.; Binder, W. H. *Macromolecules* **2010**, *43*, 1761.
38. Ma, W.-Y.; Wu, Y.-X.; Feng, L.; Xu, R.-W. *Polymer* **2012**, *53*, 3185.
39. Ren, P.; Wu, Y.-B.; Guo, W.-L.; Li, S.-X.; Chen, Y. *Chin. J. Polym. Sci.* **2013**, *31*, 285.
40. Soytaş, S. H.; Lim, G. T.; Puskas, J. E. *Macromol. Rapid Commun.* **2009**, *30*, 2112.
41. Puskas, J. E.; Soytaş, S. H.; Lim, G. T. *Macromol. Symp.* **2011**, *308*, 61.

42. Breland, L. K.; Storey, R. F. *Polymer* **2008**, *49*, 1154.
43. Zhu, Y.; Storey, R. F. *Macromolecules* **2010**, *43*, 7048.
44. Zhu, Y.; Storey, R. F. *Macromolecules* **2012**, *45*, 5347.
45. Balzano, F.; Pucci, A.; Rausa, R.; Uccello-Barretta, G. *Polym. Int.* **2012**, *61*, 1256.
46. Liston, T. V. *Lubr. Eng.* **1992**, *48*, 389.
47. Ummadisetty, S.; Storey, R. F. *Macromolecules* **2013**, *46*, 2049.
48. Liu, Q.; Wu, Y.-X.; Zhang, Y.; Yan, P.-F.; Xu, R.-W. *Polymer* **2010**, *51*, 5960.
49. Liu, Q.; Wu, Y.; Yan, P.; Zhang, Y.; Xu, R. *Macromolecules* **2011**, *44*, 1866.
50. Ummadisetty, S.; Morgan, D. L.; Stokes, C. D.; Storey, R. F. *Macromolecules* **2011**, *44*, 7901.
51. Vasilenko, I. V.; Shiman, D. I.; Kostjuk, S. V. *J. Polym. Sci., Part A: Polym. Chem.* **2012**, *50*, 750.
52. Dimitrov, P.; Emert, J.; Faust, R. *Macromolecules* **2012**, *45*, 3318.
53. Kumar, R.; Dimitrov, P.; Bartelson, K. J.; Emert, J.; Faust, R. *Macromolecules* **2012**, *45*, 8598.
54. Bartelson, K. J.; De, P.; Kumar, R.; Emert, J.; Faust, R. *Polymer* **2013**, *54*, 4858.
55. Guo, A.-R.; Yang, X.-J.; Yan, P.-F.; Wu, Y.-X. *J. Polym. Sci., Part A: Polym. Chem.* **2013**, *51*, 4200.
56. Kumar, R.; Zheng, B.; Huang, K.-W.; Emert, J.; Faust, R. *Macromolecules* **2014**, *47*, 1959.
57. De, P.; Faust, R. *Polym. Bull.* **2006**, *56*, 27.
58. Higashimura, T.; Ishihnmn, Y.; Sawamoto, M.*Macromolecules* **1993**, *26*,744.
59. Matyjaszewski, K. Cationic Polymerization of Styrenes. In *Comprehensive Polymer Science*; Pergamon Press: Oxford, 1989; Vol. 4, Chapter 41.
60. Ishihama, Y.; Sawamoto, M.; Higashimura, T. *Polym. Bull.* **1990**, *23*, 361.
61. Ishihama, Y.; Sawamoto, M.; Higashimura, T. *Polym. Bull.* **1990**, *24*, 201.
62. Lu, Y.; Larock, R. C. *Biomacromolecules* **2006**, *7*, 2692.
63. Lu, Y.; Larock, R. C. *Macromol. Mater. Eng.* **2007**, *292*, 863.
64. Camerlynck, S.; Cormack, P. A. G.; Sherrington, D. C. *Eur. Polym. J.* **2006**, *42*, 3286.
65. Gao, L.-C.; Zhang, C.-L.; Liu, X.; Fan, X.-H.; Wu, Y.-X.; Chen, X.-F.; Shen, Z.; Zhou, Q.-F. *Soft Matter* **2008**, *4*, 1230.
66. Landsmann, S.; Lizandara-Pueyo, C.; Polarz, S. *J. Am. Chem. Soc.* **2010**, *132*, 5315.
67. Kesharwani, T.; Valenstein, J. S.; Trewyn, B. G.; Li, F.; Lin, V. S. Y.; Larock, R. C. *Polym. Chem.* **2010**, *1*, 1427.
68. Hondred, P. R.; Autori, C.; Kessler, M. R. *Macromol. Mater. Eng.* **2014**, *299*, 1062–1069. DOI: 10.1002/mame.201300437.
69. Li, B.-T.; Liu, W.-H.; Wu, Y.-X. *Polymer* **2012**, *53*, 3194.
70. Yan, P.-F.; Guo, A.-R.; Liu, Q.; Wu, Y.-X. *J. Polym. Sci., Part A: Polym. Chem.* **2012**, *50*, 3383.
71. Higashimura, T.; Mitsuhashi, M.; Sawamoto, M. *Macromolecules* **1979**, *12*, 178.
72. Shohi, H.; Sawamoto, M.; Higashimura, T. *Macromolecules* **1992**, *25*, 53.
73. Shohi, H.; Sawamoto, M.; Higashimura, T. *Makromol. Chem.* **1992**, *193*, 2027.
74. Satoh, K.; Saitoh, S.; Kamigaito, M. *J. Am. Chem. Soc.* **2007**, *129*, 9586.
75. Ashida, J.; Yamamoto, H.; Yonezumi, M.; Kanaoka, S.; Aoshima, S. *Polym. Prepr.* **2009**, *50*, 156.
76. Sacristan, M.; Ronda, J. C.; Galia, M.; Cadiz, V. *Biomacromolecules* **2009**, *10*, 2678.
77. Banerjee, S.; Maji, T.; Paira, T. K.; Mandal, T. K. *Macromol. Chem. Phys.* **2014**, *215*, 440.

78. Kojima, K.; Sawamoto, M.; Higashimura, T. *J. Polym. Sci., Part A: Polym. Chem.* **1990,** *28,* 3007.
79. Miyashita, K.; Kamigaito, M.; Sawamoto, M.; Higashimura, T. *Macromolecules* **1994,** *27,* 1093.
80. Tsunogae, Y.; Kennedy, J. P. *Polym. Bull.* **1992,** *27,* 631.
81. De, P.; Faust, R. *Macromolecules* **2005,** *38,* 5498.
82. Kennedy, J. P.; Meguriya, N.; Keszler, B. *Macromolecules* **1991,** *24,* 6572.
83. De, P.; Sipos, L.; Faust, R.; Moreau, M.; Charleux, B.; Vairon, J. P.*Macromolecules* **2005,** *38,*41.
84. (a). Thomas, L.; Polton, A.; Tardi, M.; Sigwalt, P. *Macromolecules* **1992,** *25,* 5886. (b). Thomas, L.; Polton, A.; Tardi, M.; Sigwalt, P. *Macromolecules* **1993,** *26,* 4075. (c). Tsunogae, Y.; Majoros, I.; Kennedy, J. P. *J. Macromol. Sci., Pure Appl. Chem.* **1993,** *A30,* 253. (d). Kennedy, J. P.; Midha, S.; Keszler, B. *Macromolecules* **1993,** *26,* 424.
85. Ashida, J.; Yamamoto, H.; Yonezumi, M.; Kanaoka, S.; Aoshima, S. *Polym. Prep., Jpn.* **2008,** *571,* 528.
86. Yamamoto, H.; Kanaoka, S.; Aoshima, S. *Polym. Prepr., Jpn.* **2006,** *55,* 2801.
87. Shinke, Y.; Yamamoto, H.; Kanazawa, A.; Kanaoka, S.; Aoshima, S. *J. Polym. Sci., Part A: Polym. Chem.* **2013,** *51,* 4675.
88. Shohi, H.; Sawamoto, M.; Higashimura, T. *Polym. Bull.* **1989,** *21,* 357.
89. Shohi, H.; Sawamoto, M.; Higashimura, T. *Macromolecules* **1992,** *25,* 58.
90. Shimomoto, H.; Kanaoka, S.; Aoshima, S. *J. Polym. Sci., Part A: Polym. Chem.* **2012,** *50,* 4137.
91. Aoshima, S.; Kanaoka, S. *Adv. Polym. Sci* **2008,** *210,* 169.
92. Yonezumi, M.; Okumoto, S.; Kanaoka, S.; Aoshima, S. *J. Polym. Sci., Part A: Polym. Chem.* **2008,** *46,* 6129.
93. Yoshida, T.; Tsujino, T.; Kanaoka, S.; Aoshima, S. *J. Polym. Sci., Part A: Polym. Chem.* **2005,** *43,* 468.
94. Aoshima, S.; Higashimura, T. *Polym. Bull.* **1986,** *15,* 417.
95. Aoshima, S.; Higashimura, T. *Macromolecules* **1989,** *22,* 1009.
96. Ida, S.; Ouchi, M.; Sawamoto, M. *J. Polym. Sci., Part A: Polym. Chem.* **2010,** *48,* 1449.
97. Sawamoto, M. *Prog. Polym. Sci.* **1991,** *16,* 111.
98. Fukui, H.; Sawamoto, M.; Higashimura, T. *Macromolecules* **1993,** *26,* 7315.
99. Verma, A.; Nielsen, A.; McGrath, J. E.; Riffle, J. S. *Polym. Bull.* **1990,** *23,* 563.
100. Verma, A.; Nielsen, A.; Bronk, J. M.; McGrath, J. E.; Riffle, J. S. *Makromol. Chem. Macromol. Symp.* **1991,** *47,* 239.
101. Sawamoto, M.; Fujimori, J.; Higashimura, T. *Macromolecules* **1987,** *20,* 916.
102. Higashimura, T.; Deng, Y. X.; Sawamoto, M. *Polym. J.* **1983,** *15,* 385.
103. Aoki, H.; Horinaka, J.-I.; Ito, S.; Yamamoto, M.; Katayama, H.; Kamigaito, M.; Sawamoto, M. *Polym. J.* **2001,** *33,*464.
104. Lu, J.; Kamigaito, M.; Sawamoto, M.; Higashimura, T. *Macromolecules* **1997,** *30,* 22.
105. An-Long, L.; Zhang, W.; Liang, H.; Lu, J. *Polymer* **2004,** *45,* 6533.
106. Lu, J.; Kamigaito, M.; Sawamoto, M.; Higashimura, T. Deng, Y.-X. *J. Polym. Sci., Part A: Polym. Chem.* **1997,** *35,* 1423.
107. Thomas, L.; Polton, A.; Tardi, M.; Sigwalt, P. *Macromolecules* **1995,** *28,* 2105.

PART II
Advanced Synthetic Strategies for Making Biodegradable Functional Materials

CHAPTER 4

STRUCTURAL DIVERSITY IN SYNTHETIC POLYPEPTIDES BY RING OPENING POLYMERIZATION OF N-CARBOXY ANHYDRIDES

DIPANKAR BASAK and SUHRIT GHOSH

CONTENTS

4.1 INTRODUCTION

Synthetic polypeptides have attracted tremendous attention in the recent past because of their potential applications in various fields ranging from drug and gene delivery, tissue engineering, biosensing, and other biomaterials applications.[1] For most of these applications, however, it is prerequisite that the synthetic polypeptides must self-assemble into stable ordered conformations much like the natural polypeptides. It is, therefore, a challenging task for the chemists to come up with synthetic strategies that produce synthetic polypeptides with controlled molecular weight and well-defined structures. Typically, solid phase peptide synthetic strategy has been employed for successful preparation of synthetic polypeptides with specific sequences.[2] While this is a routine method for synthesis of small peptides, employing such step-by-step amino acid coupling process for synthesis of large polypeptides is impractical. In this regard, ring-opening polymerization (ROP) of γ-amino acid N-carboxyanhydrides (NCA) is proved to be more useful technique to produce high-molecular weight polypeptides (Scheme 4-1). Although this method lacks the ability to produce specific sequences of amino acids, the ease of preparation of NCA with variety of amino acids (Figure 4-1) allows preparing diverse range of synthetic polypeptides with no racemization at the chiral center. Although the synthesis of NCAs and their utility in polymerization was identified long back in 1906, only the last few decades have witnessed tremendous progress in ROP of NCA with high- and controllable molecular weights, and narrow molecular weight distributions. Especially, after integrating other recently developed controlled polymerization techniques with ROP of NCAs, the field widens the scope of producing various synthetic polypeptides with structural diversity. The following sections details the various synthetic strategies for preparation of linear polypeptides and their conjugates based on ROP of NCAs and with a brief touch upon utilization of these synthetic polypeptides for various biomedical applications.

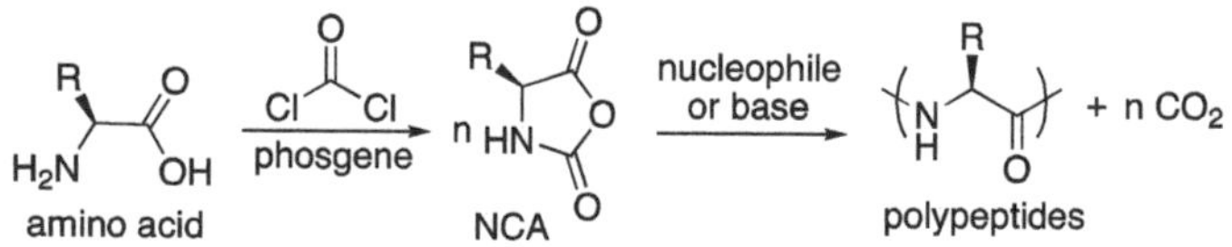

SCHEME 4-1   General synthetic scheme for polypeptides by ROP of NCAs.

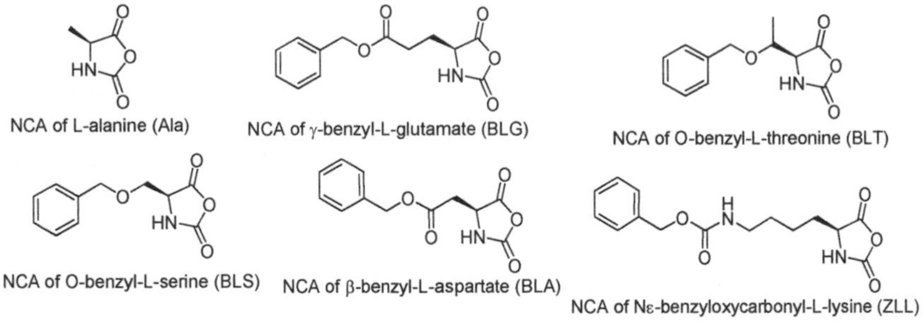

FIGURE 4-1   Structures of commonly used NCAs for polypeptide synthesis.

4.2   SYNTHESIS OF NCA

In 1906, Hermann Leuchs reported the first synthesis of NCA when he attempted to purify N-methoxycarbonyl amino acid chlorides by distillation.[3] Since then, a number of publications have appeared describing various methods for the preparation of these cyclic monomers.[4] The conventional approach involves direct addition of phosgene,[5] triphosgene,[6] or di-*tert*-butyltricarbonate[7] to γ-amino acid producing the corresponding NCA. This method, however, suffers from several drawbacks: firstly, phosgene is highly toxic; and secondly, often the NCA, prepared by this method, is contaminated with hydrogen chloride, which is difficult to remove and such contamination severely impedes subsequent polymerization reaction by favoring other side reactions. To solve this issue, NCAs were prepared where the N-center was protected with suitable protecting groups and was subsequently removed after polymerization. Two different approaches have been adopted in this regard, the first approach involves protection of N-center of the amino acid, followed by cyclization; whereas the second approach deals with attachment of suitable protecting group at the N-center of the preformed NCA. Scheme 4-2 illustrates different synthetic approaches to prepare NCA by the first route. Earlier reports in this approach include substitution of N-center of amino acid with tosyl and nitrophenylsulfonyl groups, followed by cyclization using phosgene.[5] Significant improvement was made after addition of alkoxycarbonyl group to the nitrogen, forming the urethane-protected-NCA.[8] Very recently more attractive protecting groups such as trityl and phenylfluorenyl moeities were installed on NCAs of four amino acids and the resulting NCA-derivatives were crystalline solids and exhibited excellent stability on storage at room temperature.[9] The second route involves deprotection of the N-center of NCA by a strong base, followed by N-substitution with suitable capping agent (Scheme 4-3). Since a basic reagent is necessary for deprotection, the risk of other side-reactions including NCA-polymerization is involved in this process. However, 4-nitrophenylsulfenyl chloride was found to be so reactive that complete substitution was achieved in presence of triethylamine without significant side reaction.[10] A major breakthrough in this approach occurred when benzyloxycarbonyl, [(9-fluorenylmethyl)oxy]carbonyl (Fmoc), and *tert*-butyloxycarbonyl (Boc) protected amino acid NCAs were prepared.[11] All of these protected NCAs were crystalline in nature and showed good stability upon storage at room temperature. Due to excellent stability, Boc-NCA was commercialized further and used extensively for solid-phase peptide synthesis.

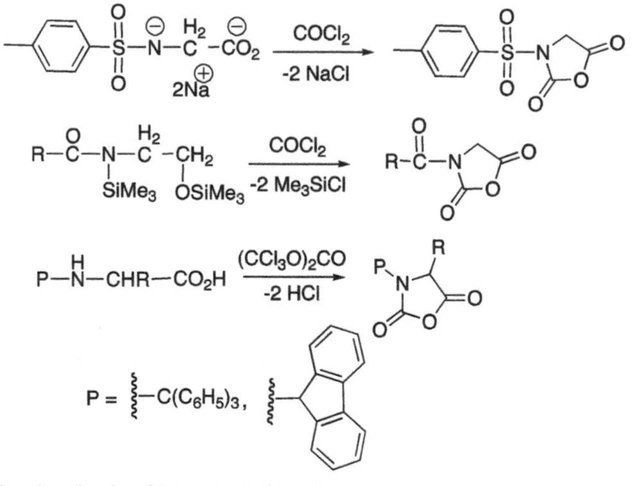

SCHEME 4-2   Synthesis of *N*-protected NCAs.

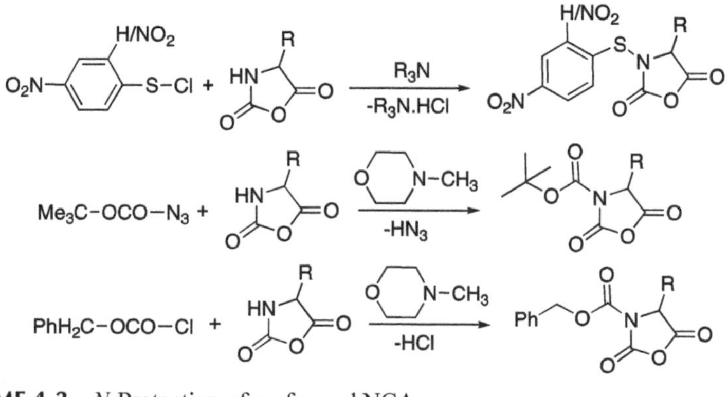

SCHEME 4-3   *N*-Protection of preformed NCAs.

4.3 ROP OF NCA

In general, there are two widely accepted mechanisms for ROP of NCA: (1) normal amine mechanism (NAM) and (2) activated monomer mechanism (AMM). NAM is usually applied for polymerizations of NCA, which are initiated by nonionic initiators such as water, alcohols, and primary or secondary amines containing at least a mobile hydrogen in their structures. Due to high nucleophilicity, primary amines such as n-hexyl amine or benzyl amine have been chosen over other protic initiators for successive ring opening polymerization. The proposed mechanism for ROP of NCA based on primary amine initiatior is shown in Scheme 4-4. The initiation step is based on nucleophilic attack of the amine on 5-CO of NCA ring, resulting in opening of the ring and formation of carbamic acid intermediate via proton transfer. Then the unstable carbamic acid decarboxylates and the newly formed amino group promote polymerization. For NAM, in general, initiation is faster than the propagation because the initiator contains primary amine, which is typically more reactive compared to the γ-amino group of the propagating chain, thus leading to well-controlled polypeptides with low polydispercity indices (PDI). It is also possible to control the degree of polymerization (DP) by simply altering the feed ratio NCA/amine (i.e., monomer/initiator or M/I). While the NAM requires primary amine for the initiation step, tertiary amine or alcoholates are the typical initiators for AMM. In this case, the initiator acts as a base rather than a nucleophile and abstracts the proton of 3-N position generating corresponding anion (Scheme 4-5). The resulting NCA anion attacks 5-CO position generating an unstable carbamate intermediate, which readily releases CO_2 to generate a dimer as well as a new NCA anion. The propagation starts when the newly formed NCA anion attacks the dimer to form a trimer and so on. Since the

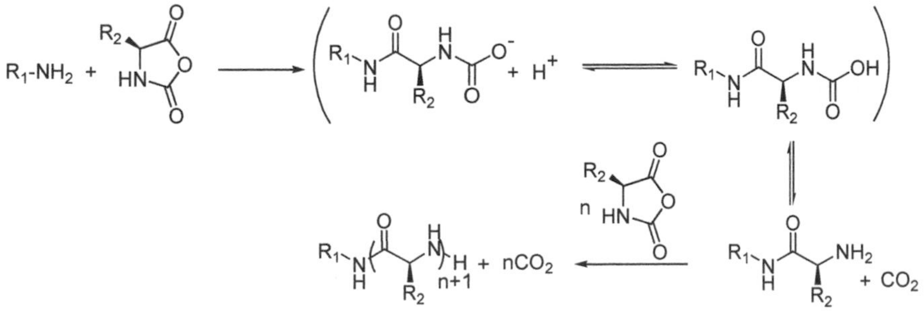

SCHEME 4-4   Normal amine mechanism (NAM).

propagation steps involves an anion, it is expected that the rate of propagation will be higher in case of AMM compared to that of NAM, often leading
to higher molecular weight polypeptides for the former case. However,
due to slow initiation and faster propagation, the polypeptides obtained via
AMM exhibit high PDI. Above all, the polymerizations following the AMM
are uncontrolled, thus restricting the synthesis of well-defined polypeptides
by this process.

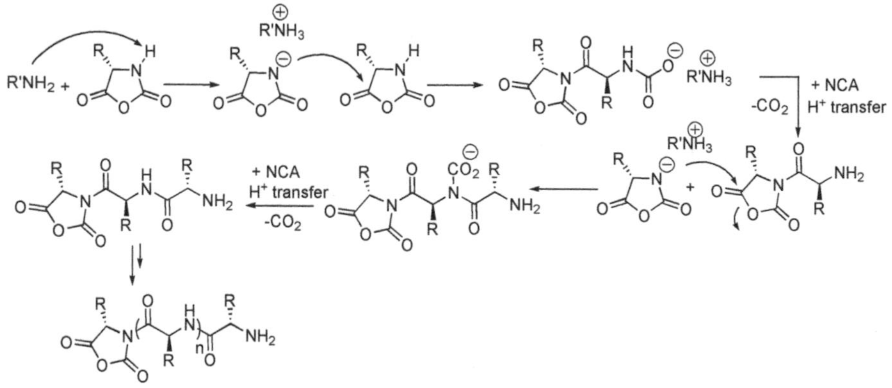

SCHEME 4-5   Activated monomer mechanism (AMM).

Although NAM offers a very facile route to synthesize well-defined
polypeptides, various side-reactions that are associated with the main
polymerization process restrict the utility of this method for polypeptide synthesis. For example, the intermediate carbamic acid plays a very
critical role in NAM. According to Ballard et al. carbamic acid forms salt
with the amino group of the propagating polymer chain and influences the
kinetics of the polymerization process.[12] They found that just by changing
the solvent from nitrobenzene to dimethylformamide (DMF), the kinetics
could be significantly altered. Another important aspect is the purity of the
NCA monomer.[13] Even the presence of trace amounts of impurities such as
water can initiate polymerization leading to undesired oligo(peptides). Other
side-reaction includes attack of the primary amine to 2-CO of NCA instead
of 5-CO resulting the formation of ureiodo acid chain end. It was found
that, as the nucleophilicity of the amine increased higher was the chance of
the attack at 5-CO position over 2-CO. For example, Goodman et al. and
others found that polymerization of ^{13}C labeled NCA that was initiated with
n-hexyl amine, which is a good nucleophile, resulted in very less (<0.15% of
the initial active sites) 2-CO attack compared to 5-CO attack.[14] Another very

common side reaction of the primary amine initiated NAM is the formation of cyclic peptide. This side reaction happens when the amino group of the growing peptide chain attacks one of the CO centers of the same polymer chain causing premature termination. In some cases the carbamate also leads to an internal nucleophilic attack to NCA ring, causing cyclization.[4b] Also the poor solubility of the growing polypeptides is another major obstacle in obtaining well-defined polypeptides. Most of the polypeptides tend to form secondary structures even at very low conversion. A major fraction of the oligomers precipitates from the reaction medium in the form of γ-sheet and only a small fraction that remains soluble in the medium, mostly in γ-helix form, tend to propagate.[4a,15] Due to this pronounced disparity in solubility between two secondary structures the polypeptides obtained by this method are usually with high PDI. In conclusion, it is very difficult to achieve well-controlled polypeptides by following classical NAM route.

To circumvent the above-mentioned problems, Deming and co-workers introduced a new class of initiators based on organonickel compounds that eliminated most of the side-reactions and maintained the growing polymer chain "living" toward new NCA monomers.[16] As a result excellent control over molecular weight as well as molecular weight distribution was observed. To formulate the initiator, NCA compound was reacted with zero-valent nickel complexes [bipyNi(COD) or (PPh$_3$)$_2$Ni(COD); bipy = 2,2'-bipyridyl, COD = 1,5-cyclooctadiene]. The first step is the formation of a metallocyclic complex by oxidative addition at the anhydride bond of NCA (Scheme 4-6). Depending upon the insertion, however, two different pathways are possible and they would give rise to two isomeric products. Reactions of bipyNi(COD) with $^{13}C_2$-L-Leu NCA (route I) and $^{13}C_5$-L-Leu NCA (route II) were carried out and after careful examination of the respective oxidative products, Deming and co-workers unambiguously concluded exclusive addition of Ni across O-C$_5$ bond of NCA. The oxidative product was then reacted with another NCA monomer and subsequently transformed to a five membered amido-amidate metallocyle (Scheme 4-7). It was proposed that elimination–reinsertion of a γ-hydrogen followed by ring contraction yielded this active intermediate. The propagation step involved nucleophilic attack of the active amido-amidate complex to 5-CO of NCA monomer resulting in a large metallocycle. This metal complex would further contract back to the amido-amidate species after a series of steps that involved probable proton migration from free amide to the tethered amidate group followed by CO$_2$ elimination. Thus, the chain-end of the propagating polypeptide always contains the metal complex and once all of the monomers are consumed the macro-initiator can be used further for growing a

second block. This particular aspect of synthesizing block copolymers will
be discussed in detail in Section 4.4.2.

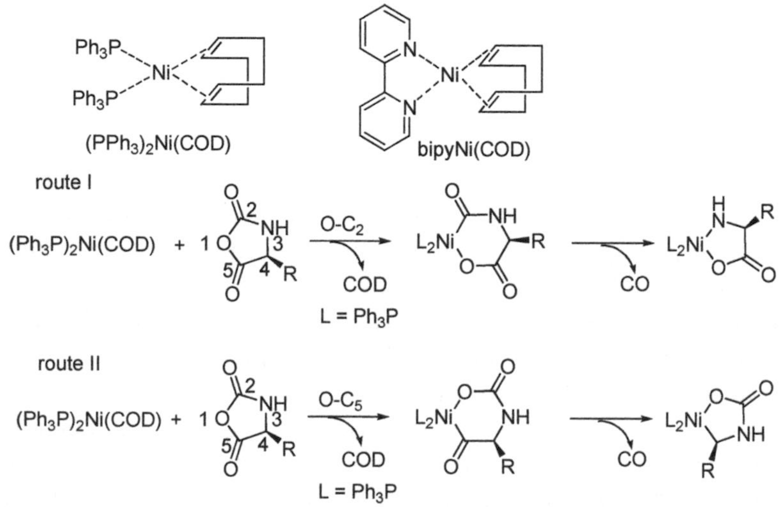

SCHEME 4-6   Proposed reaction routes (I and II) of nickel-cyclooctadiene complexes.

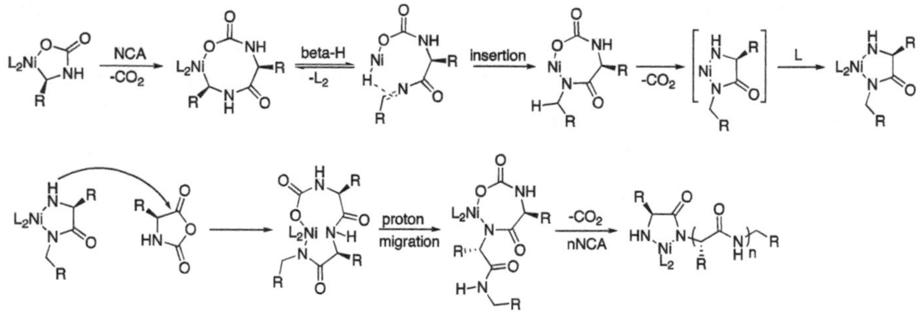

SCHEME 4-7   Mechanism of proposed living polypeptide using transition metal.

Although the method introduced by Deming and co-workers showed
high credibility, an additional step of purification of metal is required for
this approach. Also this method falls short in installing any specific initiator
functionality into the polymer chain. Therefore, the search for obtaining
a general method that would produce well-controlled polypeptide with
desired end functionality is still ongoing. Over the past years, several
researchers have reported new improved methods of NCA polymerization
using primary amine initiator. In 2004, Hadjichristidis et al. reported that

the primary-amine initiated NCA polymerization using high-vacuum technique.[17] By employing highly purified polymerization solvent and *n*-hexylamine initiator, the polymerization of γ-benzyl-L-glutamate NCA (Bn-Glu NCA) and ε-carbobenzyloxy-L-lysine NCA (Z-Lys NCA) produced homo- as well as block-polypeptides with good control over the molecular weight and narrow molecular weight distribution. It was assumed that removal of CO_2 generated during the reaction under high vacuum and suppression of the undesired side reaction between DMF solvent with the propagating polymer chain were accounted for the living nature of this NCA polymerization. Further, using this high-vacuum technique Messman et al. demonstrated the synthesis of telechelic poly(O-benzyl-L-tyrosine) with pre-determined molecular weight and low-polydispersity.[18] The rigorous end-group analysis of the resulting polymer by both matrix-assisted laser desorption/ionization time-of-flight mass spectrometry (MALDI-TOF MS) and ^{13}C NMR spectroscopy established not only the living nature of the polymerization but also demonstrated the fact that the polymerization was proceeded exclusively via NAM pathway. In 2004, Vayaboury et al. reported an important result regarding the living nature of NCA polymerization by NAM.[19] This group performed ROP of Nγ-trifluoroacetyl-L-lysine NCA using hexylamine initiator in DMF as a function of temperature and the crude polymer samples were analyzed by size exclusion chromatography (SEC) and nonaqueous capillary electrophoresis (NACE).[20] The judicious choice of employing NACE along with SEC was to quantitatively analyze the resulting polymers with different chain-ends that corresponded to either living chains (amine-terminated end group) or dead chains (carboxylate and formyl group terminated chain ends), obtained from the reaction of propagating polymer chain with NCA and DMF, respectively. They found that the population of living chain ends improved dramatically from 22% to 99% as the temperature of the polymerization was lowered from 20 to 0 °C. The suppression of the side-reactions at lower temperature was attributed to higher activation energy barrier compared to that of the chain-propagation and thus chain-propagation favored kinetically at lower temperature. However, prolonged reaction time is an immediate drawback of carrying out polymerization at 0°C.

Schlaad and co-workers introduced an innovative way to suppress the unwanted AAM in primary amine initiated NCA polymerization. NCA anions are well known to be able to rearrange to γ-isocyanocarboxylate at which nucleophilic attack of the amino group of the propagating polymer chain causes premature termination. The strategy was to use corresponding hydrochloride salt of the amine initiator, instead of the primary amine, to avoid formation of NCA anions.[21] The underlying assumption was that the acidification of the NCA anion had faster kinetics than the nucleophilic attack to another NCA molecule. Thus the nucleophilic amine terminus was converted into a dormant amine hydrochloride group with very low reactivity towards other electrophiles (Scheme 4-8). As soon as any free amine group is formed by dissociation of the salt, it would immediately react with another NCA molecule and the free amine terminus would then go back to the original dormant amine-chloride form; thus the propagation occurs only via NAM. The use of fast, reversible deactivation of the reactive terminus reminisce the working principle of any controlled radical polymerization technique and definitely has a promising future for synthesis of well-controlled polypeptides.

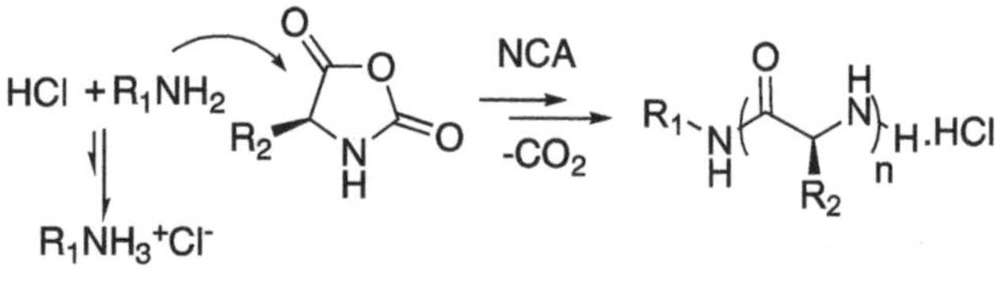

SCHEME 4-8　ROP of NCAs using primary amine hydrochloride.

In 2007, Cheng and co-workers reported controlled NCA polymerization initiated by hexamethyldisilazane (HMDS).[22] HMDS-mediated polymerization exhibited remarkable control over the polymerization with complete monomer consumption within 24 h at ambient temperature with low polydispersity. A mechanistic study of the polymerization showed that the polymerization involved trimethylsilyl carbamate (TMS-CBM) as the propagating group, which was formed by the breakage of Si-N bond during the initiation process (Scheme 4-9). The propagation step involved the transfer of the TMS group from the terminal TMS-CBM to an incoming NCA resulting a new TMS-CBM chain end. Further report from the same research group demonstrated that N-TMS amine could also be used as the initiator for controlled NCA polymerization. Use of alkene or alkyne-containing amine-TMS as the initiators could afford polypeptides that would cap with these functional groups at the C-terminus, opening the scope to conjugating this macromolecule with other polypeptides via facile chemistry.

In 2013, Wooley et al. showed that the rate of polymerization by NAM can be tuned simply by controlling the flow of N_2 during ROP of NCA (Scheme 4-10).[23] It was established from the previous studies that removal of CO_2 from the reaction medium could significantly affect the rate of polymerization. By maintaining a continuous flow of N_2 during polymerization, Wooley et al. demonstrated the following: (a) the rate of conversion of NCA monomers was higher when continuous flow of N_2 was present compared to the reactions where no N_2 flow was used, clearly indicating the continuous removal of CO_2 by N_2 flow; (b) the rate of polymerization can be tuned by changing the rate of flow of N_2; and (c) the living features of NCA polymerization was established even at very high monomer conversion.

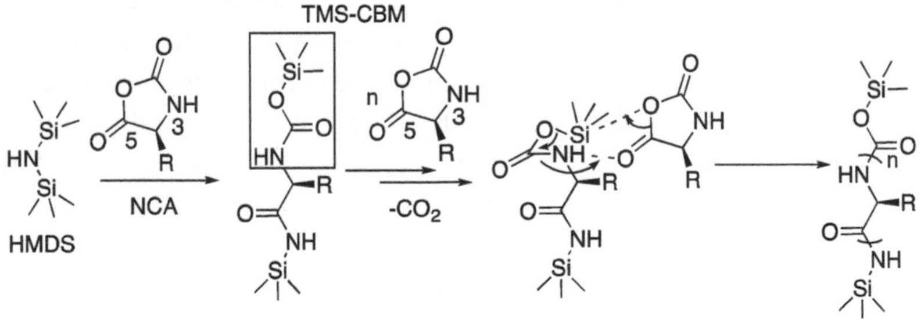

SCHEME 4-9 Synthetic scheme for HMDS-mediated NCA polymerization through TMS carbamate group.

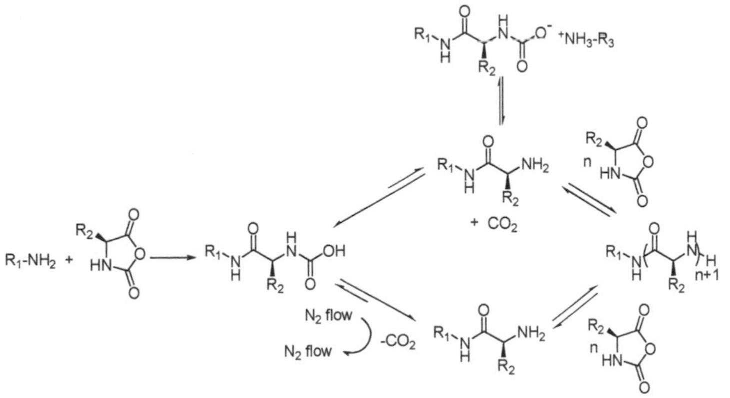

SCHEME 4-10 ROP of NCA under N_2 flow.

4.4 COPOLYPEPTIDES AND HYBRID POLYPEPTIDES

4.4.1 RANDOM COPOLYPEPTIDES

Since 1984, numerous reports have been published focusing on synthesis and characterization of random or statistical copolypeptides. Random copolypeptides can be synthesized in two ways. The first method comprises of statistical copolymerization of two or more NCA monomers, whereas the second method deals with partial removal of protecting groups on a homo-polypeptide backbone. The synthesis of random copolypeptides involves initiation by either primary/secondary amine or transition metal catalyst. The following section briefly discusses the significant advancements in this area, readers are referred to recent review articles for a detailed study.[4a–d,16b]

Deming and co-workers reported the synthesis of random copolypep-tides of Nε-carbobenzoxy L-lysine with O,O''-dicarbobenzoxy L-dihydroxy-phenylalanine by ROP of the respective NCAs using sodium *tert*-butoxide as initiator (Scheme 4-11).[24] De-protection of the copolypeptides was carried out using HBr in acetic acid. Subsequently, in presence of suitable oxidizing agent, such as O_2, $NaIO_4$, and H_2O_2, the copolymers formed cross-linked networks that were further tested as moisture-sensitive adhesive.

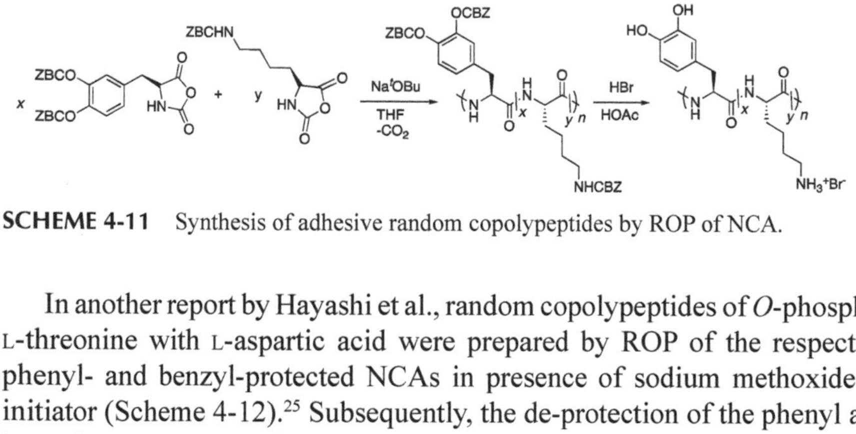

SCHEME 4-11　Synthesis of adhesive random copolypeptides by ROP of NCA.

In another report by Hayashi et al., random copolypeptides of O-phospho-L-threonine with L-aspartic acid were prepared by ROP of the respective phenyl- and benzyl-protected NCAs in presence of sodium methoxide as initiator (Scheme 4-12).[25] Subsequently, the de-protection of the phenyl and benzyl groups was carried out by catalytic hydrogenolysis over PtO_2. After-wards the effects of these synthetic copolypeptides on the growth of $CaCO_3$ crystals were examined.

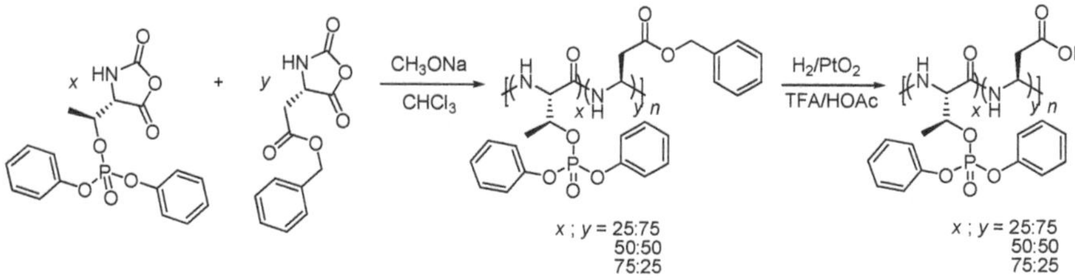

SCHEME 4-12　Synthesis of copoly[Thr(PO$_3$H$_2$)Asp].

A library of statistical copolypeptides were synthesized by Deming and co-workers by using transition metal complex, $Co(PMe_3)_4$, as initiator (Scheme 4-13).[26] The random copolypeptides were based upon combination of a hydrophilic amino acid L-lysine with one of the hydrophobic amino acids such as L-leucine, L-phenylalanine, L-isoleucine, L-valine, or L-alanine. The purpose of choosing such combinations was to mimic the "amphiphilic and cationic nature of many natural antimicrobial polypeptides."

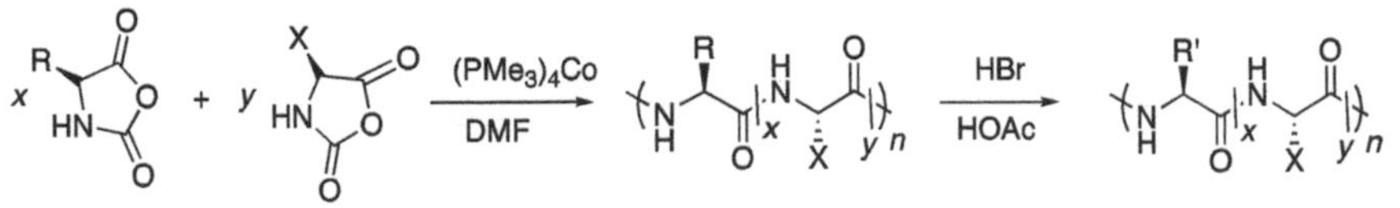

R = -$(CH_2)_4NHC(O)OCH_2C_6H_5$ X = -CH_3, -$CH(CH_3)_2$, -$CH(CH_3)CH_2CH_3$, -$CH_2CH(CH_3)_2$, or -$CH_2C_6H_5$
R' = -$(CH_2)_4NH_3^+Br^-$

SCHEME 4-13   Synthesis of random copolypeptides using $Co(PMe_3)_4$.

Hernández and Klok reported the synthesis of copolypeptides of γ-benzyl-L-glutamate N-carboxyanhydride and γ-benzyloxycarbonyl-L-lysine N-carboxyanhydride with various L-lysine NCAs that contain labile protective groups at the γ-NH_2 position (Scheme 4-14).[27] Four different L-lysine NCAs were used: N-γ-trifluoroacetyl-L-lysine N-carboxyanhydride, N-γ-(tert-butoxycarbonyl)-L-lysine N-carboxyanhydride, N-γ-(9-fluorenylmethoxycarbonyl)-L-lysine N-carboxyanhydride, and N-γ-(6-nitroveratryloxycarbonyl)-L-lysine N-carboxyanhydride. The synthesis was carried in DMF using n-hexylamine as initiator. After the polymerization, the protecting groups were removed under mild condition. Thus this method allowed the synthesis of novel polypeptides that was hitherto difficult to prepare using normal NCA ROP technique because of several side-reactions that hindered the growth of the polymer.

Higuchi and co-workers first introduced the synthesis of copolypeptides by the second approach, that is, generating a statistical copolypeptide from an already synthesized homopolypeptide by partial modification of the functional group embedded onto the homopeptide backbone (Scheme 4-15).[28] First the homopolymer poly(γ-methyl L-glutamate) was prepared by ROP of NCA of L-glutamic acid γ-methyl ester using n-hexylamine as initiator. The obtained polymer was stirred in aqueous sodium hydroxide solution (0.4 M) for 10 h causing partial hydrolysis of the methyl-ester group to produce random poly[(γ-methyl-L-glutamate)-co-(L-glutamic acid)]. From ¹H-NMR analysis, the extent of glutamic acid was estimated to be 30%.

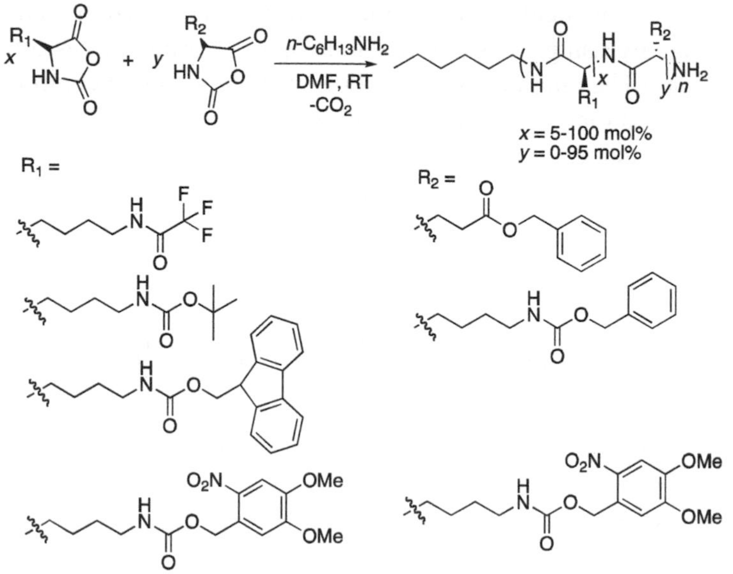

SCHEME 4-14　Synthesis of statistical copolypeptides using primary amine initiator.

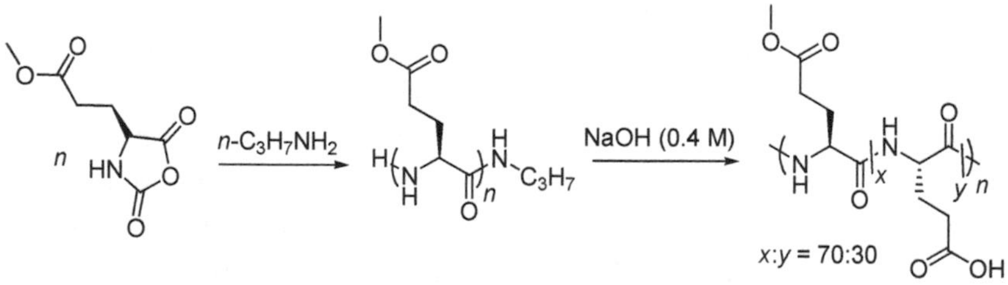

SCHEME 4-15　Synthesis of statistical copolypeptides by partial hydrolysis of L-glutamate ester.

4.4.2　BLOCK POLYPEPTIDES

In this section, various synthetic strategies that have been used to prepare linear block polypeptides will be discussed. Depending on the number of blocks that are present in the polymer structure, block polypeptides can be termed as diblock, triblock, tetrablock, and so on. All of these block polypeptides broadly fall into two major categories: (1) block copolypeptides, the block polymers those are exclusively composed of different peptide blocks; (2) hybrid block polypeptides, all other block polymers in which one of the blocks is composed of polypeptides whereas the other block(s) is made of nonpeptide chains. The standard process of preparing block copolypeptides is via sequential addition of two or more NCAs to either amine such as *n*-hexyl amine, *n*-propylamine, piperidine, and so on or metal complex, typically Ni- or Co-complex, initiator. On the other hand, different controlled polymerization techniques were employed to synthesize nonpolypeptide block(s) from which the second polypeptide block was developed by ROP of corresponding NCA.

4.4.2.1 BLOCK COPOLYPEPTIDES

Higashi et al. synthesized an amphiphilic diblock polymer of poly(γ-benzyl-L-glutamate) (PBLG)-block-poly(L-glutamic acid) (PLGA) (PBLG-*b*-PLGA). This block copolypeptide was synthesized in two steps (Scheme 4-16).[29] First step involved synthesis of PLGA block by ROP of γ-benzyl-L-glutamate-NCA using *n*-propylamine as initiator followed by removal of benzyl group through catalytic hydrogenolysis (H$_2$/Pd). Subsequently the terminal amino group of this macro-initiator initiated ROP of BLG-NCA to afford final amphiphilic polymer.

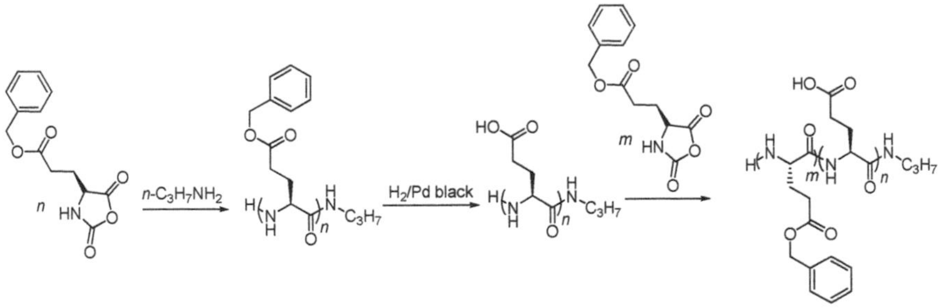

SCHEME 4-16   Synthesis of amphiphilic diblock copolypeptides PBLG-*b*-PLGA.

Using similar synthetic protocol, synthesis of homo- and block-glycopeptides was carried out by Okada et al.[30] The ROP of *O*-(tetra-*O*-acetyl-γ-D-glucopyranosyl)-L-Serine *N*-carboxyanhydride using *n*-hexylamine as initiator produced homo-glycopeptide, which was subsequently polymerized by addition of the second monomer Ala-NCA using the "living" amino group at the chain end of the first block. The exact synthetic strategy was also used by Gallot and co-workers during the preparation of various diblock copolypeptides of poly(*N*ε-trifluoroacetyl-L-lysine)-*b*-Poly(L-lysine-R); where R in the second block represents liquid crystalline groups including 11-(biphenyl-4-carboxamido)undecanamido- and $N\varepsilon$-4-phenylbenzamido-moeities.[31]

A series of reports dealing with the synthesis of block copolypeptides employing high vacuum/low temperature techniques have been documented in recent years by several researchers. The final block copolypeptides were obtained by sequential polymerizations of two or more NCAs using primary amine as initiator and in most of the cases the polymerization was well-controlled and afforded high molecular weight polymer. Another parallel and efficient method of preparation of block copolypeptides includes zero valent transition metal complexes. The polymerization initiated by either Ni or Co [bpyNy(COD) or Co(PMe$_3$)$_4$] produced the best result, furnishing great control over molecular weight and molecular weight distribution. While the elaborate discussion of every report is beyond the scope of this review, few examples of these well-controlled block copolypeptides comprised of sequential addition of different NCAs have been tabulated in Table 4-1.

TABLE 4-1   Synthesis of Different Block Copolypeptides

1st Block	2nd Block	3rd Block	4th Block	5th Block	Method	References
PBLGlu	PLAla				HV, Ni-cat, 0 °C	[32]
	PLAla	PZLLys	PBLAsp		HV/0 °C	[33]
	PZLLys				HV, Ni-Cat, Silazane	[17, 34]
	PZLLys	PLAla			0 °C	[33]
	P ᵗBocLLys				HV	[35]
	PLGly				HV	[17]
	PLTyr				HV	[17]
	PLLeu				HV, Ni-Cat	[36, 17]
	PLPro				HV, Ni-Cat, HV	[36, 37]
	PZLCys				Ni-Cat	[38]
	PBLSer				0 °C	[39]
	PtBMLCys				0 °C	[39]
	PBLCys				0 °C	[39]
	PLGly				HV	[40]
	PTFALLys				20 °C	[41]

TABLE 4-1 *(Continued)*

1st Block	2nd Block	3rd Block	4th Block	5th Block	Method	References
PMLGlu	PLLeu				20 °C, Al-Cat	[42]
PtBuLGlu	PMLGluSLGlu				0 °C	[43]
PALGlu	PCPLGlu				Bu-NH₂	[44]
PZLLys	PBLGlu				HV, Ni-Cat	[17, 39]
	PLGln				Ni-Cat	[41]
	PLLeu				Ni-Cat	[36]
	PLLeu	PZLLys			Co-Cat	[45]
	PLeu	PZLLys	PLLeu	PZLLys	Co-Cat	[46]
	PL Z2DOPA				Co-Cat	[47]
	PLAla				Ni-Cat	[32b]
	PLTyr				Ni-Cat	[32b]
	PLSer				Ni-Cat	[32b]
	PLZCys				Ni-Cat	[32b]
	PLCys				Silazane	[48]
	PLPhe				Ni-Cat, 30 °C	[49]
	Pα-manLLys				Co-Cat	[50]
	PLGly				Ni-Cat	[51]
P(EG2Lys)	PBLAsp				Ni-Cat	[52]
	PBLGlu				Ni-Cat	[53]
	PZLLys				Ni-Cat	[54]
	PLLeu				Co-Cat	[55]
Pα-manLLys L	PZLLys				Co-Cat	[50]
	Pα-galLLys				Co-Cat	[50]
PTFALLys	PLLeu				0°C	[20]
PtBocLLys	PLPro				HV	[37]
PZ2Arg	PLLeu				Co-Cat	[56]
PBLThr	PBLGlu				35 °C	[57]
	P t BocLLys				35 °C	[57]
PLPhe	PBLGlu				20 °C	[58]

Source This table is adapted from Ref. 4d.

Recently Agut et al. developed a novel synthetic methodology to prepare block copolypeptides by employing click chemistry.[59] Two separate blocks namely poly(γ-benzyl-L-glutamate) and poly(trifluoroacetyl-L-lysine) containing either an azide or an alkyne functional group were prepared using ROP of γ-benzyl-L-glutamate and trifluoroacetyl-L-lysine mono-mers, respectively (Scheme 4-17). Further these two synthetic blocks were coupled together using Cu(I)-catalyzed Huisgen 1,3 dipolar cycloaddition reaction furnishing the final block copolypeptide with a triazole group in between two blocks.

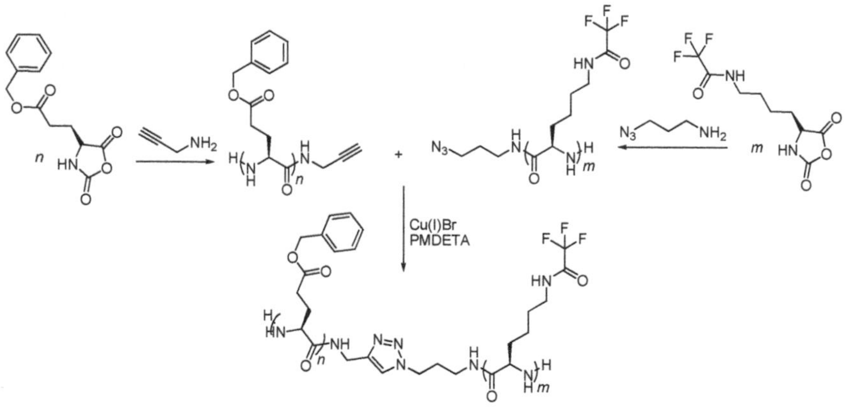

SCHEME 4-17   Synthesis of diblock copolypeptides using click chemistry.

4.4.2.2 BLOCK POLYPEPTIDE HYBRIDS

Gallot et al. carried out the first synthesis of hybrid polypeptides in the mid-1970s.[60] Since then a plethora of hybrid polypeptides consisting of a peptide block and one or more nonpeptide block(s) were synthesized. The majority of these synthetic approaches included two-step preparation of the polypep-tides. First step involves synthesis of the nonpeptide block(s) employing various controlled polymerization techniques in such a way that either one or both ends of the polymer chain would contain primary amine group or transition metal complex. These end-functionalized polymers were used as macro-initiator in the second step for ROP of NCA to install polypeptide block. In the following section, different synthetic strategies for preparation of various hybrid polymeric architectures is breifly described.

Poly(ethyelene oxide) (PEO) is perhaps the most widely used nonpep-tide block in hybrid polypeptide literature. The reasons of incorporating

PEO block are manifolds: PEO of different molecular weights are commercially available and preparation of either mono- or bi-functionalized amino end functionalized PEO, which can be used as macro-initiators for ROP of NCA, are well-established. A diblock hybrid polymer, which is composed of a hydrophilic PEO-block with a hydrophobic peptide block, shows amphiphilic character and often exhibits interesting core–shell morphology in water. Several amphiphilic A–B, A–B–A (where, A is the polypeptide block and B is PEO block) hybrid polypeptides were synthesized based on amino end group containing PEO macro-initiator (Table 4-2). Typical examples include PEO-b-PBLG,[61] PEO-b-poly(L-2-anthraquinonyl-alanine),[62] PEO-b-poly[(DL-Val)-co-(DL-Leu)],[63] and so on. Floudas and Papadopoulos carried out the synthesis of poly(γ-benzyl-L-glutamate)-poly(ethylene glycol)-poly(γ-benzyl-L-glutamate) (PBLG-b-PEG-b-PBLG) triblock copolymer using γ,ω-Diamino-PEOs as difunctional macroinitiators.[64] First the γ,ω-diamino poly(ethylene glycols) were prepared according to the synthetic scheme outlined in Scheme 4-18. Subsequently, the PEO-based macroinitiator was used for the ROP of γ-benzyl-L-glutamate N-carboxyanhydride (Bn-Glu NCA) to afford PBLG-b-PEG-b-PBLG triblock copolymers. Employing high vacuum technique Karatzas et al. synthesized triblock copolymer of PEO-b-PBLL-b-PBLG via sequential polymerization by using PEO-NH$_2$ as macroinitiator with a good control over the molecular weight and molecular weight distribution.[65] Well-defined diblock copolymer of poly(ethylene oxide)-block-poly(γ-benzyl-L-glutamate) (PEO-b-PBLG) and poly(ethylene oxide)-block-poly(γ-benzyl-L-aspartate) (PEO-b-PBLA) were synthesized by Lutz et al. using PEO-NH$_2$· HCl as the macroinitiator.[66] As shown by Schlaad and co-workers,[21] the presence of ammonium chloride salt at the chain end of the initiator facilitates NAM without other possible side-reactions and thus good control over molecular weight distribution is achieved.

Another interesting class of hybrid polypeptides includes polyester as the nonpeptide block. Aliphatic polyesters including polylactide, PLA and poly(ε-caprolactone), and PCL are particularly interesting as they are biodegradable, biocompatible, and exhibit excellent mechanical properties. When these polyesters are combined with polypeptides, the resulting hybrid displays modified stability and the semi-crystalline nature of polyesters also influences self-assembly property of the hybrid. These hybrid peptides also show high potential in drug- or gene-delivery applications, and this particular aspect of hybrid polymers will be discussed later in detail.

TABLE 4-2 Different Polymer Hybrid Block Copolymers Prepared From Amine Macroinitiator

Amine macroinitiator	Polypeptide segments (Architecture)
Polystyrene	PBLG(*AB*); PZLL(*AB*); PMDG(*ABA*)
Polybutadiene	PZLL(*AB, ABA*); PBLG (*AB, ABA*); PBL/DG (*ABA*); PML/DG (*ABA*)
Polyisoprene	PBLG(*ABA*)
Polydimethylsiloxane	PBLG(*AB, ABA*); Poly(L/D-Phe)(*AB*)
Polyethylene glycol	PZLL(*AB, ABA*); PBLG(*AB, ABA*); Poly(L-Pro)(*ABA*); PBLA(*AB, ABA*)
Polypropylene oxide	PBLG(*ABA*)
Poly(2-methyloxazoline)	PBLG(*AB*); Poly(L-Phe)(*AB*)
Poly(2-phenyloxazoline)	PBLG(AB); Poly(L-Phe)(AB)
Polymethyl methacrylate	PZLL(*AB*); PBLG(AB); PMLG(*AB*)
Polymethyl acrylate	PBLG(*AB*)
Polyoctenamer	PBLG(*ABA*)
Polyethylene	PBLG(*ABA*)
Polyferrocenylsilane	PBLG(*AB*)
Poly(9,9-dihexylfluorene)	PBLG(*ABA*)
Poly(ε-caprolactone)	BLG(*ABA*); Poly(L-Phe)(*ABA*); Poly(Gly)(*ABA*); Poly(L-Ala)(*ABA*)
Trimethyleneimine dendrimer	Poly(Sarcosine)(*Star dendrimer*)

The architectures are termed as AB (diblock) and ABA (triblock), where A is the polypeptide segment and B is the macroinitiator domain.

(**Source:** This table is adapted from the Ref. 16b and references therein.)

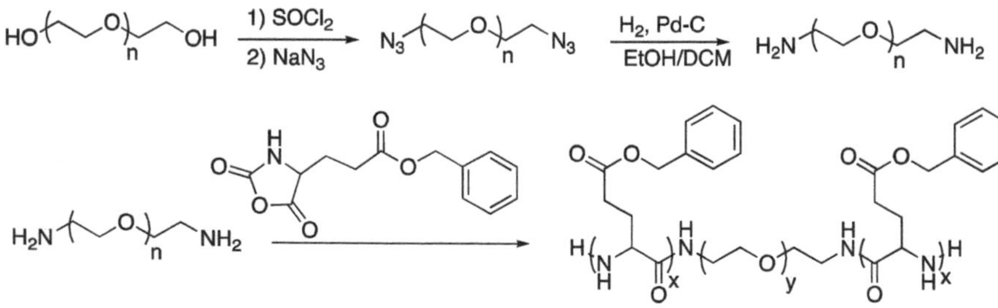

SCHEME 4-18 Synthesis of PBLG-*b*-PEG-*b*-PBLG.

The general synthetic strategy to install peptide block involves ROP of the corresponding NCA by amino-terminated polyester macroinitiator. The macroinitiator can be synthesized by following two different methods as suggested by Gotsche et al. (Scheme 4-19).[67] In the first method, reaction between N-*tert*-butoxycarbonyl-1-amino-3-propanol and Et$_2$Zn produced the corresponding Zn-alcoholate, which was used further as an initiator for ROP of L-lactide producing PLA. Subsequently, the amino group was deprotected using trifluoroacetic acid yielding the desired macroinitiator. The second method involved end capping of end-hydroxyl containing PLA with N-*tert*-butoxycarbonylphenylalanine followed by deprotection of amino group. Using this amino-terminated PLA macroinitiator, several hybrid polymers have been synthesized by reacting with NCAs of L-Ala, L-Phe, L-Leu, γ-Bzl-L-Glu, and γ-Bzl-L-Asp. Deng et al. reported the synthesis of a triblock copolymer of poly(ethylene glycol)-b-poly(L-lactide)-b-poly(L-glutamic acid) (PEG-b-PLLA-b-PLGA) by the ROP of NCA of γ-benzyl-L-glutamate (BLG-NCA) with PEG-b-PLLA-NH$_2$ as a macroinitiator.[68] This macroinitiator was synthesized by end-capping method employing a reaction between PEG-b-PLLA-OH and *tert*-Butoxycarbonyl-L-phenylalanine followed by deprotection of Boc group in presence of acid. PEG-b-PLLA-OH was prepared by the ROP of L-LA with PEG-monomethyl ether using stannous octoate as initiator.

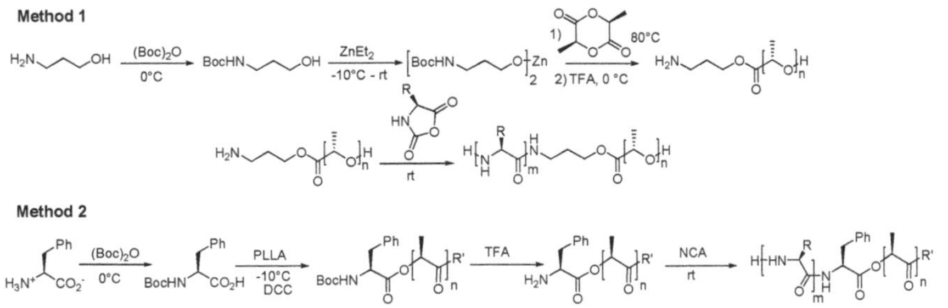

SCHEME 4-19 Synthesis of L-lactide based diblock copolymer.

Chen et al. synthesized biodegradable block copolymer of poly(ε-caprolactone)-b-poly(γ-benzyl- L-glutamic acid) (PCL-PBLGA) by ROP of NCA of γ-benzyl-L-glutamic acid with aminophenyl terminated PCL as a macroinitiator (Scheme 4-20).[69] This macroinitiator was obtained through catalytic hydrogenation of the corresponding 4-nitrophenethoxyl-teminated PCL, which was achieved through the ROP of ε-caprolactone (CL) using amino calcium 4-nitrobenzoxide as initiator. Kricheldorf and co-workers synthesized macrocyclic PCL using a seven membered cyclic initiator 2,2-dibutyl-2-stanna-1,3-dioxepane (Scheme 4-21).[70] Subsequently, the macrocycle was reacted with 4-nitrobenzoyl chloride producing 4-nitro-benzoyl-functionalized PCL, which was converted to the corresponding amino functionalities by catalytic hydrogenation. This macroinitiator was employed for the ROP of Gly-NCA, L-Ala-NCA, L-Phe-NCA, and γ-Bzl-L-Glu-NCA yielding various hybrid copolymers. It is noteworthy that despite the use of aromatic amine initiators, which are typically less reactive than the aliphatic amines, complete conversion was achieved for these hybrid polymers. An amphiphilic triblock copolymer composed of poly(ethylene glycol)-*block*-poly(L-lysine)-*block*-poly(ε-caprolactone) (PEO-*b*-PLLys-*b*-PCL) was synthesized by He et al. by combination of ROP of NCA, ROP of ε-CL and click chemistry (Scheme 4-22).[71] The amino termi-nated poly(L-lysine) was prepared via ROP of Nγ-carbobenzoxy-L-lysine *N*-carboxyanhydride, and subsequently used as a macroinitiator for ROP of ε-CL in presence of stannous octoate as a catalyst resulting poly(L-lysine)-*block*-poly(ε-caprolactone). The final triblock polymer was synthesized via copper-mediated click reaction between propargyl-terminated PLLys-*b*-PCL and azido-terminated poly(ethylene glycol) monomethyl ether (PEO-N₃).

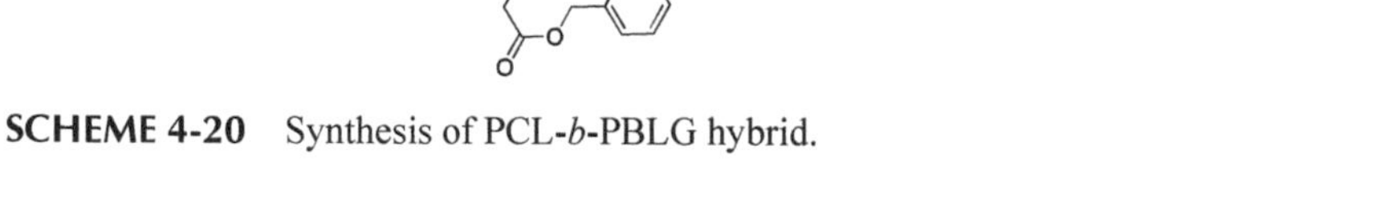

SCHEME 4-20　Synthesis of PCL-*b*-PBLG hybrid.

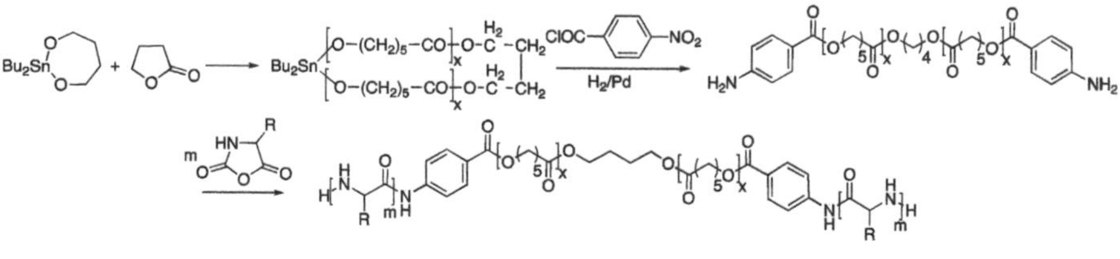

SCHEME 4-21　Synthesis of poly(peptide)-*b*-PCL-*b*-poly(peptide).

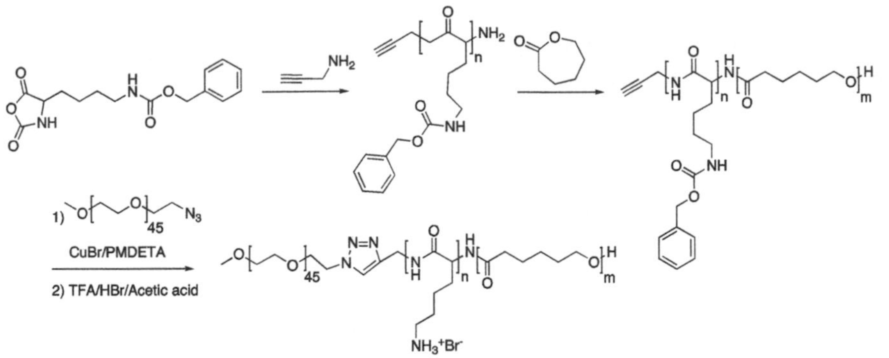

SCHEME 4-22 Synthesis of poly(ethylene glycol)-*block*-poly(L-lysine)-*block*-poly(ε-caprolactone).

The nonpeptide block can also be synthesized employing other methods such as anionic and controlled radical polymerization techniques. In this respect, Gallot and co-workers reported a diblock copolymer of poly(butadiene)-*b*-poly(γ-benzyl-L-glutamate).[31] Poly(butadiene) block was prepared by anionic polymerization followed by introduction of the amino group at the chain end using suitable chemistry. The end-amino group containing macroinitiator was subsequently used for the ROP of NCA of γ-benzyl-L-glutamate yielding the hybrid copolymer. In another report by Klok et al., oligo(styrene) was prepared by anionic polymerization using *sec*-BuLi as initiator.[72] The living end of the growing oligomer was chemically modified with 1-(3-chloropropyl)-2,2,5,5-tetramethyl-1-aza-2,5-disilacyclo-pentane followed by acidolysis installing an end-amino group (Scheme 4-23). This primary-amine terminated oligo(styrene) was subsequently used for the ROP of NCA of γ-benzyl-L-glutamate. Using similar approach Kim et al. synthesized poly(ferrocenyldimethylsilane)-*b*-poly(L-glutamic acid) (PFS-b-PLGA).[73] First dimethylsilaferrocenophane was polymerized anionically, and subsequently quenched with 1-(3-bromopropyl)-2,2,5,5-tetramethyl-1-aza-2,5-disilacyclopentane. The protecting group was removed in presence of methanol yielding amine-terminated polyferrocenylsilanes (PFS), which was used as macroinitiator for preparation of block copolymer. Nonpep-tide block having amino group at the end was also synthesized by classical free-radical polymerization technique. For example, Tanaka et al. synthe-sized poly(methyl methacrylate) employing azobisisobutyronitrile (AIBN) as initiator and 2-mercaptoethylammonium chloride as a chain-transfer reagent, which yielded amino-terminated macroinitiator for producing block copolymer with PBLG.[74] Similar strategy was used by Cheon et al.

for the synthesis of diblock copolymer composed of poly(N-isopropylacryl-amide) (PNIPAAM)-b-poly(γ-benzyl-L-glutamate) using amine-terminated PNIPAAM as a polymer initiator.[75]

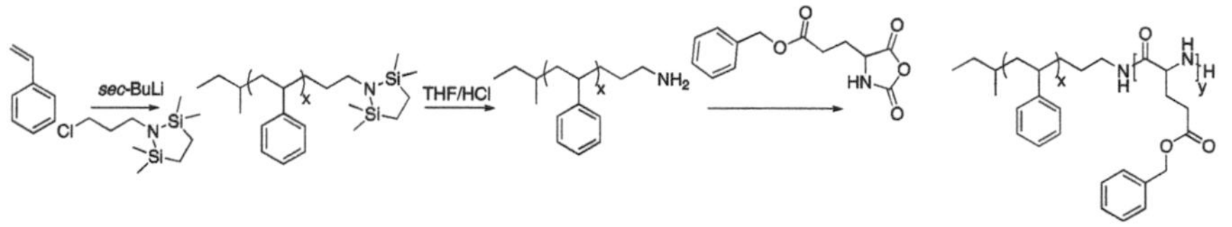

SCHEME 4-23   Synthesis of poly(styrene)-b-poly(γ-benzyl-L-glutamate).

Poly(oxazoline)-b-poly(peptide) is another interesting class of hybrid copolymer, which was synthesized by the ROP of NCAs using ω-amine-terminated poly(2-methyl-2-oxazoline) and poly(2-phenyl-2-oxazoline) macroinitiators.[76] These polymer initiators were prepared by cationic ROP of the corresponding 2-oxazoline-monomer using methyl p-toluenesulfonate as an initiator with subsequent termination with ammonia (Scheme 4-24). Using these two poly(oxazoline) macroinitiators, several hybrid copolymers consisting of poly(L-Phe), poly(γ-Bzl-L-Glu), or poly(glycosylated L-Ser) as the peptide block were prepared.

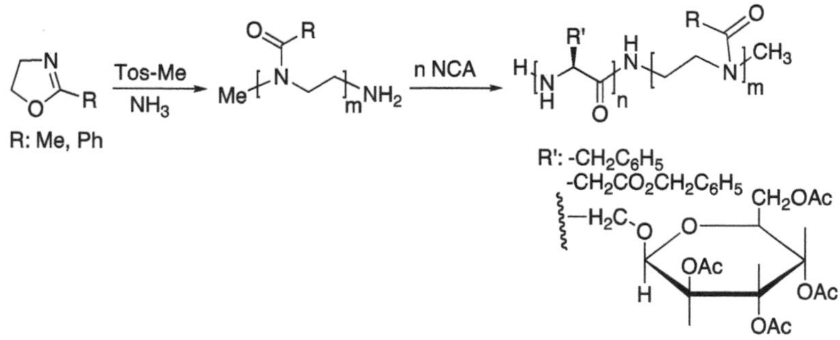

SCHEME 4-24   Synthesis of poly(oxazoline)-b-poly(peptide).

Various strategies have been adopted successfully to synthesize diverse hybrid polymers by combination of ROP of NCAs with other controlled radical polymerization techniques such as atom transfer radical polymer-ization (ATRP), nitroxide-mediated polymerization (NMP), and reversible addition-fragmentation chain transfer (RAFT). Well-defined nonpeptide blocks were prepared employing one of these techniques with amine-termi-nated chains, which were subsequently employed for ROP of NCAs. For example, Zhang et al. synthesized double-hydrophilic block copolymer of poly(L-glutamic acid)-*block*-poly(*N*- isopropylacrylamide) (PLGA-*b*-PNiPAM) by a combination of ROP of BLG-NCA and RAFT polymerization of *N*-isopropylacrylamide (NiPAM).[77] Two different strategies were employed for synthesizing this block copolymer. First strategy involves the synthesis of macro chain-transfer agent (CTA) by ROP of BLG-NCA affording the corresponding PLGA with RAFT active end group, which was then utilized for preparation of second PNiPAM block (Scheme 4-25). The second approach includes synthesis of PNiPAM block by RAFT polymerization, followed by Boc deprotection resulting in primary amine terminated block (Scheme 4.26). The amine-macroinitiator was subsequently used for ROP of BLG-NCA to form block polypeptides PNipam-*b*-PBLGA with good control over molecular weight and narrow molecular weight distribution. The removal of the benzyl group with HBr/glacial acetic acid furnished final double amphiphilic block copolymer (PNipam-*b*-PLGA). Recently, Heise et al. reported the synthesis of a library of polymer-polypeptide conjugates via ROP of NCA and RAFT polymerization techniques.[78] RAFT technique was employed to synthesize a series of polymers consisting of poly(*n*-butyl acrylate), polystyrene, and poly(*N*-isopropyl acrylamide) blocks with protected amine terminal, which after deprotection resulted in amino-end macroinitiators that were subsequently used for ROP of NCAs of γ-benzyloxycarbonyl-L-lysine or γ-benzyl-L-glutamate producing diblock hybrid polypeptides (Scheme 4-27).

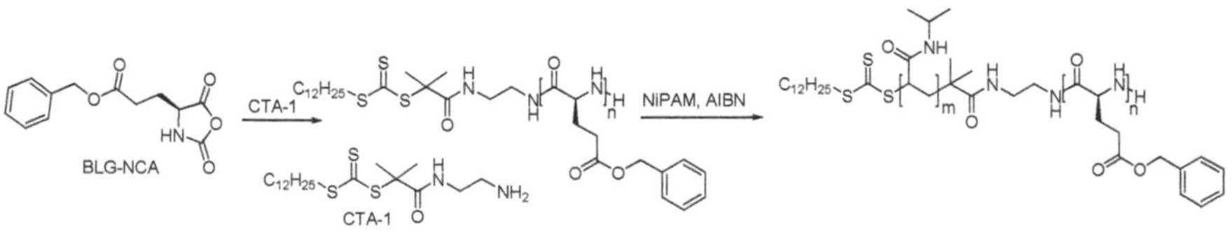

SCHEME 4-25 Synthesis of BCPs via macro-CTA route.

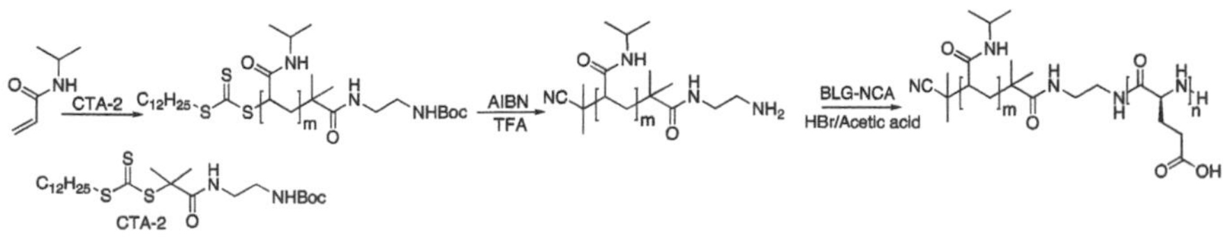

SCHEME 4-26 Synthesis of BCPs via macroinitiator route.

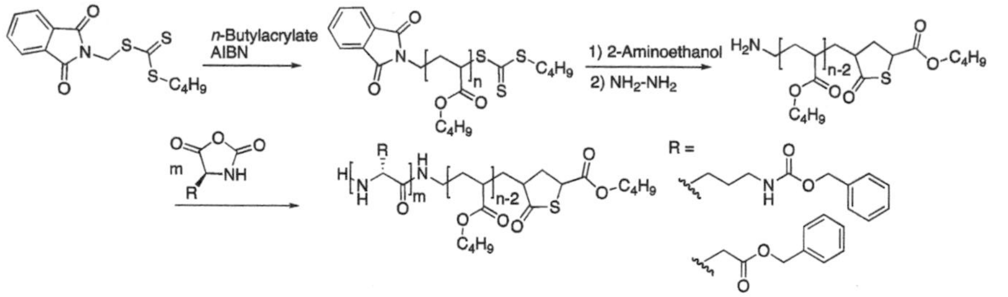

SCHEME 4-27   Synthesis of BCP combining ROP of NCA and RAFT.

Chaikof et al. synthesized glycopolymer-polypeptide hybrid composed of poly(L-alanine)-*b*-poly(2-acryloyloxyethyl-lactoside)-*b*-poly(L-alanine) via a combination of ATRP and ROP of NCAs.[79] The central glycopolymer block was synthesized by Cu-mediated ATRP of 2-*O*-acryloyl- oxyethoxyl-(2,3,4,6-*tetra-O*-acetyl-γ-D-galactopyranosyl)-(1-4)-2,3,6-*tri-O*-acetyl-γ-D-glucopyranoside, and both terminals of this glycopolymer were subsequently modified with diamino-end groups by suitable end-group transformation. The later was successively used as a macroinitiator for ROP of L-alanine *N*-carboxyanhydride yielding the final triblock copolymer. In another report, Chen et al. synthesized two graft copolymers, poly(L-glutamate)-*g*-oligo(2-(2-(2-methoxyethoxy)ethoxy)ethyl methacrylate) and poly(L-glutamic acid-*co*-(L-glutamate-g-oligo(2-(2-(2-methoxyethoxy)ethoxy)ethyl methacrylate))), via ROP of NCAs followed by ATRP of 2-(2-(2-methoxy-ethoxy)ethoxy)ethyl methacrylate (Scheme 4-28).[80]

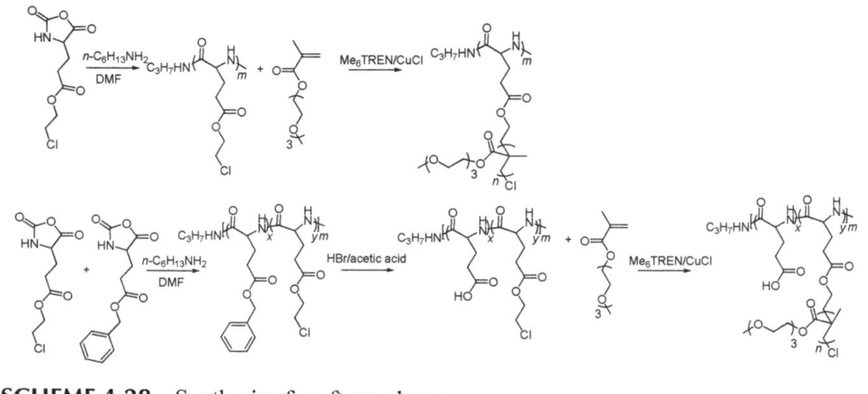

SCHEME 4-28   Synthesis of graft copolymer.

Ring opening methathesis polymerization (ROMP) was also successfully employed in combination with ROP of NCA to synthesize various hybrid copolymers. Cheng and co-workers synthesized a library of polypeptide-*b*-poly(oxa)norbornene-*b*-polypeptide triblock copolymers in two steps with well controlled molecular weights and narrow molecular weight distribution. The first step involves ROMP of a norbornene monomer containing *N*-trimethylsilyl (*N*-TMS) group at both the terminals. Subsequently, this *N*-TMS functionalized macromolecules were used as initiators for ROP of various NCAs yielding several well-defined hybrid copolymers.[81a] The same group also reported one-pot synthesis of brush-like hybrid copolymers containing polypeptide block as side chains and norbornene block as backbone by integrating ROMP with controlled ROP of NCAs.[81b]

Another approach that has drawn much attention recently is to employ transition metal initiators for preparation of hybrid block copolymers with structural precision. The detailed mechanism of ROP of NCAs using Ni(0) or Co(0) complex producing homopolypeptides has been discussed in Section 4.3. In the following section, we will focus on synthesis of several hybrid block copolymers using "living" amido-amidate metallocyle at the end of propagating nonpeptide chain as macroinitiator for subsequent ROP of NCAs. The difference from the conventional approach is that the active chain-end does not contain primary amine and thus overcomes the possibilities of side-reactions that are related to NAM. Moreover, the use of Ni(0) based initiators provided additional control over the polypeptide block. Deming and co-workers developed allyloxycarbonylaminoamides as universal precursors to amido-amidate nickelacycles, which can be successively used for ROP of NCAs producing variety of hybrid block copolymers (Scheme 4-29).[16a,b] Using this methodology, triblock copolymers composed of poly(γ-benzyl-L-glutamate)-*b*-polyoctenamer-*b*-poly(γ-benzyl-L-glutamate) and poly(-γ-benzyl-L-glutamate)-*b*-polyethylene-*b*-poly(γ-benzyl-L-glutamate) were synthesized in two steps by Deming group.[16c] The first step involved the synthesis of telechelic polyoctenamers prepared by acyclic diene metathesis (ADMET) polymerization using Grubbs' metathesis catalyst, followed by chemical modification of the chain-ends to γ,ω-bisamino-terminated poly-octenamer. The amido-amidate nickelacycles were incorporated onto amine-terminated polymers using [1,2-bis(diethylphosphino)ethane]Ni(COD), depeNi(COD) and subsequently used as difunctional macroinitiator for living ring opening polymerization of BLG-NCA producing triblock copolymers. Hydrogenation of the unsaturated block using Wilkinson's catalyst yielded the final poly(peptide)-*b*-polyethylene-*b*-poly(peptide) triblock copolymers with polyethylene segment at the middle (Scheme 4-30). In this case, since the active chain-ends of the "living" polypeptides contain amido-amidate nickelacycles, these ends can be capped, at least theoretically, by an electrophile such as isocyanate producing stable urea linkages between the capping reagent and the polypeptide block. This strategy was explored by Deming and co-workers to synthesize a series of pentablock copoly-mers composed of poly(ethylene glycol)-*b*-poly(γ-benzyl-L-glutamate)-*b*-(**polymer**)-*b*-poly(γ-benzyl-L-glutamate)-b-poly(ethylene glycol), where **polymer** = polyoctenamer, poly(ethylene glycol), or poly(dimethylsiloxane). In these cases, γ,ω-bisamino-terminated polymers (**polymer**) were used as difunctional macroinitiators for subsequent ROP of BLGA-NCA to form triblock copolymers, which were then end-capped with isocyanate-termi-nated poly(ethylene glycol) producing pentablock copolymers (Scheme 4-31).[16d] Recently, Li et al. reported molecular bottlebrush copolymers with polypeptide as backbone, polystyrene and poly(oligoethylene glycol methacrylate) as side chains by a combination of ROP of NCA followed by ATRP (Scheme 4-32). First Nε-2-bromoisobutyryl functionalized NCA of L-lysine was prepared and converted to the corresponding homopolypeptide poly(Br-L-lysine) using depeNi(COD) as the initiator. Afterwards, molec-ular bottlebrush copolymers were prepared containing polystyrene and poly(oligoethylene glycol methacrylate) as side chains using Cu-mediated ATRP polymerization.[16e]

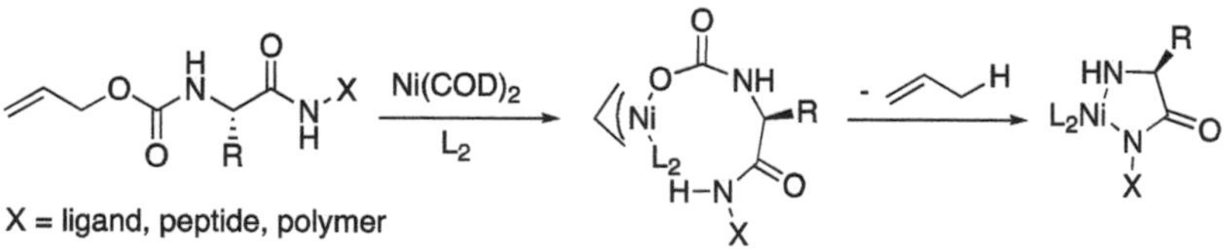

X = ligand, peptide, polymer

SCHEME 4-29 Amide-amidate nickelacycle macroinitiator for ROP of NCAs.

In the previous sections, various synthetic strategies for preparation of linear homo-, block-copolypeptides, and hybrid block copolypeptides based on ROP of NCAs are discussed. Apart from those structures, polypeptides can also adopt other architectures such as star-shaped polypeptides,[4b,82] (hyper)branched or dendritic polypeptides,[83] polypeptides showing liquid-crystalline behavior,[4b,84] surface-bound polypeptides,[4b] and polypeptides grafted on inorganic materials.[4d] All of these architectures are beyond the scope of this discussion; readers are referred to the respective references for further details.

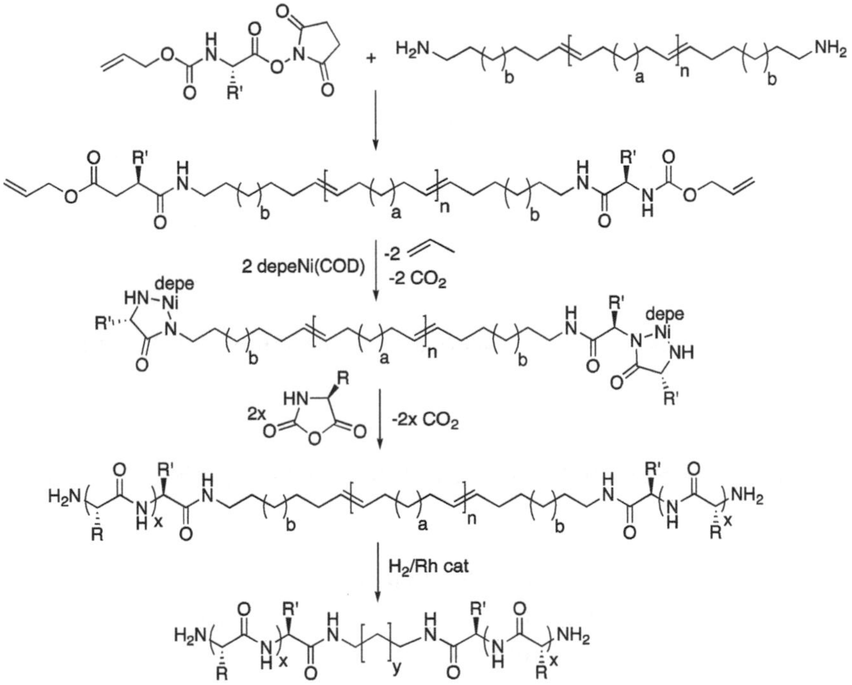

SCHEME 4-30 Synthesis of PBLG-*b*-polyethylene-*b*-PBLG triblock copolymers.

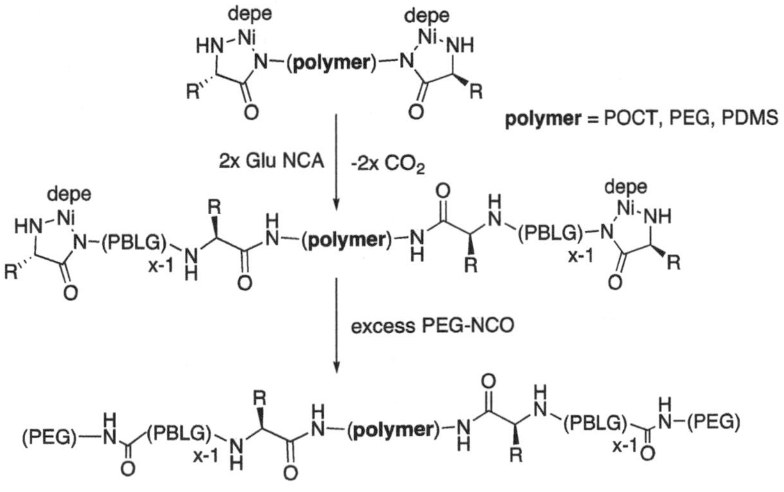

SCHEME 4-31 Synthesis of pentablock copolymers.

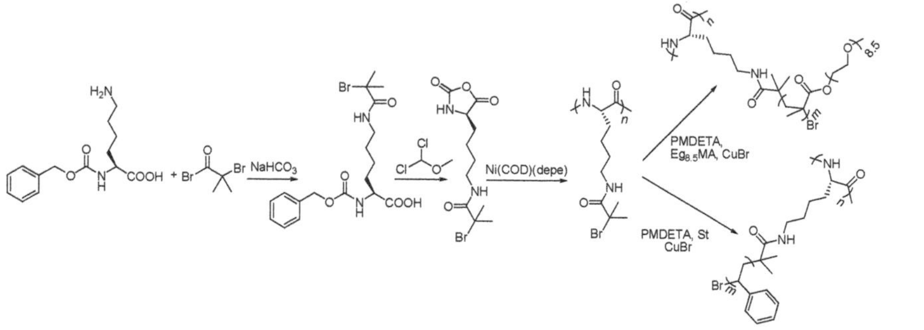

SCHEME 4-32 Synthesis of diblock copolymers by combination of ROP of NCA and ATRP.

4.5 SELF-ASSEMBLY OF BLOCK COPOLYPEPTIDES AND RELATED APPLICATIONS

The important feature of polypeptides that make them unique from other synthetic polymers is their ability to form stable well-organized assembly, which can be used for various biomedical applications. With this purpose, block copolypeptides of simple to complex architectures have been synthesized and the self-assembly of these polypeptides were studied in depth in the last two decades.[1,4d] In this regard, stimuli–responsive polypeptides, that is, the polypeptides those are capable of undergoing structural or conformational changes in response to external stimuli such as biologically relevant species (e.g., protein, enzyme, etc.), pH, light, temperature, magnetic field, redox, and so on deserve special mention.[85] Typically, ROP of NCAs initiated by either amine or metal catalyst produced these block copolymers of various architectures and their aggregates were used as polymeric nanocarriers for targeted drug/gene delivery applications. For example, amphiphilic polypeptides possessing hydrophobic core and hydrophilic corona in aqueous solution were able to encapsulate drugs, especially hydrophobic drugs, into their core, transport the drugs to targeted tissue/organ and release them in the presence of an external stimulus. In the following section, recent advancements of two stimuli responsive polypeptides, pH and redox, will be discussed in detail and readers are requested to consult with the following review articles for more elaborate discussion on this topic.[85]

Among all these stimuli-responsive polypeptides, pH-responsive polypeptides are particularly interesting and are well studied because of the presence of inherent pH gradient in body. For instance, it is known that the pH of extracellular tumor cells can be as low as 5.7 (on the average 6.8–7.0) compared to that of normal cell 7.4.[86] Thus developing a cargo that is stable at physiological pH but responds to acidic pH (i.e., the pH of cancerous cells) and releases encapsulated drug is highly desirable for targeted delivery application. Two different strategies have been employed for pH-responsive drug release studies: (1) dissociation of the polymeric nanocarrier and simultaneous release of encapsulated drug at low pH and (2) pH-triggered tunable swelling/deswelling of aggregates, which helps release of the drugs in sustained manner.[87] The second approach is advantageous to avoid any side effect that is caused by burst release of the drugs and also ensures the effective time of the drugs at certain target. Block polypeptides in which one of the blocks contains either glutamic acid[88] or lysine[89] as the pH responsive unit have been synthesized and these polymers exhibited reversible transitions in their self-assembly upon protonation/deprotonation. For example, Zhang and co-workers synthesized a series of pH-responsive amphiphilic block copolymers composed of poly(ethylene glycol)-b-poly(L-lysine)-b-poly(L-phenylalanine) (PEG-PLL-PLP), which self-assembled into micelles with PLP as hydrophobic core and PEG and PLL as hydrophilic shell.[89b] These aggregates exhibited pH-dependent swelling/deswelling due to protonation/deprotonation of the amino groups in PLL blocks; and thus pH-regulated sustained release of the encapsulated drugs was realized. Similar working principle was employed for hybrid polypeptides where poly(glutamic acid) was present as the pH-responsive block. For instance, Liu et al. synthesized star block-copolymers consisting of a hyperbranched polyethylenimine (PEI) core, a poly(L-glutamic acid) (PLG) inner shell, and a poly(ethylene glycol) (PEG) outer shell [PEI-g-(PLG-b-PEG)] and used them as polymeric nanocarriers for cationic drugs such as doxorubicin hydrochloride (DOX) as a model anticancer drug.[90] At physiological pH, at which carboxylic acid groups of PLG remains as carboxylates, DOX could be effectively entrapped inside the core of the polymer due to electrostatic interaction and as pH of the solution gradually decreases, sustained release of DOX was attained from the core of the aggregates. Lecommandoux and Rodríguez-Hernández synthesized poly(L-glutamic acid)-b-poly(L-lysine) (PGA-b-PLys) by sequential ROP of the respective NCA-monomers and investigated self-assembly behavior of the final diblock copolypeptide as a function of pH.[91] Since at acidic pH (<4) PLys block was charged and PGA block was neutralized, the conformation of PGA block changed from charged coil to compact α-helical structure; as a result the block copolypeptide aggregates exhibited vesicular morphology with PLys block at the shell and PGA block at the core of the aggregates (Figure 4-2). At basic pH (>10), PGA block was charged and remained solvated while the PLys block became neutral and hydrophobic, forming the core of the vasicular aggregates. At intermediate pH (5<pH<9), both the blocks were charged and no aggregates formation was observed.

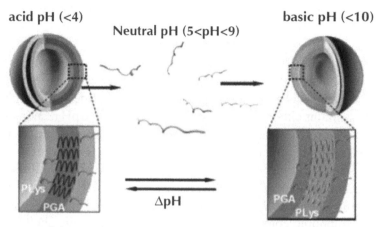

FIGURE 4-2   Self-assembly of poly(L-glutamic acid)-*b*-poly(L-lysine) (PGA-*b*-PLys) at different pH solutions.
(**Source:** Reprinted with permission from Rodríguez-Hernández, J.; Lecommandoux, S. *J. Am. Chem. Soc.* **2005,** *127,* 2026. Copyright (2005) American Chemical Society.)

Recently, Hammond and co-workers efficiently utilized Cu(I)-catalyzed click reaction as a tool for post-polymerization functionalization of a pre-polymer poly(α-propargyl L-glutamate) (PPLG), which was synthesized by ROP of corresponding NCA, to introduce thermo,[92] pH,[93] and dual[94] responsive units. For example, PPLG backbone was grafted with a combination of oligo(ethylene glycol) (as temperature responsive unit) and tertiary amine groups (as pH responsive unit) to produce dual responsive polypeptide (Scheme 4-33).

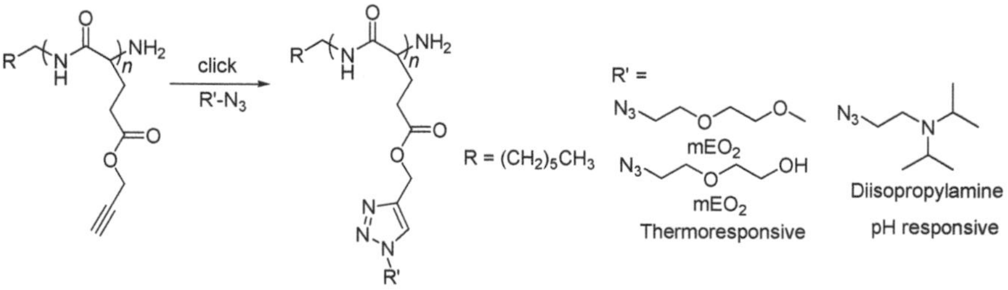

SCHEME 4-33   Postpolymerization modification of poly(propargyl-L-glutamate) with pH and thermo-responsive groups by click chemistry.

Another interesting stimuli responsive hybrid copolypeptide that has drawn much interest in recent time is block copolypeptides containing disulfide bond, a redox sensitive unit, as a part of polymer structure.[85,95] This disulfide bond cleaves in reductive environment (e.g., in intracellular environment where concentration of glutathione (GSH), a thiol-containing tripeptide, is higher (~10 mM) compared to that of extracellular environment (~2 μM)), providing an opportunity for designing thiol-responsive nanocarriers for drug or gene delivery applications.[96] Several researchers have developed series of block copolypeptides containing redox active disulfide bond as delivery vehicles for anticancer drugs. Most of these block copolymers are composed of poly(amino acid) as the hydrophobic core, which serves as the reservoir for hydrophobic drugs, linked by S-S bond with poly(ethylene glycol) (PEG) block that acts as the corona, which enables long circulation time of the encapsulated drug in blood stream. This disulfide bond is cleaved in presence of either dithiothreitol (DTT) or GSH resulting targeted release of the encapsulated drug (Figure 4-3).[97] These redox active block copolypeptides were synthesized by ROP of corresponding NCA monomer using disulfide containing PEG-amine macroinitiator (Scheme 4-34).

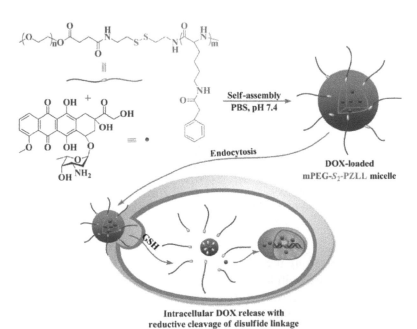

FIGURE 4-3 Schematic illustration of DOX loading and targeted intracellular release. (**Source:** Reprinted with permission from Ding, J.; Chen, J.; Li, D.; Xiao, C.; Zhang, J.; He, C.; Zhuang, X.; Chen, X. *J. Mater. Chem. B* **2013,** *1*, 69. Copyright (2013) RSC Publishing.)

Kataoka et al. developed an interesting method of preparing mono-dispersed polyion complex (PIC) micelles by self assembly of redox-sensitive block copolymers PEG-SS-P[Asp(DET)] and PEG-SS-P(Asp), where P[Asp(DET)]=poly([*N*-(2-aminoethyl)-2-aminoethyl]-γ,ω-aspartamide and P(Asp) = poly(γ,ω-aspartic acid) (Figure 4-4).[98] In this micellar aggregate, the PIC serves as the core while hydrophilic PEG block serves as the corona. Reduction of the disulfide bond by DTT causes detachment of PEG units from micellar aggregates leading to only homopolymer complex; as a result of which micelle-to-vesicle morphological transition was observed. This method of preparing nanocapsules from PEG-detachable PIC micelles opened a new avenue for developing nanocarriers for drug and gene delivery applications.[99]

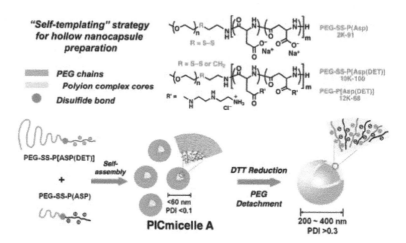

FIGURE 4-4 Schematic illustration of the preparation of hollow nanocapsules by self-templating strategy. Upon addition of reducing agent DTT into hetero-PEG-detachable PIC micelle A solution, a morphology transition occurred.
(**Source:** Reprinted with permission from Dong, W.-F.; Kishimura, A.; Anraku, Y.; Chuanoi, S.; Kataoka, K. *J. Am. Chem. Soc.* **2009**, *131*, 3804. Copyright (2009) American Chemical Society.)

Another report from the same group describes the balance between cationic charge and disulfide cross-linking densities could play huge role in targeted delivery, stability in extracellular environment and efficient release of entrapped plasmid DNA (pDNA) into intracellular medium.[100] To understand the effect of charge and disulfide cross-linking on delivery efficiency of catiomer polyplex, poly(ethylene glycol)-*b*-poly(L-lysine) (PEG-*b*-PLL) was thiolated using two different thiolation reagents N-succinimidyl

3-(2-pyridyldithio)propionate (SPDP) and 2-iminothiolane. Since thiolation by neutral SPDP decreased the charge on PLL block, while overall charge remained invariant after thiolation by 2-iminothiolane; the former system exhibited better transfection efficiency and efficient release of pDNA compared to the second system. Another PEG-detachable catiomer containing redox active disulfide bond between PEG and PLL block was reported by Cai et al. for efficient pDNA delivery.[101] The mPEG-SS-PLL/DNA complex exhibits excellent extracellular stability and efficient release of pDNA into intracellular environment upon detachment of PEG block in presence of GSH. Recently, Shi et al. reported a dual stimulus responsive catiomer system based on mPEG-SS-PLL$_{15}$-glutaraldehyde star (mPEG-SS-PLL15-star) for efficient gene delivery.[102] In reducing environment redox-sensitive disulfide bond cleaves causing detachment of outer PEG shell and intercellular uptake of mPEG-SS-PLL15-star/DNA, while acid-induced dissociation of the imine group promotes release of pDNA into targeted cells.

4.6   CONCLUSION AND FUTURE DIRECTION

Last two decades have observed tremendous progress in controlled ring opening polymerization of NCA, which result in a plethora of new synthetic precision polypeptides with structural diversity. Combined with other controlled polymerization techniques, such as ATRP, RAFT, ROMP, NMRP, and so on a series of hybrid block copolypeptides consisting of a peptide block and one or more nonpeptide block(s) are synthesized. Several interesting polypeptide architectures such as block, multi-block, star-shaped, brush-like, dendritic, and so on are evolved with precise control over the rate of polymerization, monomer conversion and end-group fidelity. In many cases, these synthetic polypeptides are found to form various types of polymer aggregates such as micelles, vesicles, nano-tube, nano-rod, and so on. When these synthetic polypeptides are combined with desired functional groups by various newly developed conjugation chemistry, a new class of functional synthetic polypeptides with tunable properties is achieved. As a result, a series of smart polypeptides that are responsive towards external stimuli such as pH, redox, light, and magnetic field are developed and their utility in various biomedical applications, for example, drug or gene delivery, are investigated. Although huge progress is achieved for last two decades, in future this field is expected to witness evolution of more complex architectures and development of next-generation smart polypeptide-based biomaterials that will be useful for biomedical and pharmaceutical applications.

REFERENCES

1. (a). Sun, J.; Chen, X.; Wei, J.; Yan, L.; Jing, X. *J. Appl. Polym. Sci.* **2010**, *118*, 1738. (b). Deming, T. J. *Adv. Mater.* **1997**, *9*, 299. (c). Deming, T. J. *Prog. Polym. Sci.* **2007**, *32*, 858–875.
2. Merrifield, R. B. *J. Am. Chem. Soc.* **1963**, *14*, 2149.
3. (a). Leuchs, H. *Chem. Ber.* **1906**, *39*, 857. (b). Leuchs, H.; Manasse, W. *Chem. Ber.* **1907**, *40*, 3235. (c). Leuchs, H.; Geiger, W. *Chem. Ber.* **1908**, *41*, 1721.
4. (a). Kricheldorf, H. R. *a-Amino acid N-Carboryanhydrides and Related Heterocycles.* Springer-Verlag: Berlin, **1987**; p. 22. (b). Hadjichristidis, N.; Iatrou, H.; Pitsikalis, M.; Sakellariou, G. *Chem. Rev.* **2009**, *109*, 5528. (c). Kricheldorf, H. R. *Angew. Chem. Int. Ed.* **2006**, *45*, 5752. (d). Habraken, G. J. M.; Heise, A.; Thornton, P. D. *Macromol. Rapid Commun.* **2011**, *33*, 272.
5. Zaoral, M.; Rudinger, J. *Collect. Czech. Chem. Commun.* **1961**, *26*, 2316.
6. (a). Daly, W. H.; Poche, D. *Terahedront. Lett.* **1988**, *29*, 5859. (b). Katakai, R.; Iizuka, Y. *J. Org. Chem.* **1985**, *50*, 715.
7. Nagai, A.; Sato, D.; Ishikawa, J.; Ochiai, B.; Kudo, H.; Endo, T. *Macromolecules* **2004**, *37*, 2332.
8. Kricheldorf, H. R. *Makromol. Chem.* **1977**, *178*, 905.
9. Sim, T. B.; Rapoport, H. *J. Org. Chem.* **1999**, *64*, 2532.
10. (a). Kricheldorf, H. R. Angew. Chem. **1973**, *85*, 86; *Angew. Chem. Int. Ed. Engl.* **1973**, *12*, 73. (b). Kricheldorf, H. R.; Fehrle, M. *Chem. Ber.* **1974**, *107*, 3533. (c). Katakai, R. *J. Org. Chem.* **1975**, *40*, 2697. (d). Katakai, R.; Nakayama, Y. **1976**, *75*, 747. (e). Halstrom, J.; Brunfeldt, K.; Kovacs, K. *J. Physiol. Chem.* **1974**, *355*, 82.
11. (a). Fuller, W. D.; Cohen, M. P.; Shabankarch, M.; Blair, R. K.; Goodman, M.; Naider, F. R. *J. Am. Chem. Soc.* **1990**, *112*, 7414. (b). Xue, C.-B.; Naider, F. *J. Org. Chem.* **1993**, *58*, 350.
12. (a). Ballard, D.; Bamford, C. *Proc. R. Soc. (London)* **1954**, *A223*, 495. (b). Ballard, D.; Bamford, C.; Weymouth, F. *Proc. R. Soc. (London)* **1954/1955**, *A227*, 155. (c). Thunig, D.; Semen, J.; Elias, H. *Makromol. Chem.* **1977**, *178*, 603.
13. (a). Ballard, D.; Bamford, C. *J. Am. Chem. Soc.* **1957**, *79*, 2336. (b). Miller, E.; Fankuchen, I.; Mark, H. *J. Appl. Phys.* **1949**, *20*, 531.
14. (a). Goodmann, M.; Hutchison, J. *J. Am. Chem. Soc.* **1966**, *88*, 3627. (b). Goodmann, M.; Hutchison, J. *J. Am. Chem. Soc.* **1965**, *87*, 3524. (c). Katchalski, E.; Shalitin, Y.; Gehatia, M. *J. Am. Chem. Soc.* **1955**, *77*, 1925.
15. Kricheldorf, H. R.; Lossow, C. V.; Schwarz, G. *Macromol. Chem. Phys.* **2004**, *205*, 918.
16. (a). Deming, T. J.; Curtin, S. A. *J. Am. Chem. Soc.* **2000**, *122*, 5710. (b). Deming, T. J. *Adv. Polym. Sci.* **2006**, *202*, 1. (c). Brzezinska, K. R.; Deming, T. J. *Macromolecules* **2001**, *34*, 4348. (d). Brzezinska, K. R.; Curtin, S. A.; Deming, T. J. *Macromolecules* **2002**, *35*, 2970. (e). Liu, Y.; Chen, P.; Li, Z. *Macromol. Rapid Commun.* **2012**, *33*, 287.
17. Aliferis, T.; Iatrou, H.; Hadjichristidis, N. *Biomacromolecules* **2004**, *5*, *1653*.
18. Deanna, L. Pickel, D. L.; Politakos, N.; Avgeropoulos, A.; Messman, J. M. *Macromolecules* **2009**, *42*, 7781.
19. Vayaboury, W.; Giani, O.; Cottet, H.; Deratani, A.; Schué, F. *Macromol. Rapid Commun.* **2004**, *25*, 1221.
20. Vayaboury, W.; Giani, O.; Cottet, H.; Bonaric, S.; Schué, F. *Macromol. Chem. Phys.* **2008**, *209*, 1628.

21. (a). Dimitrov, I.; Schlaad, H. *Chem. Commun.* **2003**, 2944. (b). Matthias Meyer, M.; Schlaad, H. *Macromolecules* **2006**, *39*, 3967.
22. (a). Lu, H.; Cheng, J. *J. Am. Chem. Soc.* **2007**, *129*, 14114. (b). Lu, H.; Cheng, J. *J. Am. Chem. Soc.* **2008**, *130*, 12562.
23. Zou, J.; Fan, J.; He, X.; Zhang, S.; Wang, H.; Wooley, K. L. *Macromolecules* **2013**, *46*, 4223.
24. Yu, M.; Deming, T. J. *Macromolecules* **1998**, *31*, 4739.
25. Hayashi, S.; Ohkawa, K.; Yamamoto, H. *Macromol. Biosci.* **2006**, *6*, 228.
26. (a). Yu, M.; Hwang, J.; Deming, T. J. *J. Am. Chem. Soc.* **1999**, *121*, 5825. (b). Wyrsta, M. D.; Cogen, A. L.; Deming, T. J. *J. Am. Chem. Soc.* **2001**, *123*, 12919.
27. Hernández, J. R.; Klok, H.-A. *J. Polym. Sci., Part A: Polym. Chem.* **2003**, *41*, 1167.
28. (a). Higuchi, M.; Takizawa, A.; Kinoshita, T.; Tsujita, Y.; Okochi, K. *Macromolecules* **1990**, *23*, 361. (b). Minoura, N.; Higuchi, M. *Macromolecules* **1997**, *30*, 1023.
29. Higashi, N.; Koga, T.; Niwa, M. *Langmuir* **2000**, *16*, 3482.
30. Aoi, K.; Tsutsumiuchi, K.; Okada, M. *Macromolecules* **1994**, *27*, 875.
31. (a). Guillermain, C.; Gallot, B. *Liq. Cryst.* **2002**, *29*, 141. (b). Guillermain, C.; Gallot, B. *Macromol. Chem. Phys.* **2002**, *203*, 1346–1356.
32. (a). Gitsas, A.; Floudas, G.; Mondeshki, M.; Spiess, H. W.; Aliferis, T.; Iatrou, H.; Hadjichristidis, N. *Macromolecules* **2008**, *41*, 8072. (b). Cha, J. N.; Stucky, G. D.; Morse, D. E.; Deming, T. J. *Nature* **2000**, *403*, 289. (c). Habraken, G. J. M.; Peeters, M.; Dietz, C. H. J. T.; Koning, C. E.; Heise, A. *Polym. Chem.* **2011**, *1*, 514.
33. Habraken, G. J. M.; Wilsens, C. H. R. M.; Koning, C. E.; Heise, A. *Polym. Chem.* **2011**, *2*, 1322.
34. Zhang, X.; Oddon, M.; Giani, O.; Monge, S.; Robin, J.-J. *Macromolecules* **2010**, *43*, 2654.
35. (a). Hanski, S.; Houbenov, N.; Ruokolainen, J.; Chondronicola, D.; Iatrou, H.; Hadjichristidis, N.; Ikkala, O. *Biomacromolecules* **2006**, *7*, 3379. (b). Karatzas, A.; Iatrou, H.; Hadjichristidis, N.; Inoue, K.; Sugiyama, K.; Hirao, A. *Biomacromolecules* **2008**, *9*, 2072.
36. Deming, T. J. *Nature* **1997**, *390*, 386.
37. Gkikas, M.; Iatrou, H.; Thomaidis, N. S.; Alexandridis, P.; Hadjichristidis, N. *Biomacromolecules* **2011**, *12*, 2396.
38. Cha, J. N.; Stucky, G. D.; Morse, D. E.; Deming, T. J. *Nature* **2000**, *403*, 289.
39. Habraken, G. J. M.; Peeters, M.; Dietz, C. H. J. T.; Koning, C. E.; Heise, A. *Polym. Chem.* **2011**, *1*, 514.
40. Papadopoulos, P.; Floudas, G.; Schnell, I.; Aliferis, T.; Iatrou, H.; Hadjichristidis, N. *Biomacromolecules* **2005**, *6*, 2352.
41. Agut, W.; Agnaou, R.; Lecommandoux, S.; Taton, D. *Macromol. Rapid Commun.* **2008**, *29*, 1147.
42. Goury, V.; Jhurry, D.; Bhaw-Luximon, A.; Novak, B. M.; Belleney, J. Biomacromolecules **2005**, *6*, 1987.
43. (a). Nguyen, L.-Y. T.; Ardana, A.; Vorenkamp, E. J.; Brinke, G. ten.; Schouten, A. J. *Soft Matter* **2010**, *6*, 2774. (b). Nguyen, L.-Y. T.; Vorenkamp, E. J.; Daumont, C. J. M.; Brinke, G. Ten.; Schouten, A. J. *Polymer* **2010**, *51*, 1042.
44. Tang, H.; Zhang, D. *Polym. Chem.* **2011**, *2*, 1542.
45. Nowak, A. P.; Sato, J.; Breedveld, V.; Deming, T. J. *Supramol. Chem.* **2006**, *18*, 423.
46. Li, Z.; Deming, T. J. *Soft Matter* **2010**, *6*, 2546.

47. Holowka, E. P.; Deming, T. J. *Macromol. Biosci.* **2010,** *10,* 496.
48. (a). Sulistio, A.; Widjaya, A.; Blencowe, A.; Zhang, X.; Qiao, G.; *Polym. Prepr. (Am. Chem. Soc., Div. Polym. Chem.)* **2010,** *51,* 121. (b). Sulistio, A.; Lowenthal, J.; Blencowe, A.; Bongiovanni, M. N.; Ong, L.; Gras, S. L.; Zhang, X.; Qiao, G. G. *Biomacromolecules* **2011,** *12,* 3469.
49. (a). Jan, J.-S.; Lee, S.; Carr, C. S.; Shantz, D. F. *Chem. Mater.* **2005,** *17,* 4310. (b). Sun, J.; Chen, X.; Deng, C.; Yu, H.; Xie, Z.; Jing, X. *Langmuir* **2007,** *23,* 8308.
50. Kramer, J. R.; Deming, T. J. *J. Am. Chem. Soc.* **2010,** *132,* 15068.
51. Gaspard, J.; Silas, J. A.; Shantz, D. F.; Jan, J.-S. *Supramol. Chem.* **2010,** *22,* 178.
52. Euliss, L. E.; Grancharov, S. G.; O'Brien, S.; Deming, T. J.; Stucky, G. D.; Murray, C. B.; Held, G. A. *Nano Lett.* **2003,** *3,* 1489.
53. Yu, M.; Nowak, A. P.; Deming, T. J.; Pochan, D. J. *J. Am. Chem. Soc.* **1999,** *121,* 12210.
54. Atmaja, B.; Cha, J. N.; Marshall, A.; Frank, C. W. *Langmuir* **2009,** *25,* 707.
55. Hanson, J. A.; Li, Z.; Deming, T. J. *Macromolecules* **2010,** *43,* 6268.
56. Holowka, E. P.; Sun, V. Z.; Kamei, D. T.; Deming, T. J. *Nat. Matter.* **2007,** *6,* 52.
57. Gibson, M. I.; Cameron, N. R. *J. Polym. Sci., Part A: Polym. Chem.* **2009,** *47,* 2882.
58. Zhuang, W.; Liao, L.; Chen, H.; Wang, J.; Pan, Y.; Zhang, L.; Liu, D. *Macromol. Rapid Commun.* **2009,** *30,* 920.
59. Agut, W.; Agnaou, R.; Lecommandoux, S.; Taton, D. *Macromol. Rapid Commun.* **2008,** *29,* 1147.
60. Perly, B.; Douy, A.; Gallot, B. *Makromol. Chem.* **1976,** *177,* 2569.
61. Choa, C.-S.; Nahb, J.-W.; Jeongc, Y.-I.; Cheonc, J.-B.; Asayamad, S.; Ised, H.; Akaiked. T. *Polymer* **1999,** *40,* 6769.
62. Matsubara, T.; Shinohara, H.; Sisido, M. Macromolecules **1997,** *30,* 2651.
63. (a). Cho, I.; Kim, J.-B.; Jung, H.-J. *Polymer* **2003,** *44,* 5497. (b). Cho, C.-S.; Kim, S.-W. *Makmmol. Chem.* **1990,** *191,* 981.
64. Floudas, G.; Papadopoulos, P. *Macromolecules* **2003,** *36,* 3673.
65. Karatzas, A.; Bilalis, P.; H. Iatrou, Pitsikalis, M.; Hadjichristidis. N. *React. Funct. Polym.* **2009,** *69,* 435.
66. Lutz, J.-F.; Schütt, D.; Kubowicz, S. *Macromol. Rapid Commun.* **2005,** *26,* 23.
67. Gotsche, M.; Keul, H.; Höcker, H. *Macromol. Chem. Phys.* **1995,** *196,* 3891.
68. Deng, C.; Rong, G.; Tian, H.; Tang, Z.; Chen, X.; Jing, X. *Polymer* **2005,** *46,* 653.
69. Rong, G.; Deng, M.; Deng, C.; Tang, Z.; Piao, L.; Chen, X.; Jing, X. *Biomacromolecules* **2003,** *4,* 1800.
70. (a). Kricheldorf, H. R.; Hauser, K. *Biomacromolecules* **2001,** *2,* 1110. (b). Kricheldorf, H. R.; Hauser, K. *Macromolecules* **1998,** *31,* 614.
71. He, X.; Zhong, L.; Wang, K.; Luo, S.; Xie, M. *J. Appl. Polym. Sci.* **2010,** *117,* 302.
72. Klok, H.-A.; Langenwalter, J. F.; Lecommandoux, S. *Macromolecules* **2000,** *33,* 7819.
73. Kim, K. T.; Vandermeulen, G. W. M.; Winnik, M. A.; Manners, I. *Macromolecules* **2005,** *38,* 4958.
74. Tanaka, M.; Mori, A.; Imanishi, Y.; Bamford, C. H. *J. Biol. Macromol.* **1985,** *7,* 173.
75. Cheon, J.-B.; Jeong, Y.; Cho, C.-S. *Polymer* **1999,** *40,* 2041.
76. (a). Tsutsumiuchi, K.; Aoi, K.; Okada, M. *Macromolecules* **1997,** *30,* 4013. (b). Naka, K.; Yamashita, R.; Nakamura, T.; Ohki, A.; Maeda, S.; Aoi, K.; Tsutsumiuchi, K.; Okada, M. *Macronol. Chem. Phys.* **1997,** *198,* 89.
77. Zhang, X.; Li, J.; Li, W.; Zhang, A. *Biomacromolecules* **2007,** *8,* 3557.
78. Jacobs, J.; Gathergood, N.; Heise, A. *Macromol. Rapid Commun.* **2013,** *34,* 1325.

79. Dong, C.-M.; Sun, X.-L.; Faucher, K. M.; Apkarian, R. P.; Chaikof, E. L. *Biomacromolecules* **2004**, *5*, 224.

80. Ding, J.; Xiao, C.; Zha, L.; Cheng, Y.; Ma, L.; Tang, Z.; Zhuang, X.; Chen, X. *J. Polym. Sci., Part A: Polym. Chem.* **2011**, *49*, 2665.

81. (a). Lu, H.; Wang, J.; Lin, Y.; Cheng, J. *J. Am. Chem. Soc.* **2009**, *131*, 13582. (b). Bai, Y.; Lu, H.; Ponnusamy, E.; Cheng, J. *Chem. Commun.* **2011**, *47*, 10830.

82. (a). Karatzas, A.; Bilalis, P.; Iatrou, H.; Pitsikalis, M.; Hadjichristidis, N. *React. Funct. Polym.* **2009**, *69*, 435. (b). Abraham, S.; Ha, C.-S.; Kim, I. *J. Polym. Sci., Part A: Polym. Chem.* **2006**, *44*, 2774. (c). Babin, J.; Leroy, C.; Lecommandoux, S.; Borsali, R.; Gnanou, Y.; Taton, D. *Chem. Commun.* **2005**, 1993.

83. (a). Klok, H.-A.; Hernadez, J. R. *Macromolecules* **2002**, *35*, 8718. (b). Hernadez, J. R.; Gatti, M.; Klok, H.-A. *Biomacromolecules* **2003**, *4*, 249. (c). Lbbert, A.; Nguyen, T. Q.; Sun, F.; Sheiko, S. S.; Klok, H.-A. *Macromolecules* **2005**, *38*, 2064. (d). Harada, A.; Kawamura, M.; Matsuo, T.; Takahashi, T.; Kono, K. *Bioconjugate Chem.* **2006**, *17*, 3. (e). Harada, A.; Nakanishi, K.; Ichimura, S.; Kojima, C.; Kono, K. *J. Polym. Sci., Part A: Polym. Chem.* **2009**, *47*, 1217.

84. Gallot, B. *Prog. Polym. Sci.,* **1996**, *21*, 1035.

85. Huang, J.; Heise, A. *Chem. Soc. Rev.* **2013**, *42*, 7373.

86. (a). Lee, E. S.; Gao, Z.; Bae, Y. H. *J. Controlled Release* **2008**, *132*, 164; (b). Ganta, S.; Devalapally, H.; Shahiwala, A.; Amiji, M. *J. Controlled Release* **2008**, *126*, 187. (c). Mellman, I.; Fuchs, R.; Helenius, A. *Annu. Rev. Biochem.* **1986**, *55*, 663.

87. Felber, A. E.; Dufresne, M. H.; Leroux, J. C. *Adv. Drug Delivery Rev.* **2012**, *64*, 979.

88. (a). Tian, H.; Chen, X.; Lin, H.; Deng, C.; Zhang, P.; Wei, Y.; Jing, X. *Chem.-Eur. J.* **2006**, *12*, 4305. (b). Huang, H. H.; Li, J. Y.; Liao, L. H.; Li, J. H.; Wu, L. X.; Dong, C. K.; Lai, P. B.; Liu, D. J. *Eur. Polym. J.,* **2012**, *48*, 696. (c). Kim, M. S.; Dayanand, K.; Choi, E. K.; Park, H. J.; Kim, J. S.; Lee, D. S. *Polymer*, **2009**, *50*, 2252. (d). Knoop, R. J. I.; Geus, M. de.; Habraken, G. J. M.; Koning, C. E.; Menzel, H.; Heise, A. *Macromolecules* **2010**, *43*, 4126.

89. (a). Yan, Y. S.; Li, J. Y.; Zheng, J. H.; Pan, Y.; Wang, J. Z.; He, X. Y.; Zhang, L. M.; Liu, D. J. *Colloids Surf., B* **2012**, *95*, 137. (b). Li, Y.-Y.; Hua, S.-H.; Xiao, W.; Wang, H.-Y.; Luo, X.-H.; Li, C.; Cheng, S.-X.; Zhang, X.-Z.; Zhuo, R.-X. *J. Mater. Chem.* **2011**, *21*, 3100. (c). Wang, C.; Kang, Y. T.; Liu, K.; Li, Z. B.; Wang, Z. Q.; Zhang, X. *Polym. Chem.* **2012**, *3*, 3056. (d). Yuan, R. X.; Shuai, X. T. *J. Polym. Sci., Part B: Polym. Phys.* **2008**, *46*, 782. (d). Krannig, K.-S.; Schlaad, H. *J. Am. Chem. Soc.* **2012**, *134*, 18542.

90. Huang, H.; Li, J.; Liao, L.; Li, J.; Wu, L.; Dong, C.; Lai, P.; Liu, D. *Eur. Polym. J.* **2012**, *48*, 696.

91. Rodríguez-Hernández, J.; Lecommandoux, S. *J. Am. Chem. Soc.* **2005**, *127*, 2026.

92. Cheng, Y.; He, C.; Xiao, C.; Ding, J.; Zhuang, X.; Chen, X. *Polym. Chem.* **2011**, *49*, 2665.

93. Engler, A. C.; Bonner, D. K.; Buss, H. G.; Cheung, E. Y.; Hammond, P. T. *Soft Matter* **2011**, *7*, 5627.

94. Chopko, C. M.; Lowden, E. L.; Engler, A. C.; Griffith, L. G.; Hammond, P. T. *ACS Macro Lett.* **2012**, *1*, 727.

95. (a). Raina, S.; Missiakas, D. *Annu. Rev. Microbiol.* **1997**, *51*, 179. (b). Meng, F.; Hennink, W. E.; Zhong, Z. *Biomaterials* **2009**, *30*, 2180.

96. (a). Elias, S. J.; Arner, A. H.; *Eur. J. Biochem.* **2000**, *267*, 6102. (b). Townsend, D. M.; Tew, K. D.; Tapiero, H. *Biomed. Pharmacother.* **2003**, *57*, 145. (c). Go, Y. M.; Jones, D. P. *Biochim. Biophys. Acta* **2008**, *1780*, 1273.

97. (a). Ding, J.; Chen, J.; Li, D.; Xiao, C.; Zhang, J.; He, C.; Zhuang, X.; Chen, X. *J. Mater. Chem. B* **2013**, *1*, 69. (b). Ren, T.; Xia, W.-J.; Dong, H.-Q.; Li, Y.-Y. *Polymer* **2011**, *52*, 3580. (c). Koo, A. N.; Lee, H. J.; Kim, S. E.; Chang, J. H.; Park, C.; Kim, C.; Park, J. H.; Lee, S. C. *Chem. Commun.* **2008**, 6570. (d). Thambi, T.; Yoon, H. Y.; Kim, K.; Kwon, I. C.; Chang Kyoo Yoo, C. K.; Park, J. H. *Bioconjugate Chem.* **2011**, *22*, 1924. (e). Wen, H.-Y.; Dong, H.-Q.; Wen-juan Xie, W.-J.; Yong-Yong Li, Y.-Y.; Wang, K.; Paulettic, G. M.; Shi, D.-L. *Chem. Commun.* **2011**, *47*, 3550.

98. Dong, W.-F.; Kishimura, A.; Anraku, Y.; Chuanoi, S.; Kataoka, K. *J. Am. Chem. Soc.* **2009**, *131*, 3804.

99. (a). Kakizawa, Y.; Harada, A.; Kataoka, K. *Biomacromolecules* **2001,** *2*, 491. (b). Matsumoto, S.; Christie, R. J.; Nishiyama, N.; Miyata, K.; Ishii, A.; Oba, M.; Koyama, H.; Yamasaki, Y.; Kataoka, K. *Biomacromolecules* **2009**, *10*, 119. (c). Christie, R. J.; Miyata, K.; Matsumoto, Y.; Nomoto, T.; Menasco, D.; Lai, T. C.; Pennisi, M.; Osada, K.; Fukushima, S.; Nishiyama N., Yamasaki, Y.; Kataoka, K. *Biomacromolecules* **2011**, *12*, 3174.

100. Miyata, K.; Kakizawa, Y.; Nishiyama, N.; Harada, A.; Yamasaki, Y.; Koyama, H.; Kataoka, K. *J. Am. Chem. Soc.* **2004**, *126*, 2355.

101. Cai, X.-J.; Hai-Qing Dong, H.-Q.; Xia, W.-J.; Wen, H.-Y.; Li, X.-Q.; Yu, J.-H.; Li, Y.-Y.; Shi, D.-L. *J. Mater. Chem.* **2011**, *21*, 14639.

102. Cai, X.; Dong, C.; Dong, H.; Wang, G.; Pauletti, G. M.; Pan, X.; Wen, H.; Mehl, I.; Li, Y.; Shi, D.-L. *Biomacromolecules* **2012**, *13*, 1024.

CHAPTER 5

ENZYMATIC AND BIOMIMETIC APPROACHES TO THE SYNTHESIS OF ELECTRICALLY CONDUCTING POLYMERS

SUBHALAKSHMI NAGARAJAN, JAYANT KUMAR, and RAMASWAMY NAGARAJAN

CONTENTS

ABSTRACT

With increasing emphasis on the development of sustainable routes to material syntheses, enzyme catalysis is rapidly emerging as a frontier tool for integration in research and industry. Enzymes catalyze a plethora of chemical transformations including polymerization reactions with unmatched specificity and selectivity. Enzyme catalysis permits the use of milder reaction conditions while maintaining high product yield and minimizing generation of by-products. This chapter reviews recent advances in the use of enzymes (oxidoreductases) and other metalporphyrins for the syntheses of conducting polymers (CP) based on aniline, pyrrole, and thiophene. The mechanism of catalysis of different enzymes belonging to the family of oxidoreductases is presented along with the role of redox potential in determining reaction feasibility. The use of aqueous and mixed solvent media and polyelectrolyte templates including biological molecules in enzymatic synthesis is presented. Strategies to influence several key reaction parameters such as temperature, nature of dopant, and pH are explained in detail, as these play a critical role in obtaining well-defined polymers with good processability and conductivity.

The high costs and limited stability of some of the naturally occurring enzymes has led to the use of modified biomimetics, which mimic the catalytic activity of the natural enzyme analogues. The synthesis and characterization of these modified biomimetics and the efficacy of these biomimetics as catalysts for the synthesis of CP are discussed. A fundamental understanding of the role of redox potential of enzymes, interaction between dopant and monomer is crucial for the development of the next generation of high performance, commercially viable CP using sustainable routes.

5.1 BIOCATALYSIS

5.1.1 ENZYMES

Enzymes catalyze biochemical reactions, critical to metabolism of all living cells, with unsurpassed efficiency under physiological conditions. Enzymes are proteins, which have evolved to perform efficiently under mild conditions required to preserve the functionality of the biological systems. Biocatalysis is the use of these naturally occurring substances for catalyzing a wide range of chemical transformations (including polymerization reactions) resulting in materials with interesting and tunable electrical, electronic, photonic, and electrochemical properties.

Enzymes have been tailored to perform exceptionally well under biological conditions. A major challenge is then to enable these catalysts to be stable and robust outside living systems. Over the last decade, advances in bioengineering have allowed for large-scale production and targeted modification of enzymes rendering them more stable for use under in vitro conditions.

Enzyme catalyzed synthesis is hence emerging as an attractive alternative to traditional chemical synthesis methodologies. This has been driven by increased knowledge on mechanism of enzymes catalysis and the imminent need to promote sustainable routes for synthesis of new materials including polymers. In addition, most of the enzyme-catalyzed reactions occur at ambient conditions in predominantly aqueous media. Enzymes are derived from renewable resources and are hence nontoxic, biocompatible, and biodegradable. The efficacy of enzymes in catalyzing polymerization reactions can have wide ramifications in terms of increasing energy efficiency, decreasing the number of steps in purification of the final product, and increasing the yield in a chemical reaction.

As a result, biocatalysis is a rapidly growing field with applications in polymers, medicine, and biology. The global market for industrial enzymes was valued at \$3.1 billion in 2009 and is expected to grow to \$6.0 billion in 2016.[1]

5.1:1.1 CLASSIFICATION OF ENZYMES

There are more than 2000 human enzymes known.[2] In order to better understand these enzymes, they are classified into six families, based on the type of chemical reaction they catalyze. The six families are further divided into several sub-classes as explained below.

i. **Oxidoreductases:** Oxidoreductases catalyze a variety of oxidation–reduction reactions. More than one-third of all documented enzymatic reactions are catalyzed by oxidoreductases.[3] There are 22 subclasses of oxidoreductases, which are used as catalysts in a number of chemical transformations.

ii. **Transferases:** Transferases catalyze the transfer of a functional moiety/group from one molecule to another. An important enzyme in this class is reverse transcriptase, which is responsible for the transcription of viral DNA to produce double stranded DNA. This enzyme is used in Reverse Transcriptase Polymerase Chain Reaction (RT-PCR) for amplification of genetic material in biological research and medical applications.[4]

iii. **Hydrolases**: Hydrolases are involved in hydrolysis of a number of chemically important bonds. Many of these enzymes are robust and stable at high temperatures and in nonaqueous media. Under proper conditions, hydrolases can catalyze esterification, transesterification, and amidation. Since some of these chemical reactions are routinely used for production of therapeutic drugs and commodity materials, these enzymes are of immense interest to both research community and industry. Lipases are an important class of enzymes belonging to this group, have been used in the food industry for flavor improvement of dairy products, in pulp manufacturing, and in cleansing of glass surfaces.[5]

iv. **Lyases**: Lyases catalyze reactions involving nonhydrolytic or nonoxidative cleavage. These typically catalyze elimination or addition reactions involving double bonds. An example of lyases is the enzyme nitrile hydratase used in the industrial production of acrylamide from acrylonitrile.[6]

v. **Isomerases:** Isomerases are known to catalyze various isomerization reactions including racemization. An important enzyme belonging to this class is "Glucose Isomerase" that is, used as a catalyst by the food industry for producing high fructose syrups from corn starch.[7]

vi. **Ligases:** Ligases catalyze coupling reactions between biomolecules/macromolecules. These enzymes are not used in large-scale industrial processes owing to their limited stability. However, some enzymes in this class are utilized in genetic engineering protocols.[8,9]

Hydrolases and oxidoreductases are the most widely used classes of enzymes clearly emphasizing the versatility of these catalysts.[10]

5.2   OXIDOREDUCTASES AS CATALYSTS FOR THE SYNTHESIS OF ADVANCED POLYMERIC MATERIALS

5.2.1   PEROXIDASES

Peroxidases are a sub-class of oxidoreductases, widely present in the plants, animals, and microorganisms. They catalyze the oxidation of a wide range of molecules including phenol and aniline by using peroxides as the substrate. This chapter focuses on use of peroxidases as catalysts for synthesis of conducting polymers (CP).

The most fascinating features of this class of enzymes are their ability to generate reactive species, which undergo nonenzymatic coupling resulting in the formation of well-defined polymers. Table 5-1 summarizes the different peroxidases, which have been used for the synthesis of CP. The synthesis and characterization of these CP are discussed in the following sections.

TABLE 5-1   Peroxidases Used for the Synthesis of Different Conducting Polymers (CP)

Commercially Available Oxidoreductases	Conjugated Polymers Synthesized
Horseradish peroxidase (HRP)	Polyaniline[11]
Soybean peroxidase (SBP)	Polypyrrole (PPy),[12] PEDOT,[13] PEDOT-co-Pyrrole[14]
Palm tree peroxidase	Polyaniline[15]
Bilirubin oxidase (BOD)	Polyaniline[16]
Laccase	Polyaniline,[17] PPy ,[18] PEDOT[19]
Glucose oxidase	Polyaniline,[20] PPy[21]
Lactate oxidase (LOD)	PPy[22]
Catalase	PEDOT[23]

The mechanism for peroxidase-catalyzed polymerization reactions is well known. As shown in Scheme 5-1, H_2O_2 oxidizes the native enzyme to an intermediate referred to as HRP-I. HRP-I in turn oxidizes the monomer by carrying out two sequential one-electron reduction steps via intermediate HRP-II. In one peroxidase cycle, two moles of radical cations are generated for every mole of hydrogen peroxide reduced to water. This catalytic cycle repeats itself until/till oligomers/polymers are formed. It has been postulated that after appreciable growth of the polymer chain, the polymer chain can no longer enter the active site of the enzyme, which terminates the polymerization reaction. However, the redox potential of a dimer is lower than that of the monomer while the redox potential of the trimer is lower than that of the dimer. This decrease in redox potential with each monomer unit being added to the growing chain facilitates chain growth even if the polymer can no longer enter the active site of the enzyme.

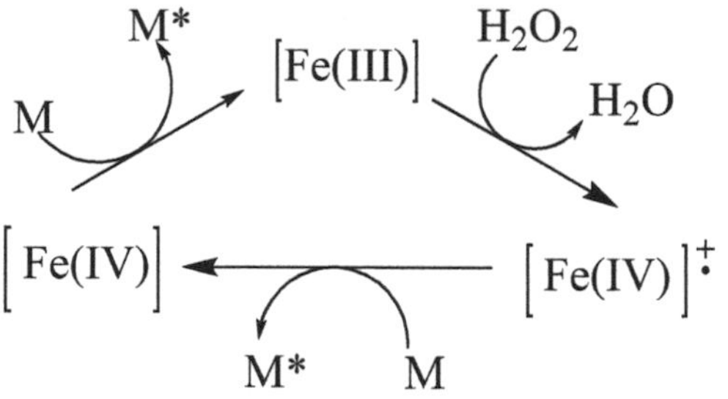

SCHEME 5-1   General catalytic cycle of peroxidase enzyme. M, Monomer; M*, radical cation of the monomer.

5.2.2   LACCASES

Laccases are copper containing oxidoreductases widely distributed in higher plants, bacteria, and fungi. These enzymes are used in pesticide/insecticide degradation and waste detoxification.[24]

The active site is shown in Figure 5-1a and consists of four copper atoms, which also confer the blue color to the enzyme owing to the intense electronic absorption of the Cu-Cu linkages. Laccase catalysis involves the four $1e^-$ oxidation of a substrate with concomitant two $2e^-$ reduction of dioxygen to water. The first step is the reduction of Cu^{2+} to Cu^+ (at the T1 site, which is the primary electron acceptor) by oxidation of the substrate. The electrons given up by the substrate are then transferred to the T2/T3 site. Dioxygen is then converted to water through two steps as illustrated in Figure 5-1b.

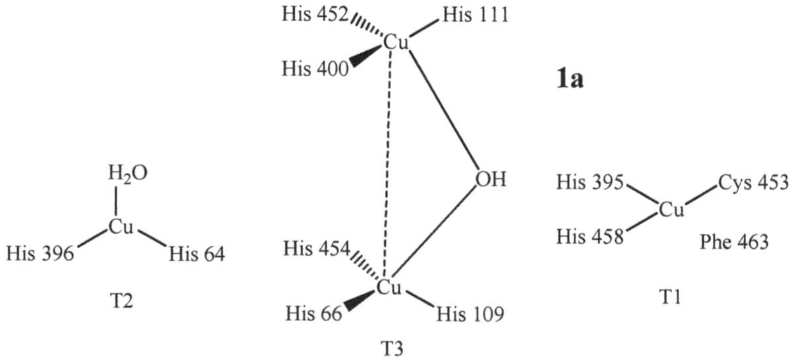

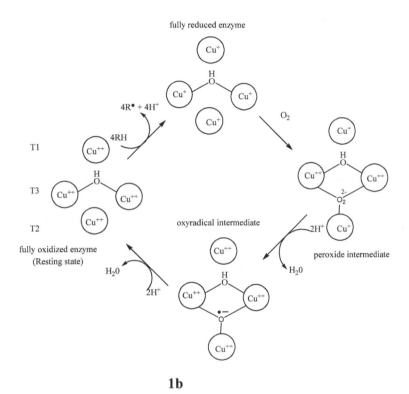

1b

FIGURE 5-1 (a) Active site of laccase. (b) Catalytic cycle of laccase.

5.2.3 OTHER ENZYMES USED IN SYNTHESIS OF POLYMERIC MATERIALS

Glucose oxidase: It is an oxidoreductase, which catalyzes the oxidation of β-D-glucose to gluconic acid by utilizing molecular oxygen as the electron acceptor with simultaneous release of hydrogen peroxide.[25] This reaction can be divided into two steps: (1) oxidation of β-D-glucose to D-glucono-δ-lactone and (2) nonenzymatically hydrolyzation of D-glucono-δ-lactone to gluconic acid. The generated hydrogen peroxide can be used to initiate polymerization reactions. The enzyme is active in the pH range of 4–7.

$$\beta\text{–D-Glucose} + O_2 \longrightarrow \text{D-glucono-δ–lactone} + H_2O_2$$

$$\text{D-glucono-δ–lactone} + H_2O \longrightarrow \text{Gluconic acid}$$

Bilirubin oxidase: These are a sub-group of multi-copper oxidases like laccases. Bilirubin oxidase (BOD) catalyzes the oxidation of bilirubin to biliverdin. BOD has been reported to display good catalytic activity at neutral pH conditions.[26]

Catalase: Catalases are oxidoreductases that catalyze the oxidation–reduction of hydrogen peroxide to oxygen and water.[27] Catalases can be broadly divided into three categories: (1) monofunctional heme containing catalases (they are the most widely studied enzymes), (2) bifunctional heme containing catalases, and (c) nonheme or manganese-containing catalases. Heme containing catalases utilize a two-stage mechanism for the degradation of hydrogen peroxide. The catalytic cycle of heme catalases is shown below. In the first step, one molecule of hydrogen peroxide oxidizes the heme to an oxyferryl species, also referred to as compound I. Heme catalases are unique in using a second molecule of hydrogen peroxide to reduce compound I, releasing dioxygen in the process. Despite the simplicity of the described reaction, there is significant variation in reactive capability of the various members of this group.[28]

$$Fe^{III} Por + H_2O_2 \longrightarrow \text{Compound I } (Fe^{IV}\!\!=\!\!O\ Por)^{\cdot+} + H_2O$$

$$\text{Compound I } Fe^{IV}\!\!=\!\!O\ Por^{\cdot+} + H_2O_2 \longrightarrow Fe^{III} Por + H_2O + O_2$$

5.2.4 LIMITATIONS OF NATURALLY OCCURRING PEROXIDASES

Peroxidase based enzymes have complex molecular structures that are intrinsically labile and costly to produce. While peroxidases can oxidize a variety of substrates,[29] the commercial applications of these enzymes, have been limited to research.[30] High cost of peroxidases and limited activity at higher temperatures or low pH conditions are major deterrents for the widespread industrial applications of these enzymes.

5.3 ENZYME MIMICS

There is clearly an imminent need for the development of efficient, low cost alternatives for peroxidases. In this context, porphyrins are considered to be attractive starting materials for development of novel biomimetic catalysts, since they bind and form complexes with metals. Among metal containing porphyrins, iron-containing porphyrins have elicited tremendous research interest due to their unique biological functions such as transport of oxygen (e.g., hemoglobin), and the catalysis of several biologically important redox reactions.[31]More importantly, since the active center in peroxidases is an Iron (III)-protoporphyrin complex, iron containing metalloporphyrins are the most rational choices for the design of biomimetic catalysts. This chapter will also cover metalloporphyrins, that have been used in the synthesis of CP. For a comprehensive review of synthesis of metalloporphyrins, the reader is referred to several review articles.[32,33,34,35]

Despite the structural diversity of biological metalloporphyrins, all of these proteins have iron protoporphyrin IX (heme) as their prosthetic group. The structure of heme *b* is shown in Figure 5-2. Oxidation of the iron to the ferric state yields hemin or ferric protoporphyrin. Anionic groups tend to co-ordinate with the axial positions occupied by iron leading to a square pyramidal geometry. When the anion bound is hydroxide, the resulting compound is referred to as hematin.

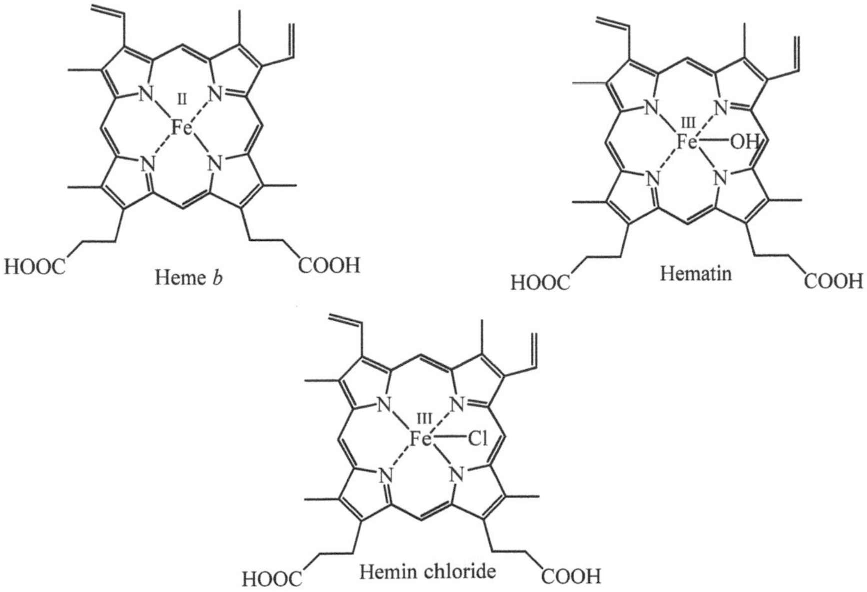

FIGURE 5-2   Structure of heme, hemin chloride, and hematin.

The insolubility of hematin at low pH conditions prevents widespread use of this catalyst. In order to overcome this limitation, hematin has been esterified with poly(ethylene glycol) (PEG) to form a pegylated hematin (PEG-hematin) using traditional chemical synthesis.[36,37] The purification process to separate the pegylated hematin from unreacted hematin is tedious. Further, the esterification reaction was done at room temperature in which hematin may remain aggregated. A more facile enzymatic route was developed[38] for the synthesis of PEG-hematin using lipase, an enzyme belonging to the family of hydrolases. Novozyme-435 is a commercially available lipase known to catalyze esterification, transesterification, and amidation reactions.[39] Both the chemically and enzymatically synthesized PEG-hematin was effective in catalyzing polymerization of several monomers, as discussed in the proceeding sections.

However, it was found that the stability of PEG-hematin was low when the pH is decreased to 1. As seen in Figure 5-3, in highly acidic conditions, the ester bond between hematin and PEG is cleaved and PEG-hematin precipitates out of solution. Since oxidative polymerization reactions are carried at fairly low pH conditions, stability of the catalyst under these conditions is important.

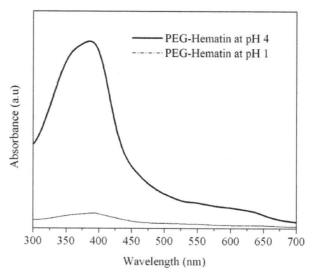

FIGURE 5-3   Stability of PEG-hematin at pH 4 and 1.

In an effort to increase the solubility of hematin at low pH conditions, amine-terminated PEG was amidated using novozyme-435 as a catalyst (Scheme 5-2).

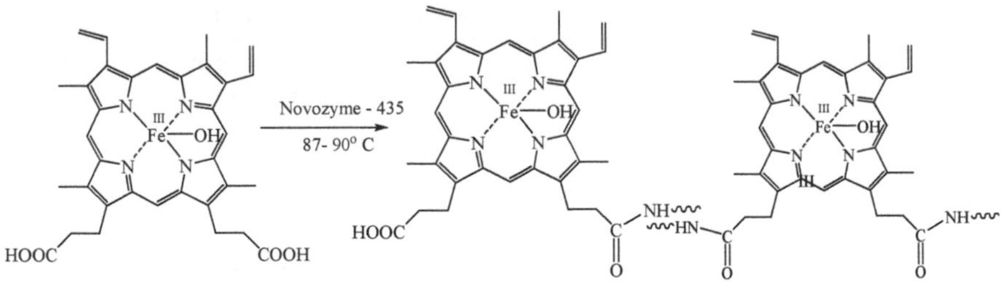

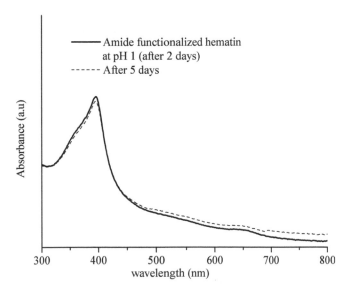

SCHEME 5-2   Lipase catalyzed amidation of hematin under solventless conditions.

The amidated hematin[40] had greater solubility and stability at lower pH conditions (pH =1) due to high stability of the amide bond over esters. As seen in Figure 5-4, the amidated hematin is stable at pH 1 after 5 days, with no change in the UV–visible spectra.

FIGURE 5-4   Stability of the amidated hematin at low pH conditions.

Besides hematin and functionalized hematin, a number of other metal containing porphyrins (synthetic and naturally occurring) have been utilized for the synthesis of CP. Table 5-2 summarizes various biomimetic catalysts, which have been utilized for the synthesis of CP.

TABLE 5-2   Biomimetic Catalysts Used for the Synthesis of Conjugated Polymers

Biomimetic Catalyst	Conducting Polymers Synthesized
Iron (III) tetrapyridyl porphyrin,	PANI[41]
Manganese (III) tetrapyridyl porphyrin	
Cobalt (III) tetrapyridyl porphyrin	
Iron (III) tetra(p-sulfonatophenyl)phthalocyanine.	PEDOT[42]
Cobalt (III) tetra(p-sulfonatophenyl)phthalocyanine.	
Manganese(III)-tetra(p-sulfonatophenyl)phthalocyanine.	
Hemoglobin (Hb)	PANI[43,44]
Hematin	PPy[45]
Pegylated hematin[36]	Polyaniline[36]
Amidated hematin[38]	Poly(sodium styrenesulphonate)[46]
	PEDOT-co-pyrrole[47]
5,10,15,20-tetraphenyl-21H,23H-porphine cobalt(II) pyridine complex [CoTPP(Py)]	Poly(indene)[48]

Since this chapter deals with the use of enzymes/enzyme mimics for the synthesis of CP, a brief introduction to CP and their applications is provided.

5.4   CONJUGATED POLYMERS

5.4.1   INTRODUCTION

Traditionally polymers have been considered to be insulators of electricity. However, pioneering work over the last few decades[49] has led to the discovery of CP which exhibit good conductivities comparable to metals while retaining other advantageous properties such as ease of processing and good environmental stability. Since their discovery in the early 1970s, the field of CP continues to expand with a wide variety of applications in photovoltaic devices,[50] optical displays,[51] and microelectronics.[52]

Figure 5-5 depicts the structure of common CPs. A key structural feature is the presence of alternating double and single bonds (conjugation) along the backbone of the polymer. However, the presence of conjugation alone is not sufficient to render a polymer conductive. Conductivity is introduced through a process known as doping.

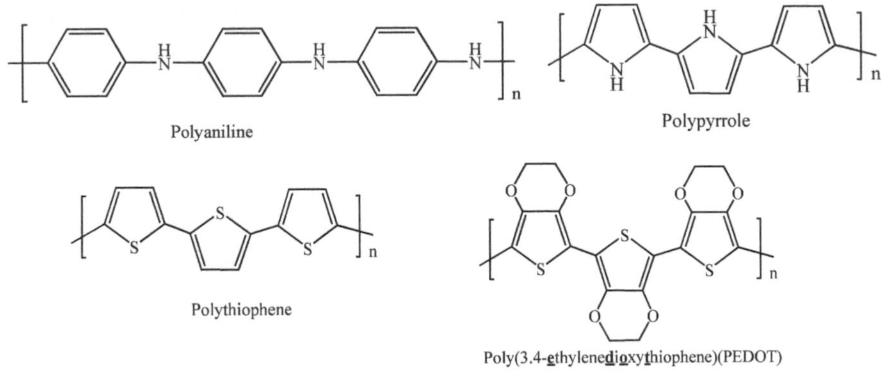

Polyaniline

Polypyrrole

Polythiophene

Poly(3,4-ethylenedioxythiophene)(PEDOT)

FIGURE 5-5   Structures of some common conducting polymers (CP).

The doping/dedoping is reversible in CP resulting in conductivities ranging from insulating to semi-conducting and conducting regions. The readers are referred to Heeger[53,54] and Bredas[55] for detailed explanations on mechanisms associated with conductivity and doping in CP. The undoped CP has a conductivity of around 10^{-10} to 10^{-8} S/cm. Upon doping, there is a rapid increase of electrical conductivity of several orders of magnitude up to values of around 10^{-1} S/cm. Generally, polymers with conductivities less than 10^{-8} S/cm are considered as insulators; materials with conductivities between 10^{-8} and 10^{-3} S/cm are considered semiconductors; and materials with conductivities greater than 10^{-3} S/cm are conductors.

Doping can be performed using chemical and/or electrochemical methods depending on the oxidation potential of the monomer. In CP, two principal types of doping can be observed. During chemical or redox doping, CP are typically partially oxidized (p-type doping) or reduced (n-type doping) often using a reducing/oxidizing agent. p-Doping was first discovered by treating trans-poly(acetylene) with iodine leading to an increase in conductivity from 10^{-5} S/cm to 10^{3} S/cm. In general, the doping process is reversible and can be accompanied by large change in conductivity/color of the polymer. Biological polymers such as DNA, chondroitin sulfate, and hyaluronic acid[56] can also be used as dopants. The use of a polymeric dopant enables the formation of a charged complex with the CP, thus mitigating any leaching out of the dopant from the polymer. Crucial to the process of doping is the presence of a π-conjugated backbone that facilitates movement of charge carriers along the polymer backbone resulting in electrical conductivity. Both n- and p-doping involves change in the number of electrons associated with the polymer backbone.

Protonic doping, on the other hand, does not involve a change in the total number of electrons associated with the polymer chain. The emeraldine form of PANI was one of the first examples of polymers produced through protonic doping. This was accomplished by treating the emeraldine base with aqueous protonic acids, which lead to an increase in conductivity (up to 3 S/cm). The average oxidation state of PANI can be varied continuously from completely reduced to half-oxidized to completely oxidized polymer. Figure 5-6 depicts the structural formulae of the PANI in different redox states.

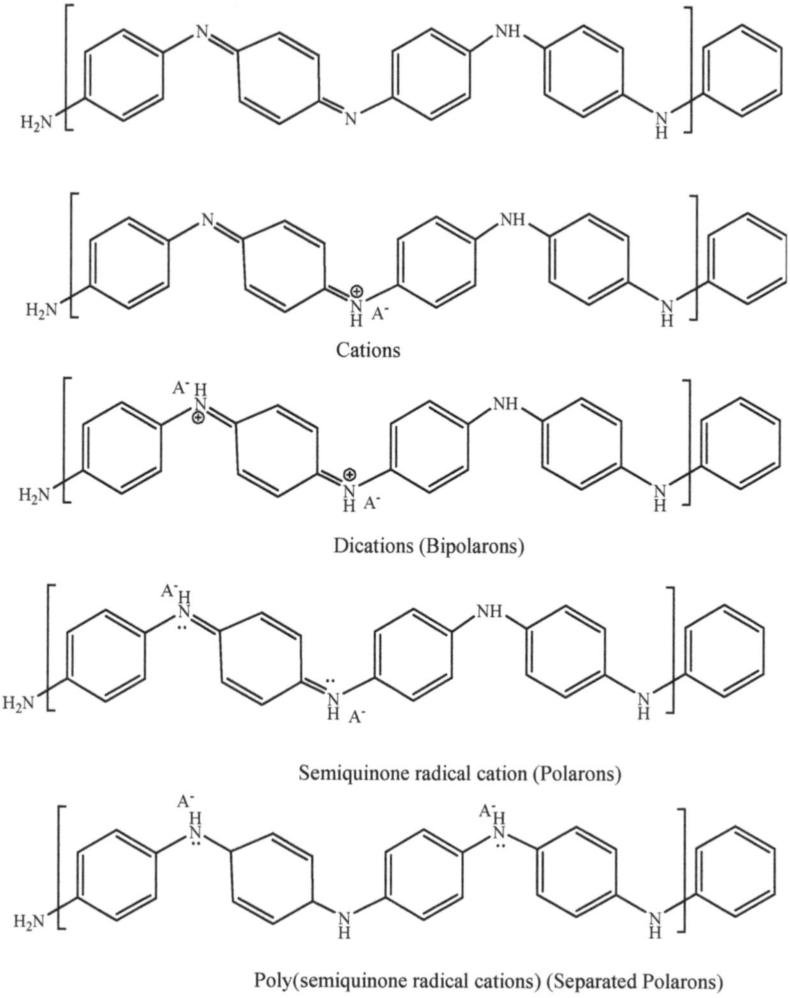

Cations

Dications (Bipolarons)

Semiquinone radical cation (Polarons)

Poly(semiquinone radical cations) (Separated Polarons)

FIGURE 5-6   Different doped forms of PANI.

The diverse array of tunable properties available with CP is based primarily on their composition and doping level.[57] The conductivity of the CP can vary depending on the degree of doping, nature of the dopant, and depending on chemical modifications to the polymer backbone.

The ability to control conductivity in CP combined with the inherent flexibility of these materials can be of immense value in biological applications. For example, studies have shown that some skeletal and tissue processes such as bone re-growth and wound healing can be susceptible to modulation via electrical stimulation.[58,59,60] While metal electrodes have been used for electrical stimulation in vivo, CP are better alternatives due to greater flexibility and biocompatibility. Hence there has been an upsurge in the use of CP for biological applications.[61,62,63,64] This has further been strengthened by research demonstrating the compatibility of CP with the biological entities.[65,66]

Among the different classes of CP, polypyrrole (PPy) and poly(3,4-ethylenedioxy thiophene) (PEDOT) have been used extensively for biological applications. CP have been used as a biological scaffold to enhance tissue degeneration[67,68,69,70] in implantable electrodes for neural probes,[71,72] in drug-delivery applications,[73] and bio-actuators.[74]

5.4.2 CONVENTIONAL METHODS OF SYNTHESIS OF CONDUCTING POLYMERS (CP)

Conventional strategies for the synthesis of CP involve chemical or electro-chemical polymerization methods. Electrochemical synthesis methods are not conducive to scale-up and the post-functionalization of the CP to introduce new functionalities is difficult. Chemical synthesis offers a variety of routes for the synthesis of CP in bulk and the ability to introduce desired functionality/groups into the polymer backbone. Several modification strategies used for the synthesis of CP have been reviewed extensively.[75,76] Chemical routes for PANI synthesis, for example, involve the use of strong oxidizers combined with low pH conditions (pH 0–2) and low temperatures. The polymer formed is not very useful in many technical applications due to its insolubility in common solvents.

5.5   ENZYME CATALYZED SYNTHESIS OF CONJUGATED POLYMERS

With the emergence of CP as frontier materials for biological applications, there exists an imminent need to develop benign methods for the synthesis of CP. Chemical synthesis often involves use of low pH conditions and/ or oxidants, both of which are not compatible with biological systems. In order for CPs to be used in biological applications, it is imperative that they are synthesized using biocompatible strategies amenable with biological systems.

Ideally, CP synthesis would be carried in biocompatible buffers/mixed solvent systems at room temperature. In addition, aqueous solvents are preferable to organic solvents due to low cost, ease of handling, safety and disposal, and the ability to utilize and incorporate biological dopants into the polymer backbone during synthesis.

Peroxidase catalyzed enzymatic polymerization reactions can be performed conveniently under mild conditions of temperature and pressure. Reactions can be performed in aqueous media or mixed solvent systems. Water-soluble and processable CP can be obtained in a single step. Since the reaction conditions are mild, biological molecules such as DNA can be readily integrated into the CP backbone.

This review will summarize efforts on enzymatic routes to the synthesis of poly(aniline) (PANI), PPy, and PEDOT. Peroxidases have also been successful in catalyzing polymerization of phenols[94] and vinyl monomers.[50] In addition to peroxidases, other interesting ecofriendly catalysts have been reported such as iron[77] and copper salts,[78] and high ionic strength solutions.[79] While these approaches are novel, they are outside of the scope of this review.

5.5.1   ROLE OF REDOX POTENTIAL IN OXIDOREDUCTASE CATALYZED POLYMERIZATION REACTIONS

Redox potential (E^0) of peroxidases plays a critical role in determining the enzyme's ability to catalyze oxidation reactions. It also serves as an important tool for identifying the range of oxidizable substrates for a particular enzyme. In theory, peroxidases can only catalyze oxidation of substrates with a lower redox potential. The reduction potential of the enzyme must be greater than the oxidation potential of the monomer to be polymerized.

Oxidoreductases typically react with hydrogen peroxide in the first step of the enzymatic cycle as shown in Scheme 5-1. From a purely thermodynamic perspective, hydrogen peroxide should be able to directly oxidize monomers such as aniline and pyrrole to initiate polymerization. Indeed polymerization reactions catalyzed by H_2O_2 have been reported.[80] However, these reactions are extremely slow and have been shown to introduce structural defects in the polymer.

Since the redox potential of the enzyme, hydrogen peroxide and substrates (in some cases)[81] are pH dependent, selection of optimum reaction conditions is important. The reduction potential of hydrogen peroxide increases with decrease in pH. The redox properties of many oxidoreductases are pH dependent; most peroxidases are effective close to neutral pH conditions. The concept of matching of redox potentials of the enzyme and substrate is further discussed in the following sections.

Other parameters such as electrostatic interactions of peroxidases with templates/solvents, local environment, and substrate orientation may also influence the redox potential of peroxidases.[82] Understanding the redox potential of heme peroxidases is fundamental in order to design enhanced biocatalysts.

5.5.2 PEROXIDASE CATALYZED SYNTHESIS OF PANI

Enzymatic polymerization of aniline was first attempted using BOD as the enzyme.[14] The polymerization of aniline was carried out on the surface of a solid matrix such as glass slide using immobilized BOD at a pH 5.5. The emeraldine salt of PANI is known to possess good stability and conductivity. Emeraldine base can be readily converted to the emeraldine salt and vice-versa by protonation/deprotonation with acid/base. The goal in most aqueous based synthetic methodologies is the synthesis of emeraldine salt of PANI. The next few sections summarizes different enzyme based approaches involving the use of mixed solvent systems, templates, and immobilized enzymes for the synthesis of PANI.

5.5.2.1 SYNTHESIS OF AROMATIC AMINE CONTAINING MONOMERS IN MIXED SOLVENT SYSTEMS

Enzymatic polymerization of a series of aniline derivatives was studied under mixed solvent systems. A series of phenylenediamines and amino-phenols were polymerized in dioxane-buffer mixtures in ambient conditions using HRP as a catalyst and a large excess of hydrogen peroxide. Polymers synthesized were soluble in organic solvents such as DMF and DMSO. Poly(2-aminophenol) and poly(4-aminophenol) were found to possess good electroactive properties.[83] HRP catalyzed oxidative coupling of 4,4'-diamino-azobenzene was carried out at pH 6 using 20% ethanol.[84] The chromophores in the resulting polymer exhibited interesting structural photoisomerization. Anilines with other azo groups were polymerized using HRP catalysis in mixed solvent systems.[85]

Kim et al.[86] polymerized a series of methoxy-derivatives of aniline using HRP in 15% ethanol using 10-camphor sulfonic acid (CSA) as the dopant. Peroxyacetic acid was used as an oxidant instead of hydrogen peroxide, for the polymerization of methoxyaniline derivatives. Figure 5-7 summarizes aniline derivatives, which have been polymerized using HRP in mixed solvent systems. Water-soluble polyanilines were enzymatically synthesized from sulphonated aniline monomers such as 2,5-diaminobenzenesulfonate.[87] A major drawback of this process was the formation of low molecular weight oligomers with a low electrical conductivity.

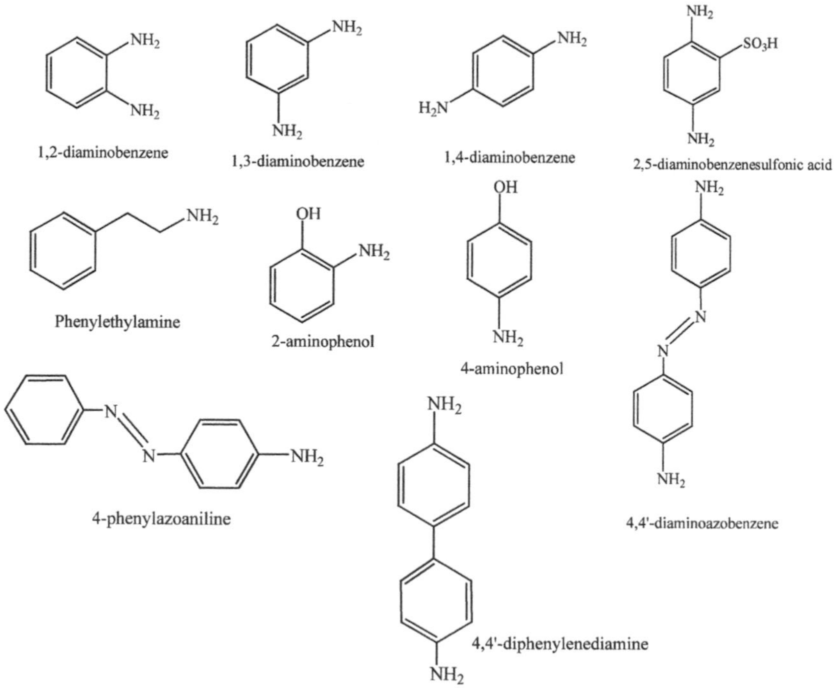

FIGURE 5-7   Structure of aniline derivatives, which have been enzymatically polymerized in mixed solvent systems.

5.5.2.2 TEMPLATE ASSISTED ENZYMATIC SYNTHESIS OF PANI

In an effort to address and improve processability, a variety of approaches were investigated including use of micelles,[88] reverse micelles,[89] polymerization at air–water interface,[90] and use of modified monomers.[91] While the processability and molecular weight improved, the resulting polymer was often a mixture of at least two structurally different types of PANIs, as shown in Figure 5-8. The presence of highly branched ortho- and para-substituted PANI limits the degree of conjugation and lowers the electrical conductivity of the resulting polymer. Solid-state NMR studies have confirmed that PANI synthesized with HRP without a template is highly branchd in structure and insulating in nature.[92]

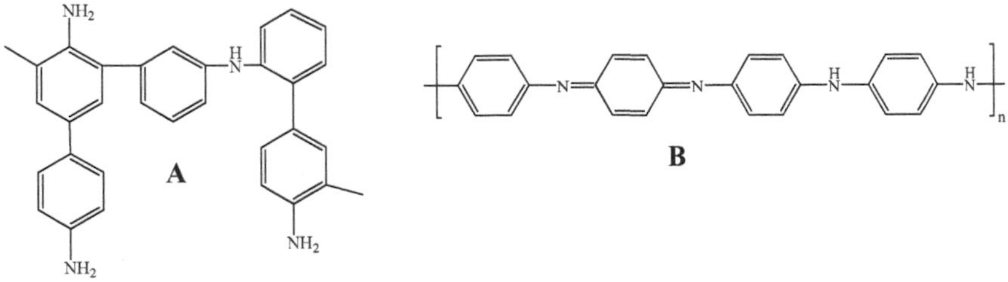

FIGURE 5-8 Two structurally different types of PANIs obtained in typical enzymatic polymerization without template: (a) benzenoid-quinoid linear structure and (b) ortho- and para-substituted branched structure.

In an effort to promote the formation of is the desired benzenoid-quinoid form, a novel template assisted polymerization was proposed by Liu et al.[9,93] A polymeric material poly(styrene-4-sulfonate) (PSSNa) was used as the polyelectrolyte template because of its commercial availability, high degree of sulfonation and low pK_a of the benzene sulfonic group which enables it to be negatively charged at most pH conditions. The anionic PSSNa couples with the protonated aniline (anilinium ion, pK_a = 4.63) thus aligning the monomer along the polymer backbone, promoting head-to-tail coupling as shown in Figure 5-9.

The charged sulfonate groups in PSSNa perform dual roles by acting as the charge-balancing dopant and help to keep the final polymer dispersed in water.

The template has been shown to provide a necessary "local" environment where the pH and charge density are vastly different from that in the bulk solution. The presence of these "nanoreactors" was critical in anchoring, aligning, and reacting with the aniline monomer to form a para directed PANI at the end of the reaction. The conductivity of the resulting PANI ranged from 6×10^{-5} S/cm to 5×10^{-3} S/cm when molar ratios of PANI-PSSNa of 0.6:1 to 2.2:1 were used, respectively.

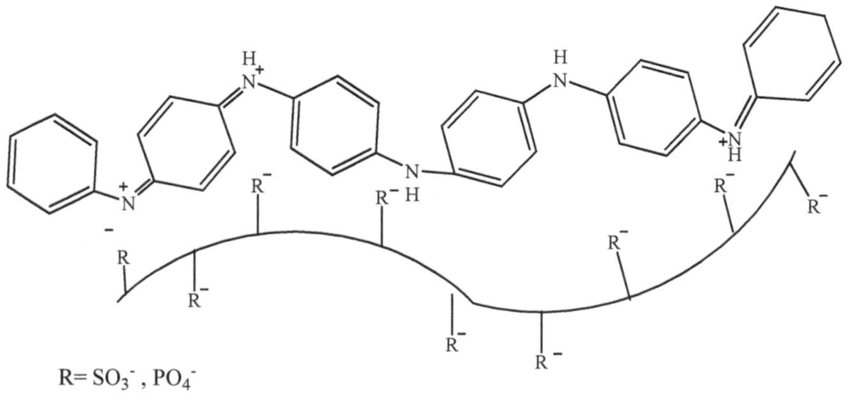

$R = SO_3^-, PO_4^-$

FIGURE 5-9 Polyaniline synthesized in the presence of polyanionic templates, which promote head-to-tail coupling.

Other anionic polymeric templates such as poly(vinylphosphonic acid) (PVP)[94] and poly(vinyl sulfonic acid, sodium salt)[95] have been shown to produce conductive PANI with a conductivity between 5.5×10^{-2} S/cm and 4.78×10^{-1} S/cm, respectively. The PANI-PVP complexes were sensitive to the ratio of aniline monomer:PVP. When the molar ratio of PANI:PVP exceeded 1:5, the free phosphonate groups could no longer solubilize the growing PANI chain and the complex is precipitated out of solution.

Small molecule strong acid dopants such as CSA were used in conjunction with weak acid polymeric templates like poly(acrylic acid) (PAA) to produce PANI with conductivity values in the range of 10^{-2} S/cm.[96] The PANI-PAA-CSA composite was produced in an enantiospecific manner by HRP regardless of the chirality of the CSA used. This suggests that the enzyme was perhaps selectively modulating the secondary structure of the newly formed polymer. A variety of other polymers and small molecules have been used as templates for the polymerization of aniline. The structure of some commonly used polymeric templates is depicted in Figure 5-10.

These research findings reinforced the central idea that strong interactions between positively charged PANI backbone and negatively charged template/dopant is necessary for the formation of conductive polymer.[98,99]

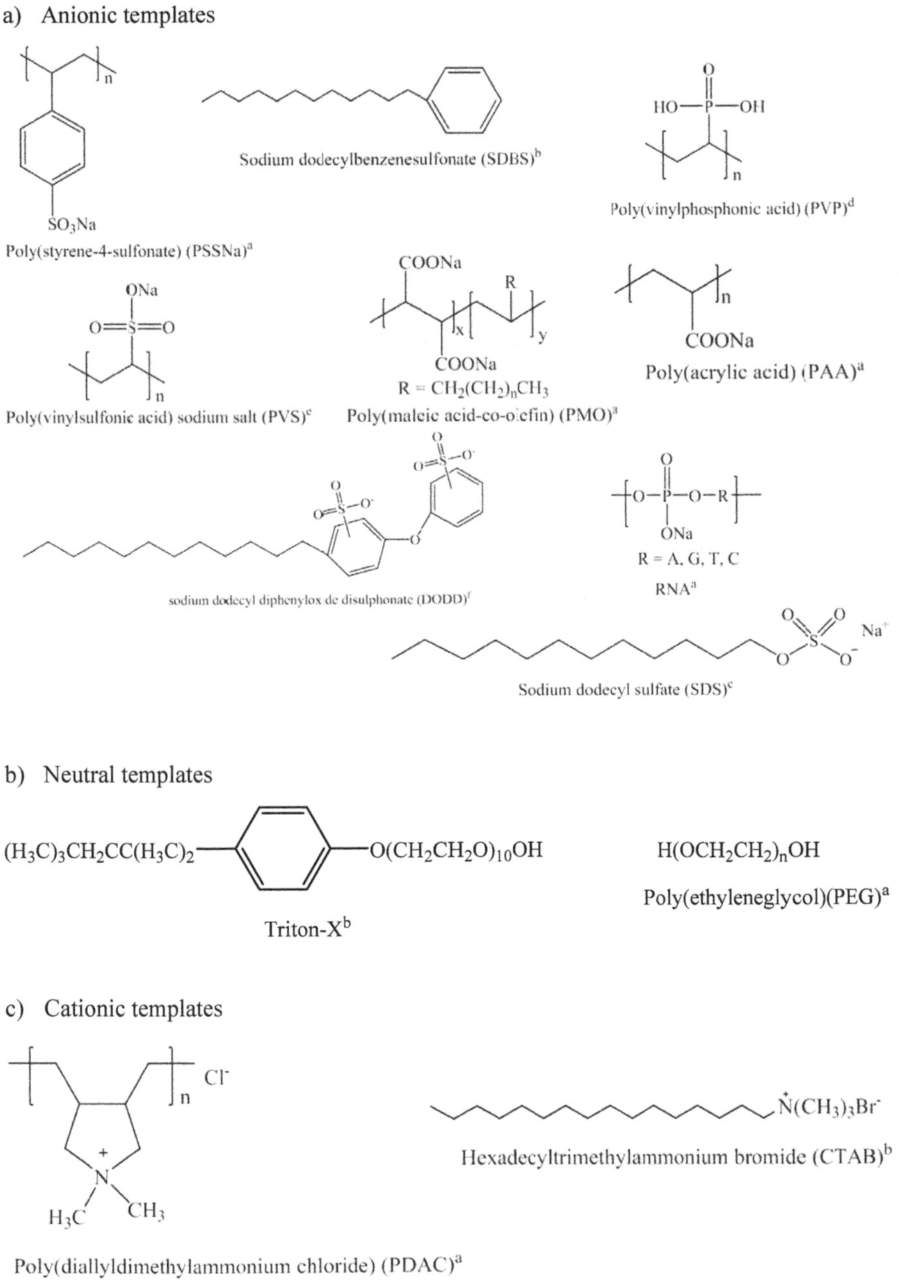

a) Anionic templates

Sodium dodecylbenzenesulfonate (SDBS)[b]

Poly(vinylphosphonic acid) (PVP)[d]

Poly(styrene-4-sulfonate) (PSSNa)[a]

Poly(vinylsulfonic acid) sodium salt (PVS)[e]

$R = CH_2(CH_2)_nCH_3$

Poly(maleic acid-co-olefin) (PMO)[a]

Poly(acrylic acid) (PAA)[a]

$R = A, G, T, C$

RNA[a]

sodium dodecyl diphenylox de disulphonate (DODD)[f]

Sodium dodecyl sulfate (SDS)[e]

b) Neutral templates

$(H_3C)_3CH_2CC(H_3C)_2$—⟨benzene⟩—$O(CH_2CH_2O)_{10}OH$

Triton-X[b]

$H(OCH_2CH_2)_nOH$

Poly(ethyleneglycol)(PEG)[a]

c) Cationic templates

Cl^-

Hexadecyltrimethylammonium bromide (CTAB)[b]

Poly(diallyldimethylammonium chloride) (PDAC)[a]

FIGURE 5-10 Structures of some commonly used templates in enzyme-assisted polymeri zation.[45c,89b,93a,94d,95e,97f]

In addition to HRP, laccase (isolated from *Cytisus hirsutus*)[100] was used as a catalyst for the polymerization of aniline in the presence of PSSNa as a dopant resulting in PANI with conductivity of around 2×10^{-4} S/cm. The role of redox potentials of laccases in the polymerization of aniline has been investigated. Laccases (*Tolmièa hirsute* and *Rhus vernicifera*) with different redox potentials at the T1 catalytic centers were evaluated for their ability to polymerize aniline. The redox potential of the T1 center of fungal laccase *T. hirsute* and *R. Vernicifera* are 0.780 V (vs. NHE)[101] and 0.42 V (vs. NHE),[102] respectively. The small difference in redox potentials between aniline (~0. 9 V vs. NHE) and *T. hirsuta* enabled the polymerization of aniline. However the difference in the oxidation potentials of aniline and the T1 center of *R. Vernicifera* is approximately 530 mV, which makes the reaction of the oxidative polymerization of aniline impossible and no absorption in the visible region is seen (Figure 5-11). Redox mediators such as potassium octocyano-molybdate (4+) have been used to accelerate the polymerization of aniline when catalyzed by *T. hirsuta.*

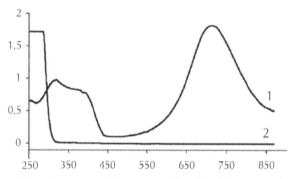

FIGURE 5-11   Absorption spectra of laccase-catalyzed aniline polymerization with different laccases: Trace 1, polymerization catalyzed by laccase *T. hirsuta* and Trace 2: polymerization catalyzed by laccase *R. vernicifera.*

Other oxidases such as Palm tree oxidase[103] and glucose oxidase[18] have also been used as catalysts for the formation of conductive PANI.

5.5.2.3   USE OF BIOLOGICAL MOLECULES AS TEMPLATES IN SYNTHESIS OF PANI

In order to realize the full potential of CP in biological applications such as biosensors, the ability to integrate bio-based molecules into CP is crucial. The mild conditions of enzymatic polymerizations are conducive to template biological entities onto the CP backbone. Molecular complexes of PANI with calf-thymus DNA[104,105] and PANI-RNA[95] have been synthesized using HRP as a catalyst. Circular dichroism (CD) studies have confirmed the formation of PANI/DNA complex. It was also interesting to note that the PANI formed induced change in the conformation of the DNA from a loosely wound "B" form to a more tightly wound polymorph. PANI formed on DNA also mimics the helicity of DNA and acquired a handness of its own (Figure 5-12).

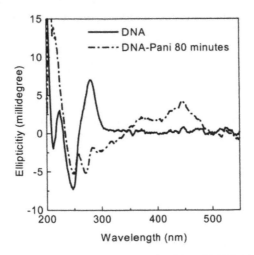

FIGURE 5-12 Circular dichroism (CD) spectra of DNA and PANI-DNA complex.

Cytosine nucleotides within larger DNA backbones have also been shown to be templates for PANI oligomer formation.[106] In an effort to template bovine serum albumin (BSA) on PANI, a double layer template approach was adopted. Since the positive charge of BSA would repel the anilinium ion, BSA was first complexed with sodium dodecylsulfate (SDS) along its backbone. The SDS could serve as a template for the polymerization of aniline to form oligo(aniline).[107]

5.5.2.4 TEMPLATE FREE SYNTHESIS OF CONDUCTIVE PANI

While use of templates for the synthesis of conductive PANI has several advantages including ease of processing and good conductivities, separation of the template from the polymer is difficult. In applications where bulk

PANI is required such as in conductive fibers, template free synthesis can be of more importance.

Conductive PANI has been synthesized without a template using HRP,[108] glucose oxidase,[109] and SBP[110] as catalysts. The higher stability of SBP at low pH conditions enabled the polymerization reactions to be carried out at pH 3.5 using toluene sulfonic acid (TSA) as a dopant. Good conductivities for PANI (2.4 S/cm) were obtained. However further chemical characterization of PANI using x-ray photoelectron spectroscopy (XPS) revealed the presence of chain defects and the existence of branched PANI. The UV–visible spectrum of PANI obtained at pH 4–7 from glucose oxidase catalysis exhibited features resembling the branched and insulating forms of the polymer.

5.5.2.5   IMMOBILIZED ENZYMES AS CATALYSTS FOR THE SYNTHESIS OF PANI

In order for enzymatic polymerization to be a commercially viable route, the high cost of enzymes needs to be mitigated. Immobilization of enzymes such as HRP is a well-established route to reuse the enzyme,[111] thus significantly lowering costs. HRP has been immobilized on other supports such as titanium dioxide,[112] collagen,[113] polyethylene,[114] and chitosan[115] for the synthesis of PANI.

HRP was encapsulated in an ionic liquid (IL) 1-butyl-3-methylimidazolium, which was insoluble in water at room temperature. Aniline, DBSA, and H_2O_2 were dissolved in water. Polymerization of aniline occurs at the IL–water interface to generate emeraldine form of PANI (conductivity: 10^{-3} S/cm).[116] Similar to HRP, laccase could be immobilized on carbozymethylcellulose to produce PANI with conductivity of approximately 10^{-2} S/cm.[117]

5.5.2.6   SYNTHESIS OF POLYANILINE DERIVATIVES VIA ENZYME CATALYSIS

A number of derivatives of aniline have also been enzymatically polymerized using HRP as a catalyst in the presence of PSSNa. The structure of these derivatives along with the appropriate references is given in Figure 5-13. In general, poly(alkyl anilines) exhibited absorption in the range between 730 and 750 nm and showed reversible electroactivity. N-substituted aniline derivatives were electroactive and exhibited absorption at wavelengths greater than 900 nm. A more systematic study with poly(alkoxyanilines).

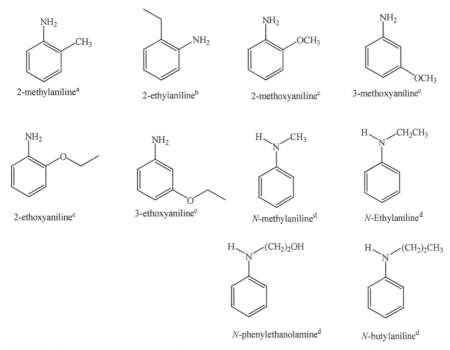

FIGURE 5-13 Structures of aniline derivatives, which have been polymerized using peroxidases in the presence of PSSNa as a template.[118,119,120,121]

5.5.2.7   SELF-DOPED POLYANILINE SYSTEMS

Self-doped polyaniline systems have negatively charged ionizable functional groups within the polymer backbone. This negatively charged moiety acts as an inner dopant anion thus effectively negating the need to add an external dopant. Self-doped PANI systems have greater thermal stability and conductivity over a wide range of pH conditions. 3-amino-4-methoxy benzene sulfonic acid (SA) was polymerized with HRP in the presence of tetradecyl trimethyl ammonium bromide (MTAB) and the polycationic template, poly(vinyl benzy dimethyl hydroxyl ethylammonium chloride) (PVAC).[122] The presence of the negative charge on the surface of the monomer led to its self-assembly along the cationic backbone of the template. Products could exist in either the expanded or collapsed coil conformations with higher conductivities (10^{-4} S/cm) seen with the expanded coil chains.

5.5.3 BIOMIMETIC ROUTES FOR THE SYNTHESIS OF POLYANILINE

Since porphyrins play a pivotal role in several biochemical and photochemical reactions, there are the most rational choices for the design of biomimetics. Metal containing porphyrins have been evaluated as catalysts for the synthesis of PANI templated on PSSNa. The structure of these porphyrins are shown in Figure 5-14. The authors provide no information on the conductivity of the PANI synthesized. Most of these metal porphyrins are synthesized using multi-step reactions and complex purification procedures.

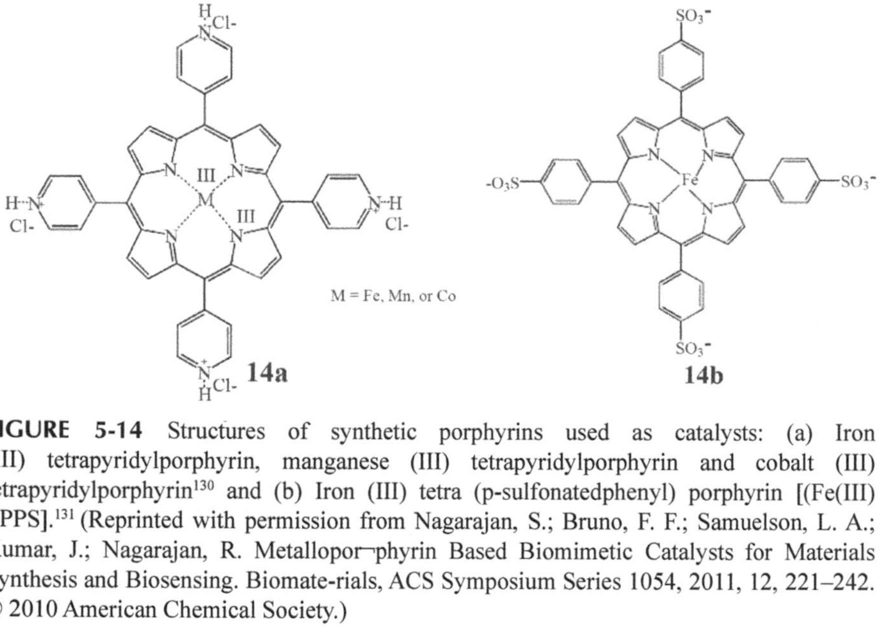

M = Fe, Mn, or Co

14a

14b

FIGURE 5-14 Structures of synthetic porphyrins used as catalysts: (a) Iron (III) tetrapyridylporphyrin, manganese (III) tetrapyridylporphyrin and cobalt (III) tetrapyridylporphyrin[130] and (b) Iron (III) tetra (p-sulfonatedphenyl) porphyrin [(Fe(III) TPPS].[131] (Reprinted with permission from Nagarajan, S.; Bruno, F. F.; Samuelson, L. A.; Kumar, J.; Nagarajan, R. Metalloporphyrin Based Biomimetic Catalysts for Materials Synthesis and Biosensing. Biomate-rials, ACS Symposium Series 1054, 2011, 12, 221–242. © 2010 American Chemical Society.)

Hematin is heme containing naturally occurring metalloporphyrin extracted from porcine blood. Hematin has exhibited good biomimetic activity and is postulated to form intermediates very similar to peroxidases.[123] The insolubility of hematin at low pH conditions prevents its use as a catalyst in the oxidative polymerization of aniline. In order to overcome this limitation, electrostatic layer-by-layer (ELBL) approach was utilized to fabricate multilayer assemblies of hematin and a polyelectrolyte, poly(dimethyl diallylammonium chloride) (PDAC) at pH 11. At pH 11, hematin is soluble in water due to carboxylate salt formation and possesses a net negative charge, while PDAC serves as the polycation. Multilayer polycomposite films prepared were then used for the polymerization of aniline in the presence of a PSSNa as a template.[124]Authors reported that it was possible to reuse assembled hematin substrates for multiple aniline polymerizations.

Another approach involved immobilization of hematin on halloysite nanotubes (HNT). HNT are aluminosilicates with aluminol (Al-OH) groups in the internal and Si-O-Si on the external surface. The authors reported that stirring a mixture of hematin and halloysite at pH 9.5 for 24 h resulted in 65% hematin adsorbed on HNT. Hematin supported HNT was then used as a catalyst for the biomimetic polymerization of aniline.[125] However, the authors provide no information on pH conditions used or on the conductivity of PANI synthesized. Since the biomimetic polymerization was carried out in the presence of p-toluene sulfonic acid (PTA), oxidation of aniline by PTA cannot be entirely ruled out.

In order to use hematin in aqueous media at lower pH conditions, PEG were tethered on to the carboxylic acid groups in hematin as explained in the preceding sections. Pegylated hematin (PEG-Hematin) obtained was water-soluble and found to be an effective catalyst for the polymerization of aniline in the presence of a number of polymeric templates such as PSSNa, PVP, synthetic oligonucleotides, and calf-thymus DNA.[104,126] PANI/PSSNa obtained using PEG-hematin catalysis has been studied extensively using solid-state ^{13}C and ^{15}N CP/MAS NMR spectroscopy. NMR studies indicate that PANI obtained resembles the chemical synthesized PANI.[127]

In an effort to develop more sustainable approaches, naturally occurring polymers such as of lignin sulfonate (LGS),[128] chondroitin sulfate, dermatin sulfate, and keratin sulfate[129] have been used as templats in the polymerization of aniline catalyzed by PEG-Hematin.

Hemoglobin (Hb) is an iron containing metalloprotein found in the red blood cells of all vertebrates. Hb has been shown to have good biomimetic properties[132] and also catalyze the formation of PANI at a pH between 1 and 3 in the presence of LGS,[133] PSSNa, and SDBS as templates. Conductivities increased with decrease in pH. Maximum conductivity was observed at around pH 1.0 (11×10^{-3} S/cm).[134]

5.5.4 BIOCATALYTIC SYNTHESIS OF POLYPYRROLE (PPY) AND DERIVATIVES

5.5.4.1 OXIDOREDUCTASES AS CATALYSTS FOR SYNTHESIS OF POLYPYRROLE (PPY)

PPy is traditionally produced by the oxidative polymerization in aqueous conditions using oxidizers such as aqueous FeCl$_3$ or other salts of trivalent iron as well as APS. The yield and conductivity of PPy is determined by many factors such as the choice of solvent, oxidizer, and temperature.[135]

However, the better biocompatibility of pyrrole with biological systems behooves the development of benign enzymatic routes for the synthesis of PPy. Since many biological processes produce hydrogen peroxide, synthesis of PPy using H_2O_2 is a clean, low-cost environmentally friendly process.[136] Hydrogen peroxide generated by enzymes such as glucose oxidase[137] and lactate oxidase (LOD)[20] has been used for the polymerization of pyrrole at pH 6–7. The rate of polymerization was very slow, confirming earlier reports that pyrrole polymerization using hydrogen peroxide proceeds very slowly in the absence of iron salts. Further direct oxidation of pyrrole using H_2O_2 lowered the electrical conductivity of the polymer, which has been attributed to the formation of 2-pyrrolidinone units along the polymer backbone due to over-oxidation of pyrrole. Besides hydrogen peroxide, oxygen can also slowly oxidize pyrrole in acidic medium,[138] but the reaction time is very slow.

Initial attempt to biocatalytically polymerize pyrrole[139] using HRP-H_2O_2 and PSSNa led to formation of PPy at pH 2. Fourier transform infrared spectroscopy (FTIR) studies indicated absorption at 1710 cm[-1], which was attributed to the carbonyl stretch of the pyrrolidinone, suggesting over-oxidation of the PPy. No data was provided on the polymerization of pyrrole in the absence of HRP. Hence nonenzymatic polymerization of pyrrole cannot be ruled out.

A further experiment by Kupriyanovich et al.[140] proved that nonenzymatic oligomerization of pyrrole was indeed possible at low pH conditions in the absence of HRP. The authors attempted enzymatic polymerization of pyrrole utilizing HRP and H_2O_2 at a higher pH of 4.5. Soluble low molecular weight reaction products isolated from the insoluble resulting polymer were analyzed by gel permeation chromatography (GPC). GPC revealed that the soluble fraction contained heptamers, octamers, and some higher molecular weight species. Both the insoluble PPy powders and low molecular weight oligomers were found to contain a carbonyl group suggesting over-oxidation.

While oxidoreductases such as HRP and lacasse can oxidize aniline based monomers using H_2O_2 the activity of these enzymes towards pyrrole between pH 3 and 6 remains low. The authors believe that this is partly due to the higher oxidation potential (E_p) of pyrrole (E_p = 1.2 V vs. Ag/AgCl)[141] compared to aniline (E_p = 0.9 V).[142] Since the oxidation potentials of laccase (E_p = 1.0 V)[143] and HRP (E_p = 1.09 V)[144] are lower than that of pyrrole, the active sites of these enzymes are unable to directly oxidize the pyrrole monomer at reasonable pH conditions to form conducting PPy free from over oxidation.

In an effort to overcome this limitation, a redox mediator such as 2,2′-azino-bis(3- ethylbenzthiazoline-6-sulfonic acid)diammonium salt (ABTS) has been used in conjunction with HRP[122] to polymerize pyrrole. The presence of sulfonate groups in ABTS enables it to play dual roles as a dopant and redox mediator. Since the oxidation potential of ABTS salt is typically much lower than pyrrole, it can be readily oxidized at the enzyme's active site to form a radical cation. ABTS radical cations then oxidize the pyrrole monomer while getting reduced. Additional secondary dopants such as I_2 have been added to increase conductivity of pyrrole. Conductivity data for PPy formed is summarized in Table 5-3.

TABLE 5-3   Dopants and Conductivity of PPy Synthesized Using Redox Mediators

Sample	Conductivity (S/cm)
Ppy (citrate buffer)	3.2×10^{-7}
PPy (ABTS)	1.2×10^{-6}
Ppy (ABTS, lithium trifluoro methane sulfonate)	6.0×10^{-5}
Ppy (ABTS, PSSNa)	2.0×10^{-4}
PPy (ABTS, PSSNa, I_2)	5.1×10^{-3}

Cruz-Silva et al.[145] employed ABTS as a redox mediator with HRP for the synthesis of PPy thin films. By adding poly(vinyl alcohol) as a stabilizer to the reaction mixture, water-dispersible PPy colloids were obtained. FTIR and XPS studies confirmed that the enzymatically synthesized PPy colloids are similar to PPy obtained through chemical oxidation.

ABTS has also been used in conjunction with laccase for the formation of PPy with conductivities of around 0.2 S/cm.[146] The rate of polymerization of pyrrole by lacasse with and without ABTS was studied. The authors report that, in presence of laccase, ABTS radical cation oxidizes pyrrole at a rate of 0.98 µM/min. In the absence of ABTS, the rate of oxidation of pyrrole by laccase alone is 0.11 µM/min. In contrast, the rate of polymerization of pyrrole using an acid catalyzed approach was reported to be around 0.068 µM/min. While redox mediators have been used successfully for the synthesis of PPy, they cannot be removed from the final product and need to be used in larger amounts to obtain conductive PPy.

Among the peroxidases, soybean peroxidase (SBP) has a reported oxidation potential of 1.2 V (Ag/AgCl).[147] SBP has previously been used to

synthesize copolymers of pyrrole and PEDOT.[14] Since SBP's redox potential
is close to pyrrole, it was hypothesized that SBP could oxidize the monomer
without the aid of a redox mediator. Indeed, PPy was successfully synthesized in high yields using SBP as catalyst, H_2O_2 as oxidant, and PSSNa as
the charge balancing dopant at pH 3.5.[12] The PPy-PSSNa complex did not
exhibit any absorption peaks were observed between 1600 and 1800 cm[-1]
suggesting that over-oxidation of Ppy did not occur. Interestingly, it was
observed that the reaction yield and the conductivity of Ppy-PSSNa obtained
was dependent on the reaction temperature. As observed in Figure 5-15, the
conductivity of the product increased with decreasing reaction temperatures.
The polymerization methodology was amenable to both small and large
molecular weight dopants.

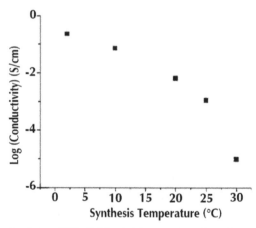

FIGURE 5-15   Conductivity of PPy-PSS complex as a function of reaction temperature.

PPy synthesized in the presence of CSA displayed the highest reported
conductivity values for any enzymatically produced PPy (>3 S/cm). The
authors report that since the reaction was conducted at low temperatures,
the generation of radical cations occurred in a more controlled fashion.
This ensured a higher concentration of dopant (compared to the radical
cations) enabling complete stabilization of all the radical cations being
generated. This stabilization led to the formation of PPy, which was
predominately α-coupled, linear, and conductive. XPS analysis indicated
that the disorder observed in the enzymatically synthesized polymer was
lower (13.1%) than both the chemically and electrochemically synthesized
PPy (Table 5-4).

TABLE 5-4   XPS Data Comparing Enzymatic vs Chemically Synthesized PPy for Structural Defects

Synthesis Method (Catalyst, Oxidant)	Degree of Structural Disorder (%)
Enzymatic PPy (SBP, H_2O_2)	13.1
Chemical PPy (APS)	22.1
Electrochemical	33.3

These results demonstrate catalyzed approaches can be employed successfully for the formation of defect-free high conductive PPy with superior properties when compared to the synthetic analogues.

In an effort to extend the applicability of enzymatic polymerization to pyrrole derivatives with substitutions at the beta position, 3-methyl pyrrole (3-MP) was polymerized using SBP-H_2O_2 (Scheme 5-3). 3-MP was enzymatically synthesized in the presence of a number of dopants, was easily dispersible in water, obtained in good yields (80%) and exhibited conductivities as high as 10^{-3} S/cm.

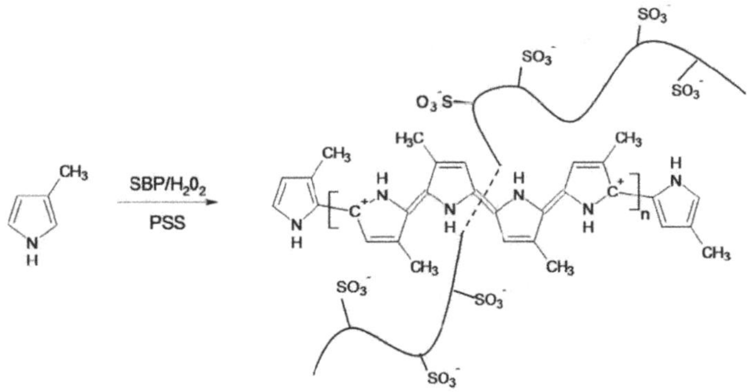

SCHEME 5-3   Enzymatic polymerization of 3-MP catalyzed by SBP in presence of H_2O_2.

PPY COMPOSITES

Composites of PPy with multi-walled carbon nanotubes (MWCNT) were prepared using HRP and ABTS as a redox mediator at a pH of 4.0.[25] The authors report a uniform coating of PPy on CNT suggesting that most of the pyrrole polymerization occurred on the surface of CNT. PPy/MWCNT composites exhibited conductivity of 2.84 S/cm.

Besides HRP, the enzyme LOD has been used to initiate the polymerization of pyrrole.[20] The LOD served dual roles: (1) LOD is known for its ability to catalyze the polymerization of lactate to pyruvic acid and hydrogen peroxide. The H_2O_2 generated was used to initiate the polymerization of pyrrole. (2) LOD also served as a dopant due to the presence of negatively charged groups in the enzyme backbone. The authors report that the LOD functions as a seed for the formation of PPy-LOD nanoparticles, which further assembled into micron, sized architectures. However UV–visible spectra of the nanoparticles show no absorption at 800 nm, suggesting the formation of reduced form of PPy. When CNT was introduced into the reaction mixture, a three-component PPy-enzyme-CNT composite was produced

More recently, Hu et al. reported the synthesis of PPy/hemin nanocomposites in presence of pluronic micelles[148] for the colorimetric detection of glucose. The authors also report that the PPy/nanocomposites were efficient in removing toxic wastes and exhibited a photothermal effect. The authors use significant amount of $FeCl_3$ that serves as an oxidant for the polymerization of pyrrole.

5.5.4.2 BIOMIMETIC APPROACHES TO THE SYNTHESIS OF POLYPYRROLE (PPY)

Bruno et al. reported the use of PEG-hematin as a catalyst for the synthesis of PPy.[149] UV and FT-IR spectra confirmed the formation of PPy with conductivities of around approximately 10^{-4} S/cm. The reaction was however performed at low pH conditions (pH =1).

Since the catalytic activity and solubility of hematin is known to increase in the presence of surfactants,[150] Ravichandran et al.[45] utilized surfactants to solubilize hematin. This micelle-encapsulated hematin was used as a catalyst for the polymerization of pyrrole. Cationic, anionic, and nonionic surfactants were evaluated for their ability to act as a charge balancing dopant in the polymerization reaction, besides solubilizing hematin. As seen in Figure 5-16, doped conducting PPy formed in presence of SLS and DBSA exhibited an absorption in the region from 800 to 1100 nm. This confirms that the presence of an anionic dopant was essential for the synthesis of electrically conducting PPy. The authors also carried out the polymerization reaction at different temperatures. PPy synthesized at lower temperatures (4 and 10 °C), displayed conductivities 3–4 orders of magnitude higher than the PPy synthesized at higher temperatures (20 and 30 °C). XPS analysis of PPy synthesized at 30 °C further indicated a decrease in ratio of α/β protons, increasing the probability of coupling at 2,3 positions. These results further corroborate with those with earlier work on conductivities of PPy/PSSNa complex increasing with decreasing synthesis temperature.[144]

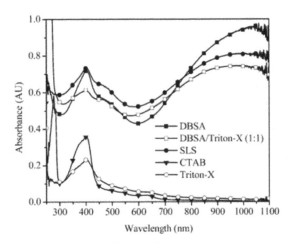

FIGURE 5-16   UV–visible spectra of PPy/surfactant complexes.

5.5.5 ENZYMATIC SYNTHESIS OF POLY(3,4-ETHYLENEDIOXY THIOPHENE) (PEDOT)

A major drawback in the synthesis of conducting polythiophenes (PTh) is the high potentials needed to oxidize the thiophene monomer. At these high potentials, PTh formed is often over-oxidized to a less conducting form.[151] The introduction of alkoxy groups at the 3- and 4-positions decreases the oxidation potential of the thiophene to form 3,4-ethylenedioxy thiophene (EDOT). EDOT can be polymerized using chemical or electrochemical methods resulting in the formation of PEDOT.

PEDOT has attracted attention for medical applications in bionics and neuronal signaling because of the high conductivities attainable and the possible advantages of improved stability. As with PPy, a number of bioactive dopants including collagen,[152] heparin,[153] and choline oxidase[154] have been incorporated into PEDOT during chemical synthesis.

The first reported enzymatic synthesis of water soluble PEDOT was performed at 4 °C at pH 2, using H_2O_2 as the oxidant and PSSNa as the dopant for 16 h. After 16 h, a blue colored polymer solution was obtained. The conductivity of the synthesized PEDOT was 2×10^{-3} S/cm. The authors observed that reaction did not proceed at higher pH conditions. Since activity of HRP is low under acidic conditions, the authors have suggested that HRP is preferentially localized in the EDOT droplets. Monomer droplets would protect the enzyme against further deactivation. The HRP solution in EDOT can be separated from the aqueous dispersion of the synthesized PEDOT-PSSNa.[155] This isolated HRP solution was reused for further PEDOT polymerization. This process could be repeated up to 10 times with the HRP solution separated out each time and the conductivity obtained for all the obtained polymers was in the order of 10^{-3} S/cm. However, nonenzymatic synthesis of PEDOT catalyzed by heme released from native peroxidase cannot be entirely ruled out.

In an effort to increase conductivity of PEDOT, Wang et al.[156] incubated the reaction mixture containing EDOT, HRP, and PSSNa at 50–60 °C. The authors report an increase in polymerization efficiency on heating besides promoting chain extension. Conductivity of around 6.55 S/cm is reported for PEDOT pellets in graphite matrix (4:1 ratio).

In an effort to synthesize PEDOT under milder conditions (pH > 2), SBP was utilized as the enzyme catalyst in the presence of terthiophene as the redox mediator and PSSNa as the dopant (Scheme 5-4).

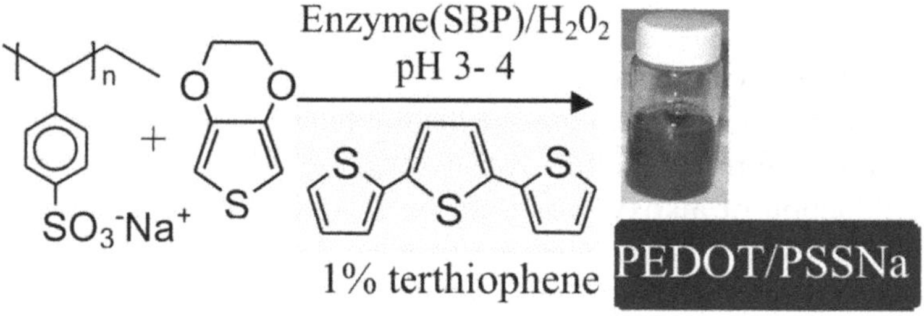

SCHEME 5-4   SBP catalyzed polymerization of EDOT using terthiophene as the redox mediator.

The oxidation potential of terthiophene was sufficiently low for initiation of the polymerization reaction catalyzed by SBP. The oxidized terthiophene helps the subsequent oxidation of EDOT, thus mediating the polymerization reaction. It was found that at least 0.5 wt.% of terthiophene was required in the reaction mixture for the polymerization to proceed. The polymerization reaction was performed at varying pH and utilizing UV–visible spectroscopy it was shown that even at pH 4 the reaction proceeded to afford

PEDOT-PSSNa. The UV–visible absorption (Figure 5-17) and the FTIR[13] spectrum of the synthesized PEDOT-PSSNa closely resembled the spectrum for the commercially available polymer.

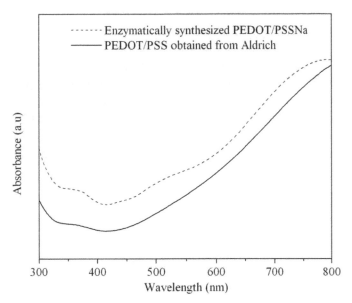

FIGURE 5-17 UV–visible spectra of the commercially available PEDOT/PSS and enzymatically synthesized PEDOT/PSSNa.

The enzymatically synthesized PEDOT/PSSNa could be reversibly dedoped (increasing pH) and re-doped (decreasing pH). The conductivities of the synthesized polymers were in the range of 10^{-3} to 10^{-4} S/cm. The use a thiophene based compound as the redox mediator ensures that any residual mediator that might be incorporated in the final polymer would still be composed of only conjugated thiophene segments and hence the electrical properties of the resulting polymers will remain unaffected.

The concept of using redox mediators in polymerization was further proved by Shumakovich et al.[117] who used laccase and a redox mediator, potassium octocyanomolybdate (4+) for the polymerization of EDOT in presence of poly(2-acrylamido-2-methyl-1-propanesulfonic) acid as the template. Fungal laccase (from *T. hirsuta*) with an oxidation potential of 0.78 V (vs. NHE) could not catalyze the polymerization of EDOT ($E_p = 1.1$ V, vs NHE). Fungal laccase catalyzes the oxidation of octocyanomolybdate (4+) ion by dioxygen to produce the octocyanomolybdate (5+) with the concomitant four electron and four proton reduction of oxygen to water. Octocyanomolybdate (5+)

formed during enzymatic reaction can oxidize EDOT monomer to form oligomeric EDOT (OEDOT). The OEDOT exhibited low conductivity in range of 2.2 10^{-5} to 3.8×10^{-5} S/cm.

More recently, it has been shown that iron containing hemeporphyrins can be used as catalysts for synthesis of PEDOT:PSSNa.[21] Catalase, like HRP, is an iron-containing enzyme on heating to 100 °C denatured the enzyme. Denatured enzyme did not exhibit any enzymatic activity as confirmed by catalase activity measurements. But both the active and the denatured enzyme were able to catalyze the polymerization of EDOT at pH 1 (Figure 5-18b). Control experiments using noniron containing proteins such as BSA and hydrogen peroxide did not result in the formation of PEDOT (Figure 5-18a). The authors concluded that enzymatic activity was not crucial; however, the presence of an iron containing protein was important for enzymatic olymerization.

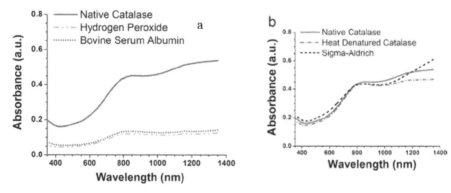

FIGURE 5-18 (a) Absorbance spectra of PEDOT:PSS synthesized using native catalase compared to the reaction product obtained from hydrogen peroxide alone and BSA. (b) Visible and near-IR absorbance spectra of PEDOT:PSS synthesized with native catalase, heat denatured catalase (dashed-dotted line) compared to commercially available PEDOT:PSS.

While the presence of polymeric templates aid in the formation of conductive PEDOT, separation of the polymer from the template is very difficult. Huo et al.[157] synthesized PEDOT in presence of silica using PEG-Hematin as a catalyst. The silica was then removed by etching with HF. PEDOT recovered after removal of silica was found to dissolve in DMSO but was dedoped.

5.5.6   ENZYMATIC AND BIOMIMETIC SYNTHESIS OF PEDOT-PYRROLE COPOLYMERS

In order to obtain enhanced electrical and optical properties, it may be beneficial to combine two materials. For example, the electrochemical properties of CP can be fine-tuned by copolymerization. Copolymers of EDOT and pyrrole have been synthesized by electrochemical methods. The first report on enzymatic synthesis of conducting copolymers of pyrrole EDOT[47] was accomplished using a biomimetic catalyst, PEG-hematin at pH 1 in the presence of PSSNa as a dopant. The electrical conductivity for the copolymers was found to be in the range of 0.1–1 S/cm, which is significantly higher than the conductivity reported for homopolymers of PPy/PSSNa and PEDOT/PSSNa.

In an effort to develop more environmentally benign routes, which are closer to neutral pH, oxidative polymerization of pyrrole/EDOT copolymers was performed using SBP and PSSNa as the charge-balancing dopant. The copolymerization reaction could be performed at higher pH conditions (85% yield at pH 5) (Figure 5-19).

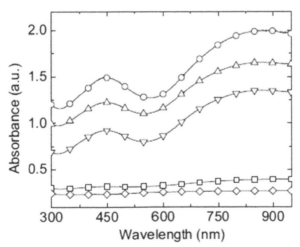

FIGURE 5-19   UV–visible absorption spectra for the copolymerization of EDOT and pyrrole at different pH conditions.

The resulting poly(pyrrole-EDOT)/PSSNa complex is not truly water-soluble, but forms a dispersion in water which is stable and processable and exhibited conductivities of around 10^{-2} S/cm.

The conductivity of the copolymers increased with increase in incorporation of pyrrole units in the copolymers. The composition of the copolymers was observed to be highly dependent on pH of the reaction, with increased pyrrole incorporation for higher pH values. The polymerization reaction was performed at different temperatures (4 and 20 °C). The authors observed that the conductivity of the copolymers was not dependent on the reaction temperature. However, since the conductivity scaled with the pyrrole content in the copolymers, higher electrical conductivities were observed for copolymers synthesized at higher pH (Figure 5-20). The lower redox potential and higher reactivity of pyrrole relative to EDOT may account for the observed trend in the product productivity.

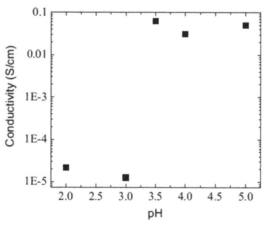

FIGURE 5-20   Conductivity for EDOT/pyrrole copolymers for varying pH.

5.6   CONCLUSIONS AND FUTURE DIRECTIONS

It is evident that oxidoreductases can catalyze polymerization of aniline, pyrrole, and EDOT under a variety of conditions as discussed. Reactions can be carried out under relatively environmentally friendly conditions (pH of around 4 in aqueous solution) even in the presence of biological entities resulting in the formation of conductive polymers, that are processable and exhibit properties comparable to the chemical/electrochemically-synthesized analogues. Enzyme catalyzed polymerizations are efficient resulting in high product yields. While redox potential of enzymes can be used as a tool for predicting polymerization, other reaction parameters such as nature of the polyelectrolyte, pH, and temperature have been shown to modulate the enzyme's redox potential and hence the enzymatic activity. Some of the reaction parameters such as temperature appear to influence the stabilities of complexes of monomer or oligomer radical intermediates with the polyelectrolyte templates. In these cases, the inverse temperature dependence of reaction rate can only be explained in terms of the stability of these complexes at lower temperatures, that increases reaction rates at these temperatures.

Current methods to measure enzyme activity of peroxidases are based on polymerization of phenols in reaction conditions that do not mimic actual polymerization conditions. Further research is required to develop new methods for the measurement of redox potential of enzymes under specific reaction conditions. In addition, new assays for measurement of enzyme activity using structurally similar substrates would be beneficial.

Advances in bioengineering had led to an increase in the commercial availability of many classes of enzymes increasing the range of monomers which can be polymerized enzymatically has increased significantly. The use of redox mediators in enzymatic synthesis has further widened the scope of enzyme-based modifications leading to exciting opportunities for the development of sustainable routes to the synthesis. Immobilization of enzymes on suitable substrates facilitates enzyme reuse, further lowering the overall cost of synthesis. However, it is also becoming apparent that hydrophilically modified metal porphyrins can serve as suitable catalysts for the synthesis of conjugated polymers. Alternatively, they can also serve as catalysts in appropriate micellar environment. Oxidoreductases and metal porphyrins differ in both in the mechanism (the catalytic cycle) and kinetics of polymerization. Further fundamental research is necessary to better understand and influence the mechanism and kinetics of metal porphyrin catalysis.

REFERENCES

1. Rajan, M. Global Markets for Enzymes in Industrial Applications. *BCC Research* March **2012**.
2. Romero, P.; Wagg, J.; Green, M. L.; Kaiser, D.; Krummenacker, M.; Karp P. D. Computational Prediction of Human Metabolic Pathways From the Complete Human Genome. *Genome Bio.* **2005**, *6*, R2.
3. Mu, F.; Pat J. Unkefer, P. J.; Unkefer, C. J.; Hlavacek, W. S. Prediction of Oxido-Reductase-Catalyzed Reactions Based on Atomic Properties of Metabolites. *Bioinformatics* **2006**, *22*, 3082–3088.
4. Bartlett J. M. S.; Stirling D. *PCR Protocols*; Humana Press: Totowa, NJ, 2006; p 556.
5. Scoville, J. R.; Novicova, I. A. Foaming Enzyme Spray Cleaning Composition and Method of Delivery. US Patent 5998342, 1999.

6. Yamada, H.; Kobayashi, M. Nitrile Hydratase and Its Application to Industrial Production of Acrylamide. *Biosci. Biotechnol. Biochem.* **1996,** *60,* 1391–1400.
7. Carasik, W.; Carroll, O. J. Development of Immobilized Enzymes for Production of High Fructose Corn Syrup. *Food Technol.* **1983,** *37,* 85–91.
8. Aslanidis, C.; Jong P. J. Ligation-Independent Cloning of PCR Products (LIC-PCR). *Nucl. Acid Res.* **1990,** *18,* 6069–6074.
9. Brenner, S.; Johnson, M.; Bridgham, J.; Golda, G.; Lloyd, D. H.; Johnson, D.; Luo, S.; McCurdy, S.; Foy, M.; Ewan, M.; Roth, R.; George, D.; Eletr, S.; Albrecht, G.; Vermaas, E.; Williams, S. R.; Moon, K.; Burcham, T.; Pallas, M.; DuBridge, R. B.; Kirchner, J.; Fearon, K.; Mao, J.; Corcoran, K. Gene Expression Analysis by Massively Parallel Signature Sequencing (MPSS) on Microbead Arrays. *Nat. Biotechnol.* **2000,** *18,* 630–634.
10. (a). Anderson, E. M.; Larsson, K. M.; Kirk, O. One Biocatalyst-ManyApplications: The Use of *Candida antartica* B-Lipase in Organic Synthesis. *Biocatal. Biotransform.* **1998,** *16,* 181–204. (b). Faber, K. *Biotransformations in Organic Chemistry: A Textbook,* 5th ed.; Springer-Verlag: Heidelberg, 2004.
11. Liu, W.; Kumar, J.; Tripathy, S.; Senecal, K. J.; Samuelson, L. A. Enzymatically Synthesized Conducting Polyaniline. *J. Am. Chem. Soc.* **1999,** *121,* 71–78.
12. Bouldin, R.; Ravichandran, S.; Kokil, A.; Garhwal, R.; Nagarajan, S.; Kumar, J.; Bruno, F. F.; Samuelson, L. A.; Nagarajan, R. Synthesis of Polypyrrole with Fewer Structural Defects Using Enzyme Catalysis. *Synthetic Metals* **2011,** *161,* 1611–1617.
13. Nagarajan, S.; Kumar, J.; Bruno, F. F.; Samuelson, L. A.; Nagarajan, R. Biocatalytically Synthesized Poly(3,4-ethylenedioxythiophene). *Macromolecules* **2008,** *41,* 3049–3052.
14. Tewari, A.; Kokil, A.; Ravichandran, S.; Nagarajan, S.; Bouldin, R.; Samuelson, L. A.; Nagarajan, R.; Kumar, J. Soybean Peroxidase Catalyzed Enzymatic Synthesis of Pyrrole/EDOT Copolymers. *Macromol. Chem. Phys.* **2010,** *211,* 1610–1617.
15. Sakharov, I. Y.; Vorobiev, A. C.; Castillo Leon, J. J. Synthesis of Polyelectrolyte Complexes of Polyaniline and SulfonatedPolystyrene by Palm Tree Peroxidase. *Enzyme Microb. Tech.* **2003,** *33,* 661–667.
16. Aizawa, M.; Wang, L. L.; Shinohara, H.; Ikariyama, Y. Enzymatic Synthesis of Polyaniline Film Using a Copper-Containing Oxidoreductase. *J. Biotechnol.* **1990,** *14,* 301–310.
17. (a). Solomon E. I.; Lowery M. D. Electronic Structure Contribution to Function in Bioinorganic Chemistry. *Science* **1993,** *259,* 1575–1581. (b). Yaropolov, A.; Skorobogat'ko, O. V.; Vartanov, S. S.; Varfolomeyev S. D. Laccase: Properties, Catalytic Mechanism, and Applicability. *Appl. Biochem. Biotechnol.* **1994,** *49,* 257–280. (c). Solomon E. I.; Sundaram U. M.; Machonkin T. E. Multicopper Oxidases and Oxygenases. *Chem. Rev.* **1996,** *96,* 2563–2606.
18. Song, H.-K.; Palmore, G. T. R. Conductive Polypyrrole via Enzyme Catalysis. *J. Phys. Chem. B.* **2005,** *109,* 19278–19287.
19. Shumakovich, G.; Otrokhov, G.; Vasil'eva, I.; Pankratov, D.; Morozova, O.; Yaropolov, A. Laccase-Mediated Polymerization of 3,4-Ethylenedioxythiophene (EDOT). *J. Mol. Catal. B: Enzym.* **2012,** *81,* 66–68.
20. Kausaite, A.; Ramanaviciene, A.; Ramanavicius, A. Polyaniline Synthesis Catalyzed by Glucose Oxidase. *Polymer* **2009,** *50,* 846–1851.
21. Ramanavicius, A.; Kausaite, A.; Ramanaviciene, A.; Acaite, J.; Malinauskas, A. Redox Enzyme—Glucose Oxidase—Initiated Synthesis of Polypyrrole. *Synth. Met.* **2006,** *156,* 409–413.

22. Cui, X.; Li, C. M.; Zang, J.; Zhou, Q.; Gan, Y.; Bao, H.; Guo, J.; Lee, V. S.; Moochhala, S. M. Biocatalytic Generation of Ppy-Enzyme-CNT Nanocomposite: From Network Assembly to Film Growth. *J. Phys. Chem. C.* **2007**, *111*, 2025–2031.

23. Hira, S. M.; Payne, C. K. Protein-Mediated Synthesis of the Conducting Polymer PEDOT: PSS. *Synthetic Metals* **2013**, *176*, 104–107.

24. Madhavi, V.; Lele, S. S. Laccase: Properties and Applications. *Bioresources* **2009**, *4*, 1694–1717.

25. Bankar, S. B.; Bule, M. V.; Singhal, R. S.; Ananthanarayan, L. Glucose Oxidase—An Overview. *Biotechnol. Adv.* **2009**, *27*, 489–501.

26. Mano, N. Features and Applications of Bilirubin Oxidases. *Appl. Microbiol. Biotechnol.* **2012**, *96*, 301–307.

27. Chelikani, P.; Fita, I.; Loewen, P. C. Diversity of Structures and Properties Among Catalases. *Cell. Mol. Sci. Life. Sci.* **2004**, *61*, 192–208.

28. Jones, P. Catalases. In *Peroxidases and Catalases*, 2nd ed.; Dunford, B. H., Ed.; Wiley and Sons: Hoboken, NJ, 2010; pp 233–256.

29. Alvarado, B.; Torres, E. Recent Patents in the Use of Peroxidases. *Rec. Pat. Biotechnol.* **2009**, *3*, 88–102.

30. Torres, E.; Ayala, M. *Biocatalysis Based on Heme Peroxidases: Peroxidases as Potential Industrial Biocatalysts*; Springer: Heidelberg, 2010.

31. Ortiz de Montellano, P. R., Ed. *Cytochrome P450 Structure, Mechanism, and Biochemistry*, 3rd ed.; Kluwer: New York, 2005.

32. Nagarajan, S.; Bruno, F. F.; Samuelson, L. A.; Kumar, J.; Nagarajan, R. Metalloporphyrin Based Biomimetic Catalysts for Materials Synthesis and Biosensing. *Biomaterials, ACS Symposium Series 1054*, **2011**, *12*, 221–242.

33. (a). Yasuda, T.; Aida, T.; Inoue, S. Living Polymerization of β-Butyrolactone Catalyzed by Tetraphenylporphinatoaluminum Chloride. *Makromol. Chem., Rapid Commun.* **1982**, *3*, 585–588. (b). Yasuda, T.; Aida, T.; Inoue, S. Living Polymerization of β-Lactone Catalyzed by (Tetraphenylporphinato)aluminum Chloride. Structure of the Living End. *Macromolecules* **1983**, *16*, 1792–1796.

34. Aida, T.; Inoue, S. Activation of Carbon Dioxide with Aluminum Porphyrin and Reaction with Epoxide. Studies on (Tetraphenylporphinato)aluminum Alkoxide Having a Long Oxyalkylene Chain as theAlkoxide Group. *J. Am. Chem. Soc.* **1983**, *105*, 1304–1309.

35. Ricoux, R.; Raffy, Q.; Mahy, J.-P. New Biocatalysts Mimicking Oxidative Hemoproteins: Hemoabzymes. *C. R. Chimie* **2007**, *10*, 684–702.

36. Bruno, F. F.; Nagarajan, R.; Roy, S.; Kumar, J.; Tripathy, S. K.; Samuelson, L. A. Use of Hematin for the Polymerization of Water-Soluble Conductive Polyaniline and Polyphenol. *Mat. Res. Soc. Symp. Proc.* **2001**, *660*, 6.1–6.6.

37. Kohri, M.; Fukushima, H.; Taniguchi, T.; Nakahira, T. Synthesis of Polyarbutin by Oxidative Polymerization Using Pegylated Hematin as a Biomimetic Catalyst. *Polymer Journal* **2010**, *42*, 952–955.

38. Nagarajan, S.; Nagarajan, R.; Tyagi, R.; Kumar, J.; Bruno, F. F.; Samuelson, L. A. Biocatalytic Modification of Naturally Occurring Iron Porphyrin. *J. Macromol. Sci.-Pure Appl. Chem.* **2008**, *45*, 951–956.

39. Roxana, I.; Katsuya, K. Lipase-Catalyzed Enantioselective Reaction of Amines with Carboxylic Acids Under Reduced Pressure in Non-Solvent System and in Ionic Liquids. *Tetrahedron Lett.* **2004**, *45*, 523–525.

40. Nagarajan, S.; Nagarajan, R.; Bruno, F. F.; Samuelson, L. A.; Kumar, J. A Stable Biomimetic Redox Catalyst Obtained by the Enzyme-Catalyzed Amidation of Iron Porphyrin. *Green Chem.* **2009,** *11,* 334–338.

41. Nabid, M. R.; Sedghi, R.; Jamaat, P. R.; Safari, N. A.; Entezami, A. Synthesis of Conducting Water-Soluble Polyaniline with Iron(III) Porphyrin. *J. Appl. Polym. Sci.* **2006,** *102,* 2929–2934.

42. Nabid, M. R.; Asadi, S.; Shamsianpour, M.; Sedghi, R.; Osati, S.; Safari, S. Oxidative Polymerization of 3,4-Ethylenedioxythiophene Using Transition-Metal Tetrasulfonated Phthalocyanine. *React. Func. Polym.* **2010,** *70,* 75–80.

43. Xing, H.; Shenggui, L.; Zhao, M.; Zou, G. Hemoglobin-Biocatalyzed Synthesis of Conducting Molecular Complex of Polyaniline and Lignosulfonate. *Journal of Wuhan University of Technology-Mater. Sci. Ed.* **2008,** *23,* 809–815.

44. Xing, H.; Yu-Ying, Z.; Kai, T.; Guo-Lin, Z. Hemoglobin-Biocatalysts Synthesis of a Conducting Molecular Complex of Polyaniline and Sulfonated Polystyrene. *Synth Met.* **2005,** *150,* 1–7.

45. Ravichandran, S.; Nagarajan, S.; Kokil, A.; Ponrathnam, T.; Bouldin, R.; Bruno, F. F.; Samuelson, L. A.; Kumar, J.; Nagarajan, R. Micellar Nanoreactors for Hematin Catalyzed Synthesis of Electrically Conducting Polypyrrole. *Langmuir* **2012,** *28,* 13380–13386.

46. Singh, A.; Kaplan, D. K. Enzyme Based Vinyl Polymerization. *J. Polym. Environ.* **2002,** *10,* 85–91.

47. Bruno, F. F.; Fosey, S.; Nagarajan, S.; Nagarajan, R.; Kumar, J.; Samuelson, L. A. Biomimetic Synthesis of Water-Soluble Conducting Copolymers/Homopolymers of Pyrrole and 3,4-Ethylenedioxythiophene. *Biomacromolecules* **2006,** *7,* 586–589.

48. Nanda, A. K.; Kishore, K. Autocatalytic Oxidative Polymerization of Indene by Cobalt Porphyrin Complex and Kinetic Investigation of the Polymerization of Styrene. *Macromolecules* **2001,** *34,* 1600–1605.

49. (a). MacDiarmid, A. G. Polyaniline and Polypyrrole: Where Are We Headed? *Synth. Met.* **1997,** *84,* 27–34. (b). Mac Diarmid, A. G.; Chiang, J. C.; Richter, A. F.; Epstein, A. J. Polyaniline: A New Concept in Conducting Polymers. *Synth. Met.* **1987,** *18,* 285–290. (c). Chinn, D.; Dubow, J.; Liess, M.; Josowicz, M.; Janata, J. Comparison of Chemically and Electrochemically Prepared Polyaniline. Films. 1. Electrical Properties. *Chem. Mater.* **1995,** *7,* 1504–1509. (d). Epstein, A. J.; MacDiarmid, A. G. The Controlled Electromagnetic Response of Polyanilines and Its Applications to Technologies. In *Science and Applications of Conducting Polymers*; Adam Hilger: Bristol, England, 1990.

50. Gurunathan, K.; Murugan, A. V.; Marimuthu, R.; Mulik, U. P.; Amalnerkar D. P. Electrochemically Synthesized Conducting Polymeric Materials for Applications Towards Technology in Electronics, Optoelectronics and Energy Storage Devices. *Mater. Chem. Phys.* **1999,** *61,* 173–91.

51. Kitani, A.; Yano, J.; Sasaki, K. ECD Materials for the Three Primary Colors Developed by Polyanilines. *J. Electroanal. Chem.* **1986,** *209,* 227–232.

52. (a). Paul, E. W.; Rico, A. J.; Wrighton, M. S. Resistance of Polyaniline Films as a Function of Electrochemical Potential and the Fabrication of Polyaniline-Based Microelectronic Devices. *J. Phys. Chem.* **1985,** *89,* 1441–1447. (c). Chen, S.-A.; Fang, Y. Polyaniline Schottky Barrier: Effect of Doping on Rectification and Photovoltaic Characteristics. *Synth. Met.* **1993,** *60,* 215–222.

53. Heeger, A. J. Polyacetylene: New Concepts and New Phenomena. In *Handbook of Conducting Polymers*; Skotheim, T. A., Ed.; Marcel Dekker: New York, 1986; Vol. 2, pp 729–56.

54. Heeger, A. J. Semiconducting and Metallic Polymers: The Fourth Generation of Polymeric Materials. *Synth. Met.* **2002**, *125*, 23–42.

55. Bredas, J. L. Electronic Structure of Highly Conducting Polymers. In *Handbook of Conducting Polymers*; Skotheim, T. A., Ed.; Marcel Dekker: New York, 1986; Vol. 2, p 859–913.

56. (a). Yang, X.; Too, C. O.; Sparrow, L.; Ramshaw, J.; Wallace, G. G. Polypyrrole-Heparin System for the Separation of Thrombin. *React. Funct. Polym.* **2002**, *53*, 53–62. (b). Wu, Y. Z.; Moulton, S. E.; Too, C. O.; Wallace, G. G.; Zhou, D. Z. Use of Inherently Conducting Polymers and Pulsed Amperometry in Flow Injection Analysis to Detect Oligonucleotides. *Analyst* **2004**, *129*, 585–588. (c). Misoska, V.; Price, W. E.; Ralph, S. F.; Wallace, G. G.; Ogata, N. Synthesis, Characterization and Ion Transport Studies on Polypyrrole/Deoxyribonucleic Acid Conducting Polymer Membranes. *Synth. Met.* **2001**, *123*, 279–286.

57. Wallace, G. G.; Spinks, G. M.; Kane-Maguire, L. A. P.; Teasdale, P. R. *Organic Conducting Polymers: Intelligent Polymer Systems*, 3rd ed.; CRC Press, Taylor & Francis: New York, 2008.

58. Kohavi, D.; Pollack, S. R.; Brighton, C. Short-Term Effect of Guided Bone Regeneration and Electrical Stimulation on Bone Growth in a Surgically Modelled Resorbed Dog Mandibular Ridge. *Biomater. Artif. Cells Immobil. Biotechnol.* **1992**, *20*, 131–138.

59. Kloth, L. C.; Mc Culloch, J. M. Promotion of Wound Healing with Electrical Stimulation. *Adv. Wound Care.* **1996**, *9*, 42–45.

60. Reger, S. I.; Hyodo, A.; Negami, S.; Kambic, H. E.; Sahgal, V. Experimental Wound Healing with Electrical Stimulation. *Artif. Organs* **1999**, *23*, 460–462.

61. Foulds, N. C.; Lowe, C. R. Enzyme Entrapment in Electrically Conducting Polymers. *J. Chem. Soc. Faraday Trans.* **1986**, *82*, 1259–1264.

62. Umana, M.; Waller, J. Protein Modified Electrodes: The Glucose/Oxidase/Polypyrrole System. *Anal. Chem.* **1986**, *58*, 2979–2983.

63. Wong J. Y.; Langer, R.; Ingber D. E. Electrically Conducting Polymers can Noninvasively Control the Shape and Growth of Mammalian Cells. *Proc. Natl. Acad. Sci.* **1994**, *91*, 3201–3204.

64. Shi, G.; Rouabhia, M.; Wang, Z.; Dao, L. H.; Zhang Z. A Novel Electrically Conductive and Biodegradable Composite Made of Polypyrrole Nanoparticles and Polylactide. *Biomaterials* **2004**, *25*, 2477–2488.

65. Shastri V. R.; Vacanti, J. P.; Langer, R. Stimulation of Neurite Outgrowth Using an Electrically Conducting Polymer. *Proc. Natl. Acad. Sci.* **1997**, *94*, 8948–8953.

66. Garner, B.; Georgevich, A.; Hodgson, A. J.; Liu, L.; Wallace, G. G. Polypyrrole-Heparin Composites as Stimulus-Responsive Substrates for Endothelial Cell Growth. *J. Biomed. Mater. Res.* **1999**, *2*, 121–129.

67. Gomez, N.; Lee, J. Y.; Nickels, J. D.; Schmidt, C. E. Micro-Patterned Polypyrrole: A Combination of Electrical and Topographical Characteristics for the Stimulation of Cells. *Adv. Funct. Mater* **2007**, *17*, 1645–1653.

68. Castano, H.; O'Rear, E. A.; McFetridge, P. S.; Sikavitsas V. Polypyrrole Thin Films Formed by Admicellar Polymerization Support the Osteogenic Differentiation of Mesenchymal Stem Cells. *Macromol. Biosci.* **2004**, *4*, 785–794.

69. Williams, R. L.; Doherty, P. J. A Preliminary Assessment of Poly(pyrrole) in Nerve Guide Studies. *J. Mater. Sci: Mater. Med.* **1994**, *5*, 429–433.

70. (a). Shastri, V. R.; Schmidt, C. E.; Kim, T.-H.; Vacanti, J. P.; Langer, R. Polypyrrole—A Potential Candidate for Stimulated Nerve Regeneration. *Mater. Res. Soc. Symp. Proc.*

1996, *414*, 113–118. (b). Chen, S. J; Wang, D. Y; Yuan, C. W; Wang, X. D; Zhang, P.; Gu, X. S. Template Synthesis of the Polypyrrole Tube and Its Bridging In Vivo Sciatic Nerve Regeneration. *J. Mater. Sci. Lett.* **2000**, *19*, 2157–2159. (c). Richardson, R. T.; Thompson, B.; Moulton, S.; Newbold, C.; Lum, M.; Cameron, A.; Wallace, G.; Kapsa, R.; Clark, G.; O' Leary, S. The Effect of Polypyrrole with Incorporated Neurotrophin-3 on the Promotion of Neurite Outgrowth from Auditory Neurons. *Biomaterials* **2007**, *28*, 512–523.

71. George P. M.; Lyckman, A. W.; LaVan D. A.; Hegde, A.; Leung, Y.; Avasare, R.; Testa, C.; Alenander, P. M.; Lander, R.; Sur, M. Fabrication and Biocompatibility of Polypyrrole Implants Suitable for Neural Prosthetics. *Biomaterials* **2005**, *26*, 3511–3519.

72. Cui, X.; Wiler, J.; Dzaman, M.; Altschuler, R. A.; Martin, D. C. In Vivo Studies of Polypyrrole/Peptide Coated Neural Probes. *Biomaterials* **2003**, *24*, 777–787.

73. (a). Entezami, A. A.; Massoumi B. Artificial Muscles, Biosensors and Drug Delivery Systems Based on Conducting Polymers: A Review. *Iranian Polym. J.* **2006**, *15*, 13–30. (b). George, P. M.; La Van, D. A.; Burdick, J. A.; Chen, C. Y.; Liang, E.; Langer, R. Electrically Controlled Drug Delivery From Biotindoped Conductive Polypyrrole. *Adv. Mater.* **2006**, *18*, 577–581. (c). Abidian, M. R.; Kim, D. H.; Martin, D. C. Conducting Polymer Nanotubes for Controlled Drug Release. *Adv. Mater.* **2006**, *18*, 405–409. (d). Li, Y.; Neoh, K. G.; Kang, E. T. Controlled Release of Heparin From Polypyrrole Poly(vinyl alcohol) Assembly by Electrical Stimulation. *J. Biomed. Mater. Res. A* **2005**, *73A*, 171–181.

74. (a). Otero, T. F.; Sansinena, J. M. Bilayer Dimensions and Movement in Artificial Muscles. *Bioelectrochem. Bioenergy* **1997**, *42*, 117–122. (b). Otero, T. F.; Cortes, M. T. A Sensing Muscle. *Sensors Actuators B* **2003**, *96*, 152–156. (c). Spinks, G. M.; Xi, B.; Troung, V.-T.; Wallace, G. G. Actuation Behavior of Layered Composites of Polyaniline, Carbon Nanotubes and Polypyrrole. *Synth. Met.* **2005**, *151*, 85–91.

75. Guimard, N. K.; Gomez, N.; Schimdt, C. E. Conducting Polymers in Biomedical Engineering. *Prog. Polym. Sci.* **2007**, *32*, 876.

76. Inzelt, G. *Chemical and Electrochemical Syntheses of Conducting Polymers in Monographs in Electrochemistry*; Springer: Berlin, Heidelberg, 2008; pp 123–148. (b). Skotheim, T. A.; Reynolds, J. Handbookof Conducting Polymers. Conjugated Polymers: Theory, Synthesis, Properties, and Characterization; CRC Press: Boca Raton, 2006.

77. Dias, H. V. R.; Fianchini, M.; Rajapakse, R. M. G. Greener Method for High-Quality Polypyrrole. *Polymer* **2006**, *47*, 7349–7354.

78. Dias, H. V. R.; Wang, X.; Rajapakse, R. M. G.; Elsenbaumer, R. L. AMild, Copper Catalyzed Route to Conducting Polyaniline. *Chem. Comm.* **2006**, 976–978.

79. Surwade, S. P.; Agnihotra, S. R.; Dua, V.; Manohar, N.; Jain, S.; Ammu, S.; Manohar, S. K. Catalyst-Free Synthesis of Oligoanilines and Polyaniline Nanofibers using H_2O_2. *J. Am. Chem. Soc.* **2009**, *131*, 12528–12529.

80. Bocchi, V.; Chierici, L.; Gardini, G. P.; Mondelli, R. On Pyrrole Oxidation with Hydrogen Peroxide. *Tetrahedron* **1970**, *26*, 4073–4082.

81. Duic, L.; Mandic, Z. Counter-Ion and pH Effect on the Electrochemical Synthesis of Polyaniline. *Electroanal. Chem.* **1992**, *335*, 207–221.

82. Cusanovich, M. A.; Hazzard, J. T.; Meyer, T. E. Electron Transfer Mechanisms in Heme Proteins. *J. Macromol. Sci. A.* **1989**, *26*, 433–443.

83. Shan, J.; Cao, S. Enzymatic Polymerization of Aniline and Phenol Derivatives Catalyzed by Horseradish Peroxidase in Dioxane. *Polym. Adv. Technol.* **2000**, *11*, 288–293.

(b). Shan, J.; Han, L.; Bai, F.; Cao, S. Enzymatic Polymerization of Aniline and Phenol Derivatives Catalyzed by Horseradish Peroxidase in Dioxane (II). *Polym. Adv. Technol.* **2003,** *14,* 330–336.

84. Alva, K. S.; Lee, T. S.; Kumar, J.; Tripathy, S. K. Enzymatically Synthesized Photodynamic Polyaniline Containing Azobenzene Groups. *Chem. Mater.* **1998,** *10,* 1270–1275.

85. Tripathy, S. K.; Kim, D. Y.; Li, L.; Viswanathan, N. K.; Balasubramanian, S.; Liu, W.; Wu, P.; Bian, S.; Samuelson, L. A.; Kumar J. Photofabrication of Electroactive Polymers for Photonics. *Synth. Met.* **1999,** *102,* 893–896.

86. Kim, S. C.; Huh, P.; Kumar, J.; Kim, B.; Lee, J. O.; Bruno, F. F.; Samuelson, L. A. Synthesis of Polyaniline Derivatives via Biocatalysis. *Green Chem.* **2007,** *9,* 44–48.

87. Alva, K. S.; Kumar, J.; Marx, K. A.; Tripathy, S. K. Enzymatic Synthesis and Characterization of a Novel Water-Soluble Polyaniline: Poly(2,5-diaminobenzenesulfonate). *Macromolecules* **1997,** *30,* 4024–4029.

88. Liu, W.; Kumar, J.; Tripathy, S. K.; Samuelson, L. A. Enzymatic Synthesis of Conducting Polyaniline in Micelle Solutions. *Langmuir* **2002,** *18,* 9696–9704.

89. (a). Rao, A. M.; John, V. T.; Gonzalez, R. D.; Akkara, J. A.; Kaplan, D. L. Catalytic and Interfacial Aspects of Enzymic Polymer Synthesis in Reversed Micellar Systems. *Biotechnol. Bioeng.* **1993,** *41,* 531–540. (b). Premachandran, R. S.; Banerjee, S.; Wu, X.-K.; John, V. T.; McPherson, G. L.; Akkara, J. A.; Ayyagari, M.; Kaplan, D. L.; Enzymic Synthesis of Fluorescent Naphthol-Based Polymers. *Macromolecules* **1996,** *29,* 6452–6460.

90. Bruno R.; Akkara, J. A.; Samuelson, L. A.; Kaplan, D. L.; Marx, K. A.; Kumar, J.; Tripathy, S. K. Enzymatic Mediated Synthesis of Conjugated Polymers at the Langmuir Trough Air–Water Interface. *Langmuir* **1995,** *11,* 889–892.

91. Alva, K. S.; Marx, K. A.; Kumar, J.; Tripathy, S. K. Biochemical Synthesis of Water-Soluble Polyanilines: Poly(*p*-aminobenzoic acid). *Macromol. Rapid Commun.* **1996,** *17,* 859–863.

92. Sahoo, S. K.; Nagarajan, R.; Chakraborty, S.; Samuelson, L. A.; Kumar, J.; Cholli, A. L. Variation in the Structure of Conducting Polyaniline With and Without the Presence of Template During Enzymatic Polymerization: A Solid-State NMR Study. *J. Macromol. Sci. Pure Appl. Chem.* **2002,** *39,* 1223–1240.

93. Liu, W.; Cholli, A. L.; Nagarajan, R.; Kumar, J.; Tripathy, S.; Bruno, F. F.; Samuelson, L. A. The Role of Template in the Enzymatic Synthesis of Conducting Polyaniline. *J. Am. Chem. Soc.* **1999,** *121,* 11345–11355.

94. Nagarajan, R.; Tripathy, S.; Kumar, J. An Enzymatically Synthesized Molecular Complex of Polyaniline and Poly(vinylphosphonic acid). *Macromolecules* **2000,** *33,* 9542–9547.

95. Shen, Y.; Sun, J.; Wu, J.; Zhou, Q. Synthesis and Characterization of Water-Soluble Conducting Polyaniline by Enzyme Catalysis. *J. App. Poly. Sci.* **2005,** *96,* 814–817.

96. Thiyagarajan, M.; Samuelson, L. A.; Kumar, J.; Cholli, A. Helical Conformational Specificity of Enzymatically Synthesized Water-Soluble Conducting Polyaniline Nanocomposites. *J. Am. Chem. Soc.* **2003,** *125,* 11502–11503.

97. Rumbau, V.; Pomposo, J. A.; Alduncin, J. A.; Grande, H.; Mecerreyes, D.; Ochoteco, E. A New Bifunctional Template for the Enzymatic Synthesis of Conducting Polyaniline. *Enzyme Microb. Technol.* **2007,** *40,* 1412–1421.

98. Bouldin, R., Kokil, A., Ravichandran, S., *Nagarajan, S.,* Kumar, J., Samuelson, L. A., Bruno, F. F., Nagarajan, R. Enzymatic Synthesis of Electrically Conducting Polymers. *ACS Symposium Series* **2010,** *1043,* 315–341.

99. Xu, P.; Singh, A.; Kaplan, D. L. Enzymatic Catalysis in the Synthesis of Polyanilines and Derivatives of Polyanilines. *Adv. Polym. Sci.* **2006,** *194*, 69–94.

100. Karamyshev, A. V.; Shleev, S. V.; Koroleva, O. V.; Yaropolov, A. I.; Sakharov, I. Y. Laccase-Catalyzed Synthesis of Conducting Polyaniline. *Enzyme Microb. Technol.* **2003,** *33*, 556–564.

101. Shleev, S. V.; Morozova, O. V.; Nikitina, O. V.; Gorshina, E. S.; Rusinova, T. V.; Serezhenkov, V. A.; Burbaev, D. S.; Gazaryan, I. G.; Yaropolov, A. I. Comparison of Physico-Chemical Characteristics of Four Laccases From Different Basidiomycetes. *Biochimie* **2004,** *86*, 693–703.

102. Reinhammar, B. R.; Vänngård, T. I. The Electron-Accepting Sites in *Rhus vernicifera* Laccase as Studied by Anaerobic Oxidation-Reduction Titrations. *Eur. J. Biochem.* **1971,** *18*, 463–468.

103. Caramyshev, A. V.; Lobachov, V. M.; Selivanov, D. V.; Sheval, E. V.; Vorobiev, A. K.; Katasova, O. N.; Polyakov, V. Y.; Makarov, A. A.; Sakharov, I. Y. Micellar Peroxidase-Catalyzed Synthesis of Chiral Polyaniline. *Biomacromolecules* **2007,** *8*, 2549–2555.

104. Nagarajan, R.; Liu, W.; Kumar, J.; Tripathy, S.; Bruno, F. F.; Samuelson, L. A. Manipulating DNA Conformations Using Intertwined Conducting Polymer Chains. *Macromolecules* **2001,** *34*, 3921–3927.

105. Nagarajan, R.; Roy, S.; Kumar, J.; Tripathy, S. K.; Dolukhanyan, T.; Sung, C.; Bruno, F. F.; Samuelson, L. A. Enzymatic Synthesis of Molecular Complexes of Polyaniline with DNA and Synthetic Oligonucleotides: Thermal and Morphological Characterization. *J. Macromol. Sci. Pure Appl. Chem.* **2001,** *38*, 1519–1537.

106. Datta, B.; Schuster, G. B. DNA-Directed Synthesis of Aniline and 4-Aminobiphenyl, Oligomers: Programmed Transfer of Sequence Information to a Conjoined Polymer Nanowire. *J. Am. Chem. Soc.* **2008,** *130*, 2965–2973.

107. Han, Y.-G.; Kusunose, T.; Sekino, T. Facile One-Pot Synthesis and Characterization of Novel Nanostructured Organic Dispersible Polyaniline. *J. Polym. Sci. B Polym. Phys.* **2009,** *47*, 1024–1029.

108. Zemel, H.; Quinn, J. F. Enzymatic Synthesis of Polyaniline. US Patent 5,420,237, 1995.

109. Kausaite, A.; Ramanaviciene, A.; Ramanavicius, A. Polyaniline Synthesis Catalyzed by Glucose Oxidase. *Polymer* **2009,** *50*, 1846–1851.

110. Cruz-Silva, R.; Romero-Garcia, J.; Angulo-Sanchez, J. L.; Ledezma-Perez, A.; Arias-Marin, E.; Moggio, I.; Flores-Loyola, E. Template-Free Enzymatic Synthesis of Electrically Conducting Polyaniline Using Soybean Peroxidase. *Euro. Poly. J.* **2005,** *41*, 1129– 1135.

111. (a). Azevedo, A. M.; Vojinovic, V.; Cabral, J. M. S.; Gibson, T. D.; Fonseca, L. P. Operational Stability of Immobilised Horseradish Peroxidase in Mini-Packed Bed Bioreactors. *J. Mol. Cat. B Enzym.* **2004,** *28*, 121–128. (b). Moeder, M.; Martin, C.; Koeller, G. Degradation of Hydroxylated Compounds Using Laccase and Horseradish Peroxidase Immobilized on Microporous Polypropylene Hollow Fiber Membranes. *J. Memb. Sci.* **2004,** *245*, 183–190.

112. Nabid, M. R.; Golbabaee, M.; Moghaddam, A. B.; Dinarvand, R.; Sedghi, R. Polyaniline/TiO_2 Nanocomposite: Enzymatic Synthesis and Electrochemical Properties. *Int. J. Electrochem. Sci.* **2008,** *3*, 1117–1126.

113. Yemini, M.; Xu, P.; Kaplan, D. L.; Rishpon, J. Collagen-Like Peptide as a Matrix for Enzyme Immobilization in Electrochemical Biosensors. *Electroanalysis* **2006,** *18*, 2049–2054.

114. Alvarez, S.; Manolache, S.; Denes, F. Synthesis of Polyaniline Using Horseradish Peroxidase Immobilized on Plasma-Functionalized Polyethylene Surfaces as Initiator. *J. Appl. Polym. Sci.* **2003**, *88*, 369–379.
115. Jin, Z.; Su, Y.; Duan, Y. A Novel Method for Polyaniline Synthesis with The Immobilized Horseradish Peroxidase Enzyme. *Synth. Met.* **2001**, *122*, 237–242.
116. Rumbau, V.; Marcilla, R.; Ochoteco, E.; Pomposo, J. A.; Mecerreyes, D. Ionic Liquid Immobilized Enzyme for Biocatalytic Synthesis of Conducting Polyaniline. *Macromolecules* **2006**, *39*, 8547–8549.
117. Vasil'eva, I. S.; Morozova, O. V.; Shumakovich, G. P.; Yaropolov, A. I. Synthesis of Electroconductive Polyaniline Using Immobilized Laccase. *Appl. Biochem. Microbiol.* **2009**, *45*, 27–30.
118. Nabid, M. R.; Entezami, A. A. Enzymatic Synthesis and Characterization of a Water-Soluble, Conducting Poly(o-toluidine). *Euro. Poly. J.* **2003**, *39*, 1169–1175.
119. Nabid, M. R.; Entezami, A. A. Synthesis of Water-Soluble and Conducting Poly(2-ethylaniline) by Using Horseradish Peroxidase. *Iran. Poly. J.* **2003**, *12*, 401–406.
120. Nabid, M. R.; Sedghi, R.; Entezami, A. A. Enzymatic Oxidation of Alkoxyanilines for Preparation of Conducting Polymers. *J. Appl. Poly. Sci.* **2007**, *103*, 3724–3729.
121. Nabid, M. R., Entezami, A. A. Comparative Study on the Enzymatic Polymerizationof N-Substituted Aniline Derivatives. *Polym. Advan. Technol.* **2005**, *16*, 305–309.
122. Kim, S. C.; Kim, D.; Lee, J.; Wang, Y.; Yang, K.; Kumar, J.; Bruno, F. F.; Samuelson, L. A. Template-Assisted Synthesis of Self-Doped Polyaniline: Morphological Effects of Templates on the Conductivity. *Macromol. Rapid Commun.* **2007**, *28*, 1356–1360.
123. Akkara, J. A.; Wang, J.; Yang, D.-P.; Gonsalves, K. E. Hematin-Catalyzed Polymerization of Phenol Compounds. *Macromolecules* **2000**, *33*, 2377–2382.
124. Ku, B. C.; Lee, S. H.; Liu, W., He, J.-A.; Kumar, J.; Bruno, F. F.; Samuelson, L. A. A New Approach to Catalyze Template Polymerization of Aniline Using Electrostatically Multilayered Hematin Assemblies. *J. Macromol. Sci.—Pure Appl. Chem.* **2003**, *A40* (12), 1335–1346.
125. Tierrrablanca, E.; Garcia, J. R.; Roman, P.; Cruz-Silva, R. Biomimetic Polymerization of Aniline Using Hematin Supported on Halloysite Nanotubes. *Appl. Catal., A: General* **2010**, *381*, 267–273.
126. Sahoo, S. K.; Nagarajan, R.; Roy, S.; Samuelson, L. A.; Kumar, J.; Cholli, A. L. An Enzymatically Synthesized Polyaniline: A Solid-State NMR Study. *Macromolecules* **2004**, *37*, 4130–4138.
127. Sahoo, S. K.; Nagarajan, R.; Roy, S.; Samuelson, L. A.; Kumar, J.; Cholli, A. L. An Enzymatically Synthesized Polyaniline: A Solid-State NMR Study. *Macromolecules* **2004**, *37*, 4130–4138.
128. Roy, S.; Fortier, J. M.; Nagarajan, R.; Tripathy, S. K.; Kumar, J.; Samuelson, L. A.; Bruno, F. F. Biomimetic Synthesis of a Water Soluble Conducting Molecular Complex of Polyaniline and Lignosulfonate. *Biomacromolecules* **2002**, *3*, 937–941.
129. Bruno, F. F.; Nagarajan, S.; Nagarajan, R.; Kumar, J.; Samuelson, L. A. *Polymer Preprints* **2007**, *48* (2),80–81.
130. Nabid, M. R.; Zamiraei, Z.; Sedghi, R.; Safari, N. Cationic Metalloporphyrins for Synthesis of Conducting, Water-Soluble Polyaniline. *React. Funct. Polym.* **2009**, *69*, 319–324.
131. Nabid, M. R.; Sedghi, R.; Jamaat, P. R.; Safari, N. A.; Entezami, A. Synthesis of Conducting Water-Soluble Polyaniline with Iron (III) Porphyrin. *J. Appl. Polym. Sci.* **2006**, *102*, 2929–2934.

132. Zhang, K.; Cai, R. X.; Chen D. H. Determination of Hemoglobin Based on Its Enzymatic Activity for the Oxidation of *o*-Phenylenediamine with Hydrogen Peroxide. *Anal. Chim. Acta* **2000,** *413*, 109–113.

133. Roy, S.; Fortier, J. M.; Nagarajan, R.; Tripathy, S.; Kumar, J.; Samuelson, L. A.; Bruno, F. F. Biomimetic Synthesis of a Water Soluble Conducting Molecular Complex of Polyaniline and Lignosulfonate. *Biomacromolecules* **2002,** *3* (5), 937–941.

134. Hu, X.; Zhang, Y.-Y.; Tang, K.; Zou, G.-L. Hemoglobin Biocatalysts of a Conducting Molecular Complex of Polyaniline and Sulfonated Polystyrene. *Synth. Met.* **2005,** *150*, 1–7.

135. (a). Armes, S. P. Optimum Reaction Conditions for the Polymerization of Pyrrole by Iron (III) Chloride in Aqueous Solution. *Synth. Met.* **1987,** *20,* 365–337. (b). Dhawan, S. K.; Trivedi, D. C. Thin Conducting Polypyrrole Film on Insulating Surface and Its Applications. *Bull. Mater. Sci.* **1993,** *16*, 371–380.

136. Bocchi, V.; Chierici, L.; Gardini, G. P.; Mondelli, R. On Pyrrole Oxidation with Hydrogen Peroxide. *Tetrahedron* **1970,** *26*, 4073–4082.

137. Ramanavicius, A.; Kausaite, A.; Ramanaviciene, A.; Acaite, J.; Malinauskas, A. Redox Enzyme—Glucose Oxidase—Initiated Synthesis of Polypyrrole. *Synth. Met.* **2006,** *156*, 409–413.

138. Reece, D. A.; Pringle, J. M.; Ralph, S. M.; Wallace, G. G. Autopolymerization of Pyrrole in the Presence of a Host/Guest Calixarene. *Macromolecules* **2005,** *38*, 1616–1622.

139. Nabid, M. R.: Entezami, A. A. A Novel Method for Synthesis of Water-Soluble Polypyrrole with Horseradish Peroxidase Enzyme. *J. App. Polym. Sci.* **2004,** *94*, 254–258.

140. Kupriyanovich, Y. N., Sukhov, B. G., Medvedeva, S. A., Mikhaleva, A. I., Vakul'skaya, T. I., Myachina, G. F., Trofimov, B. A. Peroxidase-Catalysed Synthesis of Electroconductive Polypyrrole. *Mendeleev Commun.* **2008,** *18*, 56–58.

141. Audebert, P.; Hapiot, P. Fast Electrochemical Studies of the Polymerization Mechanisms of Pyrroles and Thiophenes. Identification of the First Steps. Existence of π-Dimers in Solution. *Synth. Met.* **1995,** *75*, 95–102.

142. Matsushita, M.; Kuramitz, H.; Tanaka, S. Electrochemical Oxidation for Low Concentration of Aniline in Neutral pH Medium: Application to the Removal of Aniline Based on the Electrochemical Polymerization on a Carbon Fiber. *Environ. Sci. Technol.* **2005,** *39*, 3805–3810.

143. Claus, H. Laccases: Structure, Reactions, Distribution. *Micron* **2004,** *35*, 93–96.

144. Farhangrazi, Z. S.; Fossett, M. E.; Powers, L. S. Ellis, W. R. Jr. Variable-Temperature Spectroelectrochemical Study of Horseradish Peroxidase. *Biochemistry* **1995,** *34*, 2866–2871.

145. Cruz-Silva, R.; Amaro, E.; Escamilla, A.; Nicho, M. E.; Sepulveda-Guzman, S.; Arizmendi, L.; Romero-Garcia, J.; Castillon-Barraza, F. F.; Farias, M. H. Biocatalytic Synthesis of Polypyrrole Powder, Colloids, and Films Using Horseradish Peroxidase. *J. Coll. Inter. Sci.* **2008,** *328*, 263–269.

146. Song, H.-K.; Palmore, G. T. R. Conductive Polypyrrole via Enzyme Catalysis. *J. Phys. Chem. B.* **2005,** *109*, 19278–19287.

147. Mc Eldoon, J. P.; Pokora, A. R.; Dordick, J. S. Soybean Peroxidase Has Lignin Peroxidase-Type Activity. *Enzyme Microb. Technol.* **1995,** *17,* 359–365.

148. Hu, P.; Han, L.; Dong, S. A Facile One-Pot Method to Synthesize a Polypyrrole/Hemin Nanocomposite and Its Application in Biosensor, Dye Removal, and Photothermal Therapy. *ACS Appl. Mater. Interfaces* **2014,** *6*, 500– 506.

149. Bruno, F. F.; Nagarajan, R.; Kumar, J.; Samuelson, L. A. Biomimetic Synthesis of Water-Soluble Conducting Polypyrrole and Poly(3,4-ethylenedioxythiophene). *J. Macromol. Sci.–Pure Appl. Chem.* **2003,** *A40* (12),1327–1333.
150. Movahedi, A. A. M.; Semsarha, F.; Heli, H.; Nazari, K.; Ghourchian, H.; Hong, J.; Hakimelahi, G. H.; Saboury, A. A.; Sefidbakht, Y. Micellar Histidinate Hematin Complex as an Artificial Peroxidase Enzyme Model: Voltammetric and Spectroscopic Investigations. *Colloids Surf., A* **2008,** 320, 213–221.
151. Krische, B.; Zagorska, M. The Polythiophene Paradox. *Synth. Met.* **1989,** *28,* C263–C268.
152. Xiao, Y. H.; Li, C. M.; Wang, S. Q.; Shi, J. S.; Ooi, C. P. Incorporation of Collagen in Poly(3,4-ethylenedioxythiophene) for a Bifunctional Film with High Bio- and Electro-chemical Activity. *J. Biomed. Mater. Res. Part A* **2010,** *92A,* 766–772.
153. Asplund, M.; Thaning, E.; Lundberg, J.; Sandberg-Nordqvist, A. C.; Kostyszyn, B.; Inganas, O.; von Holst, H. Toxicity Evaluation of PEDOT/BiomolecularComposites Intended for Neural Communication Electrodes. *Biomed. Mater.* **2009,** *4,* 1–12.
154. Zeng, H. J.; Jiang, Y. D.; Yu, J. S.; Xie, G. Z. Choline Oxidase Immobilized into Conductive Poly(3,4-ethylenedioxythiophene) Film for Choline Detection. *Appl. Surf. Sci.* **2008,** *254,* 6337–6340.
155. Sikora, T.; Marcilla, R.; Mecerreyes, D.; Rodriguez, J.; Pomposo, J. A.; Ochoteco, E. Enzymatic Synthesis of Water-Soluble Conducting Poly(3,4-ethylenedioxythiophene): A Simple Enzyme Immobilization Strategy for Recycling and Reusing. *J. Polym. Sci. A: Polym. Chem.* **2009,** *47,* 306–309.
156. Wang, J.; Fang, B.-S.; Chou, K.-Y.; Chen, C.-C.; Gu, Y. A Two-Stage Enzymatic Synthesis of Conductive Poly(3,4-ethylenedioxythiophene). *Enzyme Microb. Technol.* **2014,** *54,* 45–50.
157. Huo, P. H.; Kim, S.-C.; Kim, Y.-H; Kumar, J.; Kim, B. S.; Ho, N. J.; Lee, J. O. Recovery and Characterization of Pure Poly(3,4-ethylenedioxythiophene) via Biomimetic Template Polymerization. *Polym. Eng. Sci.* **2007,** *47,* 71–75.

CHAPTER 6

THERMOREVERSIBLE ION GELS FROM SIDE-CHAIN LIQUID CRYSTALLINE BRUSHES DIBLOCK COPOLYMERS

CHI THANH NGUYEN, PRASHANT DESHMUKH, XIAORUI CHEN, SERGIO GRANADOS-FOCIL, and RAJESWARI M. KASI

CONTENTS

ABSTRACT

We report a strategy for preparing thermoreversible ion gels from a library of liquid crystalline brush block copolymers (LCBBCs). These LCBBCs were synthesized by ring-opening metathesis polymerization (ROMP) of side-chain functionalized norbonenes comprising cholesteryl ester with nine methylene spacer and poly(ethylene glycol) monomethyl ether (MPEG). The molecular composition, thermal properties, and morphologies of these brush block copolymers were investigated. These LCBBCs were used to form ion gel via self-assembly of LCBBCs in a room temperature ionic liquid (IL) 1-butyl-3-methylimidazolium hexafluorophosphate [BMIM][PF_6]. The hydrophilic PEO units are soluble in IL, while the use of cholesterol allows the formation of junction in the gel network due to the insolubility of its chain in IL, which predominantly provides mechanical robustness to the corresponding ion gels. Transparent, strong ion-gels with significant mechanical strength ($\sim$10^3 Pa) and high ionic conductivity ($\sim$10^{-2} S/cm) can be formed at approximately 20 wt.% concentration of the LCBBCs. The self-assembled ion gels have tunable thermal, mechanical, and morphological properties by virtue of their architecture and functionality is an alternative and attractive way for development of novel solid state electrolytes and offer exciting possibilities of their application in electrochemical devices.

6.1   INTRODUCTION

In recent years, self-assembling polymers and other soft materials in ionic liquids (ILs) have become an interesting topic of research.[1-3] Specifically, the development of polymer electrolytes with unique properties such as thin film formability, flexibility, and transparency in addition to high ionic conductivity has been to produce novel electrochemical devices.[4] For example, polymeric ion-gels consisting of a swollen polymeric network in an IL are among the most conductive solid state electrolytes due to their good mechanical properties and high ionic conductivity which can be used in Li-ion batteries, electrochemical devices, sensors, electromechanical actuators, gas separation membranes, and dye-sensitized solar cells.[5-12] Most of the studies on ion-gels derived from polymers are based on (1) doping of ILs with polymers, (2) in situ polymerization (or cross-linking) of vinyl monomers in Ils,[13] and (3) crystalline fluorinated copolymers. Carlin and co-workers have demonstrated the formation of stable gel electrolytes by adding poly(vinylidene fluoride)-hexafluoropropylene copolymer to

1-ethyl-3-methylimidazolium salts of triflate and tetrafluoroborate.[14] Watanabe et al. have synthesized transparent ion-gels of high ionic conductivity (~10 mS/cm) by in situ polymerization of methyl methacrylate (MMA) monomers in an imidazolium based IL.[15]

The utilization of block copolymers is of great interest since this methodology may have the potential to afford easily processable and mechanically strong ion gels and has opened unprecedented possibilities in the design of advanced materials for solid state electrolyte applications. Lodge and co-workers have demonstrated the gelation of IL through the process of self-assembly of ABA triblock copolymer in a B-block compatible IL.[16,17] Very recently, we have reported the synthesis of thermoreversible ion-gels from N-ter-butylacrylamide/ethylene oxide based triblock copolymer for solid electrolyte applications.[18]

Block copolymers containing other self-assembling units such as liquid crystals are interesting to harness mechanical features of block copolymers and stimuli responsive features of liquid crystals. Thus, systematic and comprehensive mechanical and electrical property evaluation of ion-gels produced via self-assembly of liquid crystalline block copolymers (LCBCPs) is desirable (Figure 6-1). A step forward, LCBCPs containing hydrophobic cholesterol molecules have gained importance as a platform for optoelectronics, color information technology, biotechnological, and biomedical applications.[19–23] These LCBCPs bearing cholesterol side-chains exhibit mechanical properties necessary for various applications in electrochemical devices.[24] However, most current LCBCPs are linear polymers with only a few liquid crystalline (LC) molecules, which resulted in limited defined materials properties. Brush copolymers including typical brush architecture or side-chain polymers with numerous hydrophobic groups possess a broad variety of well-defined nanoscopic morphologies, which can be controlled by type of backbone and graft density. These distinctive properties of brush block copolymers with unique untangled side chains have significant advantages in the development of nanoobjects, organic nanotubes, and photonic materials.[25–27] Several groups have established methods to prepare hydrogels using polymeric brushes.[28,29] However, the possibility of using the brush polymers for preparation of ion-gels has not been investigated.

The synthesis of cholesterol-containing LCBCPs comprising molecular brush-type semicrystalline PEG units in one block and size-chain LC units in the other block (LCBBCs) have been previously reported.[30] These polymers are known to self assemble in water into core–shell micelles and have been used for drug delivery applications. In this study, we demonstrated the gelation of and IL/LCBBC matrix through block copolymer self assembly.

The IL used is 1-butyl-3-methylimidazolium hexafluorophosphate [BMIM] [PF$_6$]. Poly(norbornene)-based side-chain liquid crystalline brush block copolymers (LCBBCPs) bearing cholesteryl mesogens in the first block and semicrystalline PEG in the second block (NBCh9-b-NBMPEG) are used as the LCBBC. These copolymers were synthesized by ring-opening metathesis polymerization (ROMP) using modified Grubbs second generation catalysts. We analyzed the relationship between composition–microstructure–rheological properties–ionic conductivity of these LCBBCs. To the best of our knowledge, this is the first report focuses on designing highly conductive IL-containing polymeric ion-gels through the self-assembly of LCBBCs. These materials are promising candidates for applications in energy storage scaffolds, dye-sensitized solar cell scaffolds, gas separation membranes, catalysis, and CO$_2$ selective sorbent coatings.[31–35]

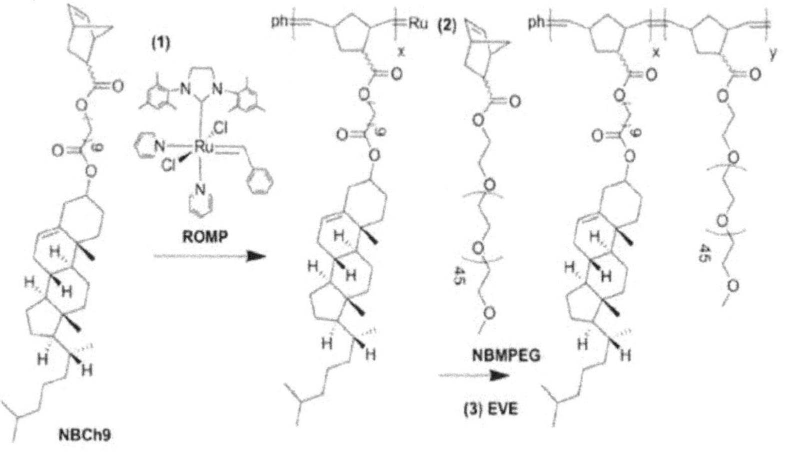

FIGURE 6-1   Synthesis of LCBBCs by sequential ROMP using side chain functionalized NBCh9 and NBMPEG monomer. (From Deshmukh, P.; Ahn, S.-K.; Geelhand de Merxem, L.; Kasi, R. M. Interplay between Liquid Crystalline Order and Microphase Segregation on the Self-Assembly of Side-Chain Liquid Crystalline Brush Block Copolymers. Macromolecules 2013, 46, 8245–8252. © Copyright 2013 American Chemical Society. Used with permission.)

6.2   EXPERIMENTAL

6.2.1   MATERIALS

[BMIM][PF$_6$] (Sigma-Aldrich Co. LLC. USA) are used as received. Modified Grubbs catalyst second generation (H$_2$IMes)(pyr)$_2$Cl$_2$RuCHPh (G3),[37] monomer norbonene bearing cholesterol with nine methylene spacer

(NBCh9), and homopolymer PNBCh9 are synthesized from previous reported procedure.[21] 5-Norbonene-2-carboxylic acid (NBCOOH, mixture of endo and exo), Grubbs catalyst second generation, poly(ethylene glycol) monomethyl ether (M_n = 2000 g/mol) are purchased from Sigma-Aldrich and used without further purification. Dry CH_2Cl_2, anhydrous Tetrahydrofuran (THF), oxalyl chloride, and ethyl vinyl ether (EVE) are obtained from Acros Organics USA. The liquid crystalline brush diblock copolymer LCBBCs was synthesized according to our published work.[30]

6.2.2 PREPARATION OF ION GELS

To prepare stable solutions and gels, co-solvent aided dissolution method was followed. In this procedure, a pre-weighed LC brush diblock copolymer is first dissolved in dichloromethane (DCM), a common solvent for both the blocks, with subsequent addition of [BMIM][PF_6] to get the desired concentration. The sample is set aside in a hood for 2 weeks at ambient temperature and DCM was removed by gradual evaporation. Thereafter, the mixture is maintained at 80 ± 5 °C for 6 h and optically clear solutions and gels are obtained. In this study the concentration is expressed in wt.%. The prepared solutions and ion-gels are stored in air-tight vials and aged for a week at room temperature prior to rheological experiments. Newly prepared samples are used for each experiment.

6.2.3 CHARACTERIZATION

^{1}H-NMR was recorded on a Bruker DMX 400 MHz NMR spectrometer using $CDCl_3$ as solvent and the peak of $CDCl_3$ at 7.24 ppm is used as internal standard. The number-average molecular weight (M_n) and polydispersity indices (PDI) of the polymers were determined by gel permeation chromatography (GPC) using a Waters 150-C ALC/GPC equipped with Evaporative Light Scattering Detector. THF was used as the eluent with a flow rate of 2.0 mL/min at 40 °C. PS is used as standard.

6.2.4 RHEOLOGY

The rheological properties of diblock copolymer solutions and ion-gels are analyzed using the AR-G2 rheometer (TA instruments, Minimum Torque

Oscillation: 0.003 μN.m and Torque Resolution: 0.1 μN.m) with peltier plate-temperature control. A cone-plate geometry with a diameter, $d = 40.0$ mm, and cone angle (deg:min:sec = 1:59:24), is used for more fluid-like samples with approximately 2 mL of sample added at experimental temperatures. Parallel plate geometry (20 mm diameter) is used for more solid-like samples. Dynamic frequency sweep experiments are performed from 10^{-2} to 10^2 rads^{-1} between 10 and 100 °C while cooling the samples. Dynamic temperature ramp experiments are performed to determine the temperature dependent rheological properties by heating the samples at a rate of 1 °C/min and oscillation frequency (ω) 1 rads^{-1}. Only linear viscoelastic properties are measured for dynamic frequency and temperature ramp experiments and the linear range is determined using strain sweep experiments. Strength of the ion-gels is qualitatively ascertained by running dynamic strain sweeps at oscillation frequency (ω) of 6.283 rads^{-1}. In each experiment, 30 min conditioning time is allowed for thermal equilibration and to get rid of any shear history introduced while transferring the copolymer solutions to the appropriate geometry.

6.2.5   IONIC CONDUCTIVITY MEASUREMENTS

The ionic conductivity (σ) measurements are carried out in an in-house designed and machined cell using an Agilent 4284A Precision LCR meter. The amplitude of the AC voltage signal is 10 mV and the applied frequency range is 20–60,000 Hz. The solutions and gels are heated at 90–100 °C and then filled into the multi-sample cell consisting of two stainless steel blocking electrodes separated by a teflon ring of diameter (d = 22.2 mm) and thickness (L = 2.05 mm). The cell is calibrated using a 0.1 N aqueous KCl standard solution at 25 °C. Before each conductivity measurement, the sample cell is equilibrated at the testing temperature for 20–30 min. The reported conductivity values for bulk [BMIM][PF$_6$] and ion-gels are based on the measurements from three repetitions in the temperature dependent experiments.

6.2.6   THERMAL ANALYSIS

The phase transitions of the LCBBCs as well as the ion gels were studied by DSC on TA instrument DSC Q-20 series. The sample was stored in a hermetically sealed alumina pan with another empty pan as reference. Both heating and cooling cycles were carried out at a ramping rate 10 °C/min. Phase transition temperature are determined in either first cooling or second heating cycle scan using TA Universal Analysis software.

Thermal gravimetric analysis (TGA) was performed on Perkin-Elmer TGA-7. The sample was heated from room temperature to 800 °C at a ramping rate of 10 °C/min under an air flow.

6.2.7 MICROSTRUCTURE ANALYSIS

2D WAXS has been used to investigate the liquid crystalline properties of the LCBBCs and ion gels. The experiment was performed on Oxford diffraction Xcalibur PX Ultra with x-ray beam of CuKα radiation ($\lambda = 1.54$ Å) and an Onyx CCD detector. Rectangular samples were prepared by compression molding of powdered samples at 60 °C and subsequently cooled to room temperature in the air. x-Ray beam is aligned along the normal the normal to the face or the edge of the film to obtain 2D diffractogram.

2D SAXS experiments were conducted with x-ray beam ($\lambda = 1.54$ Å) produced by a CuKα microsource (Rigaku) at the University of Massachusetts, Amherst, MA, USA. Silver behenate was used for calibration, which has a lamellar structure with d spacing of 58.36 Å. The scattering pattern was recorded on a gas-filled wire array detector (Molecular Metrology, INC.) with a distance of about 150 cm from the sample, providing an accessible angular range corresponding to dimensions between 4 and 100 nm.

6.3 RESULTS AND DISCUSSIONS

6.3.1 SYNTHESIS AND CHARACTERIZATION OF LCBBCS

In this paper, two monomers α-methoxy-ω-norbornenyl-PEG (NBMPEG) and 5-{9-(cholesteryloxycarbonyl)nonyloxycarbonyl}bicyclo[2.2.1]hept-2-ene (NBCh9) were synthesized according to previously reported procedures.[21,36] A series of LCBBCs with various compositions of NBCh9 and NBMPEG were synthesized by sequential ROMP using a modified Grubbs catalyst second generation (mG2nd) ((H$_2$IMes)-(pyr)$_2$(Cl)$_2$RuCHPh)[37] and their characterizations are summarized in Table 6-1. The compositions of LCBBCs are determined by integrating characteristic peaks of NBCh9 (a proton at 4.6 ppm) and NBMPEG (three protons at 3.36 ppm) in ^{1}H NMR spectra and by comparing the ratio of their integration values. The following

TABLE 6-1 Molecular Characterization of $P((NBCh9)_x\text{-}b(NBMPEG)_y)$

| Entry | Polymer[b] | M_n (kg/mol) | | PDI | Weight Fraction[e] | |
		Theory[c]	GPC[d]		NB-Ch9	NB-MPEG2000
PNBMPEG[a]	$P(NBMPEG)_{50}$	108	69	1.17	–	100
LCBBC150k-90	$P((NBCh9)_{166}\text{-}b(NBMPEG)_{10})$	150	115	1.15	94.0	6.0
LCBBC125k-85	$P((NBCh9)_{150}\text{-}b(NBMPEG)_{12})$	125	77	1.19	84.0	16.0
LCBBC120k-78	$P((NBCh9)_{135}\text{-}b(NBMPEG)_{15})$	122	74	1.06	78.0	22.0
LCBBC95k-55	$P((NBCh9)_{80}\text{-}b(NBMPEG)_{20})$	96	41	1.17	55.0	45.0
LCBBC120k-28	$P((NBCh9)_{65}\text{-}b(NBMPEG)_{35})$	119	58	1.20	28.0	72.0
LCBBC160k-16	$P((NBCh9)_{35}\text{-}b(NBMPEG)_{65})$	163	72	1.24	16.0	84.0
LCBBC200k-7	$P((NBCh9)_{25}\text{-}b(NBMPEG)_{85})$	201	92	1.20	7.0	93.0
LCBBC200k-6	$P((NBCh9)_{16}\text{-}b(NBMPEG)_{183})$	200	102	1.21	6.6	93.4
LCBBC400k-8	$P((NBCh9)_{33}\text{-}b(NBMPEG)_{367})$	400	126	1.24	7.7	92.3
LCBBC600k-6	$P((NBCh9)_{50}\text{-}b(NBMPEG)_{550})$	600	216	1.16	5.7	94.3
LCBBC600k-18	$P((NBCh9)_{120}\text{-}b(NBMPEG)_{430})$	600	118	1.16	18.0	82.0
PNBCh9[a]		84	61	1.09	100	–

[a]PNBMPEG, norbornene monomethyl ether; PNBCh9, norbornene cholesterol with nine methylene spacer; Molecular weight of MPEG is 2000 g/mol. LCBBCx–y represents LC brush block copolymer, where x represents the molecular weight of copolymer and y represents the weight percentage (wt.%) of NBCh9 in the copolymers.

[b]Subscript represents the degree of polymerization calculated based on monomer to catalyst ratio.

[c]Theoretical molecular weight calculated by $M_n = \{[MNBCh9]/[I] \times$ molar mass of NBCh9 + [MNBMPEG]/[I] $\times$ molar mass of NBMPEG\}, where [MNBCh9], [MNBMPEG], and [I] are moles of NBCh9, NBMPEG, and mG2 catalyst, respectively.

[d]Determined by GPC with RI detector, where THF was used as eluent and PS standards were used to construct a conventional calibration.

(From Deshmukh, P.; Ahn, S.-K.; Geelhand de Merxem, L.; Kasi, R. M. Interplay between Liquid Crystalline Order and Microphase Segregation on the Self-Assembly of Side-Chain Liquid Crystalline Brush Block Copolymers. Macromolecules 2013, 46, 8245–8252. © Copyright 2013 American Chemical Society. Used with permission.)

terminology is used to define the block copolymers: LCBBCx–y represents LC brush block copolymer, where x represents the molecular weight of copolymer and y represents the weight percentage (wt.%) of NBCh9 in the copolymers. Molecular weights of these polymer series are comparable and narrow polydispersity was achieved.

Thermal properties of homopolymer and LCBBCs are investigated using differential scanning calorimetry (DSC) as shown in Table 6-2. During cooling cycle, crystallization temperature (T_c) of PEG and two LC mesophase transitions (T_1 and T_2) are observed, and the intensity of these phase transitions are largely influenced by the amount of LC content in the LCBBCs.

TABLE 6-2 Thermal Properties of Homopolymers and Block Copolymers

Polymer[a]	T_c/T_m, (ΔH) (°C)[b]		Heating[c]		Cooling[c]	
	Heating	Cooling	T_1 (ΔH) (°C)	T_2 (ΔH) (°C)	T_1 (ΔH) (°C)	T_2 (ΔH) (°C)
PNBMPEG[d]	55	31	–	–	–	–
LCBBC125k-90	48	−24	92	109	88	108
LCBBC150k-85	52	−25 7	94	109	89	108
LCBBC120k-78	50.5	23.5	91.7	107.8	85.2	105.7
LCBBC95k-55	51.6	12.1	91.5	105.7	83.1	101.3
LCBBC120k-28	53.8	18.0 −24.5	90.9	106.8	81.9	101.3
LCBBC160k-16	53.2	26.8	86.0	107.5	81.6	103.6
LCBBC200k-7	52.8	26.1	–	–	–	–
LCBBC200k-6	52.9	26.0	–	–	–	–
LCBBC400k-8	53.5	25.8	–	–	–	–
LCBBC600k-6	55.8	25.4	–	–	–	–
LCBBC600k-18	54.3	22.3	90.4	112.5	80.6	102.9
PNBCh9[d]	–	–	87.6	105.6	80.2	98.3

[a]Thermal history is removed by heating to 180 °C and transition temperature and enthalpy values are recorded in the first cooling or second heating cycle with rate 10 °C/min.

[b]T_c or T_m of PEG and associated enthalpy are normalized with respect to wt.% of NBMPEG in copolymer.

[c]Two different LC transition temperatures and their associated enthalpies are normalized with respect to wt.% of PNBCh9 in copolymer.

[d]Thermal properties are reported from previous publication.

(From Deshmukh, P.; Ahn, S.-K.; Geelhand de Merxem, L.; Kasi, R. M. Interplay between Liquid Crystalline Order and Microphase Segregation on the Self-Assembly of Side-Chain Liquid Crystalline Brush Block Copolymers. Macromolecules 2013, 46, 8245–8252. © Copyright 2013 American Chemical Society. Used with permission.)

The higher LC content in the LCBBCs increases the enthalpy of LC meso-phase transitions while decreases the enthalpy of PEG crystallization (Table 6-2). In block copolymers, microphase segregated domains tend to create different population of crystalline and amorphous regions, and as a result, the crystallization temperature varies between 10 and 30 °C. However, the different populations of crystalline and amorphous regions do not have a significant impact on the melting temperature as reported previously.[38] The type of mesophase in LCBBCs will be resolved by x-ray scattering analyses, which will be discussed in the following sections.

6.3.2 MICROSTRUCTURAL ANALYSIS OF LCBBCS

The unique LCBBC architecture incorporates microphase segregation, PEG crystallization and LC order. Therefore, the interplay of these structural parameters that can be manipulated by tuning compositions of two blocks and this will determine the self-assembly of LCBBCs at multiple length scales. Hierarchical order of microphase segregated domain, PEG crystal-line lamellar and LC mesophase are examined by x-ray scattering and trans-mission electron microscopy (TEM) on melt processed film samples. The small-angle x-ray scattering (SAXS) diffractograms of LCBBCs showed three types of scattering reflections, including: (1) microphase segregation (q^* and its higher order reflections), (2) PEG crystalline lamellar (q_{PEG}), and (3) LC order (q_{LC1}, q_{LC2}).

6.3.2.1 MICROPHASE SEGREGATION (D = 40–73 NM)

All LCBBCs display microphase segregation (q^*) with higher order reflec-tions.[30] Generally, the composition of BCPs governs the type of morphology for the resulting microphase segregated domains.[39] Table 6-3 summarizes the domain (d) spacing values (40–73 nm) and speculated morphology based on the correlation between q^* and higher order reflections of the principal scattering vector. For the high LC content polymer (i.e., LCBBC120k-78), primary scattering and its higher order reflections are in the ratio of $1:\sqrt{3}:\sqrt{7}$, which suggests PEG cylinder within LC matrix with d-spacing of 43.7 nm. Based on this result, PEG cylinder to cylinder spacing can be esti-mated $d_0 = 50.5$ nm [using $d_0 = (4/3)^{1/2}d$]. The lower LC content polymer (i.e., LCBBC95k-55) exhibits primary scattering and its higher order reflec-tions in the ratio of 1:2:3 which may be attributed to lamellar morphology with d-spacing value of 54.7 nm. Meanwhile, polymers with even lower LC content (i.e., LCBBC200k-7, LCBBC160k-16, and LCBBC120k-28) show unresolved or even absent higher order reflections, indicating the absence of long-range order. The lack of higher order peaks suggest that breakout PEG spherulite crystals disrupt the microphase segregated domains leading to weak long range order.[40,41]

TABLE 6-3   Summary of SAXS Data Recorded at Room Temperature

Polymer	Smectic Layers		PEG Lamellar	Microphase Segregation	Correlation of Higher Order Peaks
	LC_1 (nm)	LC_2 (nm)	d_{PEG} (nm)	d spacing (nm)	
PNBMPEG			14.1	−	−
LCBBC120k-78	6.0	3.5		43.7	Cylinder (1:√3:√7)
LCBBC95k-55	6.1	3.5		54.7	Lamellar (1:2:3)
LCBBC120k-28	6.0	3.5		73.1	
LCBBC160k-16	6.5			49.9	
LCBBC200k-7	7.2			40.3	
PNBCh9	6.3	3.4		−	−

(From Deshmukh, P.; Ahn, S.-K.; Geelhand de Merxem, L.; Kasi, R. M. Interplay between Liquid Crystalline Order and Microphase Segregation on the Self-Assembly of Side-Chain Liquid Crystalline Brush Block Copolymers. Macromolecules 2013, 46, 8245–8252. © Copyright 2013 American Chemical Society. Used with permission.)

6.3.2.2   PEG CRYSTALLINE DOMAINS (D = 12–13 NM)

The semicrystalline PEG side-chains in LCBBCs can crystallize into lamellae and its thickness depends on the molecular weight of PEG. The periodic structure resulting from the alternate crystalline lamellae and amorphous domains of PEG is known to show a signature reflection in SAXS.[42] The q_{PEG} disappears after melting transition of PEG (T_m = ~50 °C) as shown in the temperature-dependent SAXS analyses (Figure 6-2), which suggests these peaks do not originate from microphase segregation of BCPs. The uncorrelated multiple scattering reflection imply the presence of hierarchical morphologies (structure-within-structure) in LCBBCs. Interestingly, only LCBBCs having lower LC content allows the development of periodic structure from crystalline lamellar and amorphous domains in PEG. In contrast, LCBBC120k-78 does not show this scattering reflection, probably because of the presence of PEG cylinders within LC matrix, which may inhibit the formation of periodic crystalline lamella within these cylinders.

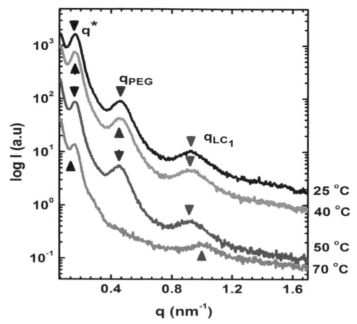

FIGURE 6-2   SAXS diffractograms of LCBBC160k-16 recorded at 25, 40, 50, and 70 °C. The q_{PEG} disappears at 70 °C (T > T_m of PEG), which suggests that q_{PEG} originates from the periodicity of PEG lamellar consisting of alternating crystalline and amorphous domains. (From Deshmukh, P.; Ahn, S.-K.; Geelhand de Merxem, L.; Kasi, R. M. Interplay between Liquid Crystalline Order and Microphase Segregation on the Self-Assembly of Side-Chain Liquid Crystalline Brush Block Copolymers. Macromolecules 2013, 46, 8245–8252. © Copyright 2013 American Chemical Society. Used with permission.)

6.3.2.3   LC ORDER (D = 3–7 NM)

LCBBC architecture also features side-chain cholesteryl mesogens, which forms smectic A (SmA) layers. In the higher q range, SmA layer reflections (q_{LC1}) are observed with layer spacing between 3 and 7 nm. The observed orthogonal arrangement indicates typical SmA mesophase. The length of cholesteryl side-chain with nine-methylene spacer is calculated to be 3.34 nm and the thickness of smectic bilayer (SmA_2) is calculated to be 6.68 nm.[43] Comparison between the calculated and the experimentally observed d-spacings of SmA layer suggests that LC orderings in LCBBCs consist of SmA_2 for lower LC content polymers or interdigitated smectic layers (SmA_d) for higher LC content polymers. For the LCBBCs having lower LC content (i.e., LCBBC200k-7 and LCBBC160k-16), loosely packed smectic layers are developed due to the dilution effect, where boarder LC scattering reflection indicates poorly defined layers.[44] In addition to q_{LC1}, the q_{LC2} peak, which does not correlate with the q_{LC1}, is also detected in higher LC content polymer (i.e., LCBBC95k-55 and LCBBC120k-78) with d-spacing of 3.5 nm. This is attributed to smectic monolayers (SmA$_1$), similar to the our previous observation for the PNBCh9.[43] Smectic polymorphism or co-existence of more than one type of smectic layer in thermotropic LC polymers has been previously reported.[45]

To summarize, the unique LCBBC scaffold consisting of semicrystal-line PEG side-chain and cholesteryl mesogen self-assembles into multi-level hierarchical nanostructures. Assuming moderate segregation between the blocks, the composition of LCBBCs governs the morphology of the resultant microphase segregated domains. Specifically, in higher PEG content copolymers (i.e., LCBBC200k-7, LCBBC160k-16, and LCBBC120k-28), periodicity of PEG crystalline lamellae is observed within microphase-segregated domains. However, in higher LC content copolymers, such as LCBBC120k-78 (i.e., lower PEG content), PEG cylinders are embedded within the LC matrix. The interplay between different types of self-organi-zation (i.e., microphase segregation, crystallization, and LC order) results in a hierarchically ordered system, which was supported by x-ray scattering and TEM analyses: microphase segregated domains (40–70 nm), PEG crys-talline regions (~13 nm), and LC mesophase (3–7 nm).

We also synthesized polymers containing even higher LC content (i.e., LCBBC125k-85 and LCBBC150k-90) to explore the composition effect of the copolymer on microphase segregation (Table 6-2). LCBBC125k-85 tends to weakly microphase-segregate (broad primary reflection and lack of higher order reflections), whereas LCBBC150k-90 does not microphase-segregate (Figure 6-3).

Due to the similar composition with LCBBC160k-16, we speculate that the LCBBC600k-18 (Figure 6-4) can form lamellar structures from side-chain microphase segregated crystalline PEG brush. In addition, the LCBBC200k-6, LCBBC400k-8, and LCBBC600k-6 with low LC content possess disrupted microsegregated structure or may not show microphase segregation.

For preparation of ion gels based on LCBBCs, the sample with high LC content (LCBBC120k-28, LCBBC95k-55, LCBBC120k-78, LCBBC125k-85, and LCBBC150k-90) are not suitable to form ion gels due to the long hydrophobic chain. Thus, the formation of ion gels using LCBBCs with high PEO content and effect of molecular weight as well as composition of block length were investigated and discussed in the following sections.

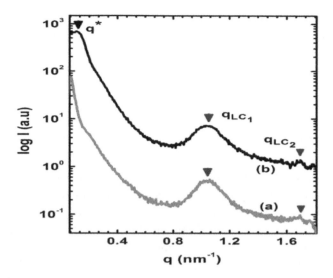

FIGURE 6-3 1D SAXS analysis at room temperature for (a) LCBBC150k-90 and (b) LCBBC125k-85. Only LCBBC125k-85 exhibits a broad microphase segregation peak (q^*), whereas LCBBC150k-90 does not microphase-segregate. (From Deshmukh, P.; Ahn, S.-K.; Geelhand de Merxem, L.; Kasi, R. M. Interplay between Liquid Crystalline Order and Microphase Segregation on the Self-Assembly of Side-Chain Liquid Crystalline Brush Block Copolymers. Macromolecules 2013, 46, 8245–8252. © Copyright 2013 Royal Society of Chemistry. Used with permission.).

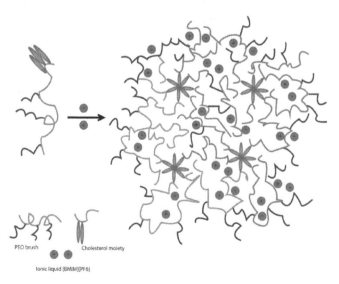

FIGURE 6-4 Self-assembled ion gels of liquid crystalline brush diblock copolymer LCBBC600k-18 in IL [BMIM][PF$_6$].

6.3.3 FORMATION OF ION GELS

6.3.3.1 MECHANICAL PROPERTIES

[BMIM][PF$_6$] is known as a good solvent for polyethylene oxide block forms ion-gels by noncovalent intermolecular aggregation of the diblock copolymer.[16a] It is interesting to explore how effectively ILs gelates with the addition of LCBBCs. The frequency and concentration dependence of the linear viscoelastic modulus of the LCBBCs solutions and gels at different concentrations and effect of norbornene cholesterol (NBCh9) block length (or copolymer composition) on the gelation characteristics were investigated. The gelation point of the block copolymers and ILs binary systems was evaluated by dynamic viscoelastic measurement. As shown in Figure 6-5, with low concentration (~5–10 wt.%), the solutions behave as viscoelastic fluids with the viscous modulus (G″) exceeds the elastic modulus (G′) over the entire frequency range, and both modulus exhibit dependence on frequency. Both G′ and G″ increases with concentration of the copolymer in ILs, and eventually with 25 wt.% concentration, an optically transparent gel is obtained, where G′ is greater than G″ and is nearly frequency independent in the terminal (or low) frequency range which is a characteristic of solid-like behavior. The results indicated that the dispersions containing 20 wt.% of LCBBC600k-18 behave as soft solid materials (ion gels). Lodge et al. reported that gelation of an IL [C4mim]PF$_6$, could be achieved by the self-assembly of as little as 5 wt.% of a triblock copolymer, polystyrene (PS)-block-poly(ethylene oxide) (PEO)-block- polystyrene (PS) (SOS), where M$_n$ of the each terminal PS is 4760 Da, M$_n$ of the middle PEO is 25,500 Da, the polydispersity index (M$_w$/M$_n$) is 1.36, and the PS weight fraction (f$_{ps}$) is 0.28.[16a] The somewhat lower gelation concentration of SOS than that of our LCBBCs may be ascribed to its higher M$_n$ of the IL-philic PEO segment.

To investigate the effect of molecular weight of LCBBCs on the elastic modulus of ion gels, we measured the dynamic viscoelastic a series of LCBBC200k-6, LCBBC400k-8, and LCBBC600k-6 with different molecular weights 200, 400, and 600 kg/mol, respectively. Figure 6-6 showed the elastic (G′) and viscous (G″) modulus of ion gels from 20 wt.% of different LCBBCs. The LCBBC200k-6 sample exhibited a larger viscous modulus (G) than the elastic modulus (G′), indicating that the formed gels are still weak. Interestingly, G′~G″ when increasing the molecular weight of LCBBCs, corresponding with LCBBC400k-8 polymer. As the molecular weight of brush polymer is increased to 600 kg/mol, corresponding to the LCBBC600k-6 sample, strong gels were observed with dynamic viscoelastic modulus approximately 50 times higher than the ion gels from 200 and 400 kg/mol block copolymer.

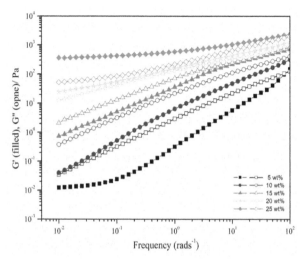

FIGURE 6-5   Storage (G′) and loss (G″) (filled and open symbols) modulus vs angular frequency of LCBBC600k-18 for various concentrations at 25 °C. The ion gel was formed at 20 wt.%.

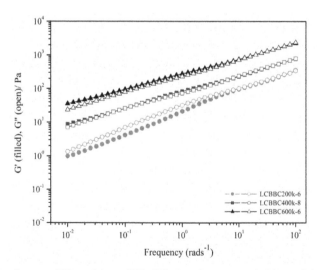

FIGURE 6-6   Storage (G′) and loss (G″) (filled and open symbols) modulus vs angular frequency of various LCBBCs polymers with 20 wt.% concentration.

The effect of NBCh9 block on the elastic modulus of ion gels was compared when changing the NBCh9 block length of LCBBCs, as shown in Figure 6-7. Increasing the length of the NBCh9 block from 50 to 120, corresponding with LCBBC600k-6 and LCBBC600k-18 sample, gel strength was improved. The improved mechanical properties of ion gel from LCBBC600k-18 polymer could be ascribed to the large number of nucleated liquid crystalline regions acting as physical cross-linkers to enhance mechanical properties. Thus, the dependence of the elastic modulus (G') on the NBCh9 length offers a straightforward way to tune the rheological response of this ion gels.

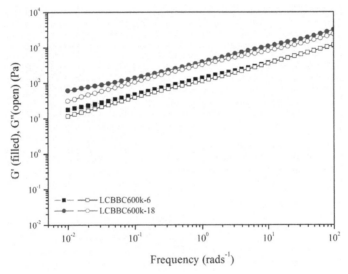

FIGURE 6-7   Storage (G') and loss (G″) moduli vs angular frequency for ion gels of 20 wt.% polymer with increasing NBCh9 block length at 25 °C.

Thermoreversibility and thermal stability are important parameters for liquid state processing and material usage. Heating–cooling cycles of the ion-gels containing 20 wt.% of LCBBCs LCBBC600k-18 in the dynamic temperature ramp measurements confirmed their thermoreversible behavior as shown in Figure 6-8. Gel-fluid phase transition shows the viscoelastic moduli (G' and G″) at the frequency of 1 rad/s as a function of temperature in the range of 20–80 °C for the ion gels. Indeed, the cross-over of G' and G″ occurs at 56 °C in the heating cycle, a lower temperature than that determined in the cooling cycle (60 °C) and gives the gel transition temperature of ion gels. The difference in the cross-over temperature between the heating and cooling cycles is due to the kinetics between the association and dissociation processes.[46–49] In addition, thermogravimetric measurements performed on this ion gel in air show a weight loss of 1–4 wt.% at 300 °C (Figure 6-9), indicated the high thermal stability of ion gel.

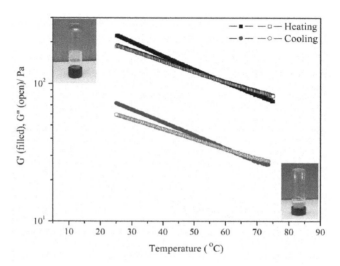

FIGURE 6-8   Storage (G′) and loss (G″) modulus of LCBBC600k-18 as a function of temperature at angular frequency 1 rads⁻¹.

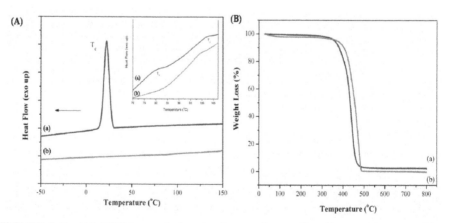

FIGURE 6-9   (A) DSC traces during the first cooling cycle with mesophase transition temperature T_1 and T_2 shown in the inset and (B) TGA traces of (a) LCBBC600k-18 polymer and (b) corresponding ion gel.

6.3.3.2 THERMAL PROPERTIES

The presence of the cholesterol in the LCBBCs is critical for the preparation of the ion gels, because the PEO block will be soluble in ILs, and the cholesterol domains can serve as junctions to uphold the integrity of the resulting ion gels. DSC was used to verify the presence of the cholesterol in the LCBBCs and their corresponding ion gels. In Figure 6-9, the DSC traces of LCBBC600k-18 and ion gel are compared. The cooling cycle at a fixed ramping rate is used to study the nonisothermal PEO crystallization behaviors for the LCBBCs and ion gel. For the plain LCBBCs polymer, the isotropic to transition temperature were observed at two peaks $T_1 = 80.6$ °C and $T_2 = 102.9$ °C. In addition, it was observed that the ion gel displayed a lower isotropic transition temperature (T_2) and isotropic transition temperature (T_1) is broadened. The broadening may also be caused by the coexistence of PNBCh9- and IL-rich region, which can be attributed to the interaction between IL and the cholesterol phase.

6.3.3.3 MICROSTRUCTURE OF ION GELS

To compare the microstructure of obtained ion-gels with plain LCBBCs, the wide- and small-angle x-ray scattering (WAXS and SAXS) were examined, as shown in Figure 6-10. The diffractogram of LCBBC600k-18 and corresponding ion gels are shown, in which the lateral distance of the cholesteryl mesogen peaks was observed at $q = 10.2$ nm^{-1} $(2\theta = 14.1°)$. This indicated that the cholesterol domains formed the junction in the ion gels network during the gelation process. Two peaks are observed at $q = 13.1$ and 16.1 nm^{-1} $(2\theta = 18.7°$ and $22°)$ in the plain LCBBCs, which are the characteristic peaks for PEO crystallization. The disappearance of the PEO crystallization peaks was observed for ion gels, indicating complete solvation of PEO chains in ILs. In the SAXS patterns, the diffractogram of LCBBC600k-18 showed a diffraction peak at $q_{LC1} = 0.86$ nm^{-1}, indicating the presence of the SmA layer reflection. The scattering peak due to the PEG-side chain segregated structure (q_{PEG}) appears at 0.42 nm^{-1}. On the contrary, the absence of PEG crystalline diffraction peaks in the diffractograms of the ion gel samples confirms that the PEG is completely soluble in ILs. From the WAXS and SAXS studies it can be indicated that, the hydrophilic PEO brush has interaction with ILs while the cholesterol domains is forming the core due to aggregation, leading to formation of ion gels with cholesterol core and surrounding soluble PEO.

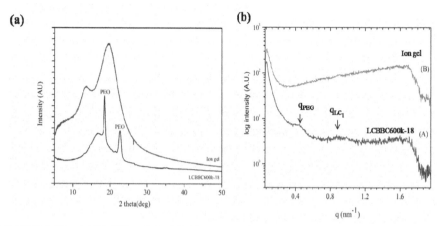

FIGURE 6-10   (a) WAXS and (b) SAXS pattern of LCBBC600k-18 and corresponding ion gel.

6.3.3.4   IONIC CONDUCTIVITY

The temperature dependent conductivity (σ) of the bulk [BMIM][PF$_6$] and the ion gels at different concentrations is shown in Figure 6-11. The conductivity of ion-gels increases with temperature, which can be explained on the basis of the Walden's rule: the product of conductivity (σ) and viscosity (η) of the pure solvent is constant for all temperatures and solvents for a given electrolyte, $\sigma*\eta$ = constant. With increasing temperature the viscosity decreases making ions more mobile, resulting in an increased σ. As seen in Figure 6-11, the conductivity of 20 wt.% ion gels is nearest to that of the bulk IL. Interestingly, the temperature dependence of σ for all concentrations of the ion-gel nearly tracks the σ of bulk IL. A similar observation of temperature dependent σ and proportionality of σ and $1/\eta$ is also seen in PS-PEO-PS/[BMIM][PF$_6$] based ion-gels.[16a] The high ionic conductivity of these ion gels can be exploited for application in development of novel solid state electrolytes.

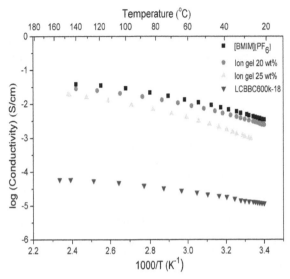

FIGURE 6-11 Ionic conductivity vs inverse temperature for the ion gels based on the LCBBC600k-18 with dissolved [BMIM][PF$_6$].

6.4 CONCLUSION

We describe a new set of ion-gels prepared via self-assembly of LCBBCs, NBCh9-b-NBMPEG in a room temperature IL, [BMIM][PF$_6$]. At concentrations of 20 wt.%, we demonstrate the formation of strong ion-gel with high gelation temperature (T$_{gel}$ ~ 56 °C) and high ionic conductivities similar to bulk [BMIM][PF$_6$]. This gives, an additional handles of copolymer concentration and block length to tune mechanical and ion transport properties to meet the design specific criteria for various electrochemical applications. Ionic conductivity studies on these gels have shown only a moderate decrease due to the structuring soft materials, as compared to bulk IL, and thus are good candidates for electrolyte applications. The hydrophilic PEO units are soluble in IL, while the use of cholesterol allows the formation of junctions in the gel network due to the insolubility of these motifs in IL, which predominantly provides mechanical robustness to the corresponding ion gels. The ion gels obtained via gelation of IL utilizing LCBBCs have tunable thermal, mechanical, and morphological properties by virtue of their architecture and functionality, which can be exploited for renewable energy scaffolds, CO$_2$ selective sorbent coatings, gas separation membranes, catalysis, and energy storage devices.

REFERENCES

1. Brazel, C. S.; Rogers, R. D., Eds. *Ionic liquids in Polymer Systems: Solvents, Additives, and Novel Applications*; ACS Symposium Series 913; American Chemical Society: Washington, DC, 2005.

2. Winterton, N. *J. Mater. Chem.* **2006,** *16*, 4281.

3. (a). Ueki, T.; Watanabe, M. *Macromolecules* **2008,** *41*, 3739. (b). Ueki, T.; Watanabe, M. *Bull. Chem. Soc. Jpn.* **2012,** *85* (1), 33–50.

4. (a). MacCallum, J. R., Vincent, C. A., Eds. *Polymer Electrolyte Reviews 1 and 2*; Elsevier: London, 1987 and 1989. (b). Gray, F. M. *Solid Polymer Electrolytes*; VCH Publishers: New York, 1991. (c). Scrosati, B., Ed. *Application of Electroactive Polymers*; Chapman & Hall: London, 1993.

5. Wang, P.; Zakeeruddin, S. M.; Comte, P.; Exnar, I.; Gratzel, M. *J. Am. Chem. Soc.* **2003,** *125* (5), 1166–1167.

6. Chen, Z. G.; Li, F. Y.; Yang, H.; Yi, T.; Huang, C. H. *Chem. Phys. Chem.* **2007,** *8* (9), 1293–1297.

7. Boswell, P. G.; Lugert, E. C.; Rabai, J.; Amin, E. A.; Buhlmann, P. *J. Am. Chem. Soc.* **2005,** *127* (48), 16976–16984.

8. Cho, J. H.; Lee, J.; Xia, Y.; Kim, B.; He, Y.; Renn, M. J.; Lodge, T. P.; Frisbie, C. D. *Nat. Mater.* **2008,** *7* (11), 900–906.

9. Lu, W.; Fadeev, A. G.; Qi, B. H.; Smela, E.; Mattes, B. R.; Ding, J.; Spinks, G. M.; Mazurkiewicz, J.; Zhou, D. Z.; Wallace, G. G.; MacFarlane, D. R.; Forsyth, S. A.; Forsyth, M. *Science* **2002,** *297* (5583), 983–987.

10. Lee, J.; Panzer, M. J.; He, Y.; Lodge, T. P.; Frisbie, C. D., *J. Am. Chem. Soc.* **2007,** *129* (15), 4532–4533.

11. Huang, J.; Riisager, A.; Wasserscheid, P.; Fehrmann, R. *Chem. Commun.* **2006,** *38*, 4027–4029.

12. Fukushima, T.; Asaka, K.; Kosaka, A.; Aida, T. *Angew. Chem. Int. Ed.* **2005,** *44* (16), 2410–2413.

13. Lu, J.; Yan, F.; Texter, J. *Prog. Polym. Sci.* **2009,** *34* (5), 431–448.

14. (a). Carlin, R. T.; Fuller, J. *Chem. Commun.* **1997,** 1345. (b). Fuller, J.; Breda, A. C.; Carlin, R. T. *J. Electrochem. Soc.* **1997,** *144*, L67. (c). Fuller, J.; Breda, A. C.; Carlin, R. T. *J. Electroanal. Chem.* **1998,** *459*, 29.

15. Susan, M. A.; Kaneko, T.; Noda, A.; Watanabe, M. *J. Am. Chem. Soc.* **2005,** *127* (13), 4976–4983.

16. (a). He, Y.; Boswell Paul, G.; Buhlmann, P.; Lodge Timothy, P. *J. Phys. Chem. B* **2007,** *111* (18), 4645–4652. (b). Lodge, T. P. *Science* **2008,** *321*, 50. (c). Zhang, S.; Lee, K. H.; Frisbie, C. D.; Lodge, T. P. *Macromolecules* **2011,** *44*, 940.

17. (a). He, Y.; Lodge, T. P. *Chem. Commun.* **2007,** *26*, 2732–2734. (b). He, Y.; Lodge, T. P. *Macromolecules* **2008,** *41*, 167.

18. Sharma, N.; Lakhman, R. K.; Zhou, Y.; Kasi, R. M. *J. Appl. Pol. Sci.* **2013,** *128*, 3982–3992.

19. Heino, S.; Lusa, S.; Somerharju, P. *Proc. Natl. Acad. Sci. USA* **2000,** *97*, 8375–8380.

20. Zhou, Y.; Sharma, N.; Deshmukh, P.; Lakhman, R. K.; Jain; M.; Kasi, R. M. *J. Am. Chem. Soc.* **2012,** *134*, 1630–1641.

21. Ahn, S.-K.; Le, L. T. N.; Kasi, R. M. *J. Polym. Sci., Part A Polym. Chem.* **2009,** *47* (10), 2690–2701.

22. Ahn, S.-K.; Kasi, R. M. *Adv. Funct. Mater.* **2011**, *21*, 4543–4549.
23. Takahashi, A.; Mallia, V. A.; Tamaoki, N. *J. Mater. Chem.* **2003**, *13*, 1582–1587.
24. Camurlu, P.; Toppare, L.; Yilmaz, F.; Yagci, Y, Galli, G. *J. Macro. Sci. Part A* **2007**, *44*, 265–270.
25. Rzayev, J. *ACS Macro Lett.* **2012**, *1*, 1146–1149.
26. Sheiko, S. S.; Sumerlin, B. S.; Matyjaszewski, K. *Prog. Polym. Sci.* **2008**, *33*, 759–785.
27. Zhang, M.; Muller, A. H. E. *J. Polym. Sci., Part A: Polym. Chem.* **2005**, *43*, 3461–3481.
28. Collett, J.; Crawford, A.; Hatton, P. V.; Geoghegan, M.; Rimmer, S. *J. R. Soc. Interface.* **2007**, *4* (12), 117–126.
29. Savina, I. N.; Tuncel, M.; Tuncel, A.; Galaev, I. Y.; Mattiasson, B. *Express Polym. Lett.* **2007**, *1*, 189–196.
30. Deshmukh, P.; Ahn, S.-K.; Geelhand de Merxem, L.; Kasi, R. M. *Macromolecules* **2013**, *46*, 8245–8252.
31. Bai, Z.; He, Y.; Lodge, T. P. *Langmuir* **2008**, *24* (10), 5284–5290.
32. Lu, J.; Yan, F.; Texter, J. *Prog. Polym. Sci.* **2009**, *34* (5), 431–448.
33. Armand, M.; Endres, F.; MacFarlane, D. R.; Ohno, H.; Scrosati, B. *Nat. Mater.* **2009**, *8* (8), 621–629.
34. Green, O.; Grubjesic, S.; Lee, S.; Firestone, M. A. *Polym. Rev.* **2009**, *49* (4), 339–360.
35. Ueki, T.; Watanabe, M. *Macromolecules* **2008**, *41* (11), 3739–3749.
36. Deshmukh, P.; Ahn, S.-K.; Gopinadhan, M.; Osuji, C. O.; Kasi, R. M. *Macromolecules* **2013**, *46*, 4558–4566.
37. Love, J. A.; Morgan, J. P.; Trnka, T. M.; Grubbs, R. H. *Angew. Chem. Int. Ed.* **2002**, *41*, 4035–4037.
38. Loo, Y.-L.; Register, R. A.; Ryan, A. J. *Macromolecules* **2002**, *35*, 2365–2374.
39. Yu, H.; Kobayashi, T.; Yang, H. *Adv. Mater.* **2011**, *23*, 3337–3344.
40. Chen, H.-L.; Wu, J.-C.; Lin, T.-L.; Lin, J. S. *Macromolecules* **2001**, *34*, 6936–6944.
41. Xu, J.-T.; Fairclough, J. P. A.; Mai, S.-M.; Ryan, A. J.; Chaibundit, C. *Macromolecules* **2002**, *35*, 6937–6945.
42. Qiu, Y.-J.; Xu, J.-T.; Xue, L.; Fan, Z.-Q.; Wu, Z.-H. *J. Appl. Polym. Sci.* **2007**, *103*, 2464–2471.
43. Ahn, S.-K; Gopinadhan, M.; Deshmukh, P.; Lakhman, R. K.; Osuji, C. O.; Kasi, R. M. *Soft Matter* **2012**, *8*, 3185–3191.
44. Verploegen, E.; Zhang, T.; Murlo, N.; Hammond, P. T. *Soft Matter* **2008**, *4*, 1279–1287.
45. Galli, G.; Chiellini, E.; Laus, M.; Angeloni, A. S.; Francescangeli, O.; Yang, B. *Macromolecules* **1994**, *27*, 303–305.
46. Rubinstein, M.; Semenov, A. N. *Macromolecules* **1998**, *31* (4), 1386–1397.
47. Li, L.; Thangamathesvaran, P. M.; Yue, C. Y.; Tam, K. C.; Hu, X.; Lam, Y. C. *Langmuir* **2001**, *17* (26), 8062–8068.
48. Afred, S. F.; Al-Badri, Z. M.; Madkour, A. E.; Lienkamp, K.; Tew, J., N. *J. Polym. Sci., Part A: Polym. Chem.* **2008**, *46*, 2640–2648.
49. Ahn, S.-K; Deshmukh, P.; Kasi, R. M. *Macromolecules* **2010**, *43*, 7330–7340.

CHAPTER 7

MALEIMIDE CONTAINING THIOL-REACTIVE POLYMERS: SYNTHESIS AND FUNCTIONALIZATION

MEHMET ARSLAN, TUGCE NIHAL GEVREK, and AMITAV SANYAL

CONTENTS

7.1 INTRODUCTION

Design and synthesis of novel functional polymeric materials has gained a widespread interest in materials and biomedical sciences since these soft materials continue to play an ever-increasing important role in enabling many of the current technologies. Polymeric materials are utilized for the design of drug/biomolecule delivery vehicles, scaffolds for tissue engineering, medical coatings for various implants and assay platforms that contains immobilized biomolecules, oftentimes need modifications to impart specific function. Rational design at the molecular level during fabrication of polymeric materials through installation of appropriate reactive groups enables their functionalization for intended applications. It is important that the functionalization should proceed with high efficiency under mild conditions. Moreover, there should not be any unwanted side reactions and introduction of undesirable contaminants due to residual reagents. Recent methods of functionalization of polymeric materials have utilized many of the reactions grouped under the umbrella of "click" reactions by Sharpless and co-workers in 2001.[1] Among these reactions, the Huisgen type alkyne-azidecyclo addition reaction has been extensively used to synthesize new polymeric materials as well as enable their modification in an efficient and modular fashion. However, in general this cycloaddition between an azide and a terminal alkyne can only be realized in the presence of a Cu(I)-based catalyst. Complete removal of the metal catalyst can be difficult and challenging and any residual copper salts can have deleterious effects on the long-term stability as well as physical and chemical properties. This potential concern has led to an increased effort towards development of metal-free methodologies for post-polymerization modifications. In recent years, addition of thiols to alkenes and alkynes via the radical-based thiol-ene and thiol-yne reactions, as well as the nucleophilic Michael-type conjugate addition of thiols to electron deficient alkenes has emerged as attractive alternatives. Notably, the versatility of thiol-maleimide addition reaction allows facile functionalization of polymers and polymeric materials for various applications. Incorporation of reactive maleimide functional groups in polymers enables the specific attachment of various thiol-containing accessories. Especially in designing soluble macromolecules for biomolecule conjugation (such as proteins and peptides), it is important to obtain well-defined systems for quantitative assessment of such conjugates. One of the important modification methods for site-specific conjugation involves reacting specific amino acids or ligand binding sites with specific functional groups on the polymers.[2,3] For such an aim, reactive cysteine residues either naturally present or genetically introduced into biomolecules provide an opportunity towards effective conjugation to maleimide groups on polymers.[4] In choosing thiol reactive functional group on polymers maleimides, vinyl sulfones, and activated disulfides are mainly preferred.[5] Among these functional groups, because of high chemoselectivity and high reaction rate with thiol groups, maleimides are preferred in preparation of bioconjugate formulations. In recent years, many synthetic strategies have emerged for fabrication of thiol-reactive maleimide containing polymers with different topologies and varying location of the maleimide functional group. It is possible to obtain polymers carrying maleimide groups at chain termini or as pendant side chains along the polymeric backbone. In the following sections, various synthetic strategies utilized to incorporate the maleimide functional group into polymeric materials have been illustrated.

7.2 RESULTS AND DISCUSSIONS

7.2.1 SYNTHESIS OF CHAIN-END MALEIMIDE FUNCTIONALIZED POLYMERS USING MODIFIED INITIATORS AND CHAIN TRANSFER AGENTS

Introducing maleimide functionality to the polymer end group can be accomplished by using controlled polymerization techniques. Appropriately designed initiators used in atom transfer radical polymerization (ATRP),[6] nitroxide-mediated polymerization (NMP),[7] and reverse-addition fragmentation chain transfer (RAFT)[8] have allowed the synthesis of well-defined polymers with desirable reactive group at the chain end. Whenever polymerization is initiated from a reactive group containing initiator, it is possible to install a functional handle at the end of each polymer chain. However, the reactive group on the initiator must not interfere with the polymerization, that is, should be preserved during the process. In case of the maleimide group, it possesses a highly activated double bond prone to participate in polymerization. A solution to this problem is to protect the reactive double bond during the polymerization process and remove the protecting group in a subsequent post-polymerization step. As a seminal contribution, Haddleton et al. reported the synthesis of maleimide end-functionalized polymers using a Diels–Alder/retro Diels–Alder reaction (rDA) based strategy (Figure 7-1).[9] In this study, copper-catalyzed living radical polymerization initiated from a protected maleimide group containing initiator was utilized to obtain thiol-reactive end-functional polymers with narrow molecular weight distributions. Maleimide terminated polymers were obtained using two different approaches. In the first approach, an amine-terminated polymer was synthesized using an initiator containing protected amine group. After the polymerization, the amine group was deprotected and thereafter reacted with maleic anhydride to install the maleimide functional group. However, in the post-polymerization functionalization reaction, even using excess amount of reagent did not convert all the polymer chain-end groups to the maleimide functionality and only a maximum of 80–85% conversion efficiency was obtained. In the second approach, a furan-protected maleimide group containing an ATRP initiator was employed to synthesize polymers with end group functionality. After obtaining the polymers, a rDA reaction was performed to remove the furan moiety to obtain the desired reactive maleimide group to the chain ends. End group activation was achieved by using a rDA reaction of polymers in a near quantitative fashion. The obtained polymers were shown to be efficiently conjugate to thiol containing bovine serum albumin (BSA) protein as demonstrated by sodium dodecyl sulfate polyacrylamide gel electrophoresis (SDS/PAGE) and fast protein liquid chromatography (FPLC) analysis.

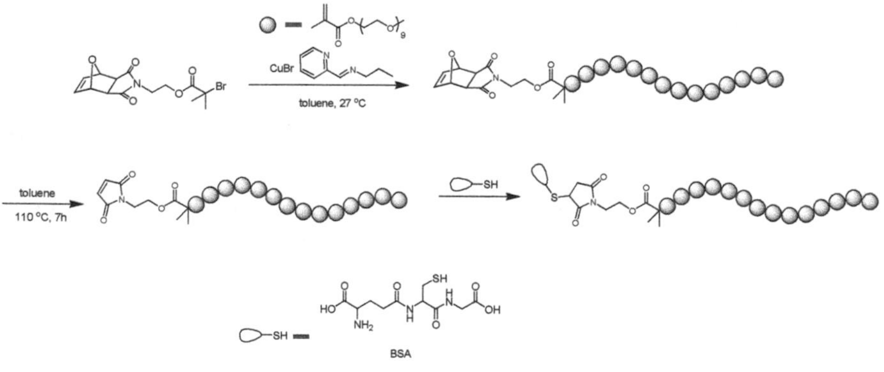

FIGURE 7-1 Synthesis of maleimide chain-end functionalized PEG-based polymers using a furan protected maleimide-bearing initiator and functionalization. (Adapted from Mantovani, G.; Lecolley, F.; Tao, L.; Haddleton, D. M.; Clerx, J.; Cornelissen, J. J.; Velonia, K. J. Design and Synthesis of NMaleimido-Functionalized Hydrophilic Polymers via Copper-Mediated Living Radical Polymerization: A Suitable Alternative to PEGylation Chemistry, Journal of the Am. Chem. Soc. 2005, 127, 2966–2973. © American Chemical Society 2005 [9].)

The concept was later extended to prepare protein-tri-block copolymer biohybrids by Velonia et al.[10] The work comprises the synthesis of masked maleimide end group containing copolymers for further attachment of the plasma proteins to prepare so called "giant amphiphilies" (Figure 7-2). In the structure of the bioconjugate, the hydrophobic polymer chain and hydrophilic protein part provides the overall amphiphilic character responsible for the bioconjugate self-assembly and aggregation. A protected maleimide group containing initiator was utilized to copolymerize solketal methacrylate monomer, silyl-protected alkyne monomer, and a monomer containing fluorescent dye via ATRP. Choice of maleimide group at chain-end was to facilitate specific coupling of polymer to the serum proteins through their reactive cysteine residues. In order to unmask the maleimide group after polymerization, a rDA reaction was performed. The site specific attachment of thiol containing proteins to the maleimide group as chain-end of a hydrophobic copolymer provided controlled building of amphiphilic structures.

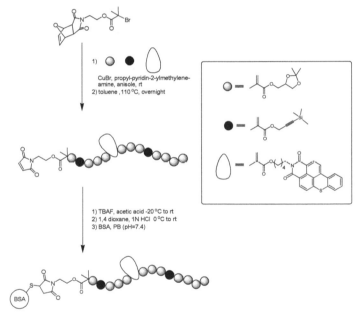

FIGURE 7-2   Synthesis of maleimide chain-end functionalized giant amphiphilic polymers and conjugation with BSA.
(**Source:** Le Droumaguet, B.; Mantovani, G.; Haddleton, D. M.; Velonia, K. J. Formation of giant amphiphiles by post-functionalization of hydrophilic protein–polymer conjugates. Journal of Materials Chemistry. 2007, 17, 1916–1922. ©Royal Society of Chemistry 2007. [10] Adapted with permission.)

Haddleton et al. also utilized the maleimide end functional polymers to fabricate glycopolymers that could conjugate to thiol containing proteins.[11] Polymerization was conducted using ATRP from the protected maleimide containing initiator and pendant clickable alkyne groups were placed along the polymer backbone to introduce sugar units (Figure 7-3). As an alternative approach, sugar moiety was clicked to alkyne containing monomer and subsequently used for polymerization. After the polymerization, maleimide group was activated by removing the furan moieties through a rDA reaction. The specific reactivity of the maleimide end functionalized glycopolymers were shown by conjugation of BSA. Successful formation of a protein-glycopolymer conjugate was confirmed by circular dichroism spectroscopy, SDS PAGE, and SEC HPLC analysis.

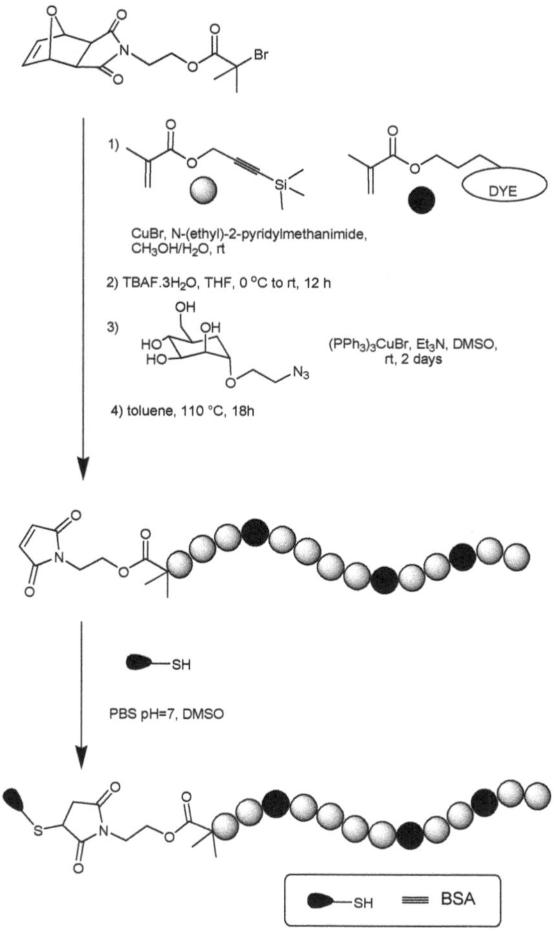

FIGURE 7-3 Synthesis of maleimide end-functionalized glycopolymer and its functionalization with thiol-containing enzyme BSA.
(**Source:** Geng, J.; Mantovani, G.; Tao, L.; Nicolas, J.; Chen, G.; Wallis, R.; Mitchell, D. A.; Johnson, B. R. G.; Evans, S. D.; Haddleton, D. M. Site-Directed Conjugation of "Clicked" Glycopolymers To Form Glycoprotein Mimics: Binding to Mammalian Lectin and Induction of Immunological Function. J. Am. Chem. Soc. 2007, 129, 15156–15163. © American Chemical Society 2007. Adapted with permission. [11])

Syntheses of maleimide end functionalized poly(PEGMA)s with various molecular weights and structures which were subsequently used for the conjugation of DNA aptamers has reported by Da Pieve et al.[12] Aptamers are oligonucleotides that can selectively bind to specific targets such as proteins on cell surfaces. Although aptamers are important macromolecules for biomedical applications, short in vivo circulating half-life and rapid renal clearance remain as the drawbacks. To improve the residence of aptamers in body, PEGylated derivatives of biomolecules were prepared. Maleimide functional group was selected as chain terminus since it has high affinity for thiol containing molecules. Synthesis of such polymers was accomplished via ATRP using a furan protected maleimide containing ATRP initiator. After the polymerization, maleimide group was activated through a rDA reaction. The coupling of disulfide-modified aptamers to maleimide terminated polymers were achieved in high yields (70–80%) with tris[2-carboxyethyl]phosphine hydrochloride (TCEP) as reducing agent at pH 4. It was observed that the conjugation efficiency was affected from the structure and conformation of the PEG polymers.

Maynard et al. introduced a novel dimethylfulvene-protected maleimide functionalized initiator for ATRP polymerization of styrene (Figure 7-4).[13] In order to compare the thermostability of novel dimethylfulvene-protected initiator, a furan-protected maleimide based initiator was also synthesized. Thermogravimetric analysis showed that the former protecting group was stable to higher temperatures than the later in the bulk phase (143 vs 125 °C). Initial attempts of polymerizations using the furan protected initiator at 70 °C resulted in polymers displaying broad and multimodal GPC traces, probably due to the undesired deprotection and subsequent copolymerization of the maleimide group with styrene monomer. In order to obtain a cycloadduct that would undergo rDA reaction at a higher temperature than the polymerization temperature, dimethylfulvene protecting group was used. Upon polymerization, it was noted that the kinetics of polymerization meet the criteria of controlled polymerization and affords polymers with low molecular weight distributions. After obtaining the polymer, the maleimide unit was deprotectedby removing the dimethylfulvene group through a rDA reaction. Modification reaction of the free maleimide groups were evaluated using benzyl mercaptan and N-acetyl-L-cysteine methyl ester as model compounds. Efficient conjugations of compounds were followed by spectroscopic techniques and it was concluded that the same strategy can be used for preparation of biohybrids from biomolecules with thiol functionality. Additionally, copper(0) catalyzed polymer chain coupling under ATR conditions was also demonstrated to prepare bismaleimide functionalized telechelicpolystyrene polymer.

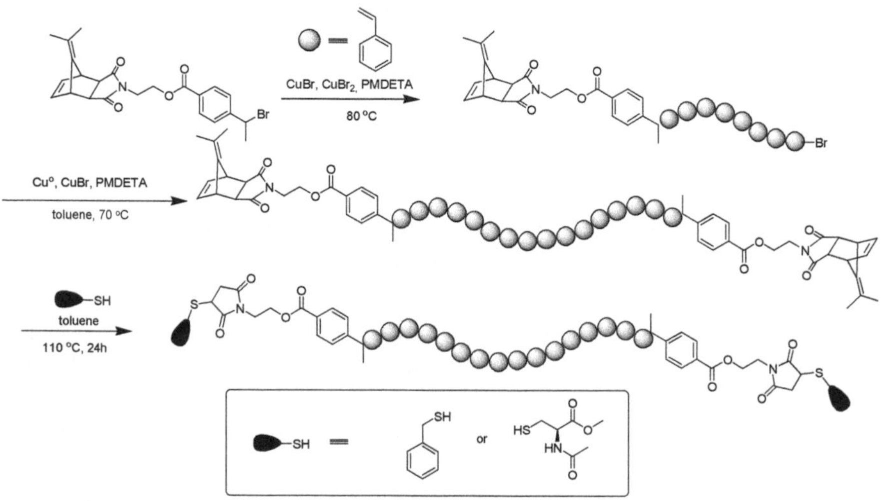

FIGURE 7-4 Synthesis of styrene-based telechelic polymers and functionalization via Michael addition.
(**Source:** Tolstyka, Z. P.; Kopping, J. T.; Maynard, H. D. Straightforward Synthesis of Cysteine-Reactive Telechelic Polystyrene. Macromolecules 2008, 41, 599–606. ©American Chemical Society 2008. Adapted with permission. [13])

Sanyal et al. extended the protection/deprotection based methodology to prepare multiarm star polymers that bear a thiol-reactive maleimide group at their core.[14] Polymers were synthesized using the ATRP technique using multiarm dendritic ATRP initiators. In order to install the maleimide group at the core, the dendritic initiators utilized in this study were designed to bear a furan-protected maleimide functional group at their focal point. The periphery of the dendrons were decorated with bromine containing initiator moieties to enable synthesis of star polymers. Multiarm polymers were synthesized using methyl methacrylate (MMA), poly(ethylene glycol) methylether methacrylate (PEGMA), and tert-butylacrylate (tBA) as model monomers. After the polymerization step, deprotection of maleimide group was achieved by heating the polymers at 100 °C under vacuum. Post-polymerization functionalization of the star polymers were demonstrated by conjugating the thiol containing tripeptide glutathione (Figure 7-5).

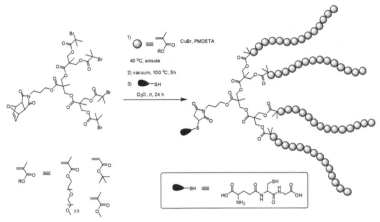

FIGURE 7-5 Synthesis and functionalization of core reactive multi-arm star polymers. (**Source:** Gok, O.; Durmaz, H.; Ozdes, E. S.; Hizal, G.; Tunca, U.; Sanyal, A. Maleimide-based thiol reactive multiarm star polymers via Diels-Alder/retro Diels-Alder strategy. J. Polym. Sci. Part A: Polym. Chem. 2010, 48, 2546–2556. ©John Wiley and Sons 2010. Adapted with permission. [14])

As an alternative to ATRP, RAFT polymerization is another important technique that is widely utilized to prepare end-functional polymers with different topologies. Since ATRP polymerization requires the use of a metal catalyst, RAFT polymerization serves as a good alternative for synthesizing polymers in a metal-free manner that allows one to obtain polymers without any undesired metal impurities. In RAFT, introducing a functional group to polymer end is accomplished by choosing a functional group containing chain transfer agent (CTA). To demonstrate this approach, Maynard et al. investigated RAFT polymerization towards the synthesis of maleimide end-functionalized polymers for site-specific conjugation to free cysteine groups of proteins (Figure 7-6).[15] A trithiocarbonate CTA containing furan protected maleimide group was used to polymerize poly(ethylene glycol) methyl ether acrylate. After the polymerization, a subsequent rDA reaction was performed to obtain thiol reactive maleimide groups at polymer chain ends. Thiol-containing proteins were shown to efficiently bind to the maleimide group at the polymer chain end to afford the protein-polymer conjugates.

As an alternative approach, Sumerlin et al. reported the direct modification of BSA protein with a RAFT CTA to prepare BSA-macro CTA (Figure 7-7).[16] The methodology established a "graft-from" strategy where the polymerization starts from the surface of protein. A maleimide group containing CTA was reacted with the free sulfhydrylcysteine residue (Cys-34) of BSA protein. Polymerization was conducted in aqueous medium using N-isopropylacrylamide (NIPAM) monomer to obtain thermoresponsive polymer–protein conjugates. The "graft-from" strategy depicted here provides an efficient installation of polymer chains onto protein. This approach circumvents the challenges encountered during conjugation of polymers to proteins that arise due to steric hindrances. Usually, an excess of polymeric component is required to achieve satisfactory conjugation, which oftentimes incurs tedious purification steps.

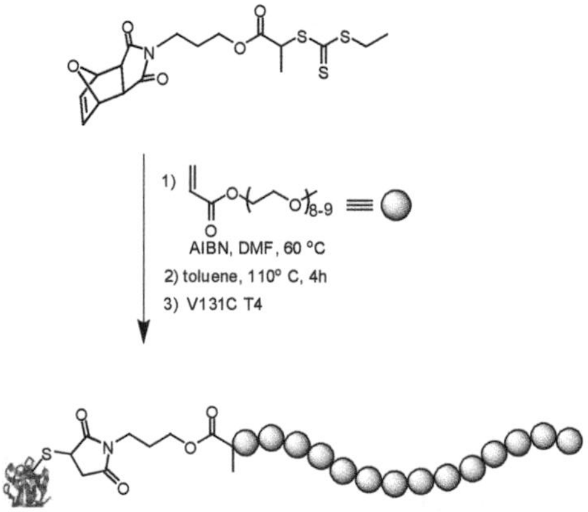

FIGURE 7-6 Synthesis of maleimide terminated polymers via RAFT polymerization and functionalization with protein.
(**Source:** Bays, E.; Tao, L.; Chang, C. W.; Maynard, H. D. Synthesis of Semitelechelic Maleimide Poly(PEGA) for Protein Conjugation By RAFT Polymerization. Biomacromolecules 2009, 10, 1777–1781. ©American Chemical Society 2009. Adapted with permission. [15].)

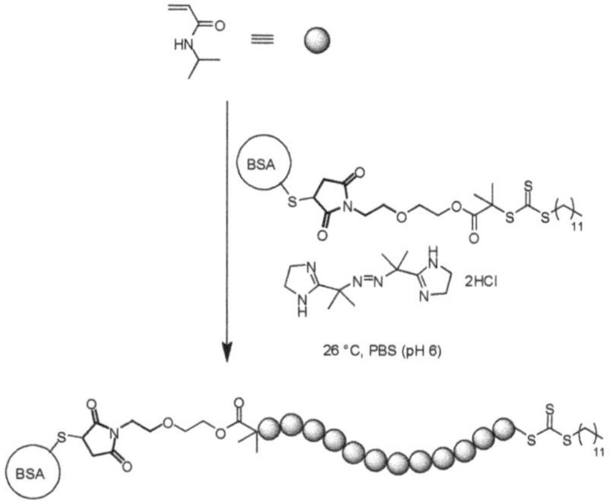

FIGURE 7-7 Synthesis and functionalization of poly NIPAM using a BSA-macro CTA.
(**Source:** De, P.; Li, M.; Gondi, S. R.; Sumerlin, B. S. Temperature-Regulated Activity of Responsive Polymer–Protein Conjugates Prepared by Grafting-from via RAFT Polymerization. J. Am. Chem. Soc. 2008, 130, 11288–11289. © American Chemical Society 2008. Adapted with permission. [16])

Maleimide end-functional biodegradable polymers were synthesized by ring opening polymerization (ROP) to obtain polymers that find potential applications in drug delivery and bioengineering. Recently Dove et al. reported the synthesis of polyester polymers via ROP of lactide monomers.[17] The organocatalytic ROP was conducted using a furan-protected maleimide containing alcohol initiator (Figure 7-8). The protecting group of maleimide was thereafter removed to obtain free maleimide groups at polymer chain ends by heating the polymers at 100 °C under vacuum. Notably, the thermal treatment did not lead to any degradation of the polyester chains. Conjugation reactions of different thiols were investigated in various conditions (solvent, catalyst) and effective functionalization was achieved without damaging the sensitive polyester backbone.

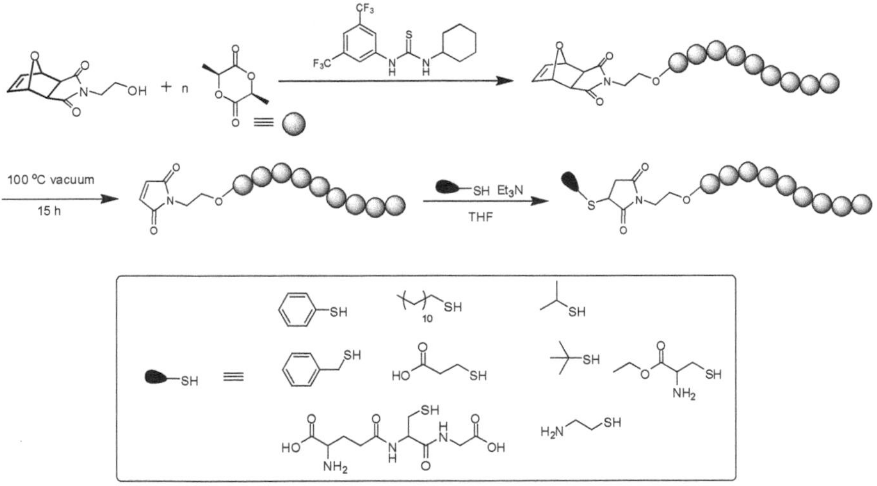

FIGURE 7-8 Synthesis and functionalization of chain end maleimide chain-end maleimide bearing biodegradable PLA polymers.
(**Source:** Pounder, R. J.; Stanford, M. J.; Brooks, P.; Richards, S. P.; Dove, A. P. Metal free thiol–maleimide 'Click' reaction as a mild functionalisation strategy for degradable polymers. Chem. Commun. 2008, 5158–5160. © The Royal Society of Chemistry 2008. Adapted with permission. [17])

Haddleton et al. demonstrated the RAFT polymerization of acrylates and acrylamides using a dibromomaleimide functionalized RAFT agent (Figure 7-9).[18] The substitution reactions of dibromomaleimides with thiol containing compounds were shown to be very efficient[19,20] and the quantitative reaction between these orthogonal groups has provided a new tool for conjugation of proteins to polymers functionalized with bromomaleimide groups. The well-defined polymer were obtained without using any protection group. Dibromomaleimide end functional RAFT polymers were efficiently functionalized with model thiophenol compound. ATRP polymerization of dibromomaleimide functionalized initiator was also investigated; however due to the participation of dibromomaleimide group to the propagating chain bimodal, ill-defined polymers were obtained.

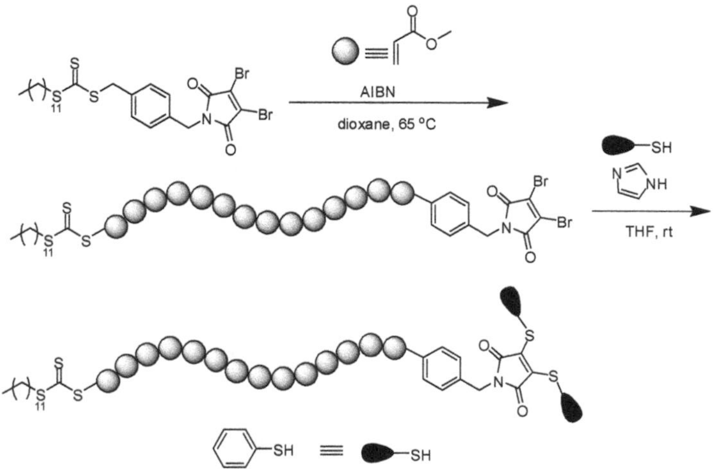

FIGURE 7-9 Synthesis and functionalization of chain-end dibromomaleimide functional polymers.
(**Source:** Robin, M. P.; Jones, M. W.; Haddleton, D. M.; O'Reilly, R. K. Dibromomaleimide End Functional Polymers by RAFT Polymerization Without the Need of Protecting Groups. ACS Macro Lett. 2012, 1, 222–226. © American Chemical Society 2012. Adapted with permission. [18])

In a later study, the methodology was extended by using dithiophenol maleimides to prepare functional polymers without the need of protecting group chemistry (Figure 7-10).[21] A linear PEG polymer (M_n = 5 kDa) bearing a dithiophenolmaleimide end group was successfully conjugated to salmon calcitonin (sCT) via disulfide bridge. Before conjugation, the single disulfide group of sCT (a 32-amino acid hormone) was effectively reduced to thiols using tris(2-carboxyethyl)phosphine (TCEP) as a reducing agent. The behavior of dithiophenolmaleimide group under living radical polymerization conditions was also investigated. An initiator functionalized with dithiophenolmaleimide groups was employed for synthesis of di(ethylene glycol) methacrylate polymer (poly(OEGMA)) by using Cu(I)/pyridine imine catalyst. Polymerization showed linear first order kinetics with low molecular weight distributions and monomodal GPC traces. Obtained dithiophenolmaleimide end group of ATRP polymer was functionalized via disulfide bridging of sCT protein.

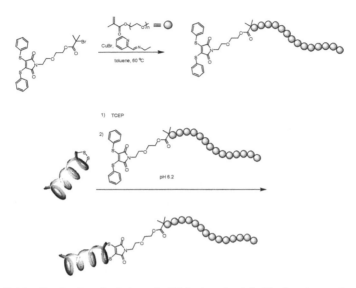

FIGURE 7-10 Synthesis of chain-end dithiophenolmaleimide bearing polymers and conjugation with sCT.
(**Source:** Jones, M. W.; Strickland, R. A.; Schumacher, F. F.; Caddick, S.; Baker, J. R.; Gibson, M. I.; Haddleton, D. M. Highly efficient disulfide bridging polymers for bioconjugates from radical-compatible dithiophenol maleimides. Chem. Commun. 2012, 48, 4064–4066. ©The Royal Society of Chemistry 2012. Adapted with permission. [21])

7.2.2 SYNTHESIS OF POLYMERS WITH SIDE CHAIN MALEIMIDE FUNCTIONALITIES USING MALEIMIDE BASED MONOMERS

The synthesis of novel functional polymers carrying reactive side chain groups has attracted wide attention in recent years since these polymers allow development of multivalent and multifunctional materials. Side-chain functional groups allow installment of desired molecules in multiple copies along the polymer backbone. The molecules attached to side chains can also act as ligands that can bind or interfere with certain receptors, especially in biological processes.[22,23] In drug delivery applications, drugs, targeting groups and/or tracking agents can be attached to reactive polymers with varying degree of functionalization. The concentration of appended side chains allows the control of attached molecules. Many functional groups including N-hydroxysuccinimide based activated ester that allows the introduction of amine containing molecules, alkyne and azide groups that allow "click" chemistries have been efficiently incorporated into polymers as pendant groups.[24,25] Maleimides constitutes another important class of functional group and recent approaches based on protection/deprotection strategies has enabled the synthesis of polymers with maleimide group bearing side chains.

Bailey and Swager reported the synthesis of maleimide-appended poly(phenyleneethynylene)s (PPEs) (Figure 7-11).[26] PPEs are useful polymersin sensor applications due to their energy funneling and transport abilities. The maleimide modified polymers were designed and synthesized for further functionalization; especially for conjugation of thiol containing molecules. In the polymerization step, masked maleimide containing aryl diiodides were cross-coupled with dialkynes under palladium catalyzed Sonogashira–Hagihara reaction conditions. Cross-coupling polymerization with unmasked maleimide monomer resulted only in short chain oligomers with side reactions. After polymerization, the masked maleimide groups were deprotected by removing furan moieties through a rDA reaction. Post-polymerization functionalization efficiency of maleimide side chain functionalized polymers were investigated by conjugating a thiol-modified carboxy-X-rhodamine (ROX) dye. Successful attachment of the thiol containing dye was demonstrated via absorbance, fluorescence spectra, and gel permeation chromatography (GPC) analyses. Maleimide based post-polymerization functionalization can be used as a powerful strategy to prepare diverse libraries of PPEs with pendant functionalities.

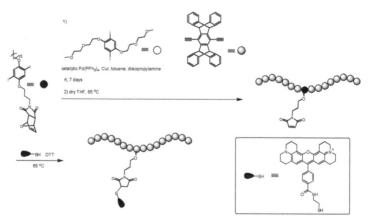

FIGURE 7-11 Synthesis and functionalization of pendant maleimide bearing poly (phenyleneethynylene)s.
(**Source:** Bailey, G. C.; Swager, T. M. Masked Michael Acceptors in Poly(phenyleneethynylene)s for Facile Conjugation. Macromolecules 2006, 39, 2815–2818. © American Chemical Society 2006. Adapted with permission. [26])

Sanyal et al. reported the utilization of a masked maleimide functional group containing monomer to obtain a polymer containing thiol-reactive maleimide groups as pendant side chains.[27] A furan protected maleimide containing methacrylate monomer was synthesized and used in AIBN initiated polymerization. The degree of maleimide functionalized monomer incorporation in the copolymer was tuned by changing the feed ratio of the monomer. After polymerization, thiol reactive maleimide groups were recovered by simply heating the polymers. In a later study, the monomer was used in the preparation of thiol-reactive polymeric micropatterns fabricated by thermal nanoimprint lithography (NIL) (Figure 7-12).[28] The furan protected maleimide monomer was copolymerized with a poly(ethyleneglycol)-based monomer and spin-coated onto a surface. During the NIL process, in situ deprotection occurs to furnish thiol-reactive polymeric micropatterns. The effective post-fuctionalization of thus obtained patterns were demonstrated by conjugating various thiol-containing molecules.

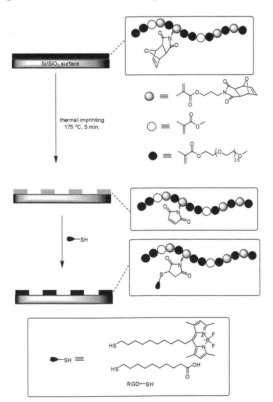

FIGURE 7-12 Synthesis and functionalization of maleimide containing polymeric patterns. (**Source:** Subramani, C.; Cengiz, N.; Saha, K.; Gevrek, T. N.; Yu, X.; Jeong, Y.; Bajaj, A.; Sanyal, A.; Rotello, V. M. Direct Fabrication of Functional and Biofunctional Nanostructures Through Reactive Imprinting. Adv. Mater. 2011, 23, 3165–3169. John Wiley and Sons 2011. Adapted with permission. [28])

Sanyal et al. extended this strategy to synthesize maleimide containing styrenic copolymers (Figure 7-13).[29] A novel styrenic monomer containing protected-maleimide unit was prepared and used to obtain reactive polymers using AIBN-initiated free radical, as well as RAFT polymerization. Masked-maleimide groups residing at the side chains were activated by removing the furan protecting group under thermal conditions. To extend the methodology, maleimide containing monomer was copolymerized with N-hydroxy-succinimide-containing or an alkene-containing styrene-based monomers to obtain orthogonally functionalizable styrenic types copolymers. Sequential click reactions were successfully employed for the functionalization of polymers with model compounds using either amine-thiol or thiol-thiol functional group based couplings.

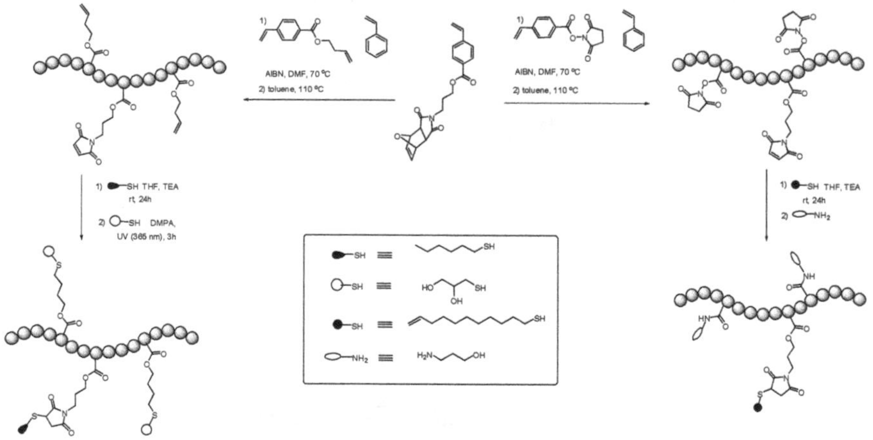

FIGURE 7-13   Synthesis and orthogonal functionalization of side-chain maleimide containing styrenic polymers.
(**Source:** Yilmaz, I. I.; Arslan, M.; Sanyal, A. Macromol. Design and Synthesis of Novel "Orthogonally" Functionalizable Maleimide-Based Styrenic Copolymers. Rapid Commun. 2012, 33, 856–862.© John Wiley and Sons 2012. Adapted with permission. [29])

Recently, Du Prez et al. reported the preparation of functional polyurethanes containing pendant maleimide functional groups. (Figure 7-14).[30] The strategy was employed by copolymerizing an appropriate functional diol and diisocyanate monomers under dibutyltindilaurate catalyzed reaction conditions. A diol monomer containing a furan-protected maleimide functionality was utilized. After activation of maleimide groups by heating polymers at 100 C under vacuum, free maleimide groups were obtained. The functionalization efficiency of maleimide groups were evaluated by attaching various thiols. The conjugation efficiency was found to be solvent and temperature dependent. For instance, in dimethylsulfoxide (DMSO) 75% coupling yield was obtained and the yield could be increased to 92% by utilizing N-methyl-2-pyrrolidone (NMP) as a solvent at a moderate temperature of 50 °C.

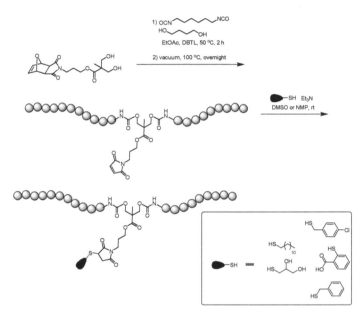

FIGURE 7-14 Synthesis and functionalization of pendant maleimide bearing polyurethanes. (**Source:** Billiet, L.; Gok, O.; Dove, A. P.; Sanyal, A.; Nguyen, L.-T. T.; Du Prez, F. Metal-Free Functionalization of Linear Polyurethanes by Thiol-Maleimide Coupling Reactions. Macromolecules 2011, 44, 7874–7878. © American Chemical Society 2011. Adapted with permission. [30])

Polycarbonate polymers have gained considerable attention because these polymers are promising candidates for various therapeutic applications due to their low toxicity, biodegradability, and biocompatibility. This necessitates the design of functionalizable polycarbonate polymers that would allow facile post-polymerization modification of these materials via conjugation of relevant molecules. Oftentimes, synthesis of polycarbonates is accomplished using the ROP of cyclic monomers such as lactides, the desired functional groups are usually incorporated into polymer backbone by choosing functional comonomers such as functional group containing lactides or carbonates. Polycarbonate based copolymers carrying various clickable functional groups have been reported.[31,32] Toward this end, Sanyal et al. reported polycarbonate based polymers carrying thiol-reactive maleimide groups along their backbone (Figure 7-15).[33] The functional polymers were obtained by ring opening (co)polymerization of a furan-protected maleimide group containing cyclic carbonate monomer using an organocatalyst, namely, 1,8-diazabicyclo[5.4.0]undec-7-ene (DBU). Obtained polymers were heated to subject the furan/maleimidecycloadduct to rDA reaction. The reaction was conducted at 100 °C under vacuum and no degradation of polymers was observed during the thermal treatment of the polymers. The maleimide groups on the polymers were efficiently conjugated with thiol containing compounds such as 6-(ferrocenyl)hexanethiol and 1-hexanethiol without any degradation of the parent polymer chains.

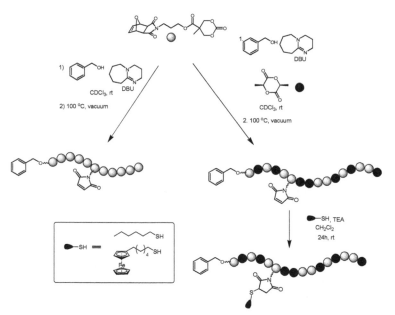

FIGURE 7-15 Synthesis and functionalization of side-chain maleimide containing polycarbonates.
(**Source:** Onbulak, S.; Tempelaar, S.; Pounder, R. J.; Gok, O.; Sanyal, R.; Dove, A. P.; Sanyal, A. Synthesis and Functionalization of Thiol-Reactive Biodegradable Polymers. Macromolecules 2012, 45 (3), 1715–1722. © American Chemical Society 2012. Adapted with permission. [33].)

Schaefer et al. recently reported the ring-opening metathesis polymerization (ROMP) of functional norbornene derivatives to synthesize multifunctional polymers (Figure 7-16).[34] Three different monomers containing a maleimide unit, a pentafluorophenyl group based activated estergroup, and a trialkylsilyl-protected alkyne group were copolymerized using the first-generation Grubbs' catalyst. The orthogonal reactivity of functional groups allowed the facile multi-functionalization of these ROMP derived copolymers that underwent efficient functionalization in a sequential manner. Firstly, the maleimide groups were conjugated with benzyl mercaptan through a nucleophilicthiol-ene reaction catalyzed by a tertiary amine. Secondly, activated pentafluorophenylester groups were functionalized with an amine. As a last step, alkyne groups were deprotected and reacted with an azide containing molecule via copper catalyzed Huisgen-type click reaction. The strategy elaborated the preparation of highly post-functionalizable ROMP terpolymers with the help of reactive group tolerance of the olefin metathesis. In particular, the maleimide functional group did not require any protection–deprotection protocol due to lack of participation of the electron deficient alkene unit during the polymerization.

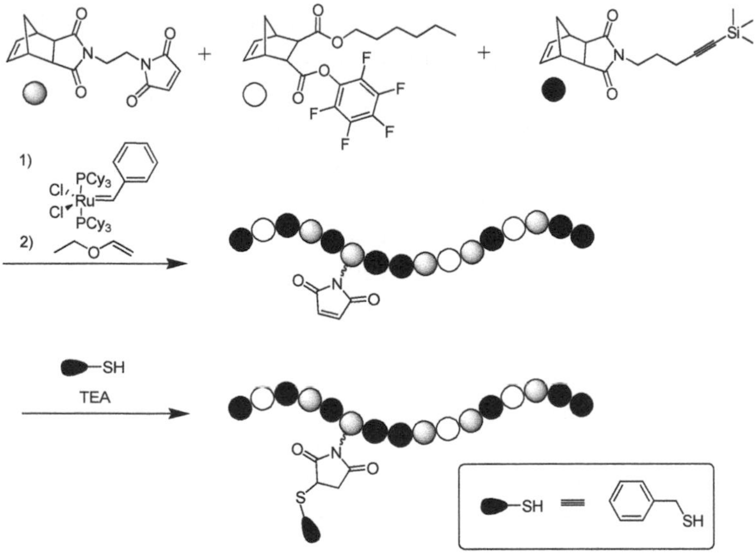

FIGURE 7-16 Synthesis of pendant maleimide bearing copolymers via ring-opening metathesis polymerization and functionalization with Michael addition.
(**Source:** 34. Schaefer, M.; Hanik, N.; Kilbinger, A. F. M. ROMP Copolymers for Orthogonal Click Functionalizations. Macromolecules 2012, 45, 6807. © American Chemical Society 2012. Adapted with permission. [34])

7.2.3 INCORPORATION OF MALEIMIDE GROUP INTO POLYMERS VIA POST-POLYMERIZATION MODIFICATION STRATEGIES

As outlined in previous sections, maleimide functional group bearing polymers are generally performed either by choosing a suitable initiator or protected maleimide group containing monomer. Beside using functional initiators and monomers that were summarized above, another strategy includes the incorporation of maleimide groups to polymers via post-polymerization modifications. In this method, after synthesis of the polymers, maleimide group containing molecules are coupled to main chain via different chemical transformations. In this section, various methods utilized to insert reactive maleimide groups onto polymers via post-polymerization modification are summarized.

The use of maleimide end functionalized polystyrene for site specific attachment of thiol containing lipase enzyme was reported by Nolte et al.[35] The synthesized well-defined polymer/enzyme hybrids constitute large amphiphilic structures and were shown to undergo self-assembly in water. The enzyme lipase B (an enzyme responsible for hydrolysis of esters) carrying a single disulfide bridge was treated with dithiothreitol (DTT) to break the disulfide bond and reveal the thiol groups. Maleimide end group containing polystyrene was synthesized by converting carboxy-terminated polystyrene to first an acid chloride derivative and this compound was coupled tomaleimide directly. The conjugation of thiol containing enzyme to maleimide-functionalized polystyrene was conducted in PBS buffer. The coupling efficiency was investigated by Ellman's analysis and it was shown that a single maleimide group was conjugated to the DTT reduced enzyme.

Synthesis of maleimide-terminated poly(N-isopropylacrylamide) (PNIPAM) polymers for conjugation of thiol containing molecules was reported by Sumerlin et al. (Figure 7-17).[36] The polymers were obtained by RAFT technique and maleimide functionality was installed by consecutive aminolysis/nucleophilicthiol-ene reactions. RAFT polymerization was performed using a trithiocarbonate CTA, 2-dodecylsulfanylthiocarbonylsulfanyl-2-methylpropionic acid. The trithiocarbonate group at the chain-end of the synthesized polymers were converted into thiol via aminolysis reaction to obtain thiol terminated PNIPAM-SH quantitatively. By treating thiol terminated polymers with 10-fold excess of bismaleimidodiethyleneglycol and removing unreacted PNIPAM-SH by immobilization onto iodoacetate-functionalized support, corresponding maleimide terminated polymers were obtained. This end-group transformation was achieved quantitatively as determined by ^{1}H NMR , ^{13}C NMR, FT-IR, and SEC analysis. The resulting maleimide end-capped polymers were efficiently functionalized with both small (4-methoxybenzyl mercaptan and dodecanethiol) and large (thiol-terminated PS) molecules. In a later study, proteins (BSA and ovalbumin) were shown to covalently attached to these maleimide terminated polymers.[37]

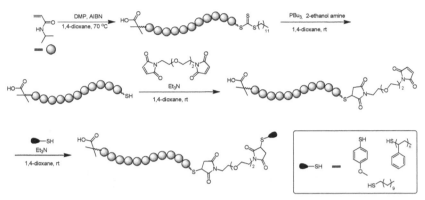

FIGURE 7-17   Incorporation of maleimide functionality to poly(N-isopropylacrylamide). (**Source:** Li, M.; De, P.; Gondi, S. R.; Sumerlin, B. S. End group transformations of RAFT-generated polymers with bismaleimides: Functional telechelics and modular block copolymers. J. Polym. Sci., Part A: Polym. Chem. 2008, 46, 5093–5100. © John Wiley and Sons 2008. Adapted with permission. [36])

Maynard and co-workers extended the manipulation of RAFT polymers to synthesize polymers that allows specific attachments of different proteins (Figure 7-18).[38] A biotin functionalized CTA was used for the polymerization of N-isopropylacrylamide to obtain polymers bearing biotin functional group at one of the chain ends. In the second step, trithiocarbonate group residing at the other end of the polymer was converted to furan-protected maleimide moiety via a radical induced coupling reaction. The protecting furan moiety was later removed by a rDA reaction to reveal biotin-maleimide functionalized hetero-telechelic polymers. The resultant polymer was efficiently functionalized by conjugation to two different proteins, streptavidin (SAv: a protein that have an extraordinarily high affinity for biotin) and BSA (carries cysteine residues which can couple to maleimide unit through thiol groups) to obtain protein–heterodimer conjugates.

In another work, a similar construct was utilized for the preparation of heterotelechelic polymers bearing cleavable disulfide bonds to modify the surfaces with proteins in a reversible manner (Figure 7-19).[39] After polymerization of N-isopropylacrylamide with a biotin containing CTA, dithioester group containing polymer chain ends were radical cross-coupled to a furan protected maleimide-containing azo initiator. Activation of maleimide group through a rDA reaction provided the heterotelechelic polymers in which biotin ligand and maleimide group placed at polymer ends. The maleimide end group was coupled with a mutant V131C T4 lysozyme and the obtained protein–polymer conjugate was bound to a streptavidin-coated surface through the biotin moiety. Since the CTA used in polymerization was decorated with a cleavable disulfide bond it enabled reversible conjugation of proteins to surfaces.

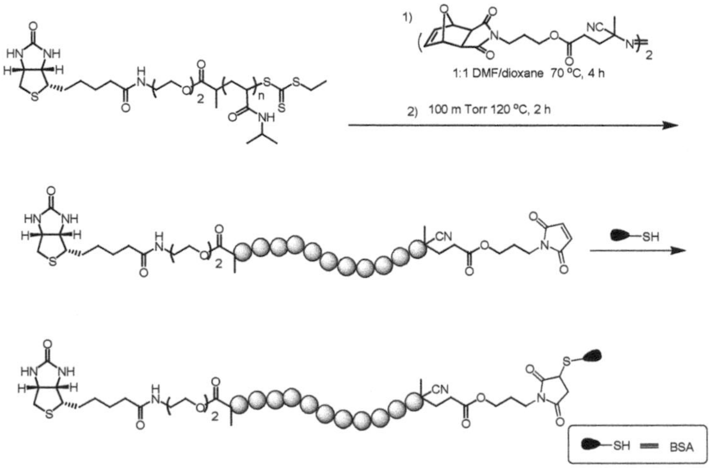

FIGURE 7-18   Modification of poly(N-isopropylacrylamide) with maleimide and functionalization with BSA.
(**Source:** Heredia, K. L.; Grover, G. N.; Tao, L.; Maynard, H. D. Synthesis of Heterotelechelic Polymers for Conjugation of Two Different Proteins. Macromolecules 2009, 42, 2360–2367. © American Chemical Society 2009. Adapted with permission. [38])

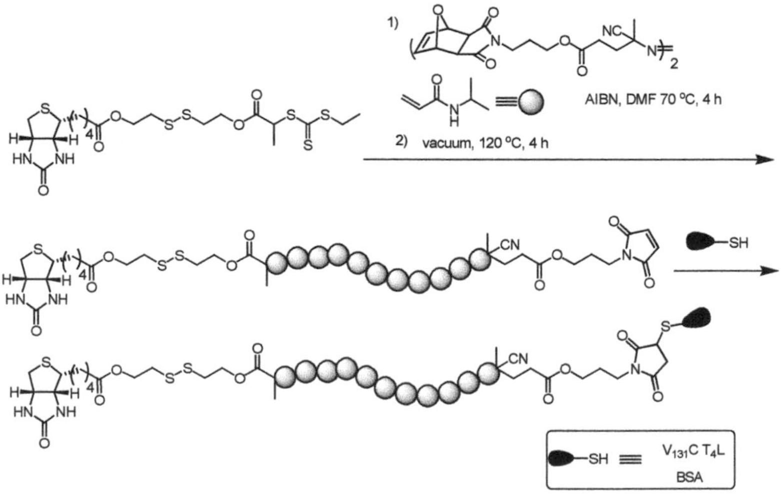

FIGURE 7-19   Synthesis and functionalization of chain-end maleimide containing heterotelechelic polymers.
(**Source:** Heredia, K. L.; Tao, L.; Grover, G. N.; Maynard, H. D. Heterotelechelic polymers for capture and release of protein–polymer conjugates. Polym. Chem. 2010, 1, 168–170. © Royal Society of Chemistry 2010. Adapted with permission. [39])

Radical induced cross-coupling methodology to replace RAFT end groups with a functionality was elaborated by Maynard et al. to synthesize maleimide-end-functionalized star polymers (Figure 7-20).[40] Four-armed poly(N-isopropylacrylamide) polymers with low polydispersity indices were synthesized using a tetrafunctional CTA. End functionalization of RAFT polymers were conducted using a maleimide functionalized azo initiator. After the radical coupling reaction, a rDA reaction was performed to remove the furan protecting group by heating the polymers at 105 °C under vacuum. The maleimide groups at polymer chain ends were functionalized via site-specific conjugation of V131C T4 lysozyme. An average of three maleimide groups were conjugated to thiol containing protein revealing the general usefulness of the methodology to prepare site-selective multimeric bioconjugates.

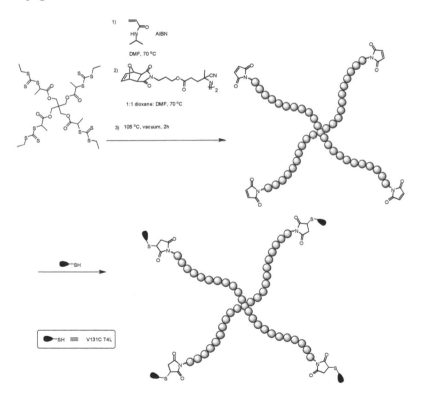

FIGURE 7-20 Synthesis and functionalization of maleimide-end-functionalized star polymers.
(**Source:** Tao, L.; Kaddis, C. S.; Loo, R. O. O.; Grover, G. N.; Loo, J. A.; Maynard, H. D. Synthesis of Maleimide-End-Functionalized Star Polymers and Multimeric Protein–Polymer Conjugates. Macromolecules 2009, 42, 8028–8033. © American Chemical Society 2009. Adapted with permission. [40])

In a similar manner, Maynard et al. exploited the strategy to prepare homodimeric protein–polymer conjugates (Figure 7-21).[41] Thermoresponsive polyNIPAAm were synthesized using RAFT polymerization by copolymerization of NIPAM and a dye-appended monomer using a bistrithiocarbonate CTA. Radical cross coupling reaction of polymer chain ends with protected maleimide containing azo compound and further deprotection of maleimide groups via a rDA reaction provided thiolreactive difunctional polymers. The resulting polymers were functionalized from both chain ends with a V131C mutant T4 lysozyme containing one free cysteine residue. According to SDS-PAGE analysis, 79% monomeric and 21% dimericprotein-polymer conjugate were obtained. Formation of monomeric conjugates were attributed to the large steric bulk of the proteins.

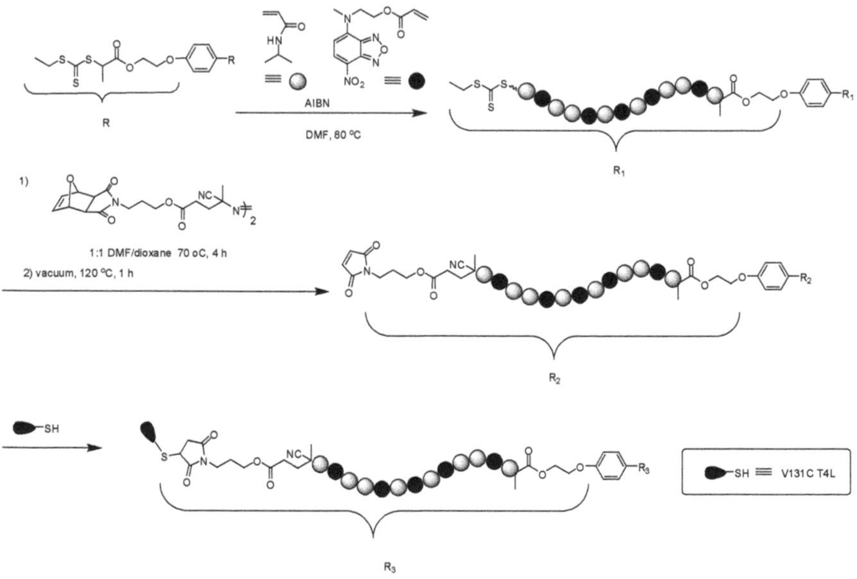

FIGURE 7-21   Synthesis of homodimeric protein-polymer conjugates.
(**Source:** Tao, L.; Kaddis, C. S.; Loo, R. R. O.; Grover, G. N.; Loo, J. A.; Maynard, H. D. Synthetic approach to homodimeric protein–polymer conjugates. Chem. Commun. 2009, 2148–2150. © The Royal Society of Chemistry 2009. Adapted with permission. [41])

Stanford and Dove reported the synthesis of maleimide end functionalized poly(lactide)s via post-polymerization functionalization of parent polymer chains (Figure 7-22).[42] The polymerization was initially conducted with aluminum methyl complexes applied in combination with isopropyl alcohol initiating species. After polymerization, while the initiating group constituted one end of polymer chain, the other chain end remains as an aluminum alkoxide species. To functionalize the ω-chain end of the polymer, polymerization was quenched with protected maleimide functional acid chloride which resulted in the formation of the desired heterobifunctional polymer. Further deprotection of furan protected maleimide at 100 °C resulted in the maleimide end-functional PLAs. To demonstrate the functionalization of maleimide group with thiols, a model compound thiophenol was used and quantitative chain end conversion was achieved as inferred by the disappearance of vinyl protons in ^{1}H NMR spectrum.

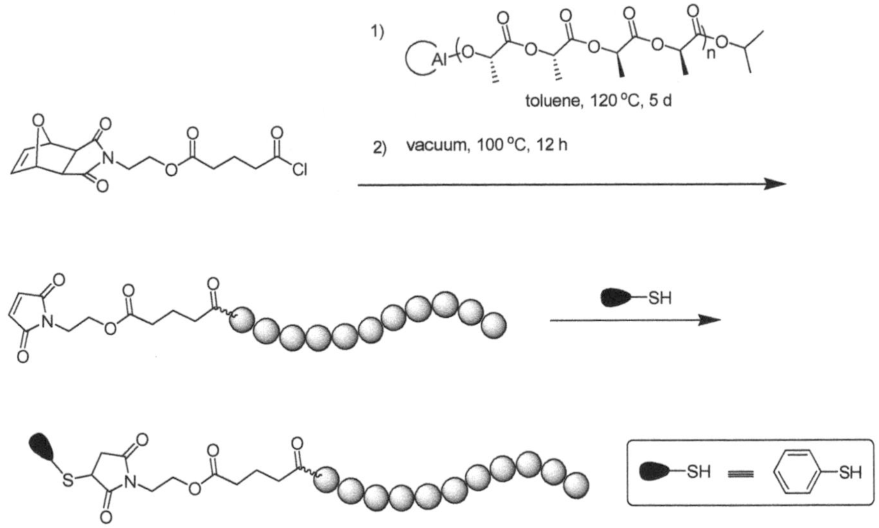

FIGURE 7-22   Synthesis of maleimide end functionalized poly(lactide)s via post functionalization modification.
(**Source:** Stanford, M. J.; Dove, A. P. One-Pot Synthesis of α,ω-Chain End Functional, Stereoregular, Star-Shaped Poly(lactide). Macromolecules 2009, 42, 141–147. © American Chemical Society 2009. Adapted with permission. [42])

In a recent study by Haddleton et al. highly efficient coupling reaction between dibromomaleimides and disulfides was utilized to prepare protein–polymer bioconjugates (Figure 7-23).[43] Dibromomaleimide end-functionalized ATRP polymers were designed and synthesized using water soluble poly(ethylene glycol) monomethyl ether methacylate (PEGMA) monomer. As mentioned previously, the dibromomaleimide group interferes with the radical polymerization. This reactivity prevents the use of unprotected dibromomaleimide group as part of initiators or monomers. Thus, the installation of dibromomaleimide group onto the polymers was investigated via post-polymerization modification routes. Two indirect routes were designed

in which azide and aniline functional initiators were used in the polymerization reactions. In the first route, after the polymerization reaction, the
azide functional group was coupled to an alkyne functional 2,3-dibro-
momaleimide via copper(I) catalyzed click reaction. Formation of new
end-group was confirmed by ^{13}C NMR spectroscopy and IR-spectroscopy.
In the second strategy, a Boc-protected aniline initiator was used for the
polymerization reaction. In the post-functionalization step, Boc protecting
group was removed and the deprotected amine of aniline was coupled to a
dibromomaleic anhydride derivative via imide formation. After obtaining
dibromomaleimide end-functionalized polymers successfully, conjugation
efficiency of disulfide containing sCT protein to both azide and aniline route
polymers were investigated. The disulfide group in sCT was reduced to free
thiol groups via tris(2-carboxyethyl)phosphine (TCEP) prior to addition to
polymer. According to RP-HPLC analysis, successful conjugation of sCT to
both polymers via re-bridged disulfide formation was observed. The conjugation reactions were performed using equimolar equivalents of the peptide
and polymer in which complete conjugation was observed in minutes. Utilization of equimolar equivalents simplifies the purification process to obtain
pure protein–polymer conjugates.

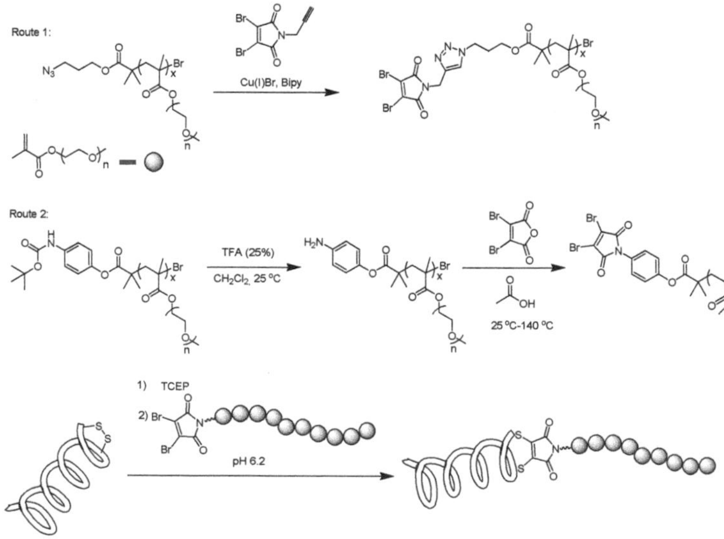

FIGURE 7-23 Synthesis of side-chain dibromomaleimide bearing polymers and conjugation
with disulfide containing protein.
(**Source:** Jones, M. W.; Strickland, R. A.; Schumacher, F. F.; Caddick, S.; Baker, J. R.; Gibson,
M. I.; Haddleton, D. M. Polymeric Dibromomaleimides As Extremely Efficient Disulfide
Bridging Bioconjugation and Pegylation Agents. J. Am. Chem. Soc. 2012, 134, 1847–1852.
© American Chemical Society 2011. Adapted with permission. [43])

7.3 CONCLUSIONS

Functionalization of polymeric materials in a desired manner, to achieve intended properties for various applications, can be realized through the adaptation of various efficient synthetic transformations available in organic chemistry. Undoubtedly, the thiol-maleimide conjugation is one of the most attractive approaches for such modifications because of the high level of efficiency and specificity that the reaction offers under mild reaction conditions. Until recently, there was a lack of efficient methods for obtaining polymers appended with reactive maleimide functional groups, due to the inherent reactivity of the electron deficient double bond in the maleimide moiety. Strategies based on various protection–deprotection sequences such as Diels–Alder/rDA reactions can be employed to incorporate the maleimide building block into various polymeric structures. The attractive nature of the thiol-maleimide based conjugations will continue to encourage the development of new methodologies for the design and synthesis of these materials as well as expand their areas of applications.

REFERENCES

1. (a). Rostovtsev, V. V.; Green, L. G.; Fokin, V. V.; Sharpless, K. B. *Angew. Chem., Int. Ed.* **2002**, *41*, 2596–2599. (b). Kolb, H. C.; Finn, M. G.; Sharpless, K. B. *Angew. Chem., Int. Ed.* **2001**, *41*, 2004–2021. (c). Lahann, J. In *Click Chemistry for Biotechnology and Materials Science*; Lahann, J., Ed.; John Wiley&Sons: Chichester, UK, 2009.
2. Heredia, K. L.; Maynard, H. D. *Org. Biomol. Chem.* **2007**, *5*, 45–53.
3. Nicolas, J.; Mantovani, G.; Haddleton, D. M. *Macromol. Rapid Commun.* **2007**, *28*, 1083–1111.
4. (a). Lundblad, R. L. *Techniques in Protein Modification*; CRC Press: Boca Raton, FL, 1995. (b). Hermanson, G. T. *Bioconjugate Techniques*; Academic Press: New York, 1996.
5. (a). Kogan, T. P. *Synth. Commun.* **1992**, *22*, 2417–2424. (b). Woghiren, C.; Sharma, B.; Stein, S. *Bioconjugate Chem.* **1993**, *4*, 314–318. (c). Herman, S.; Loccufier, J.; Schacht, E. *Macromol. Chem. Phys.* **1994**, *195*, 203–209. (d). Chilkoti, A.; Chen, G.; Stayton, P. S.; Hoffman, A. S. *Bioconjugate Chem.* **1994**, *5*, 504–507. (e). Stayton, P. S.; Shimoboji, T.; Long, C.; Chilkoti, A.; Chen, G.; Harris, J. M.; Hoffman, A. S. *Nature* **1995**, *378*, 472–474. (f). Morpurgo, M.; Veronese, F. M.; Kachensky, D.; Harris, J. M. *Bioconjugate Chem.* **1996**, *7*, 363–368.
6. Matyjaszewski, K.; Xia, J. H. *Chem. Rev.* **2001**, *101*, 2921–2990.
7. Hawker, C. J.; Bosman, A. W.; Harth, E. *Chem. Rev.* **2001**, *101*, 3661–3688.
8. Moad, G.; Rizzardo, E.; Thang, S. H. *Acc. Chem. Res.* **2008**, *41*, 1133–1142.
9. Mantovani, G.; Lecolley, F.; Tao, L.; Haddleton, D. M.; Clerx, J.; Cornelissen, J. J.; Velonia, K. *J. Am. Chem. Soc.* **2005**, *127*, 2966–2973.

10. Le Droumaguet, B.; Mantovani, G.; Haddleton, D. M.; Velonia, K. *J. Mater. Chem.* **2007**, *17*, 1916–1922.

11. Geng, J.; Mantovani, G.; Tao, L.; Nicolas, J.; Chen, G.; Wallis, R.; Mitchell, D. A.; Johnson, B. R. G.; Evans, S. D.; Haddleton, D. M. *J. Am. Chem. Soc.* **2007**, *129*, 15156–15163.

12. Da Pieve, C.; Williams, P.; Haddleton, D. M.; Palmer, R. M. J.; Missailidis, S. *Bioconjugate Chem.* **2010**, *21*, 169–174.

13. Tolstyka, Z. P.; Kopping, J. T.; Maynard, H. D. *Macromolecules* **2008**, *41*, 599–606.

14. Gok, O.; Durmaz, H.; Ozdes, E. S.; Hizal, G.; Tunca, U.; Sanyal, A. *J. Polym. Sci. Part A: Polym. Chem.* **2010**, *48*, 2546–2556.

15. Bays, E.; Tao, L.; Chang, C. W.; Maynard, H. D. *Biomacromolecules* **2009**, *10*, 1777–1781.

16. De, P.; Li, M.; Gondi, S. R.; Sumerlin, B. S. *J. Am. Chem. Soc.* **2008**, *130*, 11288–11289.

17. Pounder, R. J.; Stanford, M. J.; Brooks, P.; Richards, S. P.; Dove, A. P. *Chem. Commun.* **2008**, 5158–5160.

18. Robin, M. P.; Jones, M. W.; Haddleton, D. M.; O'Reilly, R. K. *ACS Macro Lett.* **2012**, *1*, 222–226.

19. Tedaldi, L. M.; Smith, M. E. B.; Nathani, R. I.; Baker, J. R. *Chem. Commun.* **2009**, 6583.

20. Smith, M. E. B.; Schumacher, F. F.; Ryan, C. P.; Tedaldi, L. M.; Papaioannou, D.; Waksman, G.; Caddick, S.; Baker, J. R. *J. Am. Chem. Soc.* **2010**, *132*, 1960.

21. Jones, M. W.; Strickland, R. A.; Schumacher, F. F.; Caddick, S.; Baker, J. R.; Gibson, M. I.; Haddleton, D. M. *Chem. Commun.* **2012**, *48*, 4064–4066.

22. Mammen, M.; Choi, S. K.; Whitesides, G. M. *Angew. Chem. Int. Ed.* **1998**, *37*, 2755.

23. Gestwicki, J. E.; Cairo, C. W.; Strong, L. E.; Oetjen, K. A.; Kiessling, L. L. *J. Am. Chem. Soc.* **2002**, *124*, 14922.

24. (a). Griffith, B. R.; Allen, B. L.; Rapraeger, A. C.; Kiessling, L. L. *J. Am. Chem. Soc.* **2004**, *126*, 1608–1609. (b). Drechsler, U.; Thibault, R. J.; Rotello, V. M. *Macromolecules* **2002**, *35*, 9621–9623. (c). Shunmugam, R.; Tew, G. N. *J. Polym. Sci. Part A: Polym. Chem.* **2005**, *43*, 5831–5843. (d). Savariar, E. N.; Thayumanavan, S. *J. Polym. Sci. Part A: Polym. Chem.* **2004**, *42*, 6340–6345. (e). Yanjarappa, M. J.; Gujraty, K. V.; Joshi, A.; Saraph, A.; Kane, R. S. *Biomacromolecules* **2006**, *7*, 1665–1670.

25. (a). Helms, B.; Mynar, J. L.; Hawker, C. J.; Frechet, J. M. J. *J. Am. Chem. Soc.* **2004**, *126*, 15020–15021. (b). Malkoch, M.; Thibault, R. J.; Drockenmuller, E.; Messerschmidt, M.; Voit, B.; Russell, T. P.; Hawker, C. J. *J. Am. Chem. Soc.* **2005**, *127*, 14942–14949. (c). Christman, K. L.; Maynard, H. D. *Langmuir* **2005**, *21*, 8389–8393. (d). Parrish, B.; Breitenkamp, R. B.; Emrick, T. *J. Am. Chem. Soc.* **2005**, *127*, 7404–7410.

26. Bailey, G. C.; Swager, T. M. *Macromolecules* **2006**, *39*, 2815–2818.

27. Dispinar, T.; Sanyal, R.; Sanyal, A. *J. Polym. Sci. Part A: Polym. Chem.* **2007**, *45*, 4545–4551.

28. Subramani, C.; Cengiz, N.; Saha, K.; Gevrek, T. N.; Yu, X.; Jeong, Y.; Bajaj, A.; Sanyal, A.; Rotello, V. M. *Adv. Mater.* **2011**, *23*, 3165–3169.

29. Yilmaz, I. I.; Arslan, M.; Sanyal, A. *Macromol. Rapid Commun.* **2012**, *33*, 856–862.

30. Billiet, L.; Gok, O.; Dove, A. P.; Sanyal, A.; Nguyen, L.-T. T.; Du Prez, F. *Macromolecules* **2011**, *44*, 7874–7878.

31. Tempelaar, S.; Mespouille, L.; Dubois, P.; Dove, A. P. *Macromolecules* **2011**, *44*, 2084–2091.

32. Engler, A. C.; Chan, J. M. W.; Coady, D. J.; O'Brien, J. M.; Sardon, H.; Nelson, A.; Sanders, D. P.; Yang, Y. Y.; Hedrick, J. L. *Macromolecules* **2013,** *46,* 1283–1290.
33. Onbulak, S.; Tempelaar, S.; Pounder, R. J.; Gok, O.; Sanyal, R.; Dove, A. P.; Sanyal, A. *Macromolecules* **2012,** *45* (3), 1715–1722.
34. Schaefer, M.; Hanik, N.; Kilbinger, A. F. M. *Macromolecules* **2012,** *45,* 6807.
35. Velonia, K.; Rowan, A. E.; Nolte, R. J. M. *J. Am. Chem. Soc.* **2002,** *124,* 4224–4225.
36. Li, M.; De, P.; Gondi, S. R.; Sumerlin, B. S. *J. Polym. Sci., Part A: Polym. Chem.* **2008,** *46,* 5093–5100.
37. Li, M.; De, P.; Li, H. M.; Sumerlin, B. S. *Polym. Chem.* **2010,** *1,* 854–859.
38. Heredia, K. L.; Grover, G. N.; Tao, L.; Maynard, H. D. *Macromolecules* **2009,** *42,* 2360–2367.
39. Heredia, K. L.; Tao, L.; Grover, G. N.; Maynard, H. D. *Polym. Chem.* **2010,** *1,* 168–170.
40. Tao, L.; Kaddis, C. S.; Loo, R. O. O.; Grover, G. N.; Loo, J. A.; Maynard, H. D. *Macromolecules* **2009,** *42,* 8028–8033.
41. Tao, L.; Kaddis, C. S.; Loo, R. R. O.; Grover, G. N.; Loo, J. A.; Maynard, H. D. *Chem. Commun.* **2009,** 2148–2150.
42. Stanford, M. J.; Dove, A. P. *Macromolecules* **2009,** *42,* 141–147.
43. Jones, M. W.; Strickland, R. A.; Schumacher, F. F.; Caddick, S.; Baker, J. R.; Gibson, M. I.; Haddleton, D. M. *J. Am. Chem. Soc.* **2012,** *134,* 1847–1852.

PART III
Application of Functional Materials

CHAPTER 8

METAL-CONTAINING CONJUGATED POLYMERS: PHOTOVOLTAIC AND TRANSISTOR PROPERTIES

ANJAN BEDI and SANJIO S. ZADE

CONTENTS

8.1   INTRODUCTION

The inclusion of metals into the conjugated systems may influence attractive photochemical, photophysical, and electrochemical properties. These properties can be explored for the applications in variety of area that includes electroluminescence, solar energy conversion, and nonlinear optics. The triplet-forming polymers show increased exciton diffusion length in the photovoltaic devices due to the forbidden nature of recombination from the triplet states, unlike the normal polymers.[1] It is also important in organic light emitting diodes (OLEDs) as both the conjugated organic backbone and the metal complexes can emit light upon excitation by photons and electrons. Contextually, metalloorganic complexes have been extensively used in dye-sensitized solar cell.[2]

Although organic polymers have shown dominance in the field of organic electronics, their metal-containing derivatives represent an another flexible class of polymeric semiconductors. Insertion of metal ions into conjugated polymers (CPs) benefits in many aspects:[3] (i) they may act as architectural templates to assemble the organic counterparts; (ii) they may provide redox-active and paramagnetic centres to generate active species for charge transport and may greatly influence the electronic and optical properties of organic π-systems through the involvement of the d-orbitals; (iii) the HOMO–LUMO gaps can be tuned through the interaction of the metal d-orbitals with the ligand orbitals; and (iv) structural features of these polymer may be altered depending on different coordination number, geometry, and valence shells of different metal atoms.

8.2   SCOPE OF THE CHAPTER

Polymer/organic optoelectronic devices have been the centre of attention from last two decades for both the academic and industrial research communities due to the potential for a low-cost, large-area, solution-processable technology alternative to their conventional inorganic counterparts. There are several issues related to the applications of CPs in organic electronics, which include lower efficiencies, less stabilities, and higher resistance in the devices. To address these obstacles, the significant research activities have been devoted to π-conjugated organics. One of the approaches includes the incorporation of metal complexes in the main conjugation or as a pendant substitution on that. Metalloorganic compounds with transition row elements have been extensively studied in organic electronics due to their easy tunability of electronic properties, amenable redox matching, and extended life-time of the triplet state. This successfully resulted in some high performance electronic devices containing the metalloorganic π-conjugated small molecules and polymers. Herein, we review the recent advances made in metalloorganic polymers for organic photovoltaics (OPVs) and field-effect transistors (OFETs).

8.3   ORGANIC SOLAR CELL

In a typical bulk-heterojunction (BHJ) solar cell, an active layer is sand-wiched between a transparent anode (typically indium tin oxide, ITO) and a metal cathode (Figure 8-1). To improve the electrical contact between ITO and the active layer and to adjust the corresponding energy levels, a thin layer of poly(3,4-ethylenedioxythiophene):poly(styrenesulfonate) (PEDOT:PSS) is generally applied in between the ITO and the active layer. Varieties of other interfacial layers have also been applied in polymer solar cells.

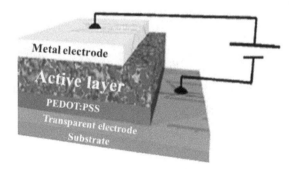

FIGURE 8-1   Structure of a typical polymer solar cell device.

Configurations of the active layer may vary depending on how the p-type semiconductor (i.e., electron-donors, such as conjugated polymers) and n-type semiconductor (i.e., electron acceptors, such as fullerene derivatives) are blended. Most successful BHJ configuration possesses electron-donors and electron acceptors, blended to form an interpenetrating network.[95] This offers two advantages: (a) it minimizes the travelling distance of excitons (electron-hole pair generated upon light absorption) to the donor/acceptor (D/A) interface, and concurrently maximizes the D/A interfacial area, thus ensures the exciton dissociation at the D/A interface to generate maximum free charge carriers and (b) it provides charge transport routes to facilitate the charge collection at electrodes, completing the conversion of the photon energy to electrical energy (i.e., photovoltaic effect).

8.3.1 IMPORTANT PARAMETERS OF POLYMER SOLAR CELLS (PSCS)

The most important performance parameter of a solar cell is the power conversion efficiency (PCE or η), which is related with the open circuit voltage (V_{OC}), short circuit current density (J_{SC}), and fill factor (FF) through eq 8-1. All of those parameters can be extracted from the current density (J)—voltage (V) curve under one sun condition (100 mW/cm^2, simulated AM1.5 solar illumination) (Figure 8-2). PCE is calculated by the following equation:

$$PCE = \frac{V_{OC} \times J_{SC} \times CE}{P_{in}} \tag{8-1}$$

And FF is defined as:

$$FF = \frac{V_{mpp} \times J_{mpp}}{V_{OC} \times J_{SC}} \tag{8-2}$$

where, V_{mpp} and J_{mpp} are the voltage and current at the maximum power point in the J–V curve, respectively.

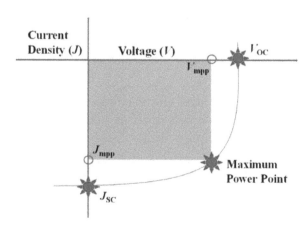

FIGURE 8-2 A general current density–voltage (J–V) graph and key parameters of device measurement.

8.4 SPACE CHARGE LIMITED CURRENTS AND FET MOBILITY

The space-charge-limited-current (SCLC) is described by the eq 8-3, where ε_r is the dielectric constant of the polymer, ε_0 is the permittivity of freespace, μ_h is the hole mobility, V is the voltage applied to the device, and L is the thickness of the polymer layer in a BHJ solar cell. The device configuration for that measurement is basically in the form of ITO/PEDOT:PSS/polymer:PCBM/Au.

$$J \cong \infty \frac{9}{8} \varepsilon_r \varepsilon_0 \mu_h \exp\left(0.891\gamma\sqrt{\frac{V}{L}}\right)\frac{V^2}{L^3} \tag{8-3}$$

On the contrary, p-Channel Field-effect transistor (PFET) devices (Figure 8-3) are constructed using the gate/dielectric/polymer/source-drain configuration for a well-documented bottom-gate top contact conformation. And the mobility is calculated from the current–voltage plot upon an applied electric field on the gate electrode. The current is measured from the eq 8-4, which provides the value of charge carrier mobility.

$$I_D = \frac{WC_{i\mu}}{2L}(V_G - V_{th})^2 \tag{8-4}$$

where W is channel width, L is channel length, C_i is the capacitance per unit area of the insulating layer, μ is the field-effect mobility, V_G is the voltage applied at the gate electrode, and V_{th} is threshold voltage, at which the channel is formed.

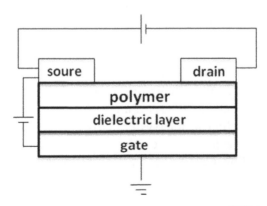

FIGURE 8-3 A general diagram for a bottom-gate top-contacts FET device.

8.5   POLYMERS CONTAINING DIFFERENT METAL CENTERS IN THE ACTIVE LAYER OF PSCS

8.5.1   POLYMERS WITH PLATINUM(II) COMPLEXES IN CONJUGATION

Materials with large triplet yields could provide increased current in organic photovoltaic devices. As the exciton diffusion length (L_D) is determined by exciton lifetime (τ) and exciton diffusivity (D) according to the equation $L_D = \sqrt{D}$, a significant increase in lifetime should lead to a longer exciton diffusion length, which would be advantageous for photovoltaic devices. Triplet-forming polymers inhibit geminate pair recombination[4] and increase the exciton diffusion length in BHJ devices. However, conjugated organometallic polymers that form triplets in the solid state are often large band gap systems (>2.5 eV), and therefore PCEs are limited by the poor overlap of their absorption spectrum with the solar irradiance. Again, low band gap (LBG) systems that demonstrate formation[5] of triplet excitons in solution does not always exhibit that in the solid state. Furthermore, performance of these LBG polymers is restricted by charge mobility.5 It is also necessary to consider that formation of triplet excitons in a BHJ device is foreseen to improve the photocurrent primarily by decreasing geminate recombination, as charge separation in a well-mixed heterojunction in the bulk occurs on the femtosecond time scale[6] while intersystem crossing occurs only in the nanosecond regime.[7] Therefore, in a BHJ with closely mixed morphology, charge separation could mostly occur before triplet excitons can form. Thus, any improvement in photovoltaic performance for triplet-forming materials in a BHJ solar cell is expected to be primarily the result of decreased geminate recombination or another factor.

Pt(II)-arylene ethynylenes are suitable building blocks for charge transportation.[3,8] In Pt-based polymetallaynes d-orbitals of Pt overlap with the p-orbitals of the alkyne unit leading to an augmentation of π-electron delocalization and intrachain charge transport along the main chain. In such metalloorganic polymers, efficient intersystem crossing facilitates the formation of triplet excitons by enhancing the spin-orbit coupling, which have longer lifetimes and thus allow extended exciton diffusion lengths.

8.5.1.1   PLATINUM IN POLYMER MAIN CHAIN

For the tuning of photophysical and charge transport properties, polymers **P1–P4** were designed with varying oligothienyl chain length in strongly absorbing polyplatinynes with solubilizing bithiazole units. The solar cell devices of **P3** (at **P3**:PCBM blend ratio of 1:5) showed PCEs of up to 2.66% and peak EQE up to 83%.[9] The solution-processable polymers **P1–P4** were obtained by Sonogashira-type dehydrohalogenation reaction between **1–4** and the platinum(II) precursor to obtain good yield and purity (Scheme 8-1). **P3** and **P4** blended with PCBM exhibited carrier mobilities in the order of 10^{-4} cm^2 V^{-1} s^{-1}, measured using space-charge-limited-current (SCLC) method and had comparable PCEs. **P1** and **P2** possessed lower mobilities than the other two polymers, which were explained by the absence of the oligothienyl chains in the polymer backbone that causes elongation of conjugation.[10] The potential of this work lies in enhancing the light-to-electricity conversion efficiencies of PSCs to a level of practical applications without the need for exploiting the triplet excited states in promoting an efficient photoinduced charge separation. Despite of using platinum polyynes with accessible triplet states for polymer solar cells, these polymers usually absorb only in the blue–violet spectral region (i.e., with wide bandgaps), and consequently the efficiency was low due to low coverage of the solar spectrum (e.g., only 0.27% in PCE for the Pt polyyne with 2,5-diethynyl-thiophene spacer). This work signifies the need for the development of not only phosphorescent polymeric materials but also the LBG counterparts to design high-efficiency photovoltaics.

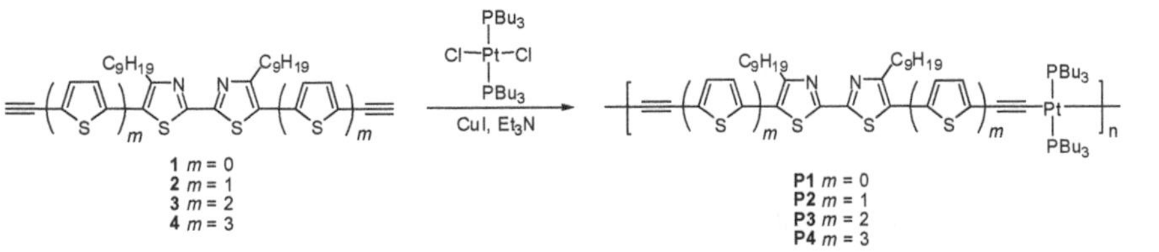

SCHEME 8-1   Synthetic route for **P1–P4**.

The maximum photocurrent quantum efficiencies in single-layer neat Pt-polyyne cells were ~0.04–1%,[11] whereas quantum efficiencies of ~1–2% were achieved after the addition of 7 wt.% C$_{60}$. More recently, external quantum efficiency (EQE) of ~9% has been achieved in BHJ cells using the **P5**/PCBM composite, resulting in a PCE of 0.27%.4 The polymer exhibited a λ_{max} of 400 nm, and consequently the low efficiency was due to limited coverage of the absorption of solar spectrum, although the triplet state of the organometallic polymer in the blend was involved in the photoinduced charge separation.4 So, the photovoltaic application of this type of material is hampered by their wide band gaps, and possibly unfavorable energy levels and charge-transport properties from blends with PCBM.

P5

Although polymetallaynes of the type *trans*-$[-Pt(PR_3)_2-C{\equiv}C-R-C{\equiv}C-]_n$ (R = carbocycles, heterocycles, main group elements) have drawn much attention in optoelectronic devices,[11,12,13] most of these Pt(II) polyynes are usually characterized by relatively large band gaps,[14,15,16,17] which compare unfavorably with those of some conjugated organic polymers comprising alternating electron donor (D) and acceptor (A) units (<1 eV).[18,19] However, successful strategies for creating LBG metallopolyynes involving D–A approach are rare.[11,20] Theoretical calculations predict that the development of novel donor materials with a lower bandgap is required to exceed the PCE of 10%. The polymer containing 4,7-di(2-thienyl)-2,1,3-benzothiadiazole as a core component was synthesized to create a new π-conjugated system with extended absorption that features unique D–A characteristics.[21] The intensely colored, novel platinum metallopolyyne **P6** showed a band gap of 1.85 eV. **P6**/PCBM blends exhibit balanced charge transport, as required for efficient solar cell performance,[22] and reached the mobility values to those measured in P3HT/PCBM cells with the optimal (1:1) blend ratio.[22] The solar cells, containing **P6**/fullerene derivative blends, showed a PCE with an average of 4.1%, without annealing or the use of spacer layers needed to achieve comparable efficiency with P3HT. This work practically explored the potential of metallated CPs for efficient photovoltaic devices.

P6   **P3HT**

In contrast to above report, Janssen et al. showed that the highest PCE of **P6**:PCBM photovoltaic devices with the reported device structure should not be higher than 2.2%.[23] Jenekhe et al.[24] further synthesized a series of 11 organometallic conjugated polymer semiconductors bearing a platinum center and studied their electronic band structures, field effect charge transport, and performance in BHJ solar cells along with structure–property relationships. Among the series of polymers including **P7**, they again synthesized **P6** and studied its OPV performance. The Pt-bridged polymers showed reversible electrochemical reduction waves from which electron affinities were found to be 2.95–3.28 eV. From the onset oxidation potentials of the polymers, ionization potentials were determined to be 4.82–5.23 eV. Optical band gaps of the D–A polymers were obtained between 1.49 and 1.97 eV. The spin coated pristine polymer thin films showed average time of flight (TOF) hole mobilities of 3.87×10^{-7} to 3.32×10^{-5} cm^2/Vs. BHJ solar cells based on of the polymers:[6,6]phenyl-C$_{71}$-butyric acid methyl ester (PC$_{71}$BM) blends produced PCEs as high as 0.68% for **P7** and 2.41% for **P6**. These results demonstrate the molecular engineering of the electronic band structures and the optical, charge transport, and photovoltaic properties of organometallic D–A conjugated polymer semiconductors.

P7

A series of polymers (**P8–P11**) were synthesized using thieno[3,2-b]thiophene instead of thiophene as the flanking spacers and ethyl instead of butyl as the solubilizing side chains.[25] Despite its amorphous nature, **P11** exhibited an encouraging FET mobility of 1×10^{-2} cm^2 V^{-1} s^{-1} with a moderate PCE value of 3.73% from the BHJ-PSCs of **P11**:PC$_{71}$BM at 1:4 (w/w) blending ratio. The significant improvement in PCEs by replacing PC$_{61}$BM with PC$_{71}$BM could be attributed to better match of energy levels. In a separate study, they reported charge carrier dynamics in solid films of a series of metalated polymers (**P8**, **P9**, and **P10**) based on Pt- and 4,7-di-2'-thienyl-2,1,3,-benzothiadiazole or 4,7-di-2'-thienothienyl-2,1,3,-benzothiadiazole upon photoexcitation of the π–π* transition using optical-pump terahertz-probe spectroscopy.[26] It was found that subpicosecond generated charge carriers recombine within 100 ps, but bound excitons persist. This is in sharp contrast to the dynamics seen in P3HT, where charge carriers persist at long delays. Application of the Drude–Smith model[27] allows estimation of the intrinsic mobility and internal quantum yield of charge carrier generation in these films. The improved conductivity in **P10** compared to **P8** is attributed to larger charge carrier yields.

P8

P9; R = C$_4$H$_9$, R$_1$ = H
P10; R = C$_4$H$_9$ and R$_1$ = C$_8$H$_{17}$
P11 ; R = C$_2$H$_5$ and R$_1$ = C$_8$H$_{17}$

Wang and Wong synthesized two stable platinum-acetylide polymers **P12** and **P13** with triphenylamine and 2,1,3-benzothiadiazole as the core components and studied their photophysical and photovoltaic properties.[28] The solvatochromic effect and narrow band gap values were attributed to ICT through the participation of strong D–A structural motif in the polymer chain. Lower band gap (E_g) of **P12** (1.85 eV) than **P13** (2.06 eV) facilitated harvesting of more solar energy. The highest PCE of 1.61% was achieved for **P13** with a V_{oc} = 0.77 V, J_{sc} = 4.9 mA cm^{-2}, and FF = 0.39 under illumination of an AM1.5 solar cell simulator in a 1:4 (**P13** : PCBM) blend ratio.

m = 0   **P12**
m = 1   **P13**

Qin et al. synthesized and studied photovoltaic properties of a novel LBG, solution-processable CP based on Pt-diacetylenide scaffolds and boron-dipyrromethene (BODIPY) chromophores.[29] To improve device performances replacement of n-Bu$_3$P ligands on Pt centers with less sterically demanding ones is expected to increase crystalline nature of these Pt-containing polymers. The highest PCE of 0.91% was obtained by using **P14**/PCBM at 1/3 weight ratio by using a typical device configuration of

ITO/MoO$_3$/active layer/Al. The low PCE were attributed to the unfavorable morphologies of the active layers and the nonoptimal energy level alignment between donor triplet states and acceptor molecules and inferior charge transfer dynamics. Miyake et al.[30] synthesized a Pt-containing copolymer (**P15**) and applied in BHJ PSC as a substitute of fullerene in combination with poly-3-hexylthiophene (P3HT) as donor. A weak photocurrent was resulted from the all-polymer solar cell which was attributed to the possible poor phase separation.

8.5.1.2 PLATINUM-CONTAINING POLYMERS WITH CYCLOPATINATED COMPLEXES

In addition to the ongoing research on platinum–aryelene CPs, Fréchet et al.[31] investigated platinum-containing LBG CPs (**P16** and P17) based on a platinum monomer with $C^{\wedge}N$ and $O^{\wedge}O$ diketonate ligands (Scheme 8-2). In these CPs the platinum atom is attached adjacent to the conjugated polymer backbone. Therefore, it docs not affect the exciton delocalization along the polymer chain, which opens up a means to study the effect of a heavy atom on diffuse excitons. The zero-field hole mobility of **P17**:PCBM and **P16**:PCBM were 2.5×10^{-9} and $1 \cdot 10^{-5}$ cm^2 V^{-1} s^{-1}, respectively. The SCLC hole mobility of **P16** is lower than that of polythiophenes[32] (10^{-4} to 10^{-3} cm^2 V^{-1} s^{-1}), but higher than the reported SCLC hole mobilities of polyplatinynes5 (10^{-8} to 10^{-7} cm^2 V^{-1} s^{-1}).

BHJ solar cells, fabricated from **P16** and **P17** yield efficiencies approaching 1.3%. Photophysical studies involved in this approach indicate that for some polymers long wavelength excitations undergo rapid nonradiative decay leading to an no observable triplet exciton formation, while other polymers showed localization of an initially delocalized exciton. This localization helps in triplet formation but is detrimental to charge transport in a photovoltaic device. Moreover, this work suggests that it is important to develop the new conjugated materials having both significant triplet yields and overlap with the solar spectrum in the solid state, which will minimize nonradiative decay paths at longer excitation wavelengths.

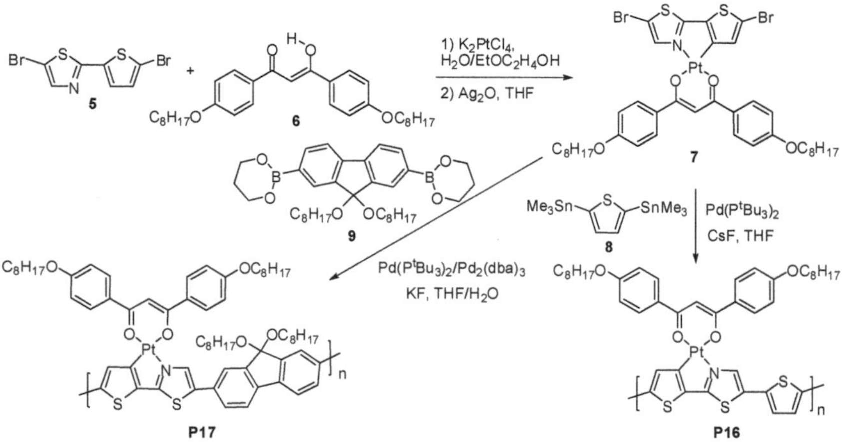

SCHEME 8-2 Synthetic route to **P16** and **P17**.

The optoelectronic properties of the cycloplatinated polymers can be tuned by adjusting the nature of the chelated ligands.[33] In pursuit of applying them in OPVs, two organometallic alternating copolymers **P20** and **P21**, comprising indaceno-dithiophene and cyclometalated Pt(II) moieties, were synthesized and compared their optical, electrochemical, and photovoltaic properties with those of related Pt-free organic polymers, **P18** and **P19**.[34] The absorption bands of the isoquinoline-derived compounds **P19**, **P21**, and BrTlqPtBr exhibited significant red shift relative to those of the pyridine-based derivatives **P18**, **P20**, and BrTPyPtBr. The absorption peaks at 430 and 480 nm were attributed to the metal-to-ligand charge-transfer (MLCT) for BrTPyPtBr and BrTIqPtBr, respectively. The shorter-wavelength absorption bands were related to $\pi-\pi^*$ transitions from the thiophene-phenylene-thiophene (TPT) core and the longer wavelength absorption bands to ICT between the TPT units and the Pt monomer. E_g of **P18–P21** were determined to be 2.17, 2.11, 1.96, and 1.82 eV, respectively. Photoinduced electron transfer (PET) from the singlet state of each polymer to its charge separated state was observed to be facile. The triplet energy states of **P20** and **P21** were too low in energy for PET to PCBM.

Large dihedral angles in polymer chain result in weak intermolecular interactions[35] and thus lower the charge mobility[36] and current density. As a result, the SCLC hole mobility of **P20** (9×10^{-6} cm²/Vs) was double that of **P21** (4.4×10^{-6} cm²/Vs). OPV device, containing **P20** as the donor material, achieved the highest PCE (2.9%) ever reported for a device fabricated with cycloplatinum-type polymers. The superior performance of Pt-containing polymers device was not attributed to the triplet contribution but the significantly red-shifted (80–90 nm) absorption bands of the Pt-containing polymers, resulting in improved overlap with the solar spectrum.

In polyplatinynes, the poor size and energetic overlap between the 5d Pt and 2pC orbitals leads to a significant smaller effective conjugation length when compared to structurally analogous poly(phenylene ethynylenes).[37] Given that exciton diffusion length is a function of both lifetime and diffusivity, this exciton localization or decreased effective conjugation length, may ultimately limit the photovoltaic performance and/or the exciton diffusion lengths observed for these polymers even if long lived triplets are realized. Organometallic conjugated polymer architectures other than polyplatinynes are therefore worthy of investigation as donor materials to better understand the role of heavy atoms on triplet formation in CPs for possible application in OPV. The performance of various metalloorganic polymers discussed in this section is summarized in Table 8-1.

TABLE 8-1 Optoelectronic Properties of the Platinum Containing Conjugated Polymers

Polymer	Band Gap (E_g)	V_{OC}(V)	J_{SC}(mA/cm2)	FF	PCE (%)	Hole Mobility (cm^2 V^{-1}s^{-1})	Reference
P1	2.40	0.73	0.91	0.32	0.21	<10^{-4}	10
P2	2.18	0.83	2.33	0.39	0.76	<10^{-4}	10
P3	2.10	0.81	6.93	0.38	2.14	~10^{-4}	10
P4	2.06	0.88	6.50	0.44	2.50 (2.66)[a]	~10^{-4}	10
P5	–	0.64	0.99	0.43	0.27	–	5
P6	1.85	0.82	13.1	0.37	4.1	–	23
P7	1.97	0.66	2.99	0.34	0.68[b]	6.27 × 10^{-6}	27
P8	1.85	0.828	4.04	0.46	1.53	6.0 × 10^{-5}	28
P9	1.84	0.844	7.33	0.39	2.45	1.5 × 10^{-3}	28
P10	1.82	0.813	8.67	0.50	3.57	5.7 × 10^{-3}	28
P11	1.81	0.787	9.61	0.49	3.73	8.9 × 10^{-3}	28
P12	2.06	0.80	4.00	0.34	1.09	–	31
P13	1.85	0.78	4.94	0.42	1.61	–	31
P14	1.70	0.86	2.23	48	0.91	–	32
P16	1.65	0.65	5.3	0.37	1.29	1.0 × 10^{-5}	34
P17	2.10	0.38	3.5	0.30	0.40	2.5 × 10^{-9}	34
P18	2.17	0.80	3.5	0.44	1.2	–	37
P19	2.11	0.69	4.0	0.32	0.9	–	37
P20	1.96	0.78	7.7	0.48	2.9	9.0 × 10^{-6}	37
P21	1.82	0.79	6.1	0.36	1.7	4.4 × 10^{-6}	37

8.5.2 ZINC COMPLEXES CONTAINING CONJUGATED POLYMERS

Porphyrin derivatives are essential light harvesting moiety in natural photo-synthetic systems. Metal porphyrins are rigid, two-dimensional planar π-conjugated structure with intense absorption in visible region, tunable optical, and redox properties.[38] The realization of ultrafast PET from porphyrins to fullerene has attracted significant attention of the current research on porphyrins as materials for organic electronics.[39,40,41,42] Efficient electron transfer was observed in porphyrin containing compounds as it undergo minimal structural changes during the uptake or release of electrons[43] to provide decent mobility of charge carriers.[44,45,46] Therefore, porphyrin containing molecules,[47,48,49] oligomers,[50] and polymers[51,52,53] have been extensively investigated in OPVs in recent years.

Unfortunately, the PCEs of the devices based on porphyrin triad,[47] liquid-crystalline porphyrin,[45] porphyrin dendrimer,[48] porphyrin-containing oligomers,[50] and main-chainporphyrin polymers[52] were obtained only upto 0.8%. Novel OPV systems using supramolecular complexes of porphyrin-peptide oligomers with fullerene clusters assembled as three-dimensional arrays onto SnO_2 films have been constructed with the PCE of 1.6%,[54] whereas photovoltaic cells using porphyrins/fullerenes composite nanoclusters with gold nanoparticles were also developed to afford a PCE of 1.5%.[44]

Wong et al.[55] developed a series of Zn-(porphyrin) chromophores containing soluble platinum metallopolyynes (Scheme 8-3). These polymers exhibited a sharp and strong Soret band near 430 nm and two weak Q-bands between 540 and 635 nm resulting in optical band gaps (E_g^{opt}) for these polymers vary from 1.93 to 2.02 eV. The best photovoltaic performance of devices based on the **P24**/PCBM blend layer was with V_{oc} = 0.77 eV, J_{sc} = 3.42 mA cm^{-2}, FF = 0.39, and PCE = 1.04%. **P23** has a lower E_g than **P22** which favors harvesting of more solar photon energy. The low PCE was attributed to the short absorption coverage in the visible region. This work suggested the need of incorporation of functional chromophores in the polymer chains to improve the absorption properties and hence enhance the photovoltaic efficiency of porphyrin-containing polymers. Such an electron rich aromatic group excludes any C–H bond which can be sterically hindered with the β-proton of the porphyrin, consequently provides further planarity to the electronic core.[56]

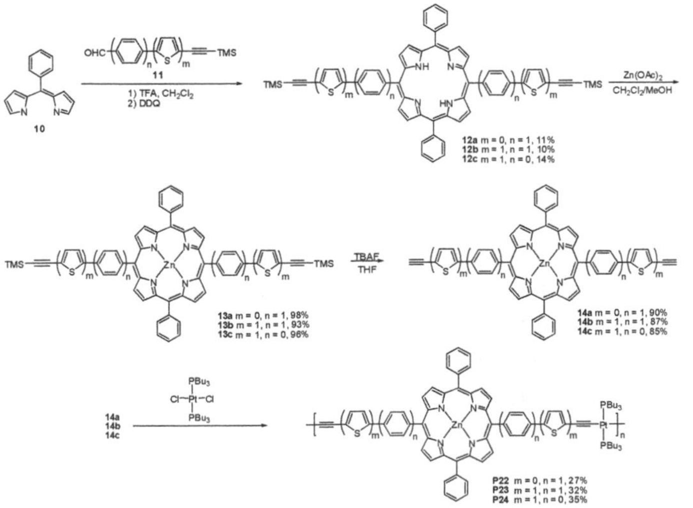

SCHEME 8-3 Synthesis of Zn-porphyrin containing conjugated alternating polymers.[55]

Porphyrin-based small molecule BHJ solar cells and dye-sensitized solar cells (DSSCs) have reached PCEs of more than 5% and 12%, respectively.[57,58] However, when applying porphyrin polymers (**P25–P32**) in solar cells, low PCEs (<1%) and poor SCLC hole mobilities were obtained.[59,60,61,62] The typical absorption spectra of porphyrin units showed a sharp and strong Soret band (410−430 nm) and weak Q-bands (530–540 nm) without absorption between these bands, and the conventional meso-aryl linkage in porphyrin polymers could not broaden the absorption effectively by intramolecular charge transfer (ICT) due to the large aryl–porphyrin dihedral angles.[63,64]

So far, the exploration of main-chain porphyrin-incorporated polymers (PPors) used in PSC applications remains at its early stage. The major obstacles come from the low J_{sc} and FF of these PPor-based PSCs, which are mainly attributed to the strong yet narrow absorption bands of the PPors, and the unfavorable morphologies in the PPor/[6,6]-phenyl-C_{61}-butyric acid methyl ester (PC$_{61}$BM) active layer. The averaged J_{sc} and FF of the PSCs was in the rage of 0.12–5.03 mA/cm^2 and 0.24–0.39, respectively, which limited the maximum PCE of the PSCs to 1%.[60] Improving J_{sc} by broadening the absorption band and light-harvesting abilities of PPors were attempted. Although broader absorption spectra were successfully demonstrated, significant enhancement in the J_{sc} of the resulted PSCs was not achieved. Thus, the mechanism of the unsatisfactory J_{sc} of PPor-based PSCs is not solely caused by the inefficient harvesting of solar radiation and remains unclear.[65] Although the polymer **P33** exhibited broad absorption range of 350–600 nm, PCE of **P33**:PCBM (1:1) blend was found to be only 0.61%.

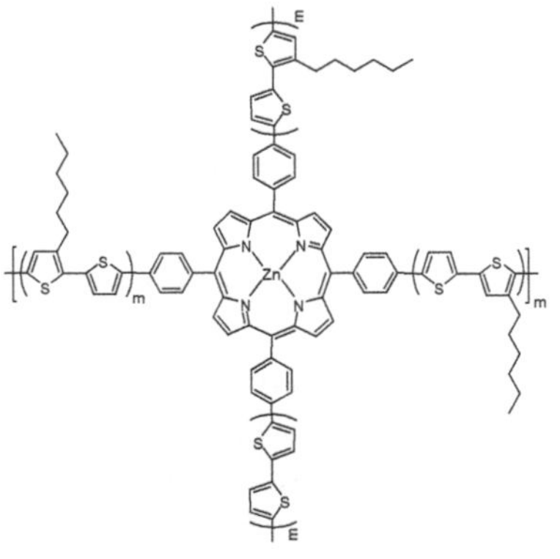

P33

Quinoxalino[2,3-b']porphyrins such as compound **15** are more π-expanded systems and[66] has flexibility towards attaching to a variety of central metal ions or selective functionalization at β-pyrrolic positions, that enable "fine tuning" of electronic properties of compounds by altering position of the frontier orbitals.[67] A new D–A copolymer (**P34**) was synthesized with quinoxalinoporphyrin as A unit, 2,7-carbazole as D unit (Scheme 8-4).[68] Thus, **P34** has laterally extended porphyrin units and not the part of polymer main chain. **P34** showed broad absorption of solar light from 400 to 750 nm. OPV device from **P34** reached a J_{sc} of 8.32 mA/cm^2, FF of 0.45, and PCE of 2.53%.

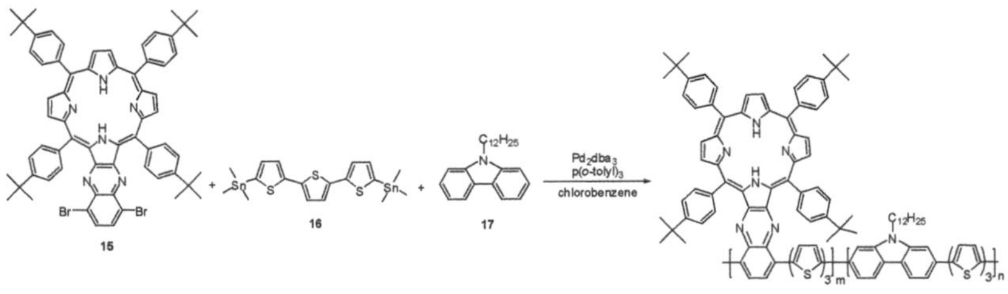

SCHEME 8-4 Synthesis of porphyrin containing polymer **P34**.

A series of quinoxalino[2,3-b']porphyrins based D–π–A polymers **P35**–**P38** were synthesized and their optoelectronic properties were tuned by the variation of oligothienyl chain π-bridge length. Structure–property relationship was studied between zinc(II)-porphyrin and free-base porphyrin.[69] The SCLC hole mobility of the polymers were measured with a device configuration of ITO/PEDOT:PSS/polymer:PC$_{71}$BM/Au (Table 8-2). The lower hole mobilities of **P35** and **P36** than **P37** can be attributed to the change of the coplanarity in the polymer backbone by the introduction of various oligo-thiophene bridges. This infers that the change of the π-bridge from thiophene to bithiophene, and then to terthiophene, results in a gradual increase of the intrachain mobility because of more extended π-conjugation and less steric hindrance. The highest hole mobility of approximately 10^{-4} cm^2 V^{-1} s^{-1} for **P38** indicates that the incorporation of zinc may be in favor of enhancing the packing structure in the blends.

The hole mobilities of **P35–P37** increased from 3.1×10^{-6} and 9.6×10^{-6} to 3.3×10^{-5} cm^2 V^{-1} s^{-1} while **P38** showed the highest p-channel mobility of 1.3×10^{-4} cm^2 V^{-1} s^{-1}, indicating that the incorporation of zinc may improve the packing structure in the blends. The PCEs of the PSCs were found to be 0.97% (**P35**), 1.97% (**P36**), 2.53% (**P37**), and 1.45% (**P38**). The increase in hole mobilities and PCEs with the number of thiophene rings was correlated with the decrease in the degree of curvature in the polymer backbone and increasing the effective conjugation length. This work provides a new insight into the design and future development of porphyrin-based CPs over the introductory research that claimed that the complexing of zinc may be a disadvantage to the improvement of PCEs due to decreasing both short-circuit current and fill factor.

Recently, a porphyrin-based polymer, **P39** featuring broad absorption in the blue-light region (400–550 nm) was reported by Lin et al.[70] Careful designing of a cascade energy levels alignment among the components guarantees efficient exciton dissociation and avoid charge trapping in a ternary-blend system.[71,72,73] The BHJ PSCs based on the binary blend of **P39**/PC$_{71}$BM at the blend ratio of 1:3 exhibited a high V_{oc} of 0.79 V, but low J_{sc} of 2.98 mA cm^{-2} and FF of 0.33. Photoluminescence (PL) quenching experiments and Atomic Force Microsopy (AFM) topology of the **P39**/PC$_{71}$BM blend indicated an efficient electron transfer from the photoexcited **P39** to PC$_{71}$BM, but an inefficient charge transport due to the nonideal morphology of the blend. This study revealed the first example of the use of porphyrine as blue-light harvester dopant in PSC applications. Although, the devices fabricated from the ternary blends **P39**:P3HT:PC$_{71}$BM and **P39**:**P40**:PC$_{71}$BM (P3HT and **P40** as hosts and **P39** as dopant) showed a cascade energy levels alignment among the components, it resulted into decrease in PCE. Low J_{sc}, FF, and PCE of the PSCs were explained by significant decrease in the crystalline nature of P3HT due to the presence of **P39** within the crystalline host (P3HT). In contrast, in the amorphous host (**P40**), the **P39** dopant effectively enhanced the power conversion in the blue-light region, maintaining the FF well at different dopant concentrations. The debt in J_{sc} in the **P39**:**P40**:PC$_{71}$BM ternary-blend PSCs was attributed to the decrease of the host concentration as indicated by the EQE experiments.

2,2':6',2''-Terpyridine (terpy) derivatives support macromolecular assemblies by chelating effect of pyridine rings to form $d\pi$-$p\pi^*$ back-bonding and high binding affinities toward many metal ions.[74] Two Zn(II)-based random (**P41** and **P42**) and two Ru(II)-based alternating (**P43** and **P44**) metallo-copolymers containing bis-terpyridyl ligands with diverse central donor (i.e., fluorene or carbazole) and acceptor (i.e., benzothiadiazole) moieties were synthesized and their thermal, optical, and electrochemical properties were investigated by Lin et al.[75] Ru(II)-based metallopolymers covered a broad range of 260–750 nm in the absorption spectra and acquired optical band gap in the range of 1.57–1.77 eV. In addition, the introduction of Ru(II)-based metallo-coplymer **P44** mixed with PC$_{60}$BM as an active layer of the BHJ solar cell device exhibited the highest PCE value up to 0.90%.

M = Zn, random polymers
P41; X = C(C$_6$H$_{13}$)$_2$
P42; X = N-CH(C$_8$H$_{13}$)$_2$

M = Ru, alternate polymers
P43; X = C(C$_8$H$_{13}$)$_2$
P44; X = N-CH(C$_8$H$_{13}$)$_2$

Wang et al.[76] reported a novel alternating D–A copolymer, **P45**, with dioctylporphyrin (Por) and 5,6-bis(octyloxy)benzo-2,1,3-thiadiazole (BDT). **P45** displayed broad absorption range of 350–950 nm with two peaks centered at 456 and 818 nm corresponding to the Soret band and Q-bands absorption of porphyrin segments, respectively. Attaching electron deficient BDT unit to with Por noticeably broadened the absorption spectrum and enhanced the Q-band absorption of the porphyrin based polymer. The HOMO and LUMO energy levels of the polymer were −5.06 eV and −3.63 eV, respectively. FETs, fabricated using this polymer with bottom gate/top-contact geometry exhibited the mobility of 4.3×10^{-5} cm^2 V^{-1} s^{-1} with a I_{on}/I_{off} of 10^4, which is one of the highest values for porphyrin-based polymers. The performance of various metalloorganic polymers discussed in this section is summarized in Table 8-2.

TABLE 8-2   Optoelectronic Properties of the Zinc Containing Conjugated Polymers

Polymer	Band Gap (E_g)	V_{OC}(V)	J_{SC}(mA/cm2)	FF	PCE (%)	Hole Mobility (cm^2 V^{-1} s^{-1})	Reference
P22	2.02	0.72	2.74	0.34	0.68	–	58
P23	2.00	0.78	3.02	0.30	0.71	–	58
P24	1.93	0.77	3.42	0.39	1.04	–	58
P27	1.75	0.58	0.34	0.24	0.048	–	62
P28	1.90	0.53	0.20	0.26	0.027	–	62
P29	1.95	0.75	2.8	0.29	0.62		64
P30	1.94	0.73	3.6	0.29	0.76		64
P31	1.85	0.46	2.03	0.34	0.32	–	63
P32	1.74	0.46	1.70	0.23	0.18	–	63
P33	1.91	0.69	8.63	0.66	3.91	–	70
P35	1.73	0.69	4.15	0.34	0.97	3.1×10^{-6}	74
P36	1.69	0.68	6.80	0.43	1.97	9.6×10^{-6}	74
P37	1.67	0.68	8.32	0.45	2.53	3.3×10^{-5}	74
P38	1.66	0.70	5.79	0.36	1.45	1.3×10^{-6}	74
P39:P40	–	0.59	7.21	0.49	2.08	–	75
P43	1.82	0.66	2.22	0.39	0.57		80
P44	1.76	0.70	3.50	0.37	0.90		80

P45

8.5.3 RUTHENIUM COMPLEXES IN METALLOORGANIC CONJUGATED POLYMERS

Recently, potential exploration of π-conjugated polyelectrolytes into photovoltaic cells (PVCs), including the easy fabrication, the ability to be used in layer-by-layer (LBL) processing, and the fact that the applications in a variety of chemical and sensory purposes, have shown promise that they could be efficiently quenched by electron acceptors.[77] Among all conjugated polyelectrolytes, terpyridyl Ru(II) complexes have been found potent materials to be used in PVCs, because the insertions of ruthenium metals into conjugated backbones have multiple advantages, such as the charge generations facilitated by the extensions of the absorption ranges because of their characteristic long-lived MLCT transitions and to exhibit reversible Ru(II,III) redox processes along with some ligand-centered redox processes.[50,78,79,80,81,82] Applications of terpyridyl Ru(II) complexes are limited by their low efficiency values in solar cell devices because of the low V_{oc} from the relatively higher HOMO levels, smaller J_{sc} caused by less sensitization ranges, and poor morphologies due to less solubilities in terpyridyl Ru(II) complexes.

Yang et al. studied poly(fluorene) (PF) and poly(thiophene) (P3HT) blended with molecular Ir(mppy)$_3$ using CdSe as an acceptor,[83] and demonstrated an increase in the triplet exciton population, attributed to the increased rate of intersystem crossing (ISC) from the PF singlet state to the triplet state. However, blends of polymers and small molecules generally suffer from local phase segregation immediately or over a period of time.[84] This was modified by chemically attaching the small molecule to the polymer backbone. One such example is the attempt by Peng et al.[85] They designed and synthesized two new Ru(II) containing supramolecular polymers (**P46** and **P47**) with D–A–D bis-terpyridyl ligands using cyclopenta[2,1-*b*:3,4-*b'*]-dithiophene as a donor block and benzodiazole/fluorinated benzodiazole as an acceptor block (Scheme 8-5). These polymers exhibited promising absorptions and deep-seated HOMO energy levels. The effect of fluorine atoms in elevation of the decomposition temperature, reduction of the optical bandgap, lowering of the HOMO level, and increase in the hole mobility was explained. Additionally, the photovoltaic performance was evaluated by PSCs with a device structure of ITO/PEDOT:PSS/polymer:PC$_{71}$BM/Ca/Al. **P46**:PC$_{71}$BM (1:2, w/w) exhibited a V_{oc} of 0.67 V, a J_{sc} of 6.17 mA cm^{-2}, and a FF of 0.48 and a PCE of 1.99%. The PCE of 2.66% in a improved device consisted of **P47**:PC$_{71}$BM (1:2, w/w), was attributed to the constructive effect of the fluorine atom on the HOMO level, band gap, hole mobility, as well as low phase separation. This is the best result of the reported supramolecular polymers applied in conventional BHJ PSCs without any additives and/or post treatment of the active films. Considering the advantages of supramolecular polymers, this approach may afford meaningful and highly efficient solution-processable polymer photovoltaic materials in PSCs.

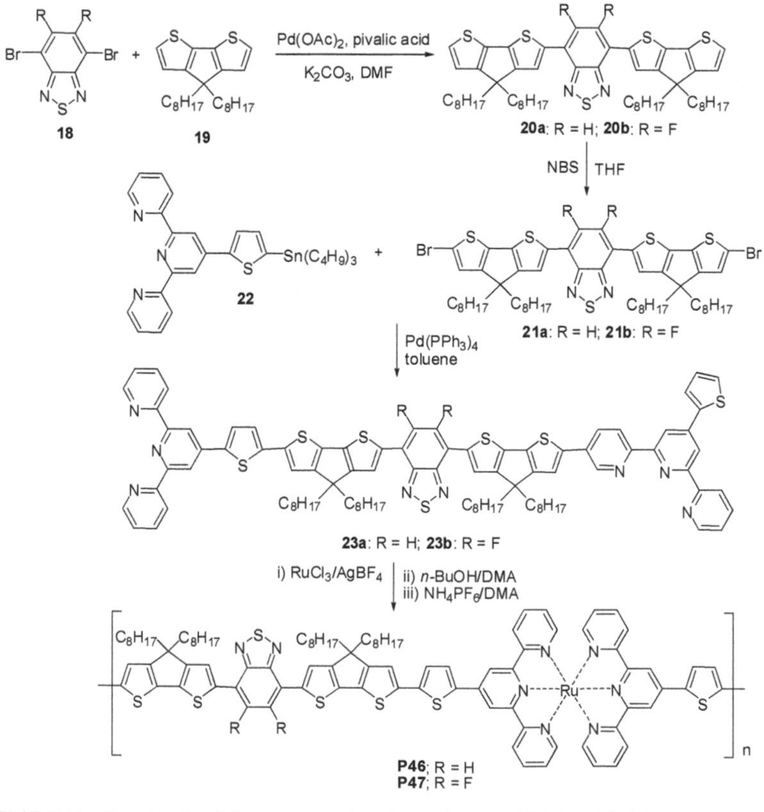

SCHEME 8-5 Synthesis of the supramolecular polymers (**P46** and **P47**).

The μ_{FET} of **P46** and **P47** were determined to be 7.5×10^{-6} and 2.8×10^{-5} cm^2 V^{-1} s^{-1}, respectively. **P47** exhibited a mobility one order of magnitude higher than **P46** because of the introduction of the fluorine atom, making the π–π stacking stronger due to the F–H, F–S, and/or F–F interactions. This has been proved to be beneficial to enhance the J_{SC}, FF, and PCE.[86,87]

Metallo-polymers are one kind of extensive supramolecular polymers, which are central to modern supramolecular chemistry, in which dynamic metal–ligand coordinative interactions serve as a reversible supramolecular polymerization motif.[88] The properties of metallo-polymers can be widely varied due to the availability of a multitude of metal ions and organic ligands, which both have effects on binding strength, reversibility, stability, and solubility.[89] Recently, Cheng et al.[90] reported the first bisterpyridyl-Ru(II)complex-containing metallopolymers (**P48** and **P49**) for fabrication of multilayer PSCs by the LBL deposition approach. Both polymers exhibits hole carrier mobilities as high as approximately 10^{-4} cm^2 V^{-1} s^{-1}. Due to low V_{OC}, these devices exhibit PCEs of approximately 0.12%. Mikroyannidis et al.[79] synthesized similar metallopolymers and used them as a buffer layer in BHJ PSCs along with a blending active layer of P3HT:PCBM (1:1 w/w). The maximum PCE value obtained from **P50** was up to 0.71%. Subsequently, Lin et al.[81] reported a series of main-chain Ru(II) metallo-polymers (**P51, P52,** and **P53**) with different donor and acceptor bis-terpyridyl ligands and used them as donor materials in BHJ PSCs. These devices exhibited PCE values in the range of 0.06–0.9%.

8.5.4 IRIDIUM COMPLEXES IN CONJUGATION

Ir-complexes may enhance the overall performance of solar cell by triplet formation and its effect on charge generation in photovoltaic devices by giving rise to longer lived excited states and longer exciton diffusion lengths. But, tethering of an Ir-complexes to the main chain of a CP was not reported till Schulz and Holdcroft[91] demonstrated that introducing triplet forming Ir-complexes into the polyfluorene-based polymer blended with an electron acceptor considerably enhances solar cell conversion efficiencies. The improved performance of **P55** over **P54** was explained by the formation of the triplet state and the longer diffusion length of the triplet exciton, compared to the singlet exciton formed in **P54**. PCEs are lower than many polymer:PCBM systems based on singlet exciton charge generation, but this was caused by the unfavorable absorption cross-section of **P55**. This effort is interesting in the premises of further improvements in conversion efficiencies, which are expected upon optimizing the iridium content in the system and choosing a main chain having a nature of covering the solar spectrum more effectively.

8.5.5 RHENIUM CONTAINING POLYMER

Rhenium(I) diimine-containing polymers have received much attention due to their desirable photophysical properties such as tunability of MLCT transition, long lifetime of the excited state, and emissive excited state. Chlorotricarbonyl rhenium(I)diimine $[Re(N^{\wedge}N)(CO)_3Cl]$ has simple coordination geometry and nice electrochemical properties, which leads to the metal containing polymer based on this kind of complexes to be extensively studied. Their photophysical properties, electrochemical properties, and charge carrier mobilities can easily be altered by using different electron donating or electron withdrawing substituents on the diimine ligand.

Metallooorganic polymers based on Re(I) and Ru(II) polypyridyl complexes demonstrated that the metal complexes of the metallopolymer facilitating the charge transport of the polymer by functioning as additional charge carriers.[92] Additionally, the emissive nature of the $Re(bipyridyl)(CO)_3Cl$ draws attention in display applications.[93] Most of the works have been concentrated on Re(I)diimine containing polymers, which are based on bipyridyl or phenanthroline unit as the coordination ligand. The emissive MLCT excited state in these polymers expelled their use in LEDs. But the use of rhenium compounds as light harvesting material in OPVs have got only scant attention.

Conjugated LBG copolymers with covalently-linked with novel rhenium(I) LBG complexes were synthesized by Chan et al.[94] by Suzuki cross-coupling reaction and Stille polymerization (Scheme 8-6). The metal-free polymers can be used as a scaffold for insertion of various other metals. This approach was direct and successful to lower the energy for the

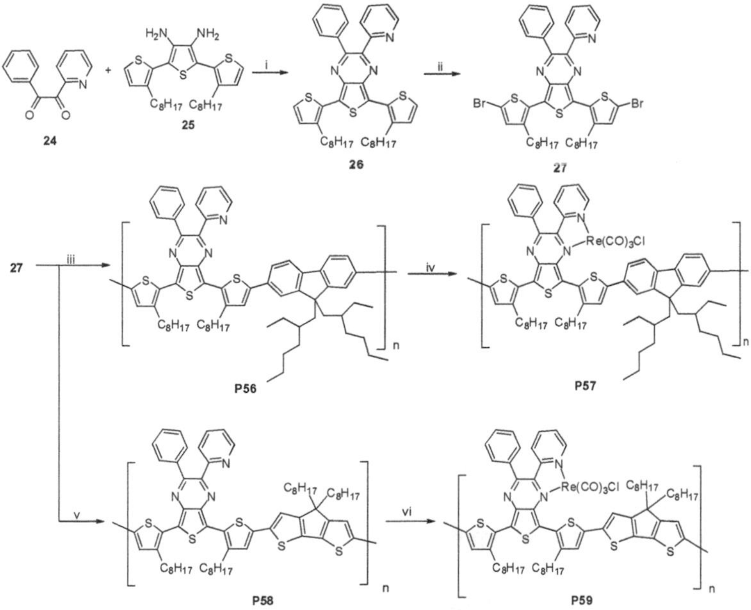

SCHEME 8-6 Synthesis of the rhenium(I)-containing CPs. Reagents and conditions: (i) EtOH, N_2, RT, 3d, 51%; (ii) NBS, DMF, RT, 16 h, 34%; (iii) 1. 2,7-di(4,4,5,5-tetramethyl-1,3,2-dioxaboralane)-9,9-diethylhexylfluorene, $Pd(PPh_3)_4$, TEAOH, Toluene, 90 °C, 16 h, 2. bromobenzene, 90 °C, 2 h, 3. phenylboronic acid, 90 °C, 2 h, 70%; (iv) rhenium(I) pentacarbonyl chloride, toluene, reflux, 16 h, 74%; (v) 1. 2,6-di(tributyltin)-4,4-dioctylcyclopenta[2,1-b:3,4-b']dithiophene, $Pd(PPh_3)_4$, THF, 60 °C, 3d, 2. 2-bromothiophene, 60 °C, 2 h, 44%; and (vi) rhenium(I) pentacarbonyl chloride, toluene, reflux, 16 h, 41%.

electronic transitions of these metal free polymers compared to their metal-free analogue by introduction of rhenium complexes into a conjugated backbone that strongly perturbed the electronic transition properties of the conjugated system of the polymer main chain. This is a new approach in designing novel metal containing LBG polymers for OPV applications. The photovoltaic performances based on these matalloorganic polymers:PCBM blend in the BHJ PSCs was studied. The charge carrier mobility (time-of-flight) was found one order higher for **P57** compared to **P56**. This is believed to be due to (a) the coordinatied metal complex to the polymer chain that provides additional charge carriers and facilitates charge transport in the system and (b) the better phase separation and superior mixing of **P57** at the metal. PCE of **P57** and **P59** were reported to be 0.03% and 0.01%, respectively. Although the hole mobility of **P57** was higher compared to **P56** by an order, the PCE for **P56** was 0.41%. This was explained on the basis of decrease of absorption intensity of **P57** in the visible region to about one-third of that of **P56** and also by the higher J_{SC} in the device for **P56** caused by efficient excition generation and dissociation.

8.6   CONCLUSION

In this chapter, we have summarized the current research outcomes on polymers for organic photovoltaics and charge transport. The chapter comprehends a quick access to the synthetic methods developed to synthesize metallooorganic polymers containing mainly the first transition series elements. The structure–property relationships have been discussed on the basis of varying metal centres with a fixed ligand and also fixed metal centre and varying ligand. The alteration in optoelectronic and device properties in OPV and transistor mobilities upon triplet state inclusion in the π-conjugated polymeric systems have been discussed with major attention. It is noticeable that after remarkable progress have been done in the organic electronic properties of conjugated metalloorganic systems, their industrial applications are still limited.

For that, the following inferences were drawn to improve the designing and application of the metalloorganic conjugated systems in OPV or transistors. Firstly, the large dihedral angles of the ligands (n-Bu3P) with steric inhibition can be detrimental for intermolecular interactions and charge mobility. In contrast, less sterically stipulating ones could provide the required crystallinity to the metallopolymers for improving device performances. Secondly, the energy-match between the frontier orbitals of the donor and acceptor in solar cells is a matter of concern. Whereas, the application of all-polymer solar cells could bring novelty in this field. Thirdly, the demonstration of efficient infrared photocurrent spectral responses for metallopolyynes would also render the design of future NIR photodetection devices but with selection of polymers which could blend well to produce larger photocurrent by introducing efficient phase separation. Finally, the metal-conjugated polymers often face the problems of inferior morphology at the electrode–semiconductor interface. Improved film forming ability of the metalloconjugated compounds could potentially address this issue to offer high charge separation and transport.

REFERENCES

1. Shao, Y.; Yang, Y. Efficient Organic Heterojunction Photovoltaic Cells Based on Triplet Materials. *Adv. Mater.* **2005,** *17,* 2841–2844.
2. Grätzel, M. Solar Energy Conversion by Dye-Sensitized Photovoltaic Cells. *Inorg. Chem.* **2005,** *44,* 6841–6851.
3. (a). Wong, W.-Y., Ho, C.-L. Di-, Oligo-, and Polymetallaynes: Syntheses, Photophysics, Structures and Applications. *Coord. Chem. Rev.* **2006,** *250,* 2627–2690. (b). Whittell, G. R.; Manners, I. Metallopolymers: NewMultifunctional Materials. *Adv. Mater.* **2007,** *19,* 3439–3468. (c). Holliday, B. J.; Swager, T. M. Conducting Metallopolymers: The Roles of Molecular Architecture and Redox Matching. *Chem. Commun.* **2005,** 23–36.
4. Guo, F.; Kim, Y.-G.; Reynolds, J. R.; Schanze, K. S. Platinum–Acetylide Polymer Based Solar Cells: Involvement of the Triplet State for Energy Conversion. *Chem. Commun.* **2006,** 1887–1889.
5. Mei, J.; Ogawa, K.; Kim, Y.-G.; Heston, N. C.; Arenas, D. J.; Nasrollahi, Z.; McCarley, T. D.; Tanner, D. B.; Reynolds, J. R.; Schanze, K. S. Low-Band-Gap Platinum Acetylide Polymers as Active Materials for Organic Solar Cells. *ACS Appl. Mater. Interfaces* **2009,** *1,* 150–161.
6. Moses, D.; Dogariu, A.; Heeger, A. J. Ultrafast Photoinduced Charge Generation in Conjugated Polymers. *Chem. Phys. Lett.* **2000,** *316,* 356–360.
7. Turro, N. J. *Modern Molecular Photochemistry*; University Science Book: Sausalito, CA, 1991; p 628.
8. Schull, T. L.; Kushmerick, J. G.; Patterson, C. H.; George, C.; Moore, M. H.; Pollack, S. K.; Shashidhar, R. Ligand Effects on Charge Transport in Platinum(II) Acetylides. *J. Am. Chem. Soc.* **2003,** *125,* 3202–3203.
9. Wong, W.-Y.; Wang, X.-Z.; He, Z.; Chan, K.-K.; Djurišić, A. B.; Cheung, K.-Y.; Yip, C.-T.; Ng, A. M.-C.; Xi, Y. Y.; Mak, C. S. K.; Chan, W.-K. Tuning the Absorption, Charge Transport Properties, and Solar Cell Efficiency with the Number of Thienyl Rings in Platinum-Containing Poly(aryleneethynylene)s. *J. Am. Chem. Soc.* **2007,** *129,* 14372–14380.
10. Harima, Y.; Kim, D.-H.; Tsutitori, Y.; Jiang, X.; Patil, R.; Ooyama, Y.; Ohshita, J.; Kunai, A. Influence of Extended π-Conjugation Units on Carrier Mobilities in Conducting Polymers. *Chem. Phys. Lett.* **2006,** *420,* 387–390.

11. Younus, M.; Köhler, A.; Cron, S.; Chawdhury, N.; Al-Mandhary, M. R. A.; Khan, M. S.; Lewis, J.; Long, N. J.; Friend, R. H.; Raithby, P. R. Synthesis, Electrochemistry, and Spectroscopy of Blue Platinum(II) Polyynes and Diynes. *Angew. Chem. Int. Ed.* **1998,** *37*, 3036–3039.

12. Wilson, J. S.; Dhoot, A. S.; Seeley, A. J. A. B.; Khan, M. S.; Köhler, A.; Friend R. H. Spin-Dependent Exciton Formation in π-Conjugated Compounds. *Nature* **2001,** *413*, 828–831.

13. Manners, I. Putting Metals into Polymers. *Science* **2001,** *294*, 1664–1666.

14. Long, N. J.; Williams, C. K. Metal Alkynyl σ Complexes: Synthesis and Materials. *Angew. Chem. Int. Ed.* **2003,** *42*, 2586–2617.

15. Manners, I. *Synthetic Metal-Containing Polymers*; Wiley-VCH: Weinheim, 2004.

16. Köhler, A.; Beljonne, D. The Singlet-Triplet Exchange Energy in Conjugated Polymers. *Adv. Funct. Mater.* **2004,** *14*, 11–18.

17. Wong, W.-Y. Recent Advances in Luminescent Transition Metal Polyyne Polymers. *J. Inorg. Organomet. Polym. Mater.* **2005,** *15*, 197–219.

18. Karikomi, M.; Kitamura, C.; Tanaka, S.; Yamashita, Y. New Narrow-Bandgap Polymer Composed of Benzobis(1,2,5-thiadiazole) and Thiophenes. *J. Am. Chem. Soc.* **1995,** *117*, 6791–6792.

19. Roncali, J. Synthetic Principle for Bandgap Control in Linear π-Conjugated Systems. *Chem. Rev.* **1997,** *97*, 173–205.

20. Wong, W.-Y., Choi, K.-H., Lu, G.-L.,Shi, J.-X. Synthesis, Redox and Optical Properties of Low-Bandgap Platinum(II) Polyynes with 9-Dicyanomethylene-Substituted Fluorene Acceptors. *Macromol. Rapid Commun.* **2001,** *22*, 461–465.

21. Wong, W.-Y.; Wang, X.-Z.; He, Z.; Djurišić, A. B.; Yip, C.-T.; Cheung, K.-Y.; Wang, H.; Mak, C. S. K.; Chan, W.-K. Metallated Conjugated Polymers as a New Avenue Towards High-Efficiency Polymer Solar Cells. *Nat. Mater.* **2007,** *6*, 521–527.

22. von Hauff, E., Parisi, J.; Dyakonov, V. Investigations of the Effects of Tempering and Composition Dependence on Charge Carrier Field Effect Mobilities in Polymer and Fullerene Films and Blends. *J. Appl. Phys.* **2006,** *100*, 43702–43708.

23. Gilot, J.; Wienk, M. M.; Janssen, R. A. J. On the Efficiency of Polymer Solar Cells. *Nat. Mater.* **2007,** *6*, 704.

24. Wu, P. T.; Bull, T.; Kim, F. S.; Luscombe C. K.; Jenekhe, S. A. Organometallic Donor–Acceptor Conjugated Polymer Semiconductors: Tunable Optical, Electrochemical, Charge Transport, and Photovoltaic Properties. *Macromolecules* **2009,** *42*, 671–681.

25. Baek, N. S.; Hau, S. K.; Yip, H. L.; Acton, O.; Chen, K. S.; Jen, A. K. Y. High Performance Amorphous Metallated π-Conjugated Polymers for Field-Effect Transistors and Polymer Solar Cells. *Chem. Mater.* **2008,** *20*, 5734–5736.

26. Cunningham, P. D.; Hayden, L. M.; Yip, H.-L.; Jen, A. K.-Y. Charge Carrier Dynamics in Metalated Polymers Investigated by Optical-Pump Terahertz-Probe Spectroscopy. *J. Phys. Chem. B* **2009,** *113*, 15427–15432.

27. Smith, N. V. Classical Generalization of the Drude Formula for the Optical Conductivity. *Phys. Rev. B* **2001,** *64*, 155106–155111.

28. Wang, Q.; Wong, W.-Y. New Low-Bandgap Polymetallaynes of Platinum Functionalized with a Triphenylamine-Benzothiadiazole Donor–Acceptor Unit for Solar Cell Applications. *Polym. Chem.* **2011,** *2*, 432–440.

29. He, W.; Jiangb, Y.; Qin, Y. Synthesis and Photovoltaic Properties of aLow Bandgap BODIPY–Pt Conjugated Polymer. *Polym. Chem.* **2014,** *5*, 1298–1304.

30. Yuan, Y.; Michinobu, T.; Oguma, J.; Kato, T.; Miyake, K. Attempted Inversion of Semi-conducting Features of Platinum Polyyne Polymers: ANew Approach for All-Polymer Solar Cells. *Macromol. Chem. Phys.* **2013,** *214,* 1465–1475.

31. Clem, T. A.; Kavulak, D. F. J.; Westling E. J.; Fréchet, J. M. J. Cyclometalated Platinum Polymers: Synthesis, Photophysical Properties, and Photovoltaic Performance. *Chem. Mater.* **2010,** *22,* 1977–1987.

32. Thompson, B. C.; Kim, B. J.; Kavulak, D. F.; Sivula, K.; Mauldin, C.; Fréchet, J. M. J. Influence of Alkyl Substitution Pattern in Thiophene Copolymers on Composite Fullerene Solar Cell Performance. *Macromolecules* **2007,** *40,* 7425–7428.

33. Brooks, J.; Babayan, Y.; Lamansky, S.; Djurovich, P. I.; Tsyba, I.; Bau, R.; Thompson, M. E. Synthesis and Characterization of Phosphorescent Cyclometalated Platinum Complexes. *Inorg. Chem.* **2002,** *41,* 3055–3066.

34. Liao, C.-Y.; Chen, C.-P.; Chang, C.-C.; Hwang, G.-W.; Chou, H.-H.; Cheng, C.-H. Synthesis of Conjugated Polymers Bearing Indacenodithiophene and Cyclometalated Platinum(II) Units and Their Application in Organic Photovoltaics. *Sol. Energ. Mat. Sol. C.* **2013,** *109,* 111–119.

35. Kim, G.; Yeom, H. R.; Cho, S.; Seo, J. H.; Kim, J. Y.; Yang, C. Easily Attainable Pheno-thiazine-Based Polymers for Polymer Solar Cells: Advantage of S,S-Dioxides into its Polymer for Inverted Structure Solar Cells. *Macromolecules* **2012,** *45,* 1847–1857.

36. Rieger, R.; Beckmann, D.; Mavrinskiy, A.; Kastler, M.; Müllen, K. Backbone Curvature in Polythiophenes. *Chem. Mater.* **2010,** *22,* 5314–5318.

37. Silverman, E. E.; Cardolaccia, T.; Zhao, X.; Kim, K.-Y.; Haskins-Glusac, K.; Schanze, K. S. The Triplet State in Pt-Acetylide Oligomers, Polymers and Copolymers. *Coord. Chem. Rev.* **2005,** *249,* 1491–1500.

38 Harvey, J. D.; Ziegler, C. J. Developments in the Metal Chemistry of N-Confused Porphyrin. *Coord. Chem. Rev.* **2003,** *247,* 1.

39. Guldi, D. M.; Luo, C.; Prato, M.; Troisi, A.; Zerbetto, F.; Scheloske, M.; Dietel, E.; Bauer, W.; Hirsch, A. Parallel (Face-to-Face) Versus Perpendicular (Edge-to-Face) Alignment of Electron Donors and Acceptors in Fullerene Porphyrin Dyads: The Impor-tance of Orientation in Electron Transfer. *J. Am. Chem. Soc.* **2001,** *123,* 9166–9177.

40. Xiao, S. Q.; Li, Y. L.; Li, Y. J.; Zhuang, J. P.; Wang, N.; Liu, H. B.; Ning, B.; Liu, Y.; Lu, F. S.; Fan, L. Z.; Yang, C. H.; Li, Y. F.; Zhu, D.[60]Fullerene-Based Molecular Triads with Expanded Absorptions in the Visible Region: Synthesis and Photovoltaic Proper-ties. *J. Phys. Chem. B* **2004,** *108,* 16677–16685.

41. Lu, F. S.; Xiao, S. Q.; Li, Y. L.; Liu, H. B.; Li, H. M.; Zhuang, J. P.; Liu, Y.; Wang, N.; He, X. R.; Li, X. F.; Gan, L. B.; Zhu, D. B. Synthesis and Chemical Properties of Conjugated Polyacetylenes Having Pendant Fullerene and/or Porphyri n Units. *Macro-molecules* **2004,** *37,* 7444–7450.

42. Liu, Y.; Wang, N.; Li, Y.; Liu, H.; Li, Y.; Xiao, J.; Xu, X.; Huang, C.; Cui, S.; Zhu, D. A New Class of Conjugated Polyacetylenes having Perylene Bisimide Units and Pendant Fullerene or Porphyrin Groups. *Macromolecules* **2005,** *38,* 4880–4887.

43. Fukuzumi, S.; Endo, Y.; Imahori, H. A Negative Temperature Dependence of the Elec-tron Self-Exchange Rates of Zinc Porphyrin π Radical Cations. *J. Am. Chem. Soc.* **2002,** *124,* 10974–10975.

44. Hasobe, T.; Imahori, H.; Kamat, P. V.; Ahn, T. K.; Kim, S. K.; Kim, D.; Fujimoto, A.; Hirakawa, T.; Fukuzumi, S. Photovoltaic Cells Using Composite Nanoclusters of Porphyrins and Fullerenes with Gold Nanoparticles. *J. Am. Chem. Soc.* **2005,** *127,* 1216–1228.

45. Sun, Q. J.; Dai, L. M.; Zhou, X. L.; Li, L. F.; Li, Q. Bilayer- and Bulk-Heterojunction Solar Cells Using Liquid Crystalline Porphyrins as Donors by Solution Processing. *Appl. Phys. Lett.* **2007**, *91*, 253505–253507.

46. Hasobe, T.; Sandanayaka, A. S. D.; Wada, T.; Araki, Y. Fullerene-Encapsulated Porphyrin Hexagonalnanorods. An Anisotropic Donor–Acceptor Composite for Efficient Photoinduced Electron Transferand Light Energy Conversion. *Chem. Commun.* **2008**, 3372–3374.

47. Walter, M. G.; Rudine, A. B.; Wamser, C. C. Porphyrins and Phthalocyanines in Solar Photovoltaic Cells. *J. Porphyrins Phthalocyanines* **2010**, *14*, 759–792.

48. Hasobe, T.; Kamat, P. V.; Absalom, M. A.; Kashiwagi, Y.; Sly, J.; Crossley, M. J.; Hosomizu, K.; Imahori, H.; Fukuzumi, S. Supramolecular Photovoltaic Cells Based on Composite Molecular Nanoclusters: Dendritic Porphyrin and C_{60}, Porphyrin Dimer and C_{60}, and Porphyrin–C_{60} Dyad. *J. Phys. Chem. B* **2004**, *108*, 12865–12872.

49. Oku, T.; Noma, T.; Suzuki, A.; Kikuchi, K.; Kikuchi, S. Fabrication and Characterization of Fullerene/Porphyrin Bulk Heterojunction Solar Cells. *J. Phys. Chem. Solids* **2010**, *71*, 551–555.

50. Hagemann, O.; Jorgensen, M.; Krebs, F. C. Synthesis of An All-in-One Molecule (For Organic Solar Cells).*J. Org. Chem.* **2006**, *71*, 5546–5559.

51. Krebs, F. C.; Hagemann, O.; Jorgensen, M. Synthesis of Dye Linked Conducting Block Copolymers, Dye Linked Conducting Homopolymers and Preliminary Application to Photovoltaics. *Sol. Energ. Mat. Sol. C.* **2004**, *83*, 211–228.

52. (a). Liu, Y.; Guo, X.; Xiang, N.; Zhao, B.; Huang, H.; Li, H.; Shen, P.; Tan, S. Synthesis and Photovoltaic Properties of Polythiophene Stars with Porphyrin Core. *J. Mater. Chem.* **2010**, *20*, 1140–1146. (b). Ravikanth, M.; Strachan, J.-P.; Li, F.; Lindsey, J. S. Trans-Substituted Porphyrin Building Blocks Bearing Iodo and Ethynyl Groups for Applications in Bioorganic and Materials Chemistry. *Tetrahedron* **1998**, *54*, 7721–7734. (c). Liu, Z.; Schmidt, I.; Thamyongkit, P.; Loewe, R. S.; Syomin, D.; Diers, J. R.; Zhao, Q.; Misra, V.; Lindsey, J. S.; Bocian, D. F. Synthesis and Film-Forming Properties of Ethynylporphyrins. *Chem. Mater.* **2005**, *17*, 3728–3742. (d). Rochford, J.; Botchway, S.; McGarvey, J. J.; Rooney, A. D.; Pryce, M. T. Photophysical and Electrochemical Properties of *Meso*-Substituted Thien-2-yl Zn(II) Porphyrins. *J. Phys. Chem. A* **2008**, *112*, 11611–11618.

53. Feng, J.; Zhang, Q.; Li, W.; Li, Y.; Yang, M.; Cao, Y. Novel Porphyrin-Grafted Poly(phenylene vinylene) Derivatives: Synthesis and Photovoltaic Properties. *J. Appl. Polym. Sci.* **2008**, *109*, 2283–2290.

54. Hasobe, T.; Saito, K.; Kamat, P. V.; Troiani, V.; Qiu, H.; Solladié, N.; Kim, K. S.; Park, J. K.; Kim, D.; D'Souza, F.; Fukuzumi, S. Organic Solar Cells. Supramolecular Composites of Porphyrins and Fullerenes Organized by Polypeptide Structures as Light Harvesters. *J. Mater. Chem.* **2007**, *17*, 4160–4170.

55. Zhan, H.; Lamare, S.; Ng, A.; Kenny, T.; Guernon, H.; Chan, W. K.; Djurišić, A. B.; Harvey, P. D.; Wong, W. Y. Synthesis and Photovoltaic Properties of New Metalloporphyrin-Containing Polyplatinyne Polymers. *Macromolecules* **2011**, *44*, 5155–5167.

56. Goudreault, T.; He, Z.; Guo, Y. H.; Ho, C.-L.; Wang, Q. W.; Zhan, H. M.; Wong, K. L.; Fortin, D.; Yao, B.; Xie, Z. Y.; Kwok, W.-M.; Wong, W.-Y.; Harvey, P. D. Synthesis, Light-Emitting, and Two-Photon Absorption Properties of Platinum-Containing Poly(arylene-ethynylene)s Linked by 1,3,4-Oxadiazole Units. *Macromolecules* **2010**, *43*, 7936–7949.

57. Matsuo, Y.; Sato, Y.; Niinomi, T.; Soga, I.; Tanaka, H.; Nakamura, E. Columnar Structure in Bulk Heterojunction in Solution-Processable Three-Layered p-i-n Organic Photovoltaic Devices Using Tetrabenzoporphyrin Precursor and Silylmethyl[60]fullerene. *J. Am. Chem. Soc.* **2009**, *131*, 16048–16050.

58. Yella, A.; Lee, H. W.; Tsao, H. N.; Yi, C.; Chandiran, A. K.; Nazeeruddin, M. K.; Diau, E. W. G.; Yeh, C. Y.; Zakeeruddin, S. M.; Grätzel, M. Porphyrin-Sensitized Solar Cells with Cobalt (II/III)–Based Redox Electrolyte Exceed 12 Percent Efficiency. *Science* **2011**, *334*, 629–634.

59. Umeyama, T.; Takamatsu, T.; Tezuka, N.; Matano, Y.; Araki, Y.; Wada, T.; Yoshikawa, O.; Sagawa, T.; Yoshikawa, S.; Imahori, H. Synthesis and Photophysical and Photovoltaic Properties of Porphyrin-Furan and -Thiophene Alternating Copolymers. *J. Phys. Chem. C* **2009**, *113*, 10798–10806.

60. Xiang, N.; Liu, Y.; Zhou, W.; Huang, H.; Guo, X.; Tan, Z.; Zhao, B.; Shen, P.; Tan, S. Synthesis and Characterization of Porphyrin-Terthiophene and Oligothiophene π-Conjugated Copolymers for Polymer Solar Cells. *Eur. Polym. J.* **2010**, *46*, 1084–1092.

61. Lee, J. Y.; Song, H. J.; Lee, S. M.; Lee, J. H.; Moon, D. K. Synthesis and Investigation of Photovoltaic Properties for Polymer Semiconductors Based on Porphyrin Compounds as Light-Harvesting Units. *Eur. Polym. J.* **2011**, *47*, 1686–1693.

62. Lamare, S.; Aly, S. M.; Fortin, D.; Harvey, P. D. Incorporation of Zinc(II) Porphyrins in Polyaniline in its Perigraniline Form Leading to Polymers with the Lowest Band Gap. *Chem. Commun.* **2011**, *47*, 10942–10944.

63 Gouterman, M. Spectra of Porphyrins. *J. Mol. Spectrosc.* **1961**, *6*, 138–163.

64. Anderson, H. L. Building Molecular Wires From the Colours of Life: Conjugated Porphyrin Oligomers. *Chem. Commun.* **1999**, 2323–2330.

65. Liu, Y.; Guo, X.; Xiang, N.; Zhao, B.; Huang, H.; Li, H.; Shen, P.; Tan, S. Synthesis and Photovoltaic Properties of Polythiophene Stars with Porphyrin Core. *J. Mater. Chem.* **2010**, *20*, 1140–1146.

66. (a). Crossley, M. J.; Burn, P. L.; Langford, S. J.; Pyke, S. M.; Stark, A. G. A New Method for the Synthesis of Porphyrin-α-Diones That is Applicable to the Synthesis of Trans-Annular Extended Porphyrin Systems. *J. Chem. Soc., Chem. Commun.* **1991**, 1567–1568. (b). Crossley, M. J.; Burn, P. L. An Approach to Porphyrin-Based Molecular Wires: Synthesis of a Bis(porphyrin)tetraone and Its Conversion to a Linearly Conjugated Tetrakisporphyrin System. *J. Chem. Soc., Chem. Commun.* **1991**, 1569–1571. (c). Crossley, M. J.; Burn, P. L.; Langford, S. J.; Prashar, J. K. Porphyrins with Appended Phenanthroline Units: A Means by Which Porphyrin π-Systems can be Connected to an External Redox Centre. *J. Chem. Soc., Chem. Commun.* **1995**, 1921–1923. (d). Crossley, M. J.; Govenlock, L. J.; Prashar, J. K. Synthesis of Porphyrin-2,3,12,13- and -2,3,7,8-tetraones: Building Blocks for the Synthesis of Extended Porphyrin Arrays. *J. Chem. Soc., Chem. Commun.* **1995**, 2379–2380.

67. (a). Fukuzumi, S.; Ohkubo, K.; Zhu, W.; Sintic, M.; Khoury, T.; Sintic, P. J.; E, W.; Ou, Z.; Crossley, M. J.; Kadish, K. M. Androgynous Porphyrins. Silver(II) Quinoxalinoporphyrins Act as Both Good Electron Donors and Acceptors. *J. Am. Chem. Soc.* **2008**, *130*, 9451–9458. (b). Ee, W.; Kadish, K. M.; Sintic, P. J.; Khoury, T.; Govenlock, L. J.; Ou, Z.; Shao, J.; Ohkubo, K.; Reimers, J. R.; Fukuzumi, S.; Crossley, M. J. Control of the Orbital Delocalization and Implications for Molecular Rectification in the Radical Anions of Porphyrins with Coplanar 90° and 180° β,β'-Fused Extensions. *J. Phys. Chem.*

A **2008,** *112,* 556–570. (c). Hutchison, J. A.; Sintic, P. J.; Crossley, M. J.; Nagamura, T.; Ghiggino, K. P. The Photophysics of Selectively Metallated Arrays of Quinoxaline-Fused Tetraarylporphyrins. *Phys. Chem. Chem. Phys.* **2009,** *11,* 3478–3489.

68. Shi, S.; Wang, X.; Sun, Y.; Chen, S.; Li, X.; Li, Y.; Wang, H. Porphyrin-Containing D–π–A Conjugated Polymer with Absorption Over the Entire Spectrum of Visible Light and Its Applications in Solar Cells. *J. Mater. Chem.* **2012,** *22,* 11006–11008.

69. Shi, S.; Jiang, P.; Chen, S.; Sun, Y.; Wang, X.; Wang, K.; Shen, S.; Li, X.; Li, Y.; Wang, H. Effect of Oligothiophene π-Bridge Length on the Photovoltaic Properties of D–A Copolymers Based on Carbazole and Quinoxalinoporphyrin. *Macromolecules* **2012,** *45,* 7806–7814.

70. Wu, Y.-C.; Chao, Y.-H.; Wang, C.-L.; Wu, C.-T.; Hsu, C.-S.; Zeng, Y.-L.; Lin, C.-Y. Porphyrin–Diindenothieno[2,3-b]thiophene Alternating Copolymer–ABlue-Light Harvester in Ternary-Blend Polymer Solar Cells. *J. Polym. Sci. Part A: Polym. Chem.* **2012,** *50,* 5032–5040.

71. Koppe, M.; Egelhaaf, H. J.; Dennler, G.; Scharber, M. C.; Brabec, C. J.; Schilinsky, P.; Hoth, C. N. Near IR Sensitization of Organic Bulk Heterojunction Solar Cells: Towards Optimization of the Spectral Response of Organic Solar Cells. *Adv. Funct. Mater.* **2010,** *20,* 338–346.

72. Huang, J. H.; Velusamy, M.; Ho, K. C.; Lin, J. T.; Chu, C. W. A Ternary Cascade Structure Enhances the Efficiency of Polymer Solar Cells. *J. Mater. Chem.* **2010,** *20,* 2820–2825.

73. Khlyabich, P. P.; Burkhart, B.; Thompson, B. C. Efficient Ternary Blend Bulk Hetero-junction Solar Cells with Tunable Open-Circuit Voltage. *J. Am. Chem. Soc.* **2011,** *133,* 14534–14537.

74. (a). Constable, E. C., 2,2′:6′,2″-Terpyridines: From Chemical Obscurity to Common Supramolecular Motifs. *Chem. Soc. Rev.* **2007,** *36,* 246–253. (b). Wild, A.; Winter, A.; Schlütter, F.; Schubert, U. S. Advances in the Field of π-Conjugated 2,2′:6′,2″-Terpyri-dines. *Chem. Soc. Rev.* **2011,** *40,* 1459–1511.

75. Padhy, H.; Ramesh, M.; Patra, D.; Satapathy, R.; Pola, M. K.; Chu, H.-C.; Chu, C.-W.; Wei, K.-H.; Lin, H.-C. Synthesis of Main-Chain Metallo-Copolymers Containing Donor and Acceptor Bis-Terpyridyl Ligands for Photovoltaic Applications. *Macromol. Rapid Commun.* **2012,** *33,* 528–533.

76. Wang, X.; Wen, Y.; Luo, H.; Yu, G.; Li, X.; Liu, Y.; Wang, H. Donor–Acceptor Copo-lymer Based on Dioctylporphyrin: Synthesis and Application in Organic Field-Effect Transistors. *Polymer* **2012,** *53,* 1864–1869.

77. Hoven, C. V.; Garcia, A.; Bazan, G. C.; Nguyen, T.-Q. Recent Applications of Conju-gated Polyelectrolytes in Optoelectronic Devices. *Adv. Mater.* **2008,** *20,* 3793–3810.

78. (a). Cheng, K. W.; Mak, C. S. C.; Chan, W. K.; Ng, A. M. C.; Djurišić, A. B. Synthesis of Conjugated Polymers with Pendant Ruthenium Terpyridine Trithiocyanato Complexes and Their Applications in Heterojunction Photovoltaic Cells. *J. Polym. Sci. Part A: Polym. Chem.* **2008,** *46,* 1305–1317. (b). Pan, Y.; Tong, B.; Shi, J.; Zhao, W.; Shen, J.; Zhi, J.; Dong, Y. Fabrication, Characterization, and Optoelectronic Properties of Layer-by-Layer Films Based on Terpyridine-Modified MWCNTs and Ruthenium(III) Ions. *J. Phys. Chem. C* **2010,** *114,* 8040–8047.

79. Vellis, P. D.; Mikroyannidis, J. A.; Lo, C. N.; Hsu, C. S. Synthesis of Terpyridine Ligands and Their Complexation with Zn^{2+} and Ru^{2+} for Optoelectronic Applications. *J. Polym. Sci. Part A: Polym. Chem.* **2008,** *46,* 7702–7712.

80. Man, K. K. Y.; Wong, H. L.; Chan, W. K.; Djurišić, A. B.; Beach, E.; Rozeveld, S. Use of a Ruthenium-Containing Conjugated Polymer as a Photosensitizer in Photovoltaic Devices Fabricated by a Layer-by-Layer Deposition Process. *Langmuir* **2006**, *22*, 3368–3375.

81. Padhy, H.; Sahu, D.; Chiang, I. H.; Patra, D.; Kekuda, D.; Chu, C. W.; Lin, H. C. Synthesis and Applications of Main-Chain Ru(II) Metallo-Polymers Containing Bis-Terpyridyl Ligands with Various Benzodiazole Cores for Solar Cells. *J. Mater. Chem.* **2011**, *21*, 1196–1205.

82. (a). Wild, A.; Schlutter, F.; Pavlov, G. M.; Friebe, C.; Festag, G.; Winter, A.; Hager, M. D.; Cimrova, V.; Schubert, U. S. π-Conjugated Donor and Donor–Acceptor Metallo-Polymers. *Macro Mol. Rapid Commun.* **2010**, *31*, 868–874. (b) Siebert, R.; Hunger, C.; Guthmuller, J.; Schlutter, F.; Winter, A.; Schubert, U. S.; Gonzalez, L.; Dietzek, B.; Popp, J. Direct Observation of Temperature-Dependent Excited-State Equilibrium in Dinuclear Ruthenium Terpyridine Complexes Bearing Electron-Poor Bridging Ligands. *J. Phys. Chem. C* **2011**, *115*, 12677–12688.

83. Yang, C.-M.; Wu, C.-H.; Liao, H.-H.; Lai, K.-Y.; Cheng, H.-P.; Horng, S.-F.; Meng, H.-F.; Shy, J.-T. Enhanced Photovoltaic Response of Organic Solar Cell by Singlet-to-Triplet Exciton Conversion. *Appl. Phys. Lett.* **2007**, *90*, 133509–13511.

84. Noh, Y.-Y.; Lee, C. L.; King, J.-J.; Yase, K. Energy Transfer and Device Performance in Phosphorescent Dye Doped Polymer Light Emitting Diodes. *J. Chem. Phys.* **2003**, *118*, 2853–2864.

85. Feng, V.; Shen, X.; Li, Y.; He, Y.; Huang, D.; Peng, Q. Ruthenium(II) Containing Supramolecular Polymers with Cyclopentadithiophene–Benzothiazole Conjugated Bridges for Photovoltaic Applications. *Polym. Chem.* **2013**, *4*, 5701–5710.

86. Peng, Q.; Liu, X. J.; Su, D.; Fu, G. W.; Xu J.; Dai, L. M. Novel Benzo[1,2-b:4,5-b'] dithiophene–Benzothiadiazole Derivatives with Variable Side Chains for High-Performance Solar Cells. *Adv. Mater.* **2011**, *23*, 4554–4558.

87. Yang, F.; Shtein M.; Forrest, S. R. Controlled Growth of aMolecular Bulk Heterojunction Photovoltaic Cell. *Nat. Mater.* **2005**, *4*, 37–41.

88. Friese, V. A.; Kurth, D. G. Soluble Dynamic Coordination Polymers as a Paradigm for Materials Science. *Coord. Chem. Rev.* **2008**, *252*, 199–211.

89. Burnworth, M.; Rowan, S. J.; Weder, C. Structure–Property Relationships in Metallosupramolecular Poly(*p*-xylylene)s. *Macromolecules* **2012**, *45*, 126–132.

90. Cheng, K. W.; Mak, C. S. C.; Chan, W. K.; Ng A. M. C.; Djurišić, A. B. Synthesis of Conjugated Polymers with Pendant Ruthenium Terpyridine Trithiocyanato Complexes and Their Applications in Heterojunction Photovoltaic Cells. *J. Polym. Sci., Part A: Polym. Chem.* **2008**, *46*, 1305–1317.

91. Schulz, G. L.; Holdcroft, S. Conjugated Polymers Bearing Iridium Complexes for Triplet Photovoltaic Devices. *Chem. Mater.* **2008**, *20*, 5351–5355.

92. (a). Chan, W. K.; Ng, P. K.; Gong, X.; Hou, S. Synthesis and Electronic Properties of Conjugated Polymers Based on Rhenium or Ruthenium Dipyridophenazine Complexes. *J. Mater. Chem.* **1999**, *9*, 2103–2108. (b). Ng, P. K.; Gong, X.; Chan, S. H.; Lam, L. S. M.; Chan, W. K. The Role of Ruthenium and Rhenium Diimine Complexes in Conjugated Polymers That Exhibit Interesting Opto-Electronic Properties. *Chem. Eur. J.* **2001**, *7*, 4358–4367.

93. Zhang, M.; Lu, P.; Wang, X.; He, L.; Xia, H.; Zhang, W.; Yang, B.; Liu, L.; Yang, L.; Yang, M.; Ma, Y.; Feng, J.; Wang, D.; Tamai, N. Synthesis and Photophysical Properties

of π-Conjugated Polymers Incorporated with Phosphorescent Rhenium(I) Chromophores in the Backbones. *J. Phys. Chem. B* **2004,** *108,* 13185–13190.

94. Mak, C. S. K.; Cheung, W. K.; Leung, Q. Y.; Chan, W. K. Conjugated Copolymers Containing Low Bandgap Rhenium(I) Complexes. *Macromol. Rapid Commun.* **2010,** *31,* 875–882.

CHAPTER 9

POLYMER RESIST TECHNOLOGY IN LITHOGRAPHY

RAMAKRISHNAN AYOTHI and BRIAN OSBORN

CONTENTS

9.1   INTRODUCTION

The word *lithography* comes from the Greek and simply means *writing on stones*. In a lithography printing process, a master pattern is formed on a stone or metal surface using a hydrophobic material, then this pattern is wetted with an oil-based ink, and finally the inked pattern is transferred to paper, producing copies of the master. Today, lithography has evolved into a variety of techniques which allow transfer of complex micro- and nano-patterns from a single master to a number of identical reproductions on a variety of substrates, including silicon. Photolithography is the transferring of integrated circuit (IC) patterns using projected ultraviolet light onto a silicon substrate. Both photolithography and particle beam lithography are also known as conventional lithography techniques.[1-4] These techniques are mainstream processes in the semiconductor industry for printing micro- and nano-scale features.

Alternative lithography techniques[4,5] such as imprint lithography (thermal and UV),[5-7] scanning probe lithography,[8,9] edge lithography,[10] and self-assembly[11,12] offer alternative solutions to photolithography and create the possibility of low cost, large area patterning on a variety substrates including non-planar surfaces.[4,5] The combination of both conventional lithography and alternative techniques have also been investigated for nanofabrication.[13,14] Critical parameters such as resolution, critical dimension control, throughput, and overlay accuracy must be considered in developing new lithography techniques. In the semiconductor industry, photolithography is the method of choice to manufacture devices made of billions of transistors, and possessing features less than 1/100 the width of a human hair, onto a single silicon substrate. In today's photolithography, patterns are still printed on "stone" (single crystal silicon) using a hydrophobic material. The hydrophobic material in the lithography printing process is called *photoresist*.

A photoresist is a polymeric material that undergoes a solubility change upon irradiation (photo), replicating the desired pattern in the film. Photoresist must have sufficient etch resistance (resist) when transferring the pattern into the substrate. Resist, as photoresist is commonly called, is predominantly made from polymer, and may contain other compounds as additives.[15-20] Polymers that have specific structure and functionalities are the major portion of any effective resist film. Polymers are the preferred choice in lithography due to their effective length scale, processability, availability, tunable properties, and diverse functionalities. Advances in lithographic technologies have always been enabled by advances in resist polymer materials.[4,15-20]

This chapter reviews several forms of semiconductor lithography techniques and materials—in particular, the polymers used in them. The first section focuses on the lithographic patterning process and technological trends in lithographic exposure techniques. Next, the focus lies on the properties and types of polymers used in resist. In the following section, discussion is on how controlling the structure and functionality of polymers has helped to enable the field of photolithography since its inception. Then a brief account on emerging lithography techniques is given, and on the polymers used in these processes. Finally, an outlook for the future of photoresist technology is provided.

9.2 LITHOGRAPHIC PATTERNING PROCESS AND TECHNOLOGICAL EVOLUTION IN LITHOGRAPHY

The key elements of a typical lithographic system are essentially the same for all technologies (Figures 9-1 and 9-2).[19,21] In a modern lithography process, electromagnetic radiation is projected through a set of lenses (projection lithography) and a mask containing the target patterns, onto a semiconductor substrate which is coated with a thin layer of polymer resist (Figure 9-1). Irradiation and optional substrate bake steps induce chemical changes that modify the solubility of the exposed region of resist film. The exposed resist film is immersed in aqueous base or organic solvent developer that dissolves the exposed (positive tone resist) or unexposed (negative tone resist) regions, providing access to the surface of the substrate. The resist patterns are only the template to transfer this pattern to the underlying material, so they are removed or stripped after the image has been transferred into the substrate.

In the semiconductor industry, the ability to manufacture billions of transistors and devices on the nanometer size scale onto a single silicon substrate is critical. The industry finds ways to shrink devices even further in accordance with Moore's Law.[22] In Figure 9-1, the finite resolution of the lens results in a light intensity distribution that does not have such clearly defined edges due to diffraction.[21] The edge slope of the light intensity profile become more sloped and the peak intensity decreases for higher resolution (R) features. The minimum R achievable with conventional projection lithography tool is governed largely by the well-known Rayleigh's scaling (eq 9-1). The corresponding depth of focus (DoF) of the same tool is estimated by q. 2.

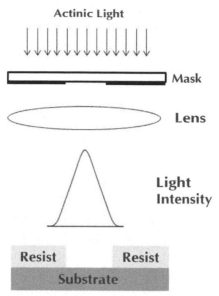

FIGURE 9-1 A simplified illustration of a typical exposure optical system. Light passes through a mask and lens, with the resulting pattern imaged onto a resist-covered substrate.

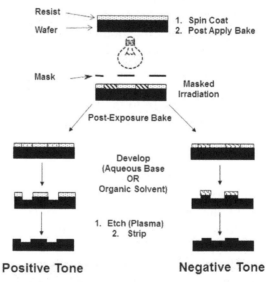

FIGURE 9-2 The key steps in the lithography process. The resist is first spin-coated onto a silicon wafer, and then baked to densify the film and remove residual solvent. Exposure results in a solubility change in the exposed region, rendering it soluble (positive tone) or less soluble (negative tone) in developer. The pattern is transferred into the substrate by etching, and finally the residual resist is stripped from the wafer.

$$R = \frac{k_1 . \lambda}{NA} \tag{1}$$

$$DoF = \frac{k_2 . \lambda}{(NA)^2} \tag{2}$$

In eqs 9-1 and 9-2, λ is the exposure wavelength, NA is the numerical aperture of the optical system, and k_1 and k_2 are constants dependent upon the resist materials, the process technology, and the image formation technique used. According to eqs 9-1 and 9-2, higher resolution can be achieved by reducing exposure wavelength (λ) while maintaining acceptable DoF.[21] However, new resist materials, light sources, and lens systems suitable for any new wavelength are required.

The development of new resist materials has been advanced in large part by major advances in photolithography. Photolithography exposure tools operating at 400–450 nm (g-line), 365 nm (i-line), 248 nm (KrF excimer laser), and 193 nm (ArF excimer laser) wavelengths were introduced into the semiconductor industry for high volume manufacturing (HVM) at regular intervals.[2,4,23] The most advanced ArF lithographic tools today can pattern 37 nm wide resist patterns. To extend the resolution capabilities of ArF exposure tools, the semiconductor industry found another route to scale the effective wavelength: immersion lithography.[23,24] In immersion lithography, imaging resolution is improved by increasing the refractive index of the medium between the imaging lens and the imaging plane. The use of water with a refractive index (R.I. is 1.44) greater than that of air (R.I. is 1), between the last lens element and the photoresist provides an increase in the DoF and also allows the use of larger lenses (higher NA) to be used. Today, all leading edge IC are fabricated using ArF immersion lithography for critical device levels.

Lithographic tools operating at 157 nm (F_2 excimer laser) wavelength[25] were evaluated as a successor to 193 nm tools, but that effort was abandoned due to various technical difficulties. Extreme ultraviolet (EUV) exposure technology operating at 13.5 nm wavelength[26] has been the front-runner to replace 193 nm for many years, but has been delayed due to many technical challenges. It is currently scheduled for first use in the 2015–2016 time frames. Several other lithographic techniques such as nano-imprint lithography and charged particle lithography have been proposed and demonstrated for high resolution patterning (<100 nm). Serial writing with charged particles (electrons or ions) is a high resolution (~10 nm or less) lithographic technique with low throughput and high cost, well-suited for small area nano-fabrication. Electron beam lithography is used extensively in the manufacture of photomasks but the serial nature of the exposure limits its throughput. A substantial effort is underway to develop multibeam electron beam tools[27] and UV-assisted nano-imprint tools[28] with faster throughput that is required for high volume device manufacturing.

9.3   POLYMER RESIST SYSTEM FOR LITHOGRAPHY

9.3.1   *PROPERTIES OF RESIST POLYMERS AND TYPES OF RESIST*

Photoresists must satisfy several requirements, which have become more demanding as feature sizes continue to decrease. The most important lithographic properties of a resist are resolution, process window (DoF and exposure latitude), and etch resistance. The resist is also expected to have good substrate adhesion, high transmittance to the exposure wavelength, thermal stability, compatibility with standard developer, and good shelf-life stability. Several lithographic properties are influenced by the polymers in the photoresist, and some are also dependent on the tools and processes used.

Resist films must have a high enough glass transition temperature (T_g) to withstand post-application and post-exposure annealing processes, and they should undergo a solubility switch upon exposure. Resist films comprised of polymers with specific structures and functionalities can act as effective photoresists. In photolithography, the resolution of smallest features obtained is directly related to the wavelength of light used during patterning (eq 9-1). To enhance transmittance, different functional groups can be built in to the photoresist polymer to ensure transparency[29–32] at the exposure wavelength used (Table 9-1). In general, the polymer's contribution to the absorbance should be kept as low as possible at the required film thickness. Photoresist must be resistant to the plasma etch process that transfers the pattern into the silicon substrate. This etch resistance is largely determined by the structure of the polymer in the resist.[33–35] These specific requirements led to the design and synthesis of new polymer platforms for evolving lithographic exposure technologies.

Resist are broadly classified as positive or negative (Figure 9-2) based on the imaging tone, and can further be divided on the basis of their design into one-component and multi-component systems.[15,19] A one component resist is composed solely of radiation sensitive polymers. Poly(methyl methacrylate) (PMMA) is a positive tone, one component resist that is frequently used in e-beam lithography. Modern, advanced photoresists are exclusively based on a multi-component design, where resist functionality is provided by more than one component. In such systems, the polymers are typically radiation insensitive, functioning as either an etch resistant binder or a participant in the reaction caused by exposure of a radiation sensitive component. Another frequently used resist classification is one based on the radiation source, such as UV or DUV (248 and 193 nm) resists, EUV resists, and electron beam (e-beam) resists. The chemistries of the most commonly used resists are described in the following section.

TABLE 9-1 Summary of the Major Polymer Platforms Developed for Successive Exposure Technologies

Exposure Wavelength	I-line (365 nm)	KrF (248 nm)	ArF (193 nm)	
Polymer Platform				
Absorption Coefficient $(1/\mu m)$ at Exposure Wavelength	0	<0.2	<0.3	
Relative high density Oxide $(C_2F_6;$ 5 mT) Etch Rate Vs APEX-E$^\$$	0.73	1	1.70	0.84

[#]Absorption coefficient of various polymers (Ref. 29).

[$]Etch rates of KrF, ArF polymers and I-line resist (polymer + additives) (Ref. 34).

9.3.2 NOVOLAC-DNQ G-AND I-LINE RESIST

The first important development that enabled the patterning of sub-micron features was carried out using diazonaphthoquinone (DNQ)-novolac based photoresists[36] (Figure 9-3) for 436 nm (G-line) and 365 nm (I-line) UV exposure technologies. Novoac resin is soluble in 0.26 N aqueous base developer [tetramethylammonium hydroxide (TMAH)], but DNQ is insoluble in aqueous TMAH. In unexposed areas, DNQ acts as a dissolution inhibitor preventing the free hydroxyl groups in the novolac resin from dissolving during the development process. However, upon UV exposure, DNQ undergoes the Wolff rearrangement,[36,37] converting the DNQ into a indenecarboxylic acid (dissolution promoter) and allowing the exposed areas to be dissolved quickly, producing a positive tone relief image (Figure 9-3). The photoconversion of the DNQ from a compound that is insoluble in TMAH, to one that is TMAH soluble, is the foundation for g- and i-line resist photochemistry.

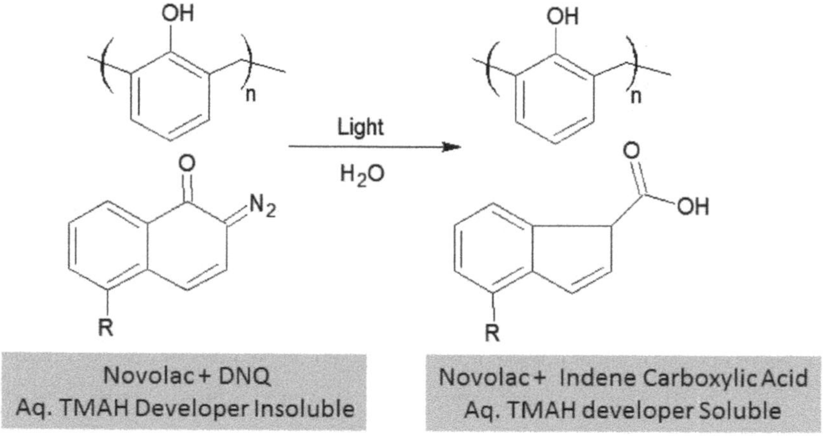

| Novolac + DNQ | Novolac + Indene Carboxylic Acid |
| Aq. TMAH Developer Insoluble | Aq. TMAH developer Soluble |

FIGURE 9-3　Novolac/DNQ i-line photoresist. UV-induced rearrangement of DNQ from a dissolution inhibitor to a dissolution promoter.

Suppliers of novolac-DNQ resists distinguish their materials by using different forms of the novolac resin and DNQ structures.[36] The role of novolac resin in i-line resists is much broader than the typical role of being the base soluble binder and high etch resistant polymer. The novolac structure and properties (ratio of *o*-cresol to *m*-cresol, the ratio of *ortho* to *para* backbone linkages, and the molecular weight characteristics) affect the dissolution behavior, thermal flow resistance, and lithographic performance. The position nature and number of R group present in the DNQ molecule alter absorption and dissolution characteristics.[36,38] The development of advanced optimized novolac-DNQ resists in conjunction with improved exposure tools, has pushed the resolution limit of i-line photolithography to the sub-500 nm regime.

9.3.3　ACID CATALYZED DEEP-UV (254 OR 248 NM) RESIST

The demand for smaller device features continued, and the industry required a shift to deep-UV wavelengths (254 or 248 nm) for patterning.

The novolac-DNQ system was not sensitive enough at DUV wavelength, resulting in sloping sidewalls because the resist polymer absorbed UV light too strongly.[29] In addition, the DUV light sources (254 nm) were typically much weaker than g- and i-line light sources. KrF excimer lasers (248 nm) were invented in the late 1990s but the relative intensity available at the wafer plane is still weaker than their 365 nm counterparts. This necessitated the development of a different type of photoresist, one more sensitive to maintain sufficient exposure tool throughput. This problem was addressed with the advent of the *chemical amplification* concept.[39–42]

Chemically amplified (CA) resists relied on the use of protected polymer and photoacid generators (PAGs) that were developed. The majority of 248 nm polymer systems for CA resists have been based on low molecular weight poly(4-hydroxystyrene) (PHS) copolymers. The 0.26 N TMAH used for i-line resist has become standard developer. PHS has good solubility in aqueous base developer, and the aromatic rings impart high plasma etch resistance. The first example of the CA concept was demonstrated using poly(*tert*-butoxycarbonyloxystyrene) (PTBOCST) polymer and onium salt PAG.[39,42] This process is illustrated in Figure 9-4.

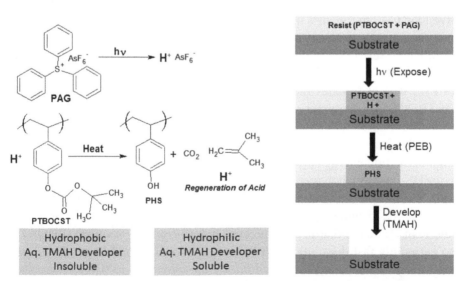

FIGURE 9-4 The process of chemical amplification, shown here with tert-butoxycarbonyl protected poly(4-hydroxstyrene). Exposure to an appropriate wavelength radiation activates a photoacid generator, releasing catalytic amounts of protons. These diffuse thorough the resist and invoke the solubility switch in the exposed regions of the resist, forming the desired pattern.

A PAG generates a proton upon absorption of a photon at a DUV wavelength. During the post exposure bake step, the exposed regions of a photoresist film are deprotected and the resulting polymer is PHS, which is more aqueous base soluble. This result in a positive tone resist system, and the dissolution contrast between the exposed and unexposed area is very high. The PAGs used in photoresist films are primarily sulfonium or iodonium salts, and many different PAG structures exist based on the wavelength and performance required.[43-48] The acid production mechanisms are complex and the fact that one proton is capable of catalyzing hundreds of subsequent reactions resulted in these resists being termed CA resists. Consequently, CA resists are approximately two orders of magnitude more sensitive than novolac-DNQ resists. Resists based on this acid catalyzed deprotection mechanism have played an important role in realization of 248 nm lithography, and continue to play a major role in the extension of lithography to smaller dimensions.[49]

PTBOCST-based resists showed good dissolution contrast, but an effective photoresist needs to satisfy several requirements as described in Section 9.3.1. Therefore, several PHS-based copolymer platforms were developed for 248 nm lithography (Figure 9-5). In general, all deep UV, positive tone CA photoresists used today are copolymers comprised of different monomers with different properties.[49,50] IBM APEX,[49] IBM ESCAP,[51] and IBM KRS[52] resists are typical examples for 248 nm lithography (Figure 9-5).

The APEX family of copolymers were developed by the partial protection of PHS with carbonates such as tert-butylcarbonate (TBOC). The poly(HS-co-TBOCST) copolymer is more hydrophilic, and has better adhesion properties than its fully protected version (PTBOCST). Poly(hydroxystyrene-co-tert-butyl acrylate) [poly(HS-co-tBA)] based resists were developed as a high activation energy (E_a) resist to improve lithographic performance. The poly(HS-co-tBA) polymer does not undergo thermal decomposition until 180 °C, and the tert-butylester group requires higher post-exposure bake temperature to undergo deprotection reaction. The deprotected methacrylate is more acidic that the deprotected poly(4-hydroxy styrene) (PHOST), so extremely fast dissolution rate in the exposed area and larger developer selectivity were observed. In addition, the poly(HS-co-tBA) resist also showed very good process stability in the fabrication (fab) environment, and became to known as environmentally stable chemical amplification photoresist (ESCAP).[51] The ESCAP resist has become the workhorse resist for 248 nm lithography, and ESCAP concept is under investigation for EUV lithography. KRS is a low activation energy type resist developed for 248 nm lithography. It was evaluated for e-beam lithography applications due to its sensitivity.[52] The KRS deprotection mechanism relies on the more labile *acetal* protecting group, which can deprotect at room temperature. High resolution Light Source (LS) patterning (<100 nm) with KRS resist has been demonstrated with EUV and E-beam lithography. Several other PHS copolymers, including tetrahydropyranyl or ether protecting groups were investigated (Figure 9-5).[53,54]

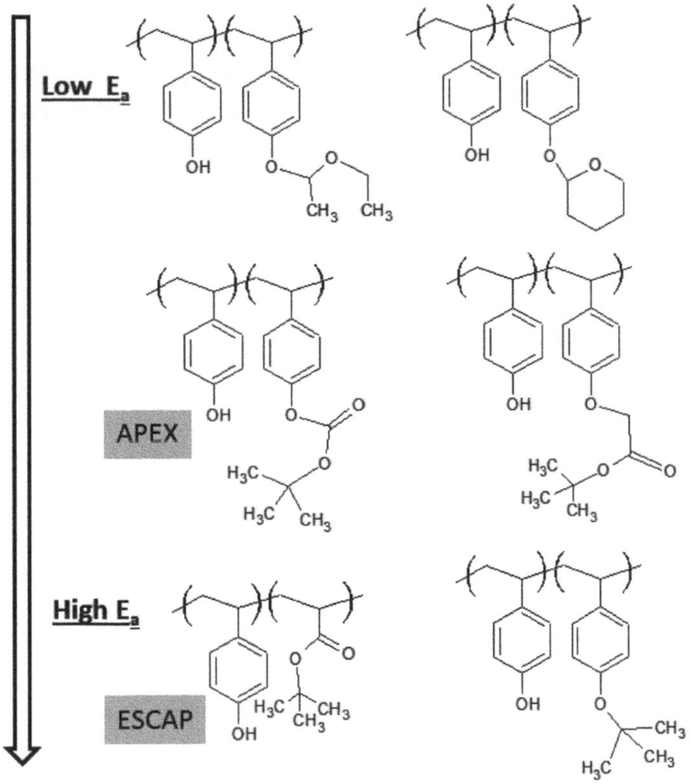

FIGURE 9-5   Poly(hydroxystyrene)-based copolymers for 248 nm lithography.

Several PHS-based negative tone photoresists were also developed in parallel. PHS-based negative tone resists become less soluble through acid-induced cross-linking reactions in the exposed region.[55–57] Negative tone imaging through cross-linking is a most widely studied mechanism, but polarity differences between the protected and deprotected regions can also be used to produce negative tone imaging. In latter case, the developer is an organic solvent.[58] In the case of PTBOCST resist, organic solvent, such as anisole, produces negative tone images since organic solvent dissolves the hydrophobic PTBOCST readily, but not the hydrophilic PHS (Figure 9-4). Narrow trenches (spaces) are best printed with negative tone imaging. However PHS-based positive tone CA resists are predominantly used in 248 nm lithography, and features smaller than 150 nm in width have been achieved.[49]

9.3.4　ARF (193 NM) RESISTS

In early 2000, the semiconductor industry required features smaller than 150 nm, which is below what can be patterned with 248 nm lithography. This required the transition to the smaller 193 nm (ArF excimer source) wavelength. However, PHS-based aromatic polymers used at 248 nm are highly absorbing at 193 nm, so a different polymer platform was needed in order to create suitable ArF resists (Table 9-1). (Meth)acrylate-based polymer systems showed high transparency at the ArF wavelength,[59] but they exhibited poor plasma etch resistance. Methyacrylate units such as tert-butyl methacrylate was introduced as an acid-labile group for chemical amplification.[59,60] The carboxylic acid functionality has become the acid group of choice for 193 nm lithography.

Early on, a terpolymer of methyl methacrylate (MMA), t-butyl methacrylate (t-BMA), and methacrylic acid (MAA) based CA resist showed excellent imaging at 193 nm when developed with dilute TMAH (0.01 N) solution.[60] Good imaging with dilute aqueous base developer can be achieved with simple methacrylic photoresists, but the plasma etch resistance is poor. Methacrylate polymers containing pendant alicyclic groups including adamantane, lactone, cyclohexane, and norbornane groups were developed to improve the etch resistance;[61–65] the incorporation of an alicyclic structure in the methacrylate polymer improved etch resistance without sacrificing transparency. However, the incorporation of such highly lipophilic groups resulted in poor aqueous base development and the formation of cracks. It was also difficult to control the hydrophobic/hydrophilic balance and T_g, resulting in poor adhesion and resolution.[63]

The right balance of imaging quality and plasma etch resistance was managed by providing hydrophilicity to these lipophilic alicyclic structures.[64,65] Thus, ArF resist polymer development has been centered around methacrylate polymers bearing pendant alicyclic groups. Examples of methacrylate polymers developed for ArF resist are provided in Figure 9-6. Several other copolymers such as norbornenes, norbornene-maleic anhydride, norbornene-sulfur dioxide, and vinyl ether-maleic anhydride were also developed for ArF resists.[66-68] Benchmarking studies were performed to understand the advantages and issues of these ArF polymer platforms. Each ArF platform has specific issues and advantages but process optimization and polymer design can lead to sub-100 nm resolution.[66]

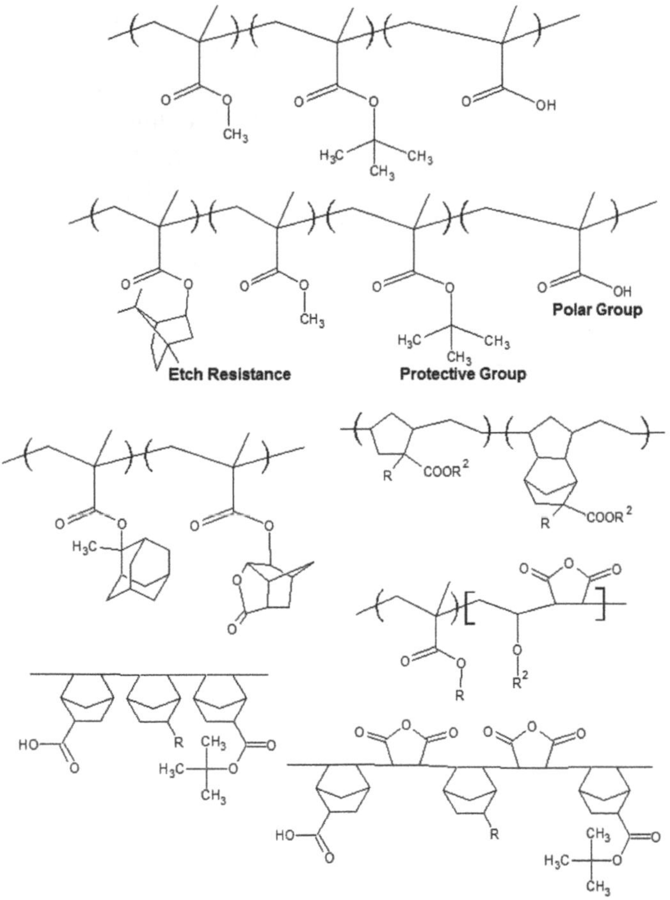

FIGURE 9-6 Methacrylate and alicyclic polymer platforms developed for 193 nm lithography.

In addition to alicyclic chemistries, silicon- and fluorine-containing polymers were also investigated (Figure 9-7) for ArF resists. Silicon-containing polymers were expected to show improved etch resistance. Poly-silsesquioxane-based 193 nm positive bilayer resists were developed and investigated.[69] Silsesquioxane resists showed superior etch resistance than alicyclic-based COMA resists. Siloxane- or fluorine-containing polymers, showed good solubility in supercritical CO_2 solvent. Resist containing fluorinated or siloxane polymers were successfully developed using environmentally friendly supercritical CO_2 (sCO_2). Poly(tBMA-*block*-3-methacryloxypropylpentamethyldisiloxane) polymer, prepared by group transfer polymerization (GTP), showed better development characteristics in aqueous base than that of the corresponding random copolymer.[70] Thus, block copolymers with higher silicon content are developable as negative resists using the environmentally friendly sCO_2 process.[71] Hexafluoroalcohol (HFA) pendant methacrylate polymers developed for ArF resist showed swelling-free dissolution properties due to the presence of the HFA functionality. However, fluorine-containing polymers and their resulting resists showed poor etch stability under a variety of etch processes. This problem was mitigated by insertion of acyclic and aliphatic spacers between the main chain and the HFA functionalities.[72] HFA methacrylate-based ArF resists resolved sub-100 nm resolution and showed excellent process windows for trench applications. Both fluorine and silicon based resist materials were investigated for ArF immersion lithography.

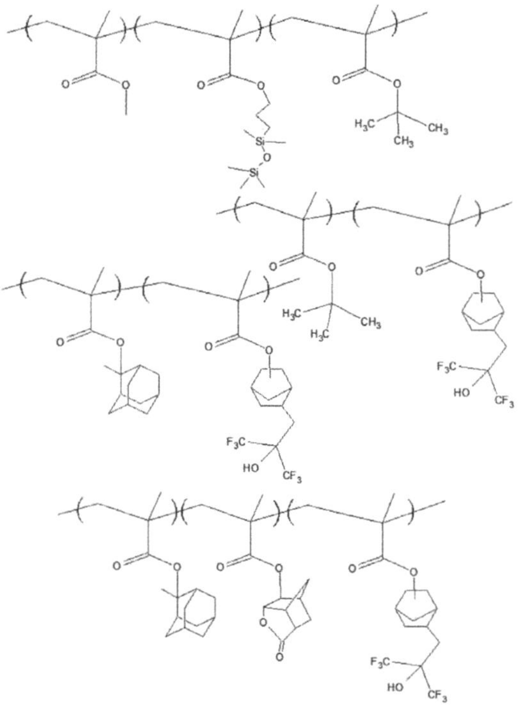

FIGURE 9-7   Silicon and HFA-based polymer platforms for 193 nm lithography.

CA, positive-tone single layer resists dominate the 193 nm lithography landscape. Several attempts were made to develop high performance, single layer, negative tone resists which can be processed using industry standard 0.26 N tetramethylammonium hydroxide (TMAH) aqueous developer. Several of the 193 nm negative tone resist designs studied were based on aqueous base soluble alicyclic polymers containing a reactive functional unit (Figure 9-8). Negative tone resist polymers can undergo either a cross-linking reaction or a polarity change via rearrangement in presence of photo-generated acid.[73–75] Cho et al. developed negative tone resist systems via polarity change through an acid catalyzed pinacol rearrangement.[73]

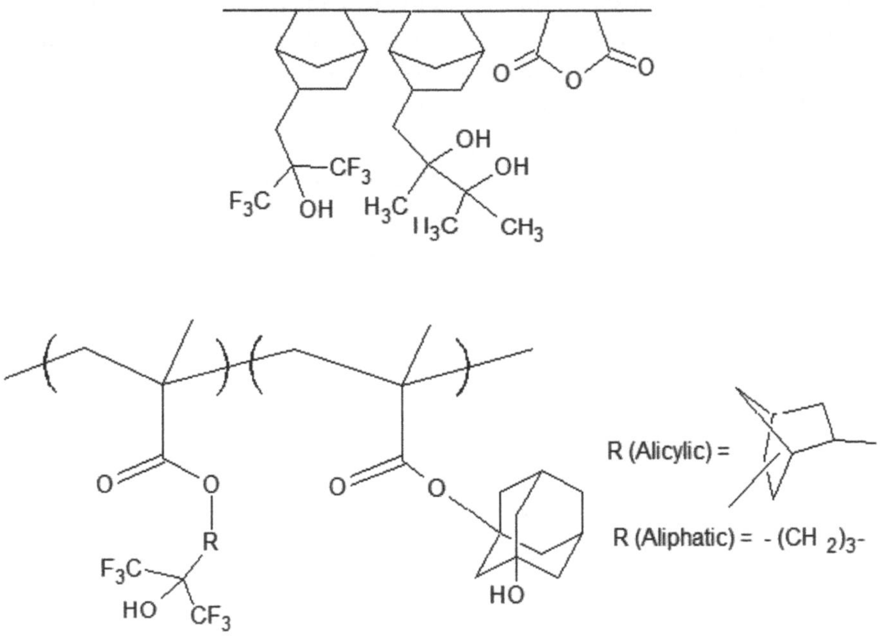

FIGURE 9-8   Representative polymer platforms for a 193 nm negative tone resist.

For IBM and JSR, a high performance negative tone resist, using a base soluble methacrylate polymer platform with pendent nucleophilic units, was developed.[75] These polymers reacted with an electrophilic crosslinker in the presence of photo-generated acid. The resulting increase in molecular weight of the polymer decreased its solubility in aqueous base developer, enabling negative-tone imaging. The optimized negative resist was capable of printing sub-100 nm patterns and provided a process window comparable to leading positive-tone resists. Several high resolution 193 nm negative tone resists were developed, but the transition to a negative tone process in the HVM has yet to be achieved. Recently, a great deal of progress has been made in the design of organic solvent developable negative tone resists based on a polarity change mechanism. Organic solvent-based, negative tone develop 193 nm imaging for patterning high resolution trenches and contact holes will be described later.

CA resists are well suited for high-throughput manufacturing processes due to their high sensitivity. However, excessive acid diffusion during the post-exposure bake may reduce resolution and increase line edge roughness (LER) at smaller dimensions (sub-50 nm). Polymers with low activation energy (E_a) protecting groups[76] and photoacid generator (PAG) bound polymers[77,78] were developed to minimize acid diffusion and achieve higher resolution. Lower bake CA resists based on methacrylate polymers and containing 1-ethylcyclooctyl protecting groups resolved sub-50 nm LS patterns at 60 °C PEB.[76] CA resist-based polymer-bound PAG systems showed lower LER than typical CA resists.[78] One alternative to CA resists are non-chemically amplified (non-CA) materials based on chain scission (like PMMA) and dissolution inhibition (e.g., i-line type, diazo group) (Figure 9-9); such a mechanism was investigated as it eliminates acid diffusion.

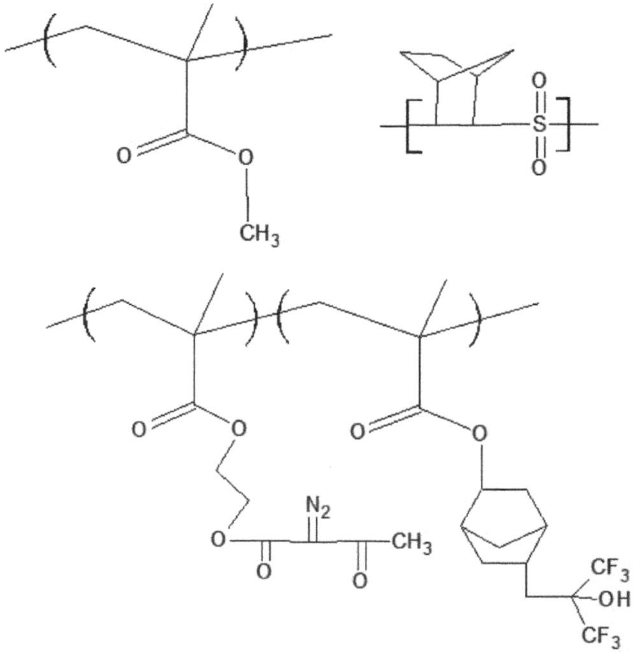

FIGURE 9-9   Polymer platforms investigated for 193 nm non-CA resists.

PMMA was the first resist to be imaged using an ArF excimer laser, but it showed poor sensitivity.[79] Sensitivity was improved by reducing the PMMA film thickness to approximately 20 nm.[80] Polysulfones that are more sensitive than PMMA were also investigated for 193 immersion lithography.[81] Non-CA resists based on a chain scission process showed higher outgassing and also required a special development process. Methacrylate copolymers containing dicyclodiazo (DCD) side chain units as well as a HFA unit were developed and investigated.[82] Resist compositions containing DCD/HFA polymers have imaging occur via a dissolution inhibition mechanism. Optimized resists based on DCD/HFA polymers showed good lithography performance and Line width roughness (LWR).

9.3.5 RESIST MATERIALS FOR ARF IMMERSION (193I NM) LITHOGRAPHY

ArF lithography also has its eventual resolution limits, but the technology was successfully extended by the use of immersion lithography.[24,83–85] The basic idea of immersion lithography is to fill the space between the final lens element and the resist with a fluid which has a much higher refractive index (n) than air ($n = 1$) so that the resolution and the DOF can be increased. Immersion (ArFi) lithography tools with NA > 1.0 were successfully developed and implemented for semiconductor device manufacturing. The increased numerical aperture and greater DOF provided by water immersion lithography has enabled the extension of 193 nm optical lithography beyond the dry ArF resolution limit. The photoresists developed for ArFi lithography must be water-resistant and suffer minimal leaching of components into the water. A number of interactions between the water and the photoresist materials must be understood since these interactions affect the lithographic profiles, process window and defect formation mechanisms.[86–88] (Meth)acrylic-based CA resist systems developed for ArF dry lithography has shown an excellent stability to water, but tend to show poor performance under immersion conditions.

Several materials strategies have been employed to mitigate component leaching and improve the immersion compatibility of conventional ArF resist materials.[88] The use of protective polymeric coatings (or topcoats) atop photoresist coatings has been proposed (Figure 9-10). Both organic solvent and aqueous base removable topcoat materials were investigated. Organic solvent removable topcoats required additional process steps such as coat and removal. To reduce additional processing steps, aqueous base-soluble topcoats that can be stripped during the resist develop step were developed and implemented for HVM processes.[89,90] The topcoat structures are typically random linear polymers containing saturated hydrocarbon, fluorocarbon, or organosilicon-containing groups to reduce surface energy, increase water contact angles, and enable high wafer scan rates (>500 mm/s) without resist component leaching (Figure 9-11A and B).[88] Recently, topcoat-free immersion-compatible resists have been developed (Figure 9-10).[91] Topcoat-free resists contains either surface-segregating additives or hydrophobic resist materials with non-leaching PAGs.[88] Surface segregating additives based on random linear polymers have been the most widely reported. A variety of other polymer architectures have also been explored (Figure 9-11C).[92]

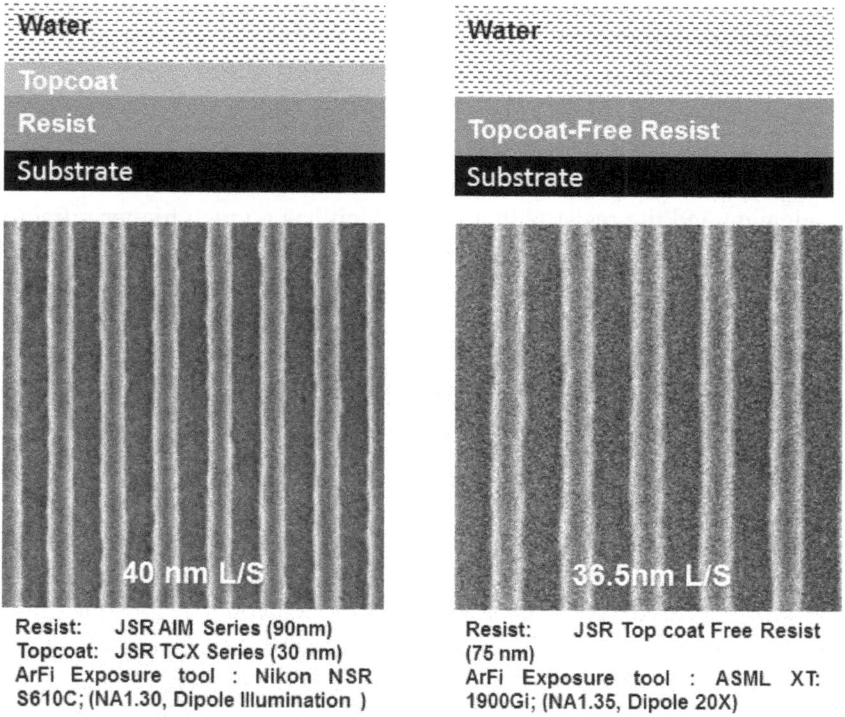
Water
Topcoat
Resist
Substrate
Water
Topcoat-Free Resist
Substrate
40 nm L/S
36.5nm L/S
Resist: JSR AIM Series (90nm)
Topcoat: JSR TCX Series (30 nm)
ArFi Exposure tool : Nikon NSR S610C; (NA1.30, Dipole Illumination)
Resist: JSR Top coat Free Resist (75 nm)
ArFi Exposure tool : ASML XT: 1900Gi; (NA1.35, Dipole 20X)

FIGURE 9-10 Resist with topcoat and topcoat free resist process (top). Lithographic performance of topcoat/resist and topcoat-free resist with an ArF immersion lithography exposure tool (bottom). (Courtesy: JSR Corporation.)

ArF immersion lithography with a NA of between 1.3 and 1.35 can print $\leq$40 nm LS patterns with either topcoat and resist, or by using a topcoat free resist (Figure 9-10). To extend 193 nm immersion lithography to 32 nm LS, an NA of about 1.7 is required.[88] Immersion with a high refractive index fluid (n = 1.64) could achieve 32 nm patterning or better resolution. The application of high index fluids (HIF) 193 nm immersion lithography was also proposed. This led to the development of high refractive index lenses and high refractive index resist materials.[88,93–95] Several second and third generation HIF candidates with good transmittance have been reported by several research groups.[95,96] Immersion exposure using a second generation HIF such as JSR HIL-001 achieved 32 nm LS patterns with two beam interferometric exposure tool.[95] Many commercial 193 nm photoresists have refractive indices around 1.66–1.70, which is sufficient for use with second-generation immersion fluids. Therefore, work focused next on the development of resists (Figure 9-11D) with refractive indices greater than 1.8, to enable 1.7 NA imaging.[97] In addition to high NA imaging, photoresists with higher refractive indices ($n \approx$ 1.8–1.9) has also increased process window and reduced mask error enhancement factor (MEEF) for conventional 193 nm water immersion lithography. In late 2008, high index immersion lithography activities were abandoned due to delays in the development of the high index lenses—particularly the final lens element. However, high index immersion material research is expected to impact future resist materials development.

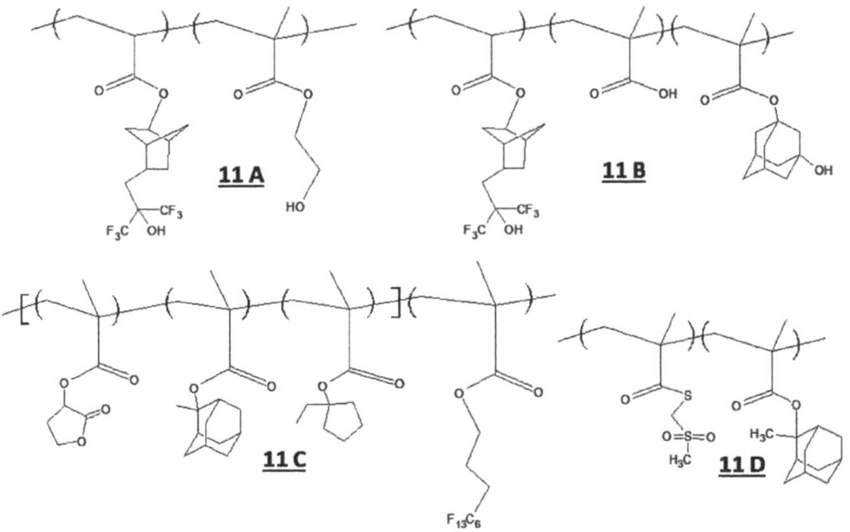

FIGURE 9-11 Representative polymer platforms investigated for topcoat and topcoat-free resists (11A–C) and high index resists (11D).

9.3.6 157 NM RESISTS

Attempts were made to reduce the exposure wavelength from 193 to 157 nm (F_2 laser) for higher resolution (≤70 nm). At this vacuum ultraviolet (VUV) wavelength, light is absorbed by water, air, simple organic materials, existing DUV resist materials, and fused silica (the traditional lens material).[31,98] The need for polymers semi-transparent to VUV radiation were necessary for application in photoresists.[31,98] Fluoro-polymers and polysilsesquioxanes were found to be transparent at this wavelength.[98] In addition to high transparency, compatibility with aqueous base development is critical. This led to structures similar to 193 nm resists being used for 157 nm resists, even though the acidic carboxylic groups strongly absorbs at 157 nm. Fortunately, the acidic hexafluoroisopropanol group incorporated into polymers showed good transparency, so the hexafluoroisopropanol functionality became the acid group of choice for 157 nm lithography (Figure 9-12).[31,99-104] Fluorinated compounds have larger molecular volumes than those expected from their van der Waals volumes, which further reduces the absorbance.

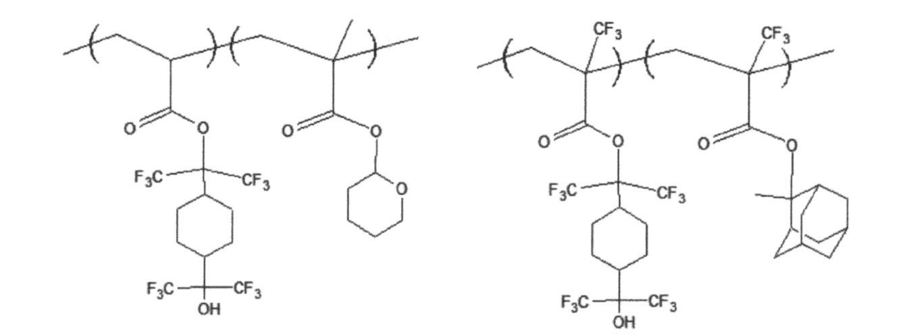

FIGURE 9-12   HFA-based polymer platform developed for 157 nm photoresist.

Photoresists containing fluorinated monomers reduce absorption while increasing the acidity of the hydroxyl group, and the cyclic moiety improves etch resistance. Etch resistance of these systems can be improved by introducing more alicyclic groups; this was achieved by using photoresist polymers with a polynorbornene backbone and pendent acid labile moieties. In 2003, 157 nm lithography development activities were abandoned due to lower than expected improvement from the smaller wavelength, as well as several technical challenges created by working with VUV light. However, the fluorinated materials originally developed for 157 nm lithography eventually found use in 193 nm immersion topcoats and topcoat-free resists. In addition, the materials developed for 157 nm were also investigated for extreme ultraviolet (EUV) lithography and e-beam lithography.

9.3.7 EXTREME ULTRAVIOLET (EUV OR 13.5 NM) RESISTS

The ultimate advanced optical lithography technique is EUV lithography, which operates at 13.5 nm, the range of soft X-rays.[26] EUV lithography is expected to produce feature sizes in the sub-20 nm regime due to its extremely short wavelength by single exposure. Therefore, EUV lithography is the leading candidate to succeed 193 nm immersion lithography but EUV lithography presents new challenges—in the nature of the EUV radiation and the development of new infrastructure radically different from traditional optical lithography systems.[105–108] At the 13.5 nm wavelength, almost any material absorbs energy to some degree, and energy absorption is dictated by atomic composition (Figure 9-13).[32] EUV absorption depends on the atomic composition and the density of each element in the absorbing material. Since all materials absorb at 13.5 nm, EUV exposure have to be performed under vacuum condition with reflective types of photomasks instead of the usual transmission photomask.[105–108]

The Sn plasma sources used for EUV lithography are either plasma-produced or discharge produced.[109] Current EUV plasma light sources are not spectrally pure and emit out-of-band (OOB) radiation over a broad range including the DUV bands.[110] In addition, EUV sources are relatively weak, and every spare photon must be utilized to maximize the imaging capability. Resist designers have a new set of requirements due to the change in wavelength. EUV resists must simultaneously achieve sub-22 nm resolution [R] with low LER (<2 nm) [L] and high sensitivity (10–15 mJ/cm^2) [S].[111] EUV resists must have low outgassing of resist components during exposure, as well as low sensitivity to OOB radiation. EUV lithographic processes are to expected use ultrathin films (<80 nm) to avoid pattern collapse. Resist absorption must be in the acceptable range. High sensitivity CA resists developed for 248 and 193 nm lithography are the leading candidates under investigation for EUV lithography.[112–115]

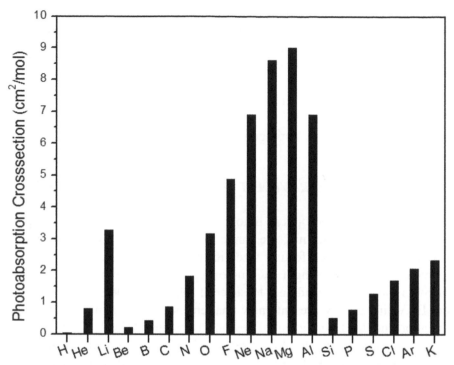

FIGURE 9-13 Photoabsorption cross section of elements at 13.5 nm wavelength. (Courtesy: Prof. Ober Cornell University.)

Light absorption at ionizing radiation (13.5 nm or e-beam) involves core–shell rather than valence electrons, so the CA resist acid-generation mechanism is very different at these ionizing wavelengths (13.5 nm or e-beam) than at 193 and 248 nm.[116–118] Several research groups use e-beam exposure tools to screen EUV resists, due to limited access to EUV exposure tools. Representative CA resist polymer systems studied for EUV resist are summarized in Figure 9-14. Resist polymers containing carbon, hydrogen, boron, and silicon were investigated due to their high transparency (Figure 9-14A and B).[119] Ultrathin 248 nm positive resists based on the PHS polymer platform (Figure 9-14C) were also investigated for EUV lithography. High activation energy (ESCAP; Figure 9-5)[120] and low activation types (ketol protecting group; Figure 9-5)[121] of resist were evaluated for EUVL, but a tradeoff in RLS was observed. ESCAP-based resists showed high resolution capability down to 19 nm LS with EUV interferometry lithography, but LER is a major issue.[122]

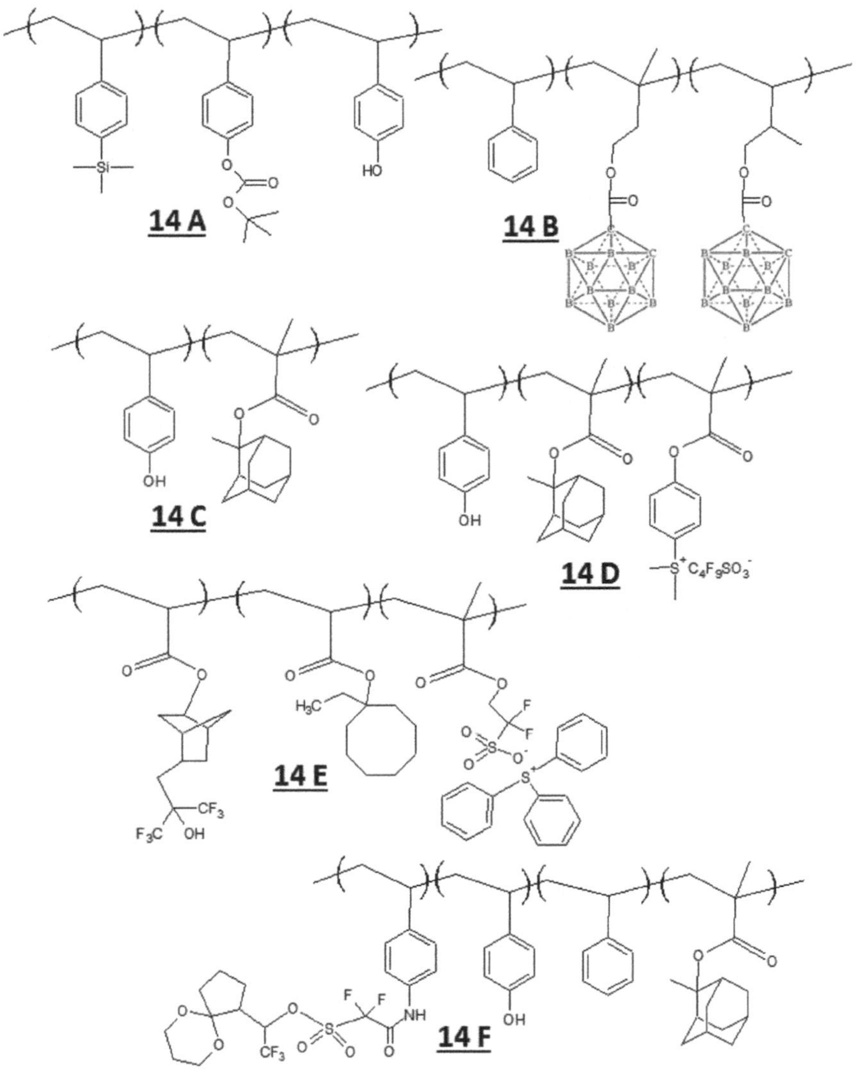

FIGURE 9-14 Polymer platforms investigated for EUV lithography.

To improve RLS, low acid diffusion resists based on polymer-bound PAG systems (Figure 9-14D and E), where the photoacid generator is bound to the polymer main were studied.[123–126] PAG anion-bound (Figure 9-14E) polymer resists showed improvement in resolution and LER. Resists based on high absorption polymers and polymer-bound acid amplifiers[127] (Figure 9-14F) were explored for improving RLS performance. Acid amplifiers[128,129] offer a means to improve the photospeed of a CA resist by introducing a second channel for acid-catalyzed reactions. Several CA EUV resists showed promising performance at 22 nm LS resolution,[126,130,131] but further optimization is in progress to achieve the desired RLS performance for sub-20 nm patterning. Post-lithographic process optimizations such as special rinses during resist develop were introduced to improve RLS performance.[132] Optimized EUV CA resists with post process optimization were able to resolve 15 nm LS using EUV microexposure field tools (Figure 9-15).[133,134]

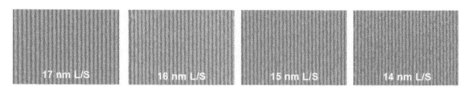

FIGURE 9-15   EUV lithography performance of CA EUV rResist. (Courtesy: JSR Corporation).

In addition to CA resists, non-CA resist materials which undergo chain scission or depolymerization upon exposure EUV radiation, were investigated to improve EUV RLS. PMMA was exposed using EUV interferometric lithography and showed clean profiles down to about 20 nm.[135] The resolution of non-CA resists such as PMMA is clearly superior to that of the CA resist materials, but showed poor sensitivity. Fluorine, which has higher absorption at EUV (Figure 9-13), was incorporated into PMMA to improve sensitivity,[136] and indeed, fluorinated PMMA showed higher sensitivity due to its increased resist absorbance. In addition to linear polymers, star polycarbonate-*block*-PMMA copolymers and polysulfone-*comb*-PMMA have also been investigated.[137] These polymers showed improved sensitivity compared to PMMA. Non-CA resists based on chain scission or depolymerization mechanisms need to address outgassing and special process (develop) concerns for use in HVM.

Predominantly, most resist systems developed for optical lithography are based on organic components. Nanocomposite resists (organic polymers with inorganic components as additives) and inorganic resists are interesting candidates for ultrathin film imaging processes, due to their superior mechanical properties and high etch resistance. These resists also showed high contrast and improved resolution when evaluated with ionizing radiation (e-beam). Nanocomposite resists based on silica particles incorporated KRS-XE (a PHS polymer platform with ketol protecting groups), resulting in a CA resist that was evaluated with 75 keV e-beam lithography.[138] KRS-XE nanocomposite resists showed improved resolution (75 nm LS) and etch resistance in O_2. Hydrogen silsesquioxane (HSQ) is a commercially available inorganic polymer that works well with e-beam lithography, and was evaluated for EUV lithography. HSQ as a negative-tone photoresist resolved patterns as small as 20 nm with EUV interference lithography.[139] Inorganic resists derived from hafnium and zirconium oxide sulfates, or polytungstic acids containing peroxo groups, have been reported.[140] For these metal oxide materials, imaging was achieved via negative tone imaging, which in turn depends on the presence of peroxide. The hafnium and zirconium oxide sulfate materials have imaged sub-20 nm LS resolution with a EUV microexposure tool and achieved 8 nm LS resolution using EUV interference lithography.[141]

Significant progress has been made in the development for EUV lithography technology.[142,143] EUV lithography full field tools (NXE3330) from ASML are available now. CA resist materials, and hafnium and zirconium oxide-based resists, have showed 18 nm LS resolution with the NXE3300.[142] However, EUV resist technology incorporating new chemistries and process schemes must be explored to achieve the RLS and process window necessary for large-scale manufacturing of 16 nm and smaller patterns, and beyond.

9.4 EMERGING PATTERNING METHODS: STATUS AND CHALLENGES

Deep ultraviolet lithography can pattern features below 22 nm in size, and resolution enhancement techniques should help to achieve even higher resolution features. However, a point will be reached where optical lithography will need even more stringent requirements to achieve the required features. This creates a need for research into alternative patterning methods, ones with capabilities that can complement those of DUV lithography and other established approaches. Several alternative patterning methods and materials are under investigation to overcome the resolution limits. A brief account on the status and challenges associated with these emerging patterning methods is provided below.

9.4.1 NEGATIVE TONE DEVELOPMENT (NTD) PROCESS WITH ORGANIC SOLVENT DEVELOPER

Recently, NTD processes with organic solvent developer was proposed for small contact hole and trench patterning using ArF immersion, since higher optical image contrast is provided by NTD compared to traditional aqueous-based positive tone develop (PTD).[144,145] The dissolution contrast comes from the polarity change, which occurs during the acid catalyzed deprotection reaction of polymer, similar to the positive tone process. This concept is not new and was already explored for PTBOCST (refer to Section 9.3.3), the first CA resist used in 1986 for device manufacturing.[58] The logistics of handling large volumes of organic solvents in a manufacturing environment is a concern, but the semiconductor industry recently implemented ArFi NTD in HVM process due to its technological benefits.

Several 193 nm resist polymers (Figures 9-6 and 9-7) showed good solubility in ester (e.g., n-butyl acetate) and ketone (e.g., methyl amyl ketone) organic solvents.[144,146,147] These solvents dissolve the nonpolar, unexposed film and leave behind the deprotected, more polar carboxylic acid-containing polymer. High resolution trench patterning performance was observed for ArF resists when developed with these solvents.[147] Polymeric CA resist platforms designed for a PTD process were successfully used for NTD process with minor adjustments to polymer structure or molecular weight characteristics. The applicability and advantages of NTD for EUV lithography has also been investigated. New polymer platforms were developed for EUV NTD that showed good isolated trench (<22 nm) as well as dense contact hole patterning (22 nm) performance.[148]

9.4.2 MULTIPLE PATTERNING

Multiple patterning is the repeated combination of patterning and process steps used for manufacturing ICs. The simplest form of multiple patterning is double patterning (DP). By patterning a single mask layer more than once, patterning resolution will improve because the k_1 factor of eq 9-1 (refer to Section 9.2) is effectively reduced. The effective k_1 will be k_1/number of the patterning passes (N), if the same lithographic system is used. DP requires minimal change in the equipment infrastructure to extend the resolution capability of current lithography tools. To extend resolution below the 32 nm node, the semiconductor industry has focused on enabling DP technology to circumvent the limitations of Rayleigh scaling (eq 9-1). However, DP means half the throughput and double the cost for the critical layers where it is used, in addition to increased process cost and complexity.

There are several forms of DP technologies such as Litho-Etch-Litho-Etch (LELE), Litho-Process-Litho-Etch, Litho-Litho-Etch (LLE), the dual tone develop process, and self-aligned double patterning (SADP).[149–159] LELE is the most complex and expensive process to implement, but does not require developing new materials or chemical treatments.[149,150] Litho-Process-Litho-Etch requires hardening of the first resist pattern, so it is unaffected during the second resist coat-expose-develop cycle.[151–156] Litho-Freeze-Litho-Etch (LFLE) is one variant of Litho-Process-Litho-Etch, where the first lithography pattern is stabilized by a chemical freezing material, or by a post develop bake to prevent damage by the second lithography process (Figure 9-16).[155,156] The LFLE process with ArF immersion exposure resolved sub-32 nm LS features (Figure 9-16). LLE is the simplest variant of the DP approaches. It can be carried out using either two positive tone resists as the two patterning layers, or a combination of both positive and negative tone resist.[157] In a dual tone develop process, the photoresist is developed twice. A conventional development step (PTD) with aqueous base removes the high exposure dose areas of the patterns. Subsequent development with an organic solvent developer (NTD) removes the unexposed or low exposure dose areas.[158] This leads to removal of all materials except the intermediate dose areas. The main challenge is to not only show successful positive and negative tone development process windows, but also to ensure the windows overlap sufficiently.

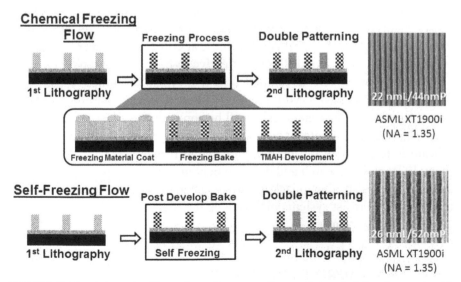

FIGURE 9-16 Schematic illustration of process flows of chemical-freezing and self-freezing process. ArF immersion exposure were performed with ASML XT1900i (NA = 1.35). (Courtesy: JSR Corporation.)

SADP is the most promising technique for patterning sub-22 nm features.[159] It has many advantages such as accurate CD control, relaxed overlay requirements, and low LER. Recently, self-aligned multi-patterning (SAMP) techniques including self-aligned quadruple-patterning (SAQP) and self-aligned octuple-patterning (SAOP) were investigated for much higher resolution (≤10 nm LS).[160] SADP was also investigated using EUV lithography to further reduce feature sizes. The combination of EUV lithography and SADP achieved 10 nm LS patterns.[161]

9.4.3   ELECTRON BEAM DIRECT WRITE (EBDW) LITHOGRAPHY

Electron beam (e-beam) lithography systems use a fine scanning e-beam to directly write on a radiation sensitive resist. Electron beam (e-beam) lithography for high resolution patterning was demonstrated as early as 1960. E-beam lithography operating at 10–100 keV has higher spatial resolution and a wider DoF than photolithography, but single-beam writing has never been able to compete with the massively parallel optical systems in throughput and cost. E-beam lithography has been widely used in the semiconductor industry to make mask and small scale writing of exploratory devices or prototyping. A multiple beam solution overcomes the throughput limitation of EBDW. Several multiple e-beam direct write (MEB DW) have been conceived and are being developed for semiconductor industry applications.[162,163] IMS PML MEB DW, multiple aperture pixel by pixel enhancement of resolution (MAPPER), reflective e-beam lithography (REBL) MEB DW systems are developed and their high resolution capabilities were demonstrated.[163–165] Multiple EUV resist systems such as CA resist, non-CA resist, and inorganic resist will be applicable to EBDW, because the secondary electron generated either by high energy photon or by high-energy e-beam is responsible for the radiation-induced reactions.[116–118] At present, many EUV resist research and development groups use e-beam exposure to screen EUV resists, due to lack to EUV exposure tools. Therefore, developments of EUV resist systems may reduce the preliminary R&D to develop e-beam resist systems.

9.4.4   NANOIMPRINT LITHOGRAPHY (NIL)

An imprint lithography process to fabricate nanometer sized IC features was proposed in 1996. This process is now known as nanoimprint lithography

(NIL). In simple terms, NIL creates patterns by mechanical deformation of an imprint resist, followed by an etch process.[166] In NIL, the imprint resist, which is typically a monomer or polymer formulation, is applied onto a wafer surface, and then the template is pressed into the resist. The resist is cured by applying heat or UV light. Finally, the template is released from the wafer surface and a subsequent etch process transfers the resist pattern to the substrate.

There are several forms of NIL but the two widely studied are thermal-NIL and photo-NIL. For thermal-NIL, a thin layer of imprint resist (thermoplastic polymer) is spin-coated on the substrate, which is then stamped with a heated mold. After cooling down, the mold is removed and the resist pattern is left on the substrate. An etch step transfers the resist pattern into the underlying substrate. The formation of feature sizes <10 nm has been reported.[167,168]

Photo-NIL[28,169] uses a low viscosity photocurable liquid monomer or polymer as the imprint material. The transparent mold and the substrate are pressed together and the resist is formed with UV light exposure. After mold separation, the resist pattern remains on the substrate, and is then etched into the substrate. Acrylate monomer mixtures, vinyl ether compositions, and silicon-hydrocarbon based resists have been reported. Resolutions of <11 nm with very low LER have been demonstrated with photo-NIL.[170] At present, photo-NIL is the preferred method for high resolution manufacturing due to its exposure process advantages. NIL is high resolution, low LER with a low cost of ownership technique, but further improvements in template defects, overlay, throughput and management of its working templates are required to apply NIL into high volume production.

9.4.5 BLOCK COPOLYMER DIRECTED SELF-ASSEMBLY (DSA)

Block copolymer (BCP) directed self-assembly (DSA) is an alternative method to achieve a desired pattern by striking a balance between chemistry and thermodynamics. Block copolymers microphase separate to dimensions ranging from 10 to 100 nm, depending on the Flory-Huggins interaction parameters (χ) between the two blocks, and their degree of polymerization (N).[171] The resulting morphology can be altered by changing the volume fraction of the components, keeping the interaction parameters constant. Spheres, cylinders, and lamellae are the commonly encountered morphologies, even though the microstructure can be complicated by the number of blocks (di-, tri-, and multi-block) and the connectivity between the blocks.

The thermodynamics of these morphologies have been extensively studied in bulk.[171,172] The cylinder and lamellae morphologies are of particular interest for lithographic applications as they closely resemble the patterns of contact holes (CH) and lines-space (LS) arrays.[173]

In thin films, microdomain formation is influenced by the interfacial energies and the geometrical constraints introduced by the interfaces. In general, self-assembly is limited to a few periodic forms of high symmetry such as spheres or cylinders, but some degree of intelligent guidance or DSA can be used to achieve the preferred structures and long range ordering depending upon the application.[174-179] Three different types of mechanisms are generally used for DSA of block copolymer thin films: (1) control-ling the topography of the substrate (Grapho-epitaxy; Figure 9-18)[174,175]; (2) chemical modification of the surface to change the substrate–polymer interactions (Chemo-epitaxy)[176,177]; and (3) applying external stimuli (elec-tric, thermal, crystallization, and solvent evaporation, etc.) to induce long-range ordering.[178] The guide pattern for grapho or chemo-epitaxy BCP DSA process can be prepared using any of the lithography techniques discussed in the previous sections. Compatibility of BCP DSA with ArFi,[179] e-beam,[180] and EUV[133] lithography processes has been demonstrated. Though the guide patterns are larger than those of the final microdomain, BCP DSA has the ability to perform a function similar to the DP techniques described previ-ously. DSA has been investigated for line, space, or contact hole creation, contact hole shrink, pattern repair, and pattern doubling.[178–183] BCP DSA is an economical process compared to the advanced lithographic techniques discussed in the previous sections, and is emerging as an alternative route for leading edge device fabrication.

In order to complement block copolymer DSA with classical top-down lithographic processes, the block copolymer material must be compatible with existing semiconductor fabrication processes and equipment.[13,184,185] Optimi-zation of a BCP material for DSA applications requires careful balancing of all functional properties. The composition and architecture of block copo-lymers play an important role in forming and transferring nanoscale struc-tures into an underlying substrate. Controlled polymerization methods were generally used to synthesize well-defined block copolymers. To enhance wet and dry etch resistance, selective metal decoration of one block, a copo-lymer with a metal-containing block, or photochemical crosslinking was used. Self-assembly of a wide range of block copolymer materials has been reported (Figure 9-17).[178] Poly(styrene-*block*-methyl methacrylate) [PS-*b*-PMMA] has been adopted for initial evaluations of block copolymer DSA in a manufacturing environment. However, PS-b-PMMA is not an ideal material, so development of high-χ-block copolymers is now in progress. Very high resolution performance has been demonstrated for the BCP DSA process (Figure 9-18).[183]

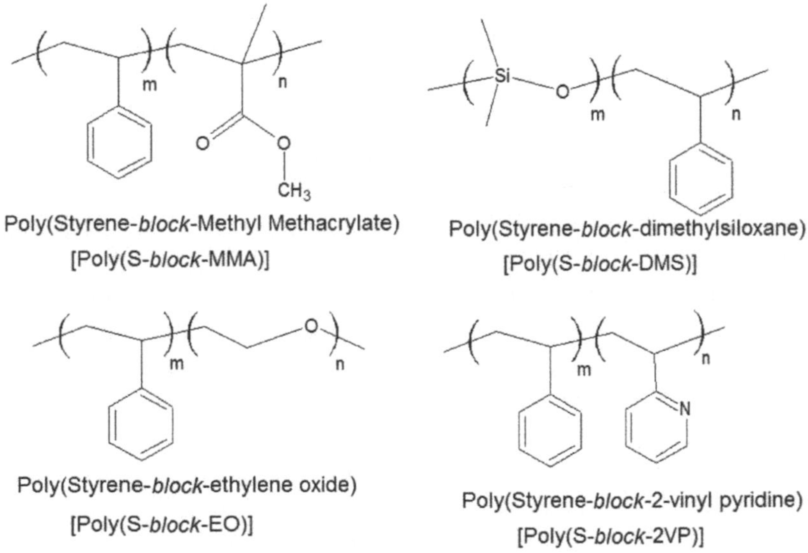

Poly(Styrene-*block*-Methyl Methacrylate)

[Poly(S-*block*-MMA)]

Poly(Styrene-*block*-dimethylsiloxane)

[Poly(S-*block*-DMS)]

Poly(Styrene-*block*-ethylene oxide)

[Poly(S-*block*-EO)]

Poly(Styrene-*block*-2-vinyl pyridine)

[Poly(S-*block*-2VP)]

FIGURE 9-17 Examples of block copolymers investigated for DSA process.

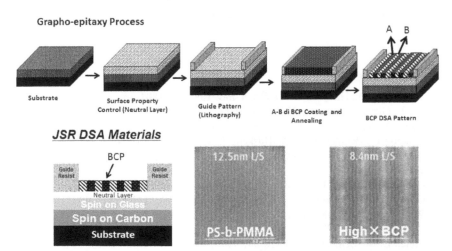

FIGURE 9-18 Block copolymer DSA process. Grapho-epitaxy process for forming oriented micro domains using a lithographically-generated pattern to control alignment and orientation (top); SEM images of self-assembled patterns of microdomains formed with grapho-epitaxy process (bottom). (Courtesy JSR Corporation.)

9.5 SUMMARY AND OUTLOOK

Photolithography has been the dominant technique used in the semiconductor industry for the fabrication of ICs, with feature sizes ranging from several tens of nanometers to the micrometer scale. CA resists are the preferred patterning material for all advanced patterning processes. Photolithography continues to overcome obstacles and achieve new resolution limits; however, some techniques increase the process complexity and cost. Similarly, polymer-based CA resists are challenged by the RLS trade off in the sub-50 nm patterning regime. While CA resists continue to show excellent resolution, several new materials including non-CA polymeric resists are under investigation. Patterning of features to 20 nm and below has been demonstrated by a variety of patterning techniques. Many reports believe that EUV lithography will be more economical despite the complexity and cost of the exposure tool. But alternative lithography techniques such as NIL, MEBDW, and DSA are very attractive for semiconductor IC manufacturing. These options come with additional process complexities and costs, as well as other unforeseen complications. New patterning techniques will require development of new, dedicated materials to satisfy increasingly stringent lithography requirements. The semiconductor industry is planning to achieve sub-20 nm resolution in the near future. The complexities and requirements for 20 nm resolution and beyond will require more patterning options, processes, and new patterning materials.

REFERENCES

1. Wallraff, G M.; Hinsberg, W. D. *Chem. Rev.* **1999,** *99,* 1801.
2. Ito, T.; Okazaki, S. *Nature* **2000,** *406,* 1027.
3. Gamo, K. *Microelectron. Eng.* **1996,** *32,* 159.
4. Acikgoz, C.; Hempenius, M. A.; Huskens, J.; Vancso, G. J. *Eur. Polym. J.* **2011,** *47,* 2033.
5. Gates, B. D.; Xu, Q. B.; Stewart, M.; Ryan, D.; Willson, C. G. *Chem. Rev.* **2005,** *105,* 1171.
6. Chou, S. Y.; Krauss, P. R.; Renstrom, P. J. *Science* **1996,** *272,* 85.
7. Colburn, M.; Johnson, S. C.; Stewart, M. D.; Damle, S.; Bailey, T. C.; Choi, B.; Wedlake, M.; Michaelson, T. B.; Sreenivasan, S. V.; Ekerdt, J. G.; Willson, C. G. *Proc. SPIE* **1999,** *3676,* 379.
8. Eigler, D. M.; Schweizer, E. K. *Nature* **1990,** *344,* 524.
9. Piner, R. D.; Zhu, J.; Xu, F.; Hong, S. H.; Mirkin, C. A. *Science* **1999,** *283,* 661.
10. Rogers, J. A.; Paul, K. E.; Jackman, R. J.; Whitesides, G. M. *Appl. Phys. Lett.* **1997,** *70,* 2658.
11. Melosh, N. A.; Boukai, A.; Diana, F.; Gerardot, B.; Badolato, A.; Petroff, P. M.; Heath, J. R. *Science* **2003,** *300,* 112.

12. Whitesides, G. M.; Mathias, J. P.; Seto, C. T. *Science* **1991**, *254*, 1312.
13. Hawker, C. J.; Russell, T. P. *MRS Bull.* **2005**, *30*, 952.
14. Jeong, S. J.; Kim, J. Y.; Kim, B. H.; Moon, H. S.; Kim, S. O. *Materials Today* **2013**, *16* (12), 468.
15. Moreau, W. M. *Semiconductor Lithography: Principles and Materials*; Plenum: New York, 1988.
16. Reichmanis, E.; Novembre, A. *Annu. Rev. Mater. Sci.* **1999**, *23*, 1.
17. Hinsberg, W. D.; Wallraff, G. M. *Lithography Resists, Encyclopedia of Polymer Science and Technology*; John Wiley and Sons Inc.: New York, 2012.
18. Reichmanis, E.; Thompson, L. F. *Chem. Rev.* **1989**, *89*, 1273.
19. Willson, C. G.; Thompson, L.; Bowden, M. *Introduction to Microlithography*, 2nd ed.; American Chemical Society: Washington, DC, 1994.
20. Bratton, D.; Ayothi, R.; Felix, N.; Ober, C. K. Applications of Controlled Macromolecular Architectures to Lithography. In *Macromolecular Engineering: Precise Synthesis, Materials Properties, Applications*; Matyjaszewski, K., Gnanou, Y., Leibler, L., Eds.; Wiley-VCH Verlag GmbH & Co. KgaA: Weinheim, Germany, 2007.
21. Levinson H. *Principles of Lithography*, 2nd ed.; SPIE Press: Bellingham, Washington, 2010.
22. Moore, G.E. *Proc. SPIE* **1995**, *2438*, 2.
23. Lin, B. J. *Microelectron. Eng.* **2006**, *83* (4–9), 604.
24. Lin, B. J. *J. Micro/Nanaolith. MEMS MOEMS* **2004**, *3* (3), 377.
25. Bloomstein, T. M.; Rothschild, M.; Kunz, R. R.; Hardy, D. E.; Goodman, R. B.; Palmacci, S. T. *J. Vac. Sci. Techn. B.* **1998**, *16* (6), 3154.
26. Bjorkholm, J. E. *Intel Technology Journal,* **1998**, Q3.
27. Lin, B. J. *J. Micro/Nanaolith. MEMS MOEMS* **2012**, *11* (3), 033011.
28. Stewart, M. D.; Johnson, S. C.; Sreenivasan, S. V.; Resnick, D. J.; Willson, C. G. *J. Micro/Nanaolith. MEMS MOEMS* **2005**, *4* (1), 011002.
29. Allen, R. D.; Wallraff, G. M.; Hofer, D. C.; Kunz, R. R. *IBM J. Res. Dev.* **1997**, *41* (1/2), 95.
30. Opitz, J.; Allen, R. D.; Wallow, T. I.; Wallraff, G. M.; Hofer, D. C. *Proc. SPIE* **1998**, 3333, 571.
31. Patterson, K.; Yamachika, M.; Hung, R.; Brodsky, C.; Yamada, S.; Somervell, M.; Osborn, B.; Hall, D.; Dukovic, G.; Byers, J.; Conley, W.; Willson, C. G. *Proc. SPIE* **2000**, *3999*, 365.
32. Kwark, Y. J.; Bravo-Vasquez, J. P.; Chandhok, M.; Cao, H.; Deng, H.; Ober, C. K. *J. Vac. Sci. Techn. B* **2006**, *24* (4), 1822.
33. Gokan, H.; Esho, S.; Ohnishi, Y. *J. Electrochem. Soc.* **1983**, *130*, 143.
34. Wallow, T.; Brock, P.; Di Pietro, P.; Allen, R. D.; Opitz, J.; Sooriyakumaran, R.; Hofer, D.; Meute, J.; Byers, J.; Rich, G.; McCallum, M.; Schuetze, S.; Jayaraman, S.; Hullihen, K.; Vicari, R.; Rhodes, L.; Goodail, B.; Shick, R. *Proc. SPIE* **1998**, *3333*, 92.
35. Oehrlein, G. S.; Phaneuf, R. J.; Graves, D. B. *J. Vac. Sci. Techn. B.* **2011**, *29* (1), 010801.
36. Dammel, R. *Diazonapthoquinone-Based Resists (Tutorial Texts in Optical Engineering, Vol. TT 11)*; SPIE Optical Engineering Press: Bellingham, Washington, 1993.
37. Pacansky, J.; Lyerla, J. R. *IBM J. Res. Dev.* **1979**, *23* (1), 42.
38. Trefonas, P.; Daniels, B. K. *Proc. SPIE* **1987**, *771*, 194.
39. Ito, H.; Willson, C. G.; Frechet, J. M. J. *Digest of Technical Papers of 1982 Symposium on VLSI Technology* 1982, 86.

40. Ito, H.; Willson, C. G. *Technical Papers of SPE Regional Technical Conference on Photopolymers* 1982, 331.
41. Ito, H.; Willson, C. G. *Polym. Eng. Sci.* **1983,** *23*, 1012.
42. Ito, H.; Willson, C. G *Polymers In Electronics. ACS Symposium Series 242*; American Chemical Society: Washington, DC, 1984; p 11.
43. Crivello, J. *Radiation Curing in Polymer Science and Technology, Vol. II: Photoinitiating Systems*; Elsevier Science Publishing Co: New York, 1993; Chapter 8.
44. Crivello, J. V. *J. Polym. Sci. Part A: Polym. Chem.* **1999,** *37*, 4241.
45. Shirai, M.; Tsunooka, M. *Prog. Polym. Sci.* **1996,** *21* (1), 1.
46. Cameron, J. F.; Ablaza, S. L.; Xu, G.; Wang, Y. *J. Photopolym. Sci. Technol.* **1999,** *12* (4), 607.
47. Ayothi, R.; Yi, Y.; Cao, H. B.; Wang, Y.; Putna, S.; Ober, C. K. *Chem. Mater.* **2007,** *19* (6), 1434.
48. Yi, Y.; Ayothi, R.; Wang, Y.; Li, M.; Barclay, G.; Sierra-Alvarez, R.; Ober, C. K. *Chem. Mater.* **2009,** *21* (17), 4037.
49. Ito, H. *Adv. Polym. Sci.* **2005,** *172*, 37.
50. Reichmanis, E.; Macdonald, S. A.; Iwayanagi, T. Eds., *Polymers in Microlithography: Materials and Processes, ACS Symposium Series 412*; American Chemical Society: Washington, DC, 1989; p 27.
51. Ito, H.; Breyta, G.; Hofer, D.; Sooriyakumaran, R.; Petrillo, K.; Seeger, D. *J Photopolym. Sci. Technol.* **1994,** *7* (3), 433.
52. Medeiros, D. R. *J. Photopolym. Sci. Technol.* **2002,** *15* (3), 411.
53. Kim, J. H.; Kim, Y.-H.; Chon, S. N.; Nagai, T.; Noda, M.; Yamaguchi, Y.; Makita, Y.; Nemoto, H. *J. Photopolym. Sci. Technol.* **2004,** *17* (3), 379.
54. Murata, M.; Takahashi, T.; Koshiba, M.; Kawamura, S. *Proc. SPIE* **1990,** *1262*, 8.
55. Thackeray, J. W.; Adams, T.; Cronin, M. F.; Denison, M.; Fedynyshyn, T. H.; Georger, J.; Mori, J. M.; Orsula, G. W.; Sinta, R. *J. Photopolym. Sci. Technol.* **1994,** *7* (3), 619.
56. Lee, S. M.; Frechet, J. M. J. *Macromolecules* **1994,** *27*, 5160.
57. Shaw, J. M.; Gelorme, J. D.; LaBianca, N. C.; Conley, W. E.; Holmes, S. J. *IBM J. Res. Develop.* **1997,** *41* (1/2), 81.
58. Maltabes, J. G.; Holmes, S. J.; Morrow, J.; Barr, R. L.; Hakey, M.; Reynolds, G.; Brunsvold, W. R.; Willson, C. G.; Clecak, N. J.; MacDonald, S. A.; Ito, H. *Proc. SPIE* **1990,** *1262*, 2.
59. Allen, R. D.; Wallraff, G. M.; Hinsberg, W. D.; Simpson, L. L. *J. Vac. Sci. Technol. B.* **1991,** *9* (6), 3357.
60. Kunz, R. R.; Allen, R. D.; Hinsberg, W. D.; Wallraff, G. M. *Proc. SPIE* **1993,** *1925*, 167.
61. Wallraff, G. M.; Allen, R. D.; Hinsberg, W. D.; Larson, C. F.; Johnson, R. D.; DiPietro, R.; Breyta, G.; Hacker, N.; Kunz, R. R. *J. Vac. Sci. Technol. B.* **1993,** *11* (6), 2783.
62. Allen, R. D.; Sooriyakumaran, R.; Optiz, J.; Wallraff, G. M.; Hinsberg, W. D.; Kunz, R. R.; Jayaraman, S.; Shick, R.; Goodall, B.; Okoroanyanwu, U.; Willson C. G. *Proc. SPIE* **1996,** *2724*, 334.
63. Takahashi, M.; Takechi, S.; Nozaki, K.; Kaimoto, Y.; Abe, N. *J. Photopolym. Sci. Technol.* **1994,** *7* (1), 31.
64. Allen, R. D.; Wallraff, G. M.; DiPietro, R. A.; Kunz, R. R. *J. Photopolym. Sci. Technol.* **1994,** *7* (3), 507.
65. Allen, R. D.; Wang, I. Y.; Wallraff, G. M.; DiPietro, R. A.; Hofer, D. C.; Kunz, R. R. *J. Photopolym. Sci. Technol.* **1995,** *8* (4), 623.

66. Kajitaa, T.; Nishimura, Y.; Yamamotoa, Y.; Ishii, H.; Soyano, A.; Kataoka, A.; Slezak, M.; Shimizu, M.; Varanasi, P. R.; Jordahamo, G.; Lawson, M. C.; Chen, R.; Brunsvold, W. R.; Li, W.; Allen, R. D.; Ito, H.; Truong, H.; Wallow. T. *Proc. SPIE* **2001**, *4345*, 712.

67. Okoroanyanwu, U.; Shimokawa, T.; Byers, J.; Willson, C. G. *Chem. Mater.* **1998**, *10* (11), 3319.

68. Okoroanyanwu, U.; Byers, J.; Shimokawa, T.; Willson. C. G. *Chem. Mater.* **1998**, *10* (11), 3328.

69. Ito, H.; Truong, H. D.; Burns, S. D.; Pfeiffer, D.; Huang, W.-S.; Khojastehc, M. M.; Varanasi, P. R.; Lercel, M. *Proc. SPIE* **2005**, *5753*, 109.

70. Gabor, A. H.; Ober, C. K. *Silicon-Containing Block Copolymers as Microlithographic Resists, Microelectronics Technology, ACS Symposium Series 614*; American Chemical Society: Washington, DC, 1995; p 281.

71. Sundararajan, N.; Yang, S.; Ogino, K.; Valiyaveettil, S.; Wang, J.; Zhou, X.; Ober C. K.; Obendorf, S. K.; Allen, R. D. *Chem. Mater.* **2000**, *12* (1), 41.

72. Varanasi, P. R.; Kwong, R. W.; Khojasteh, M.; Patel, K.; Chen, K.-J.; Li, W.; Lawson, M. C.; Allen, R. D.; Sorriyakumaran, R.; Brock, P.; Sundberg, L. K.; Slezak, M.; Dabbagh, G.; Liu, Z.; Nishimura, Y.; Shimokawa, T. *J. Photopolym. Sci. Technol.* **2005**, *18* (3), 381.

73. Cho, S.; Heyden, A. V.; Byers, J.; Willson, C. G. *Proc SPIE* **2000**, *3999*, 62.

74. Pugliano, N.; Bolton, P.; Barbieri, T.; King, M.; Reilly, M. *Proc. SPIE* **2003**, *5039*, 698.

75. Patel, K.; Lawson M.; Varanasi P.; Medeiros M.; Wallraff G.; Brock, P.; DiPietro, R.; Nishimura, Y.; Chiba T.; Slezak, M. *Proc. SPIE* **2004**, *5376*, 94.

76. Sooriyakumaran, R.; DiPietro, R.; Truong, H.; Brock, P.; Allen, R.; Bozano, L.; Popova, I.; Huang, W.-S.; Chen, R.; Khojasteh, M.; Varanasi, P. R. *Proc. SPIE* **2008**, *6923*, 69230C.

77. Stewart, M. D.; Tran, H. V.; Schmid, G. M.; Stachowiak, T. B.; Becker, D. J.; Willson C. G. *J. Vac. Sci. Technol. B.* **2002**, *20* (6), 2946.

78. Wang, M.; Jarnagin, N. D.; Lee, C.-T.; Henderson, C. L.; Yueh, W.; Roberts, J. M.; Gonsalves, K. E. *J. Mater. Chem.* **2006**, *16*, 3701.

79. Kawamura, Y.; Toyoda, T.; Namba, S. *J. Appl. Phys.* **1982**, *53* (9), 6489.

80. Blakey, I.; Chen, L.; Goh, Y.-K.; Piscani, E.; Zimmerman, P. A.; Whittaker, A. K. *The International Symposium on Immersion Lithography Extensions* 2008.

81. Blakey, I.; Chen, L.; Goh, Y.; Lawrie, K.; Chuang, Y.; Piscani, E.; Zimmerman, P. A.; Whittaker, A. K. *Proc. SPIE* **2009**, *7273*, 72733X.

82. Nishimura, I.; Heath, W. H.; Matsumoto, K.; Jen, W.-L.; Lee, S. S.; Neikirk, C.; Shimokawa, T.; Ito, K.; Fujiwara, K.; Willson C. G. *Proc. SPIE* **2008**, *6923*, 69231C.

83. Smith, B. W.; Bourov, A.; Kang, H.; Cropanese, F.; Fan, Y.; Lafferty, N.; Zavyalova, L. *J. Micro/Nanaolith. MEMS MOEMS* **2004**, *3* (1), 44.

84. Owa, S.; Nagasaka, H. *J. Micro/Nanaolith. MEMS MOEMS* **2004**, *3* (1), 97.

85. Mulkens, J.; Flagello, D.; Streefkerk, B.; Graeupner, P. *J. Micro/Nanaolith. MEMS MOEMS* **2004**, *3* (1), 104.

86. Hinsberg, W.; Wallraff, G.; Larson, C.; Davis, B.; Deline, V.; Raoux, S.; Miller, D.; Houle, F.; Hoffnagle, J.; Sanchez, M.; Rettner, C.; Sundberg, L.; Medeiros, D.; Dammel, R.; Conley, W. *Proc. SPIE* **2004**, *5376*, 21.

87. Taylor, J. C.; Chambers, C. R.; Deschner, R.; LeSuer, R. J.; Conley, W.; Burns, S. D.; Willson, C. G. *Proc. SPIE* **2004**, *5376*, 34.

88. Sanders, D. P. *Chem. Rev.* **2010**, *110* (1), 321.

89. Nakagawa, H.; Hoshiko, K.; Shima, M.; Kusumoto, S.; Shimokawa, T.; Nakano, K.; Fujiwara, T.; Owa, S. *Proc. SPIE* **2006**, *6153*, 61530D.

90. Nakagawa, H.; Nakamura, A.; Dougauchi, H.; Shima, M.; Kusumoto, S.; Shimokawa, T. *Proc. SPIE* **2006**, *6153*, 61531R.

91. Matsumura, N.; Sugie, N.; Goto, K.; Fujiwara, K.; Yamaguchi, Y.; Tanizaki, H.; Nakano, K.; Fujiwara, T.; Wakamizu, S.; Takeguchi, H.; Arima, H.; Kyoda, H.; Yoshihara, K.; Kitano, J. *Proc. SPIE* **2008**, *6923*, 69230D.

92. Sheehan, M. T.; Farnham, W. B.; Chambers, C.R.; Tran, H. V.; Okazaki, H.; Brun, Y.; Romberger, M. L.; Sounik J. R. *Proc. SPIE* **2011**, *7920*, 89210T.

93. Sewell, H.; Mulkens, J.; Wagner, C.; McCafferty, D.; Markoya, L.; Lipson, M.; Samarakone, N. *J. Photopolym. Sci. Technol.* **2007**, *20* (5), 651.

94. Ohmura, Y.; Nakashima, T.; Nagasaka, H.; Sukegawa, A.; Ishiyhama, S.; Kamijo, K.; Shinkai, M.; Owa, S. *Proc. SPIE* **2007**, *6520*, 652006.

95. Kusumoto, S.; Shima, M.; Wang, Y.; Shimokawa, T.; Sato, H.; Hieda, K. *Polym. Adv. Technol.* **2006**, *17*, 122.

96. Furukawa, T.; Kishida, T.; Miyamatsu, T.; Kawaguchi, K.; Yamada, K.; Tominaga, T.; Slezak, M.; Hieda, K. *Proc. SPIE* **2007**, *6519*, 65190B.

97. Matsumoto, K.; Costner, E.; Nishimura, I.; Ueda, M.; Willson, C.G. *Proc. SPIE* **2008**, *6923*, 692305.

98. Kunz, R. R.; Bloomstein, T. M.; Hardy, D. E.; Goodman, R. B.; Downs, D. K.; Curtin, J. E. *Proc. SPIE* **1999**, *3678*, 13.

99. Hung, R. J.; Tran, H. V.; Trinque, B. C.; Chiba, T.; Yamada, S.; Sanders, D. P.; Connor, E. F.; Grubbs, R. H.; Klopp, J.; Frechet, J. M. J.; Thomas, B. H.; Shafer, G. J.; DesMarteau, D. D.; Conley, W.; Willson, C. G. *Proc. SPIE* **2001**, *4345*, 385.

100. Trinque, B. C.; Osborn, B. P.; Chambers, C. R.; Hsieh, Y.-T.; Corry, S.; Chiba, T.; Hung, R. J.; Tran, H. V.; Zimmerman, P.; Miller, D.; Conley, W.; Willson, C. G. *Proc. SPIE.* **2002**, *4690*, 58.

101. Crawford, M. K.; Farnham, W. B.; Feiring, A. E.; Feldman, J.; French, R. H.; Leffew, K. W.; Petrov, V. A.; Schadt, F. L., III.; Zumsteg, F. C. *J. Photopolym. Sci. Technol.* **2002**, *15* (4), 671.

102. Bae, Y. C.; Douki, K.; Yu, T.; Dai, J.; Schmaljohann, D.; Kang, S. H.; Kim, K. H.; Koerner, H.; Conley, W.; Miller, D.; Balasubramanian, R.; Holl, S.; Ober, C. K. *J. Photopolym. Sci. Technol.* **2001**, *14* (4), 613.

103. Vohra, V.; Douki, K.; Kwark, Y.; Liu, X.; Ober, C.K.; Bae, Y. C.; Conley, W.; Miller, D.; Zimmerman, P. *Proc. SPIE* **2002**, *4690*, 84.

104. Bae, Y. C.; Douki, K.; Yu, T.; Dai, J.; Schmaljohann, D.; Koerner, H.; Ober, C. K.; Conley, W. *Chem. Mater.* **2002**, *14* (3), 1306.

105. Meiling, H.; Banine, V.; Kuerz, P.; Harned, N. *Proc SPIE* **2004**, *5374*, 31.

106. Naulleau, P.; Goldberg, K. A.; Anderson, E. H.; Bradley, K., Delano, R.; Denham, P.; Gunion, B.; Harteneck, B.; Hoef, B.; Huang, H.; Jackson, K.; Jones, G.; Kemp, D.; Liddle, J. A.; Oort, R.; Rawlins, A.; Rekawa, S.; Salmassi, F; Tackaberry, R.; Chung, C.; Hale, L.; Phillion, D.; Sommargren, G.; Taylor, J. *Proc. SPIE* **2004**, *5374*, 881.

107. Benschop, J.; Banine, V.; Lok, S.; Loopstra, E. *J. Vac. Sci. Technol. B.* **2008**, 26 (6), 2204.

108. Wagner, C.; Harned, N. *Nature Photonics* **2010**, *4*, 24.

109. Tomie, T. *J. Micro/Nanaolith. MEMS MOEMS* **2011**, *11* (2), 021109.

110. Roberts, J. M.; Bristol, R. L.; Younkin, T. R.; Fedynyshyn, T. H.; Astolfi, K.; Cabral, A. *Proc. SPIE* **2009**, *7273*, 72731W.

111. Naulleau, P.; Anderson, C.; George. *J. Photopolym. Sci. Technol.* **2011,** *24* (6), 637.
112. Thackeray, J. W. *J. Micro/Nanaolith. MEMS MOEMS* **2011,** *10* (3), 033009.
113. Itani, T. *Microelectron. Eng.* **2009,** *86,* 207.
114. Maruyama, K.; Shimizu, M.; Hirai, Y.; Nishino, K.; Kimura, T.; Kai, T.; Goto, K.; Sharma, S. *Proc. SPIE* **2010,** *7636,* 76360T.
115. Ayothi, R.; Singh, L.; Hishiro, Y.; Pitera, J. W.; Sundberg, L. K.; Sanchez, M. I.; Bozano, L.; Virwani, K.; Truong, H. D.; Arellano, N.; Petrillo, K.; Wallraff, G. M.; Hinsberg, W. D.; Hua, Y. *J. Vac. Sci. Technol. B.* **2012,** *30* (6), 06F506.
116. Kozawa, T.; Yoshida, Y.; Uesaka, M.; Tagawa, S. *Jpn. J. Sppl. Phys.* **1992,** *31,* 4301.
117. Tagawa, S.; Nagahara, S.; Iwamoto, T.; Wakita, M.; Kozawa, T.; Yamamoto, Y.; Werst, D.; Trifunac, A.D. *Proc. SPIE* **2000,** *3999,* 204.
118. Torok, J.; Re, R. D.; Herbol, H.; Das, S.; Bocharvoa, I.; Paolucci, A.; Ocola, L. E.; Ventrice, C., Jr; Lifshin, E.; Denbeaux, G.; Brainard, R. L. *J. Photopolym. Sci. Technol.* **2013,** *26* (5), 625.
119. Dai, J.; Ober, C. K.; Kim, S. O.; Nealey, P. *Proc. SPIE* **2003,** *5039,* 1164.
120. Gonsalves, K. E.; Thiyagarajan, M.; Choi, J. H.; Zimmerman, P.; Cerrina, F.; Nealey, P.; Golovkina, V.; Wallace, J.; Batina, N. *Microelectron. Eng.* **2005,** *77,* 27.
121. Naulleau, P.; Rammeloo, C.; Cain, J. P.; Dean, K.; Denham, P.; Goldberg, K. A.; Hoef, B.; Fontaine, B. L.; Pawloski, A. R.; Larson, C.; Wallraff, G. *Proc. SPIE* **2006,** *6151,* 61510Y.
122. Hinsberg, H.; Houle, F.; Sanchez, M.; Hoffnagle, J.; Wallraff, G.; Medeiros, D.; Gallatin, G.; Cobb, J. *Proc. SPIE* **2003,** *5039,* 1.
123. Thiyagarajan, M.; Dean, K.; Gonsalves, K. E. *J. Photopolym. Sci. Technol.* **2005,** *18* (6), 737.
124. Allen, R. D.; Brock, P. J.; Na, Y.-H.; Sherwood, M. H.; Truong, H. D.; Wallraff, G. M.; Fujiwara, M.; Maeda, K. *J. Photopolym. Sci. Technol.* **2009,** *22* (1), 25.
125. Bozano, L. D.; Brock, P. J.; Truong, H. D.; Sanchez, M. I.; Wallraff, G. M.; Hinsberg, W. D.; Allen, R. D.; Fujiwara, M.; Maeda, K. *Proc. SPIE* **2011,** *7972,* 797218.
126. Thackeray, J. W.; Jain, V.; Coley, S.; Christianson, M.; Arriola, D.; LaBeaume, P.; Kang, S.-J.; Wagner, M.; Sung, J. W.; Cameron, J. *J. Photopolym. Sci. Technol.* **2011,** *24* (2), 179.
127. Kruger, S.; Hosoi, K.; Cardineau, B.; Miyauchi, K.; Brainard, R. *Proc. SPIE* **2012,** *8325,* 832514.
128. Arimitsu, K.; Kudo, K.; Ichimura, K. *J. Am. Chem. Soc.* **1998,** *120* (1), 37.
129. Kruger, S. A.; Revuru, S.; Higgins, C.; Gibbons, S.; Freedman, D. A.; Yueh, W.; Younkin, T.; Brainard, R. *J. Am. Chem. Soc.* **2009,** *131* (29), 9862.
130. Nishino, K.; Maruyama, K.; Kimura, T.; Kai, T.; Goto, K.; Sharma, S. *Proc. SPIE* **2011,** *7969,* 79692I.
131. Nakagawa, H.; Fujisawa, T.; Goto, K.; Kimura, T.; Kai, T.; Hishiro, Y. *Proc. SPIE* **2011,** *7972,* 79721I.
132. Chandhok, M.; Frasure, K.; Putna, E. S.; Younkin, T. R.; Rachmady, W.; Shah, U.; Yueh, W. *J. Vac. Sci. Technol. B.* **2008,** *26* (6), 2265.
133. Maruyama, K.; Nakagawa, H.; Sharma, S.; Hishiro, Y.; Shimizu, M.; Kimura, T. *Proc. SPIE* **2012,** *8325,* 83250A.
134. Maruyama, K.; Ayothi, R.; Hishiro, Y.; Inukai, K.; Shiratani, M.; Kimura, T. *Proc. SPIE* **2013,** *8682,* 86820B.
135. Gronheid, R.; Solak, H. H.; Ekinci, Y.; Jouve, A.; Roey, F. V. *Microelectron. Eng.* **2006,** *83,* 1103.

136. Gronheid, R.; Fonseca, C.; Leeson, M.; Adams, J. R.; Strahan, J. R.; Willson, C. G.; Smith, B. W. *Proc. SPIE* **2009,** *7273,* 727332.

137. Whittaker, A. K.; Blakey, I.; Blinco, J.; Jack, K. S.; Lawrie, K.; Liu, H.; Yu, A.; Leeson, M.; Yeuh, W.; Younkin, T. *Proc. SPIE* **2009,** *7273,* 72732.

138. Merhari, L.; Gonsalves, K. E.; Hu, Y.; He, W.; Huang, W.-S.; Angelopoulos, M.; Bruenger, W. H.; Dzionk, C.; Torkler, M. *Microelectron. Eng.* **2002,** *63,* 391.

139. Ekinci, Y.; Solak, H. H.; Padeste, C.; Gobrecht, J.; Stoykovich, M. P.; Nealey, P. F. *Microelectron. Eng.* **2007,** *84,* 700.

140. Stowers, J.; Keszler, D. A. *Microelectron. Eng.* **2009,** *86,* 730.

141. Ekinci, Y.; Vockenhuber, M.; Hojeij, M.; Wang, L.; Mojarad, N. *Proc. SPIE* **2013,** *8679,* 867910.

142. Peeters, R.; Lok, S.; Alphen, E.; Harned, N.; Kuerz, P.; Lowisch, M.; Meijer, H.; Ockwell, D.; Setten, E. V.; Schiffelers, G.; Van der Horst, J.-W.; Stoeldraijer, J.; Kazinczi, R.; Droste, R.; Meiling, H.; Kool, R. *Proc. SPIE* **2013,** *8679,* 86791F.

143. Setten, E. V.; Schiffelers, G.; Toma, C.; Finders, J.; Oorschot, D.; van Dijk, J.; Lok, S.; Peeters, R. *Proc. SPIE* **2013,** *8886,* 888604.

144. Tarutani, S.; Tsubaki, H.; Kanna, S. *Proc. SPIE* **2008,** *6923,* 69230F.

145. Bekaert, J.; Van Look, L.; Truffert, V.; Lazzarino, F.; Vandenberghe, G.; Reybrouck, M.; Tarutani, S. *J. Micro/Nanaolith. MEMS MOEMS* **2010,** *9* (4), 043007.

146. Landie, G.; Xu, Y.; Burns, S.; Yoshimoto, K.; Burkhardt, M.; Zhuang, L.; Petrillo, K.; Meiring, J.; Goldfarb, D.; Glodde, M.; Scaduto, A.; Colburn, M.; Desisto, J.; Bae, Y.; Reilly, M.; Andes, C.; Vohra, V. *Proc. SPIE* **2011,** *7972,* 797206.

147. Tarutani, S.; Fujii, K.; Yamamoto, K.; Iwato, K.; Shirakawa, M. *Proc. SPIE* **2012,** *8325,* 832505.

148. Tarutani, S.; Nihashi, W.; Hirano, S.; Yokokawa, N.; Takizawa, H. *Proc. SPIE* **2013,** *8682,* 868214.

149. Finders, J.; Dusa, M.; Vleeming, B.; Hepp, B.; Maenhoudt, M.; Cheng, S.; Vandeweyer, T. *J. Micro/Nanaolith. MEMS MOEMS* **2009,** *8* (1), 011002.

150. Drapeau, M.; Wiaux, V.; Hendrickx, E.; Verhaegen, S.; Machida, T. *Proc. SPIE* **2007,** *6521,* 652109.

151. Rex Chen, K. J.; Huang, W.-S.; Li, W.-K.; Varanasi, P. R. *Proc. SPIE* **2008,** *6923,* 69230G.

152. Hori, M.; Nagai, T.; Nakamura, A.; Abe, T.; Wakamatsu, G.; Kakizawa, T.; Anno, Y.; Sugiura, M.; Kusumoto, S.; Yamaguchi, Y.; Shimokawa, T. *Proc. SPIE* **2008,** *6923,* 69230H.

153. Ando, T.; Takeshita, M.; Takasu, R.; Yoshi, Y.; Iwashita, J.; Matsumaru, S.; Abe, S.; Iwai, T. *Proc. SPIE* **2008,** *7140,* 71402H.

154. Bekiaris, N.; Cerveraa, H.; Dai, J.; Kim, R.-H.; Acheta, A.; Wallow, T.; Kye, J.; Levinson, H. J.; Nowak, T.; Yu, J. *Proc. SPIE* **2008,** *6923,* 692321.

155. Wakamatsu, G.; Anno, Y.; Hori, M.; Kakizawa, T.; Mita, M.; Hoshiko, K.; Shioya, T.; Fujiwara, K.; Kusumoto, S.; Yamaguchi, Y.; Shimokawa, T. *Proc. SPIE* **2009,** *7273,* 72730B.

156. Fujisawa, T.; Anno, Y.; Hori, M.; Wakamatsu, G.; Mita, M.; Ito, K.; Tanaka, H.; Hoshiko, K.; Shioya, T.; Goto, K.; Ogawa, Y.; Takikawa, H.; Kozuma, Y.; Fujiwara, K.; Sugiura, M.; Yamaguchi, Y.; Shimokawa, T. *Proc. SPIE* **2010,** *7639,* 76392Y.

157. Crouse, M.; Uchida, R.; van Dommelen, Y.; Ando, T.; Schmitt-Weaver, E.; Takeshita, M.; Wu, S.; Routh, R. *J. Micro/Nanaolith. MEMS MOEMS* **2009,** *8* (1), 011006.

158. Fonseca, C.; Somervell, M.; Scheer, S.; Kuwahara, Y.; Nafus, K.; Gronheid, R.; Tarutani, S.; Enomoto, Y. *Proc. SPIE* **2010**, *7640*, 76400E.
159. Bencher, C.; Chen,Y.; Dai, H.; Montgomery, W.; Huli, L. *Proc. SPIE* **2008**, *6924*, 69244E.
160. Natori, S.; Yamauchi, S.; Hara, A.; Yamato, M.; Oyama, K.; Yaegashi, H. *Proc. SPIE* **2013**, *8682*, 86821F.
161. Hermans, J. V.; Dai, H.; Niroomand, A.; Laidler, D.; Mao, M.; Chen, Y.; Leray, P.; Ngai, C.; Cheng, S. *Proc. SPIE* **2013**, *8679*, 86791K.
162. Pfeiffer, H. C. *Proc. SPIE* **2008**, *7823*, 782316.
163. Lin, B. J. *J. Micro/Nanaolith. MEMS MOEMS* **2012**, *11* (3), 033011.
164. Wieland, M. J.; Derks, H.; Gupta, H.; van de Peut, T.; Postma, F. M.; van Veen, A. H. V.; Zhang, Y. *Proc. SPIE* **2010**, *7637*, 76371Z.
165. Freed, R.; Gubiotti, T.; Sun, J.; Kidwingira, F.; Yang, J.; Ummethala, U.; Hale, L. C.; Hench, J. J.; Kojima, S.; Mieher, W. D.; Bevis, C. F.; Lin, S.-J.; Wang, W.-C. *Proc. SPIE* **2012**, *8323*, 83230H.
166. Chou, S. Y.; Krauss, P. R.; Renstrom, P. J. *J. Vac. Sci. Technol. B.* **1996**, *14* (6), 4129.
167. Chou, S. Y.; Krauss, P. R.; Renstrom, P. J. *Science* **1996**, *272*, 85.
168. Chou, S. Y.; Krauss, P. R. *Microelectron. Eng.* **1997**, *35*, 237.
169. Colburn, M.; Johnson, S.; Stewart, M.; Damle, S.; Bailey, T.; Choi, B.; Wedlake, M.; Michaelson, T.; Sreenivasan, S. V.; Ekerdt, J.; Willson, C. G. *Proc. SPIE* **1999**, *3676*, 379.
170. Malloy, M.; Litt, L. C. *Proc. SPIE* **2010**, *7637*, 763706.
171. Bates, F. S. *Science* **1991**, *251*, 898.
172. Bates, F. S.; Fredrickson, G. H. *Phys. Today* **1999**, *52*, 32.
173. Park, M.; Harrison, C.; Chaikin, P. M.; Register, R. A.; Adamson, D. H. *Science* **1997**, *276*, 1401.
174. Segalman, R.; Yokoyama, H.; Kramer, E. *Adv. Mater.* **2001**, *13*, 1152.
175. Cheng, J.; Ross, C.; Thomas, E.; Smith, H.; Vansco, G. *Appl. Phys. Lett.* **2002**, *81*, 3657.
176. Cheng, J. Y.; Sanders, D. P.; Truong, H. D.; Harrer, S.; Friz, A.; Holmes, S.; Colburn, M.; Hinsberg, W. D. *ACS Nano* **2010**, *4* (8), 4815.
177. Stoykovich, M. P.; Nealey, P. F. *Mater. Today* **2006**, *9* (9), 20.
178. Kim, H.-C.; Park, S.-M.; Hinsberg, W. *Chem. Rev.* **2010**, *110*, 146.
179. Bencher, C.; Smith, J.; Miao, L.; Cai, C.; Chen, Y.; Cheng, J. Y.; Sanders, D. P.; Tjio, M.; Truong, H. D.; Holmes, S.; Hinsberg, W. D. *Proc. SPIE* **2011**, *7970*, 79700F.
180. Cheng, J. Y.; Rettner, C. T.; Sanders, D. P.; Kim, H.-C.; Hinsberg, W. D. *Adv. Mater.* **2008**, *20*, 3155.
181. Minegishi, S.; Anno, Y.; Namie, Y.; Nagai, T. *J. Photopolym. Sci. Technol.* **2012**, *25* (1), 21.
182. Minegishi, S.; Namie, Y.; Izumi, K.; Anno, Y.; Buch, X.; Naruoka, T.; Hishiro, Y.; Nagai, T. *J. Photopolym. Sci. Technol.* **2013**, *26* (1), 27.
183. Shimokawa, T.; Hishiro, Y.; Yamaguchi, Y.; Shima, M.; Kimura, T.; Takimoto, Y.; Nagai, T. *Proc. SPIE* **2013**, *8682*, 868202.
184. Black, C. T.; Ruiz, R.; Breyta, G.; Cheng, J. Y.; Colburn, M. E.; Guarini, K. W.; Kim, H.-C.; Zhang, Y. *IBM J. Res. Dev.* **2007**, *51*, 605.
185. Rathsack, B.; Somervell, M.; Muramatsu, M.; Tanouchi, T.; Kitano, T.; Nishimura, E.; Yatsuda, K.; Nagahara, S.; Hiroyuki, I.; Akai, K.; Ozawa, M.; Negreira, A. R.; Tahara, S.; Nafus, K. *Proc. SPIE* **2013**, *8682*, 86820K.

CHAPTER 10

POLYNORBORNENE BASED SENSORS FOR "IN-FIELD" HEAVY METAL AND NERVE AGENT SENSING APPLICATIONS

SANTU SARKAR, SOURAV BHATTACHARYA, and RAJA SHUNMUGAM

CONTENTS

ABSTRACT

The main threats to human health from heavy metals are associated with exposure to lead, cadmium, mercury, and arsenic. To provide sensitive and selective sensor systems, fluorophores (e.g., rhodamine and hydroxyquinoline) with binding moieties are attached to norbornene backbone. It is found that hydroxyquinoline modified polynorbornene behaves as a fluorometric responsive molecule for Cd^{2+}, whereas rhodamine based polynorbornene acts as chemodosimeter for As^{3+}. Norbornene based monomers are polymerized by ring-opening metathesis polymerization (ROMP) technique using Grubbs' Catalyst. Using UV and fluorescence method, the sensing ability of the new molecules is confirmed. The response of the sensing event is instantaneous. These novel designs are able to sense ppb level of As^{3+} and Cd^{2+}, very selectively in aqueous environment. Paper strips are prepared to demonstrate in-field application for sensing both heavy metals and nerve agent.

PART A: HEAVY METAL SENSING BY ROMP BASED POLYMER

Photoluminescence techniques are considered to be one of the most effective tools for the detection of a variety of analytes due to their high sensitivity and selectivity. The fine instrument manipulability, commercial availability, lower detection limit, and in vivo detection ability make the techniques extremely attractive in biological and environmental sciences.[1-6]

Metal coordination-induced alteration in photophysical, electrophysical, or biological properties are essential for the related applications in photoluminescent sensing and detection. Most fluorescent probes for metal ions are developed based on metal coordination-induced alteration in emission intensity, lifetime, or wavelength of organic dyes. The fluorescent emission of the metal complex comes from the radiative relaxation of π–π^* excited states of organic dyes. Photoluminescent probes based on Ln^{3+} and transition metal cations of d^6, d^8, and d^{10} configurations are emitted differently from organic dyes, in which the luminescent emission comes from the radiative relaxation of metal-centered (MC, via Ligand to Metal Charge Transfer (LMCT) process) or Metal to Ligand Charge Transfer (MLCT) excited states.[7-11]

Integrating an organic fluorophore as a fluorescence signal reporter with a specific chelator (receptor) is a common approach to design fluorescence probes for metal ions. In this case, metal coordination to the receptor will alter the fluorescence intensity, lifetime, or excitation/emission maxima, reporting the presence of metal cations. The intramolecular interaction between the fluorophore and receptor is essential for the design of these fluorescent probes. Conventional mechanisms such as photo-induced electron transfer (PET),[12,13] photo-induced charge transfer (PCT),[14] fluorescence resonance energy transfer (FRET),[15] and photo-induced excimer/exciplex formation have been frequently adopted for the construction of probe molecules.[12,13] On the other hand, a number of new rationales, such as metal ion coordination inhibited excited-state intramolecular proton transfer (ESIPT) and aggregation-induced emission (AIE), have also been explored to devise probes.[16]

PET is the most commonly employed mechanism for the design of fluorescence probes.[13] Conventional PET probes for metal cations contain three parts: fluorophore, spacer, and ionophore (Figure 10-1). Ionophores are normally electron donors (e.g., amino-containing group), while fluorophores are electron acceptors. For free probe, the electron occupying the highest occupied molecular orbital (HOMO) of the fluorophore can be promoted to the lowest unoccupied molecular orbital (LUMO) by absorbing an excitation photon. If the energy of the ionophore HOMO is just higher than that of the fluorophore, the electron of the ionophore HOMO will transfer to the HOMO of the excited fluorophore through space, which blocks the emission transition of the excited electron occupying the fluorophore LUMO to fluorophore HOMO. This fluorescence quenching effect is termed as PET.

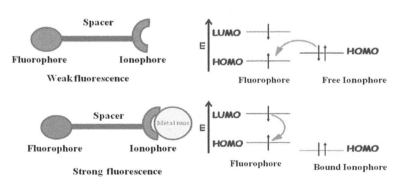

FIGURE 10-1   PET fluorescent probe for metal cations and their "turn-on" sensing mechanism.

Internal charge transfer (ICT) fluorophores are a conjugated system of electron-donating/electron-withdrawing groups (donor/acceptor, D/A). The ICT effect decreasing the basicity of the donor amine, and therefore it would be applicable in biological environment as the system is pH independent.

Metal coordination to the donor of an ICT fluorophore decreases the HOMO energy and induces the hypsochromic shift of excitation or emission maxima, whereas a reverse change is observed when there is metal coordination to its acceptor. Ratiometric probes are extremely useful in sensing because self-calibration effect of two excitation/emission bands can eliminate the interference of photobleaching and deviated microenvironments, local probe concentration, and experimental parameters. Therefore, these probes are extremely useful for the determination of targeted metal ions in biological system (Figure 10-2).[17,18] PCT mechanism is therefore, an efficient strategy for the construction of ratiometric metal ion probes.

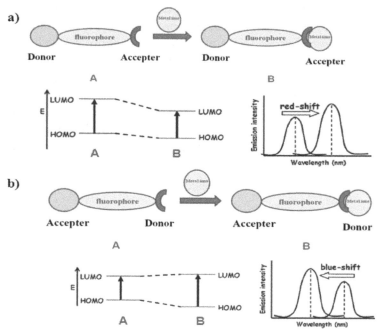

FIGURE 10-2   PCT fluorescent probes for metal cations and their ratiometric sensing mechanisms.

Fluorescence resonance energy transfer, is a nonradiative process which is also known as Förster resonance energy transfer (FRET) where the resonance energy transfers from an excited state of a donor fluorophore (D-F) to the ground state of an acceptor fluorophore (A-F) via a nonradiative dipole–dipole interaction. FRET occurs through the effective interaction distance (Förster distance) typically is 6 nm. Typically, FRET requires the spectral overlap between the emission band of donor and the absorption band of acceptor.[19] Ono has pioneered the development of FRET sensors for heavy metal detection.[20,21] Ono has linked an organic dye (fluorophore) and a quencher to two ends of a molecular system to form a FRET sensor (Figure 10-3).

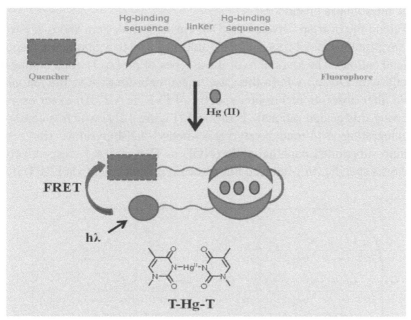

FIGURE 10-3 Sketch showing the molecular beacon-based FRET sensor. The presence of Hg^{2+} ions leads to quenching of fluorescence by the emission of fluorescein.

The presence of Hg^{2+} ions induced a hairpin structure due to the formation of the $T-Hg^{2+}-T$ sandwich structure, which brought the quencher close to the fluorophore and thus enabled the FRET process, leading to fluorescence quenching of the organic dye. This sensor achieved a limit of detection (LoD) of 40 nM towards Hg^{2+} in a buffer solution and excellent selectivity toward Hg^{2+} ions coexisting with other metal ions.

FRET sensor can also be created by linking a nonemissive moiety such as rhodamine spirolactam derivatives to a donating fluorophore. After binding to metal ions FRET will be switched on and it will convert the nonemissive acceptor into a fluorophore. For instance, compound **1** is a ratiometric fluorescent Cr^{3+} probe based on FRET, in which 1,8-naphthalimide and rhodamine derivatives are selected as the D-F and A-F, respectively. There is no energy transfer between the two because of the nonemissive nature of rhodamine spirolactam derivatives. An efficient ring-opening reaction induced by

Cr^{3+} generates fluorescent rhodamine, which induces an effective FRET process from naphthalimide to switch-on rhodamine.[22] In this case, the probe normally functions as an emission ratiometric probe. If the metal-induced switch on is devised for D-F, then the probe is normally an excitation ratiometric one. It is clear that the selectivity comes from the selective response of the fluorophore precursor.

Tuning the overlap between the D-F emission spectrum and A-F absorption spectrum via metal coordination could also lead to a ratiometric response to metal cations. The ionophore of these types of probes is often conjugated directly with D-F or A-F. In this case, metal coordination to the ionophore would alter absorption/emission spectra of D-F or A-F. An exact example following this design rationale is probe **2** (Figure 10-4) which is composed by integrating 5-(4-methoxystyryl)-5′-methyl-2,2′-bipyridine (bpy) with diamino-substituted naphthalimide (NDI) as D-F and A-F, respectively. It displays a specific Zn^{2+}-induced fluorescence enhancement via FRET. It was

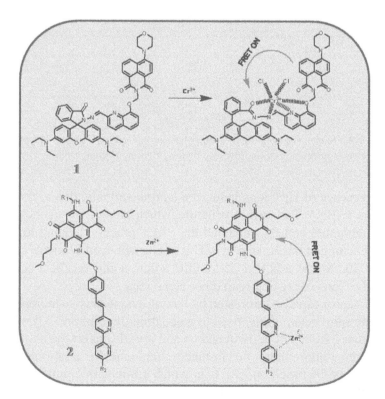

R₁, R₂ = Alkyl group

FIGURE 10-4   Fluorescent probes **1** and **2** based on a FRET mechanism.

proposed that there is very small overlap between the emission spectrum of D-F and the absorption spectrum of A-F before coordination to Zn^{2+}. The coordination makes its emission band undergo a red-shift which enables a significant spectral overlap between the emission of the bpy/Zn^{2+} complex and the absorption band of NDI. This facilitates the occurrence of FRET.

Most transition metal ions have electron withdrawing ability and act as Lewis acids, which can promote or catalyze certain chemical reactions. Therefore, chemodosimeters for metal cations can be constructed based on a mechanism of metal-induced formation of an emissive fluorophore from a nonfluorescent precursor.[23] In fact, this approach is frequently used in the design of turn-on probes for metal cations with an emission quenching nature, such as Hg^{2+}, Cu^{2+}, and Fe^{3+}/Fe^{2+}. Most of these probes are based on metal induced-opening of the nonfluorescent spirolactam of xanthenes. The first example based on this reaction is reported in 1997 by Czarnik and co-workers, on a rhodamine B based fluorescent probe for Cu^{2+} (Figure 10-5).[24,25] Other metal coordination induced reactions, such as the hydrolysis of esters/hydrazides/ethers, rearrangement, and ring formation

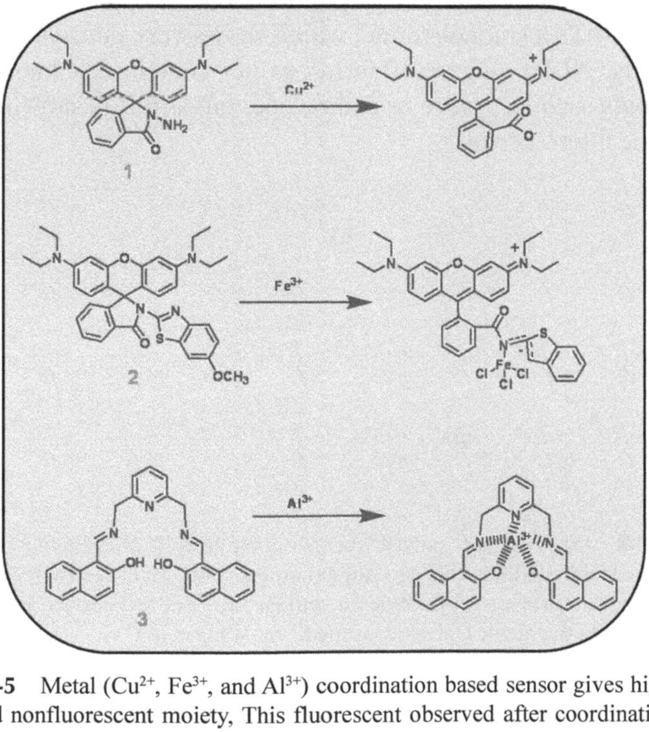

FIGURE 10-5 Metal (Cu^{2+}, Fe^{3+}, and Al^{3+}) coordination based sensor gives high emission intensity and nonfluorescent moiety, This fluorescent observed after coordination of metal with the molecule.

have also been successfully employed for the development of other chem-dosimeters.[23] In the second example, coordination of Fe(III) with the N atom of the benzothiazole moiety is accompanied by the transfer of electrons of the benzothiazole, resulting in the opening of the spiro-ring. Benzothiazole moieties selectively and sensitively sensing Fe(III) over the other metal ions.[26] The third example demonstrates a turn-on and reversible Schiff-base fluorescence based sensor for Al^{3+} ion through chelation enhanced fluorescence coordination with nitrogen and oxygen of the corresponding probe.[27]

10.1 POLYMER BASED SENSORS

Herein we are demonstrating two examples regarding Fe(III) sensing.

Examples given in Figure 10-6 demonstrates the sensing of Fe(III) in aqueous environment spontaneously and selectively through a turn-off method, but making of these molecule is really difficult and sensing mechanism is not at all simple in nature.[28,29] But in this case, a very simple molecule **NDP1** just by coupling of exo-5-norbornene carboxylic acid with alcohol derivative of diethyl(hydroxylmethyl)phosphonate in presence p-Toluene-sulfonic acid (pTSA) monohydrate, which shows very unusual strong emission property. **NDP1** (Figure 10-7a) does not contain any known organic chromophores such as pyrene or anthracene, but still it is showing a strong unusual blue fluorescence.[30]

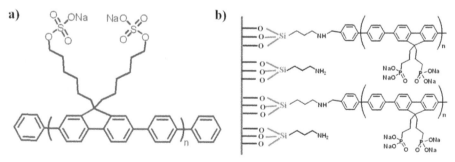

FIGURE   10-6   (a)   Polymer   poly(9,9-bis(6′-sulfate)hexyl)   fluorene-alt-1,4-phenylene sodium salt used for Fe(III) sensor in aqueous environment by Iyer et al.[28] (b) Poly (9,9-bis(30-phosphatepropyl)fluorene-alt-1,4-phenylene) sodium salt (PFPNa) surface is also used for Fe(III) sensing in both organic and aqueous media by Wang et al.[29]

The rare phenomenon of **NDP1** has prompted us to design **NDP2** and **NDP3** molecules to explore the reason behind the unusual fluorescence

(Figure 10-7b). From the DFT calculations, it is observed that the reason behind fluorescence of **NDP1** is the difference between the positions of HOMO and LUMO wave functions that lead to the prevention of nonradiative relaxation pathways. However, in the case of **NDP2** the difference in HOMO and LUMO is decreased drastically, hence weak emission is observed; but in case of **NDP3**, both HOMO and LUMO wave functions are located on norbornene, which results in an efficient nonradiative relaxation pathway and hence no emission is observed.

a)   b)

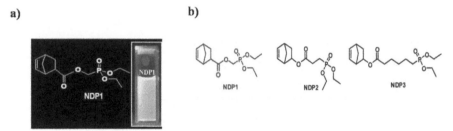

FIGURE 10-7   (a) Sensor molecule NDP1 and it's unusual blue emission. (b) Schematic representation of the monomers **NDP1**, **NDP2**, and **NDP3**.

For the comparative fluorescence, UV–visible spectroscopy of **NDP1**, **NDP2**, and **NDP3** were studied which shows that maximum absorbance values are found to be 350 nm for **NDP1**, 320 nm for **NDP2**, and 306 nm for **NDP3** in MeOH (Figure 10-8a). The fluorescence spectrum shows a strong emission for **NDP1** with emission maximum at 430 nm in MeOH. **NDP2** shows an emission maximum with a lesser intensity and **NDP3** shows a very weak emission (Figure 10-8b). The molar absorptivity (ε) of **NDP1**, **NDP2**, and **NDP3** are calculated as 0.257×10^{-2}, $0.069 \; 10^{-2}$, and 0.224×10^{-2} M^{-1} cm^{-1} respectively. Quantum yield (F) 0.212 of compound **NDP1** is calculated from its absorption and emission spectrum. Similarly, the quantum yield of **NDP2** (0.08) as well as **NDP3** (0.02) were calculated from its absorption and emission spectrum using quinine sulfate as standard. To investigate the unusual emission, the HOMO and the LUMO of the optimized molecular geometries are calculated at B3LYP/6-31G level using the Gaussian 03 package (Figure 10-9). For **NDP1**, it is observed that the difference between the positions of HOMO and LUMO wave functions led to the prevention of nonradiative relaxation pathways. This hypothesis is supported by the optimized molecular geometries of **NDP2** and **NDP3**. It is observed that the difference in HOMO and LUMO decreased drastically, hence weak emission is observed in the case of **NDP2**; but in case of **NDP3**, both HOMO and

LUMO wave functions are located on norbornene itself, resulting in a nonradiative relaxation pathway and hence no emission is observed. These results confirmed the importance of the interaction between the norbornene double bond and the phosphonate functionality for the observed unusual fluorescence. To the best of our knowledge, this is the first report on the unusual fluorescence of molecules that contain the phosphate ester functionality.

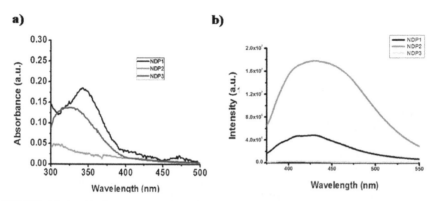

FIGURE 10-8 (a) Comparative absorbance spectra of norbornene derived phosphonate molecules **NDP1** to **NDP3** in MeOH. (b) Comparative fluorescence spectra of norbornene derived phosphonate molecules (**NDP1** to **NDP3**) (concentration 2 mg mL^{-1}).

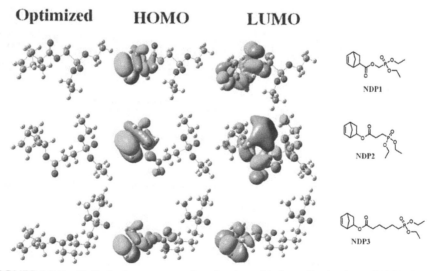

FIGURE 10-9 Optimized structure and molecular orbital amplitude plots of HOMO and LUMO energy levels of (a) **NDP1**, (b) **NDP2**, and (c) **NDP3** calculated with the use of B3LYP/6-31G basis set.

Having studied the unusual emission using control molecules, the interesting phenomenon is further explored with **NDP1** molecule. The fluorescence measurements of **NDP1** in THF are done starting from the lowest concentration (5 mM), where there is very weak emission, but it started to show stronger emission with an increasing concentration of **NDP1**, as shown in Figure 10-10 suggesting the formation of aggregates which could be due to high local concentration of the species.

a)   b)

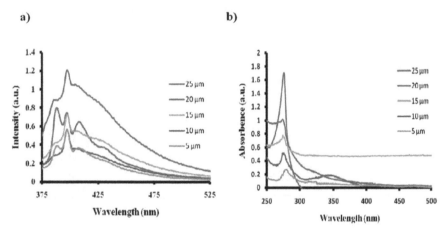

FIGURE 10-10   (a) Fluorescence and (b) absorption spectra measured with increasing concentration of **NDP1** in THF starting from 5 to 25 μM.

Further support for this hypothesis is from three other observations. Similarly, the same solutions are used to measure the absorbance of **NDP1**, where a similar trend of increasing absorbance with increasing concentrations is observed, which supported the fluorescence studies (Figure 10-10). The formation of aggregates with increasing concentration of **NDP1** is also ascertained by dynamic light scattering (DLS) and transmission electron microscope (TEM) analysis of these solutions. The DLS data of a 5 mM solution of **NDP1** in THF showed particles of 10 nm (r, nm) in size (Figure 10-11a) while the 25 mM solution had particles of about 60 nm (r, nm) in size (Figure 10-11c), which clearly demonstrated the formation of larger objects due to aggregation of **NDP1** with increasing concentration. These aggregates are further investigated by TEM analysis. When the solution is drop cast onto a carbon coated TEM grid, the grid from the 5 mM solution does not show larger objects while 25 mM solution shows individual objects with diameters of approximately 120 nm, as shown in Figure 10-11b and d, respectively. The importance of the intimate spatial arrangement provided by the aggregation and hence an increase in intensity at higher concentrations is clear from these studies. Finally, the aggregation behavior of **NDP1** is tested in both nonpolar solvents (DCM) and polar solvents (methanol). It is observed that the aggregation behavior is greater in methanol and less in DCM in comparison with THF. Due to this, an increase in fluorescence intensity is observed in methanol while a much less emission is observed in DCM. Hence, the difference in fluorescence spectra of **NDP1** in methanol and THF can be clearly attributed to its greater aggregation behavior in methanol.

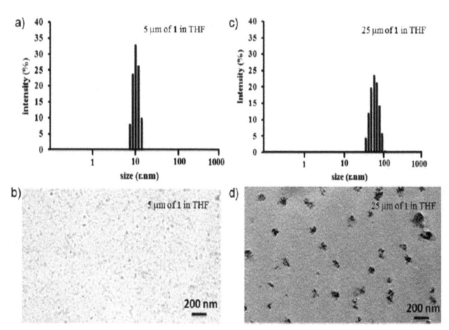

FIGURE 10-11   (a) DLS data of 5 mM solution of NDP1 in THF. (b) TEM micrograph of 5 mM solution of **NDP1**. The solution is drop cast onto a carbon coated TEM grid. (c) DLS data of 25 mM solution of NDP1 in THF. (d) TEM micrograph of 25 mM solution of **NDP1**. The solution is drop cast onto a carbon coated TEM grid (RuO_4 is used as staining agent).

The optical events generated by inducing changes to the optical properties of fluorophores through associating metal ions with the molecule **NDP1** is chosen to study the viability of metal binding since the phosphate ester terminated side chains would facilitate its association with metal ions. Salts of Pb^{II}, Ba^{II}, Mn^{II}, Hg^{II}, Fe^{II}, Fe^{III}, Co^{II}, Ni^{II}, Cu^{II}, and Cd^{II} (with concentrations of 0.015 M in water) are titrated with **NDP1** (5×10^{-5} M in THF) and monitored by fluorescence spectroscopy (Figure 10-12). The addition of $FeCl_3$ in water to a solution of **NDP1** in THF/water (1:1) caused the immediate quenching.

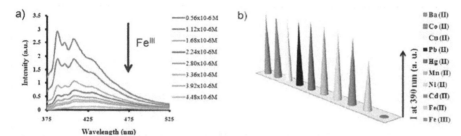

FIGURE 10-12 Emission spectra of (a) **NDP1** as a function of increasing Fe(III) concentration. The spectra are collected in THF with an excitation wavelength (Ex) of 350 nm. (b) Bar diagram of Fe(III) sensing where **NDP1** selectively sense in presence of other metal ions.

In conclusion, herein synthesis of **NDP1** molecule which has no traditional organic chromophore showing unusual blue fluorescence, clearly explained through DFT calculation and AIE. Because of the presence of terminal phosphate ester functionality it selectively sense Fe(III) over other metal ions in aqueous environment.

10.1.1 CADMIUM SENSOR PART

Our development of sensor system still continued as we are looking for a cadmium sensor system. Cadmium, being used in many things of our daily life such as batteries, fertilizers, military affairs, agriculture, and metallurgy, its exposure to human health is quite possible. Excessive exposure of cadmium leads to health hazards like lung, breast, and prostate cancer.[31] So, detection of cadmium is of great importance to the society. Our aim is to develop a polymer based sensor system that could be used for in-field detection as paper strip model. Though some literature examples of cadmium sensing are available, polymer based system is yet to be developed.[32] The most general problem of cadmium sensing is the selectivity of the sensor between cadmium.[33] They possess similar chemical properties being in the same group.

Here we designed a norbornene based sensor system that can act as a selective as well as sensitive detector for cadmium.

First we developed our scheme based upon existing literature technique as shown in Scheme 10-1. Here compound **3** behaved as fluorometric cadmium sensor reported by Yao et. al.[31g] 8-hydroxyquinoline along with 1,3,4 oxadiazole captured cadmium in aqueous media (Figure 10-13).

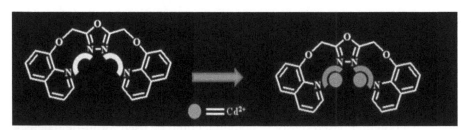

SCHEME 10-1   Our approach to develop a norbornene based cadmium sensor molecule.

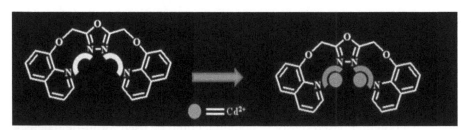

FIGURE 10-13   Schematic demonstration of cadmium sensing by Yao et al.[31g]

A prominent fluorescence enhancement of **3** is found after binding to cadmium.

This detection technique is selective to cadmium only. Based upon that, we planned to replace the one hand of 8-hydroxyquinoline of **3** with norbornene functionalized alcohol **4** which leads to **5**. **5**, if formed, expected to act as a cadmium sensor and we could polymerize that through a ring-opening metathesis polymerization (ROMP). But after our several effortless trials we could not make it. The possible reason is to control the one side replacement of **3**.

Continuing with our intense search for cadmium sensor we developed a simple molecule, that is, norbornene attached 8-hydroxyquinoline (**N8HQ**) which fulfilled our aim.[34] Here, we have shown the scheme of synthesis of **N8HQ** and its polymer (Scheme 10-2). As synthesized norbornene acid is coupled to 8-hydroxyquinolne through N,N'-Dicyclohexylcarbodiimide (DCC) coupling. Formation of product is confirmed through NMR and mass spectroscopy technique.

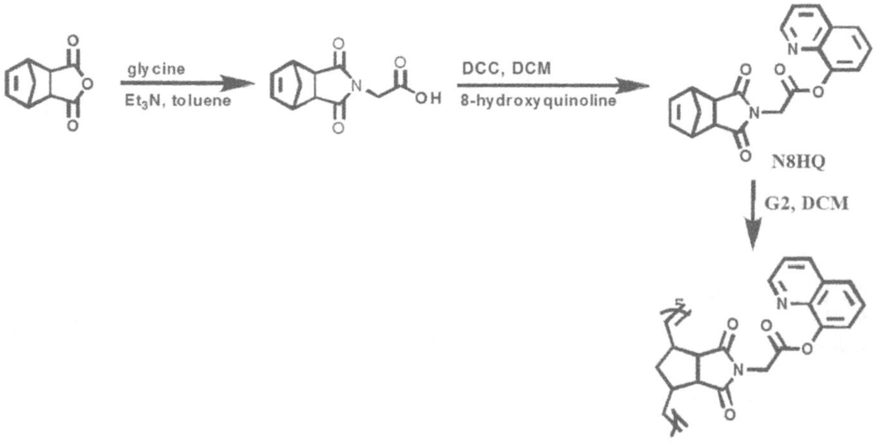

SCHEME 10-2   Schematic representation of synthesis of **N8HQ** and its polymer **PN8HQ**.

Sensing behavior of **N8HQ** towards cadmium ion is explored in MeOH-H_2O (1:1) system. Absorbance spectrum of **N8HQ** exhibited three bands at 288, 300, and 313 nm. While titrating **N8HQ** with Cd^{2+}, two new peaks appeared gradually at 335 and 373 nm with an isobestic point at 326 nm (Figure 10-14a). A plot of absorbance of **N8HQ** at 373 nm vs Cd^{2+} emerged a linear behavior initially and then remained constant which proposed a 1:1 binding between **N8HQ** and Cd^{2+}. It was also proved by Job's plot (Figure 10-14c).

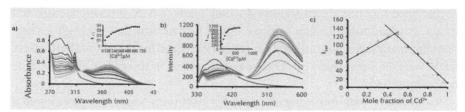

FIGURE 10-14 (a) Absorbance spectra of **N8HQ** upon titration with Cd²⁺ in in MeOH-H₂O (1:1). Inset: titration curve based on absorbance at 373 nm. (b) Emission spectra of **N8HQ** (4 × 10⁻⁴ M) in MeOH-H₂O (1:1) with gradual addition of Cd²⁺ (4 × 10⁻⁴ M). Inset: titration curve based on emission at 540 nm. λ_{Ex}: 310 nm. (c) Job's plot showing 1:1 binding stoichiometry between **N8HQ** and Cd²⁺ in MeOH-H₂O (1:1).

From emission spectrum of **N8HQ**, weak emission is observed at 372 nm with excitation at 310 nm. With addition of cadmium a new peak is found to rise at 536 nm while peak at 372 nm decreased. A huge red shift of 164 nm is found with cadmium addition (Figure 10-14b). We hypothesized that the huge red shift is might be for two reasons: (1) the prevention of nonradiative relaxation pathways of N lone electron pairs in the molecule by Cd²⁺ binding and (2) PET between 8-hydroxyquinoline and norbornene motif (Figure 10-15a).

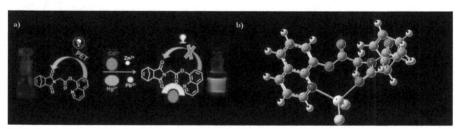

FIGURE 10-15. (a) A cartoon representation for the proposed binding site and PET. (b) Calculated energy minimized structure of N8HQ with proposed binding sites for CdCl₂

The response behavior is noticed under UV lamp. **N8HQ** is found to be colorless while cadmium addition made it green color under UV lamp.

Selectivity of **N8HQ** towards other metal ions especially zinc ions is tested. **N8HQ** is found to be very selective to cadmium among other metals. Only cadmium but no other metal made the red shift in emission spectroscopy. The bar chart clearly showed the selectivity (Figure 10-16a). Under UV lamp also it is observed that only cadmium made the green color after addition to **N8HQ** (Figure 10-15a).

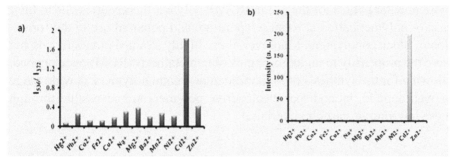

FIGURE 10-16 (a) Emission ratio at 536 and 372 nm (I_{536}/I_{372}) of **N8HQ** in the presence of indicated metal ions in MeOH:H$_2$O (1:1). Ex: 310 nm. (**b**) A representation of fluorescence intensity of **PN8HQ** film on quartz slide in presence of Cd^{2+} and different metal ions. Λ_{Ex}: 310 nm.

DFT calculation is carried out to get an explanation for this binding. Calculation on **N8HQ** without Cd^{2+} at the B3LYP/6-31G(d) level using the Gaussian 09 package is done. Structure optimization with Cd^{2+} is done for two proposed structures, **I** and **II**. In proposed structure **I**, the coordination of Cd is through carboxylate oxygen while in the case of **II**, the coordination is through norbornene anhydride oxygen. It could be seen from the optimized structure that a Cd^{2+} ion suitably fit between the nitrogen of 8-hydroxyquinoline and norbornene anhydride oxygen (Figure 10-15b).

ROMP is a chain growth polymerization process where a mixture of cyclic olefins is converted to a polymeric material (Figure 10-17). The mechanism of the polymerization is based on olefin metathesis, a unique metal-mediated carbon–carbon double bond exchange process (Figure 10-18).[35] As a result, any unsaturation associated with the monomer is conserved as it is converted to polymer. This is an important feature that distinguishes ROMP from typical olefin addition polymerizations (e.g., ethylene- polyethylene). In ROMP, strained rings are being opened by a ruthenium carbene-catalyzed reaction with a second alkene following the mechanism of the Cross Metathesis. The driving force is the relief of ring strain. As the products contain terminal vinyl groups, further reactions of the Cross Metathesis variety may occur. Therefore, the reaction conditions (time, concentrations) must be optimized to favor the desired product. We are doing ROMP in our laboratory by using simple norbornene materials. Norbornene is a bicyclic olefin, possess ring strain so the molecule contains high reactive double bonds. It is prepared via the Diels–Alder reaction of cyclopentadiene and ethylene. Functionalized norbornenes exists as two isomers endo (major) and exo (minor) (Figure 10-19). Derivatives of norbornene, tend to be a more popular choice for the use in ROMP polymer therapeutics due to their ease of polymerization at room temperature and potential use in the formation of block copolymers. Moreover, these highly strained monomers do not have the propensity to undergo ring closing metathesis (RCM) once opened, allowing for the synthesis of high molecular weight polymers. A wide range of well defined, monodisperse, bioactive polymers are accessible through these few simple starting materials.[36,37]

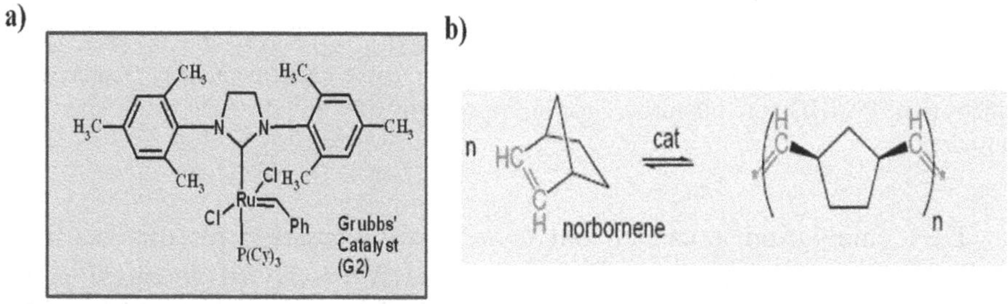

a)   b)

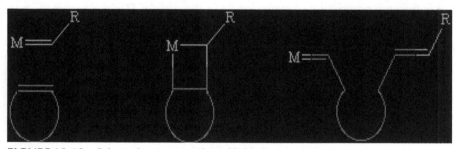

FIGURE 10-17   Schematic reprsentation (a) Grubbs' catalyst and (b) ROMP.

FIGURE 10-18   Schematic representation of ROMP.

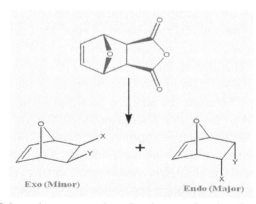

FIGURE 10-19   Schematic representation of norbornene functionalization.

After the successful explanation of the binding towards cadmium of **N8HQ** we proceeded further to develop polymer based sensor. **N8HQ** is polymerized by using Grubbs' second generation catalyst[38] to form **PN8HQ**. Characterization of polymer is done with NMR and GPC technique. Polymer coated paper strip model is developed and tested under UV lamp for cadmium sensing. Under UV lamp colorless paper strip showed green emission after dipping into cadmium containing solution. In paper strip level also it is sensitive to cadmium only as showed in Figure 10-20.

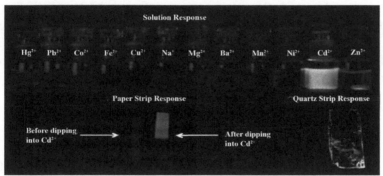

FIGURE 10-20 Image of solutions of **N8HQ**-metal complexes of various metal ions in MeOH-H₂O (1:1) taken under UV lamp (Top). Demonstration of the Cd^{2+} sensing in water by polymer (**PN8HQ**) coated paper strip (bottom left). **PN8HQ** film on quartz slide after dipping into cadmium solution exposed under UV lamp (bottom right).

10.1.2 COLORIMETRIC SENSOR

Though fluorometric sensors are more sensitive than colorimetric sensors, colorimetric sensors are preferable for practical use. As a signaling event, color changes can be detected by the naked eye and hence can be widely used. These chemosensors are generally developed with a receptor linked with a chromophore in such a way that binding of a specific analyte to the receptor produce color change of the chromophore. This color event can occur either due to the conformational changes in the receptor or formation of a charge-transfer complex. Most of the metal sensors are fluorometric and very few are colorimetric as found in the literature. Large part of these colorimetric sensors is for copper and mostly based upon ICT principle.[39] Here, some examples are discussed.

As a only colorimetric sensor Govindaraju and Maity have reported in 2011 copper sensor ligand 4-(diethylamino)salicylaldehyde conjugated urea/thiourea.[40]

The (diethylamino)phenol moiety in ligands **1** and **2**, acts as the chromophore core. The extended π conjugation and the stronger electron-withdrawing ability of the ketone/thioketone group produces an ICT mechanism. These highly conjugated compounds **1** and **2** are able to sense Cu^{2+} by colorimetry showing red shifts in their absorbance spectra in CH_3CN and aqueous medium. These sensors are selective to copper in presence of other transition metal ions. A characteristic near-infrared (NIR) shift is found in case of ligand **2** and can be used for biological imaging. Different spectral behavior of ligands **1** and **2** is explained from their XRD data. Compound **2** got additional stability by conjugation as well as coordination with copper (Figure 10-21).

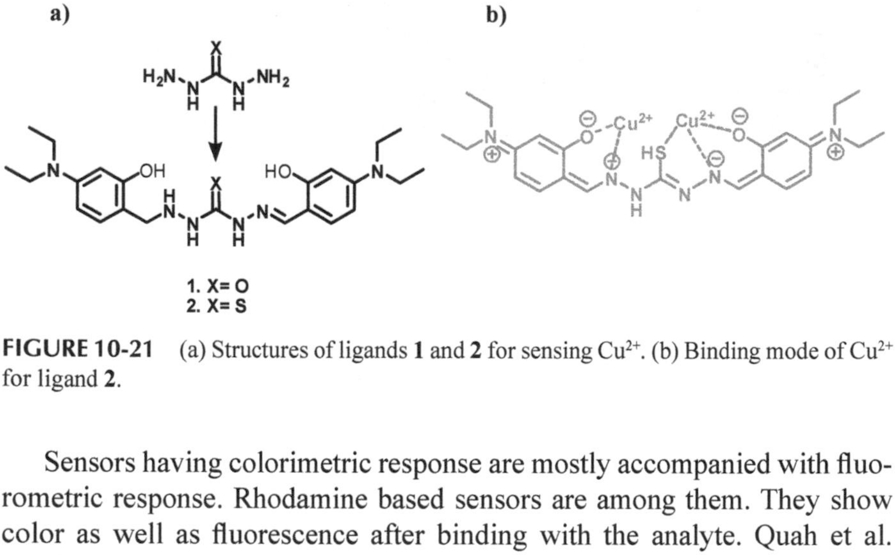

FIGURE 10-21   (a) Structures of ligands **1** and **2** for sensing Cu^{2+}. (b) Binding mode of Cu^{2+} for ligand **2**.

Sensors having colorimetric response are mostly accompanied with fluorometric response. Rhodamine based sensors are among them. They show color as well as fluorescence after binding with the analyte. Quah et al. developed a rhodamine derivative bearing an 8-aminoquinoline moiety that selectively sense Pb^{2+} colorimetrically and fluorometrically.[41]

Here, the role of the quinoline nitrogen is very effective to bind Pb^{2+}. With addition of Pb^{2+} a pink color is developed which is due to the spirolactum ring opening of the rhodamine, as shown in Figure 10-22. In absorbance spectra strong absorption band at 540 nm is found to be increasing with gradual addition of Pb^{2+} to **3** in aqueous solution. In emission spectra also a turn-on response is found with addition of Pb^{2+} which is explained as a chelation enhanced fluorescence (CHEF) effect. Among other transition metal this sensing system selectively detects Pb^{2+}. This is proved to be a reversible sensing system as addition of S^{2-} to the pink colored metal-ligand complex produces a colorless solution.

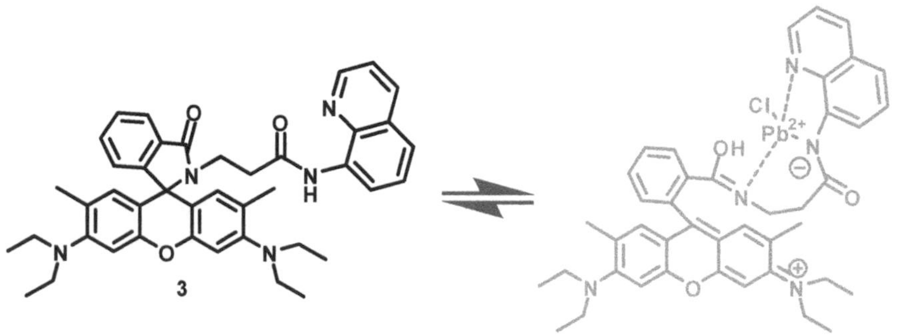

FIGURE 10-22 Pb^{2+} sensing and binding modes by Quah et al.[41]

Qian et al. synthesised a napthalimide based probe for the detection of Cu^{2+} that produce fluorescent and colorimetric response.[42] This probe **4** produces a large red shift in absorbance spectra with addition of Cu^{2+} in HEPES buffer. Among all other transition metal ions, selectively it detects Cu^{2+} and produce primrose yellow to pink color. Even in the emission spectra also **4** shows red shift after binding with Cu^{2+}. The strategy used to produce this red shift in spectra is well explained. In structure **4** the deprotonation occurs in the two secondary amines conjugated to 1,8-naphthalimide. Hence, the electron donating ability of these nitrogen to naphthalene ring would be enhanced and that produces the red shift in absorbance and fluorescence spectra (Figure 10-23).

Here we are mainly focusing on development of a colorimetric sensor for the toxic metal Arsenic(III). Arsenic is a well-known toxic metal and is present mainly as oxyanion compounds in groundwater. Arsenite (As^{III}) interferes with the function of enzymes by bonding to the –OH and –SH groups leading to the inactivation of the enzymes. Therefore, our ultimate goal is (1) to attach ligand which have specific affinity towards arsenic along with a dye in a norbornene moiety, (2) making its polymer, and (3) finally sense it either colourimetrically or fluoremetrically. From literature survey, the stability constant of arsenic-thiol is more than arsenic-oxygen.[43–45] This prompted us to choose thiol functionlized ligand to sense arsenic. To achieve the goal we prepared unique redox-based "in-field" sensor of As(III).[46] Very few examples are available in the literature regarding spectrophotometric determination of arsenic in water.

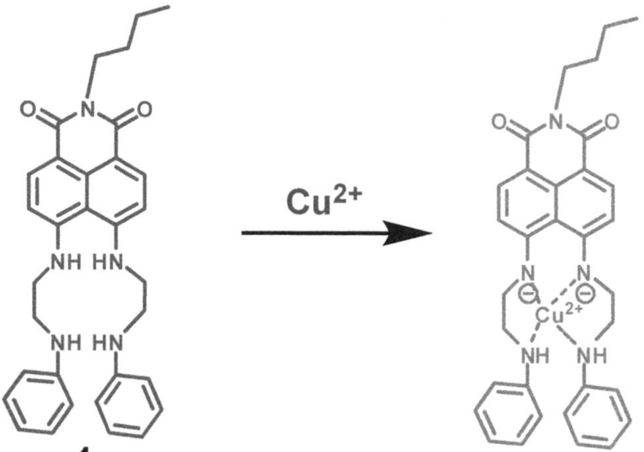

FIGURE 10-23   N-butyl-4,5-di[2-(phenylamino)ethylamino]-1,8-naphthalimide probe synthesised by Qian et al.[42] for Cu^{2+} sensing.

Pillai et al. (2000) reported a new simple and reliable spectrophotometric method to determine total arsenic in environmental and biological samples. It involves bleaching the pinkish-red dye rhodamine-B (measured at 553 nm) by the action of iodine released from the reaction between potassium iodate and arsenic in a slightly acidic medium. But the main draw back is that they did not mention the specific color change after sensing As(III).[47]

A sensitive and selective fluorescence sensor for the detection of Arsenic(III) in organic media by Ezeh and Harrop have not discussed this sensing method in aqueous environment.[48]

Roy et al. invented selective detection of arsenic in groundwater by gold nanoparticles in a simple colorimetric and ultrasensitive DLS assay. But main drawback is that without the need of special equipment they cannot sense As(III). Moreover nanoparticles could not be stabilized longer time in aqueous enviorenment.[49]

Banerjee et al. (2012) discussed the AS-prepared gold cluster-based fluorescent sensor for the selective detection of As(III) ions in aqueous solution. But this water-soluble fluorescent gold clusters (AuCs) only stabilized at room temperature using a dipeptide L-cysteinyl-L-cysteine as a capping ligand. Moreover without the help of instrument they can not conclude wheather sensing is occurring or not.[50]

Before going to the discussion of unique redox-based "in-field" sensor of As(III), nature of the failure reactions regarding sensing As(III) in norbornene backbone (Scheme 10-3) are discussed.

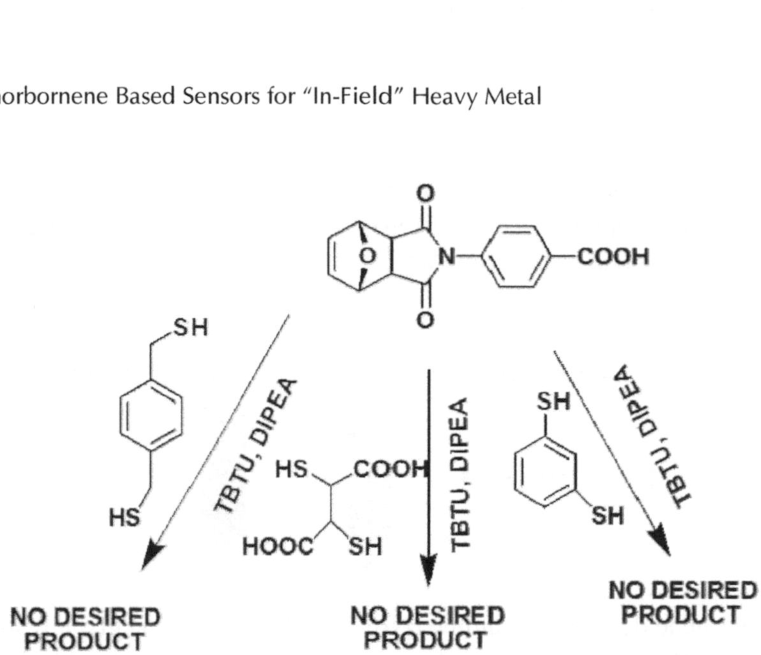

SCHEME 10-3   Failure reactions of thiol-functionlized ligand with nadic acid

We failed to attach the above mentioned thiol functionlised ligand because exo-nadic acid is not soluble in ethyl acetate and the reaction is only reported in only that solvent. Next we successfully attached the above mentioned thiol-functionalized ligand with norbonene motif and which is thoroughly characterized with ^{1}H and ^{13}C NMR spectroscopic technique (Scheme 10-4). But drawback of this polymer is not forming because of crosslinking (free thiol functionality) and the absence of dye which is not giving any kind of response through UV–visible and fluorescence spectroscopy.

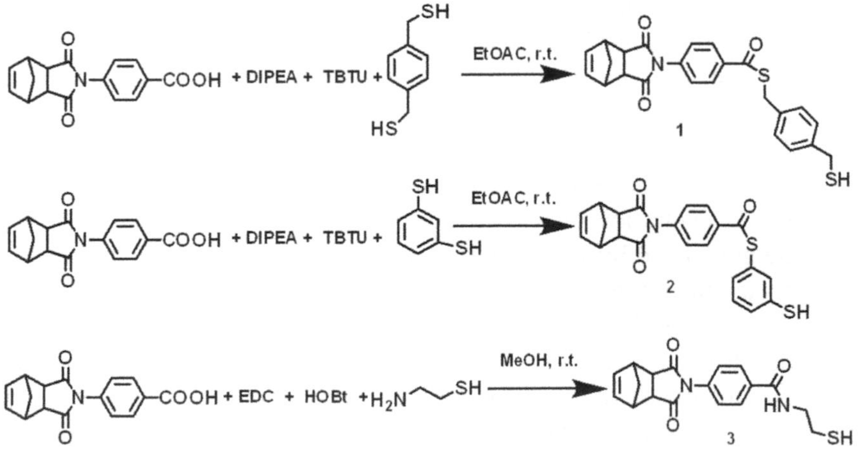

SCHEME 10-4   Synthesis of thiol-functionlized norbornene monomer.

To overcome this problem we have chosen rhodamine as dye and protected thiol functionality for ROMP. Herein we are discussing about redox-based "in-field" sensor for As(III).

Sensor Components:

R + KIO₃ + HCl = 1, where R = Nor-Rh or PNor-Rh

Sensor Response:

1 + As(III) ⟶

Colorimetric   Fluorimetric

Norbornene derived rhodamine (Nor-Rh) (Figure 10-24) givve response against As(III) in the presence of potassium iodate and hydrochloric acid which is monitored through UV–visible and fluorescence spectroscopy. Sensing properties of the Nor-Rh monomer towards As(III) are explored in presence of KIO₃ and HCl. A stock solution of Nor-Rh is prepared in methanol–water mixtures, whereas As(III) and KIO₃ are prepared in water. First we monitored the response of sensor 1 with As(III) by UV–visible spectroscopy. As soon as a drop of HCl is added to the mixture of Nor-Rh and KIO₃ an immediate pink color is developed. UV spectroscopy confirmed the characteristic absorbance band of rhodamine at 560 nm is due to spirolactam ring opening. Then the As(III) solution is gradually added dropwise to the sensing solution 1. A new absorbance band is observed at 288 and 356 nm with increasing intensity whereas the absorbance at 560 nm gradually decreased with increased concentration of the As(III) solution. A characteristic isobestic point is observed at 520 nm and the color of the solution became brown. The observed color change is very unique and interesting since the very strong pink color of rhodamine completely vanished due to the As(III) addition (Figure 10-25).

The emission properties of sensing solution 1 are explored with the increasing concentration of As(III), the change in the emission of solution 1 with the gradual addition of As(III). Solution 1 initially showed a characteristic rhodamine emission at 640 nm when it is excited at 560 nm. With each addition of As(III), there is a shift of 5 nm from the initial emission wavelength towards the blue region. Finally, a blue shift of 40 nm from its initial emission of solution 1, (from 640 to 600 nm) is observed with gradual addition of As(III). The change of emission can be clearly visible from red-orange to green (before and after the addition of As(III)) under the UV lamp (Figure 10-25).

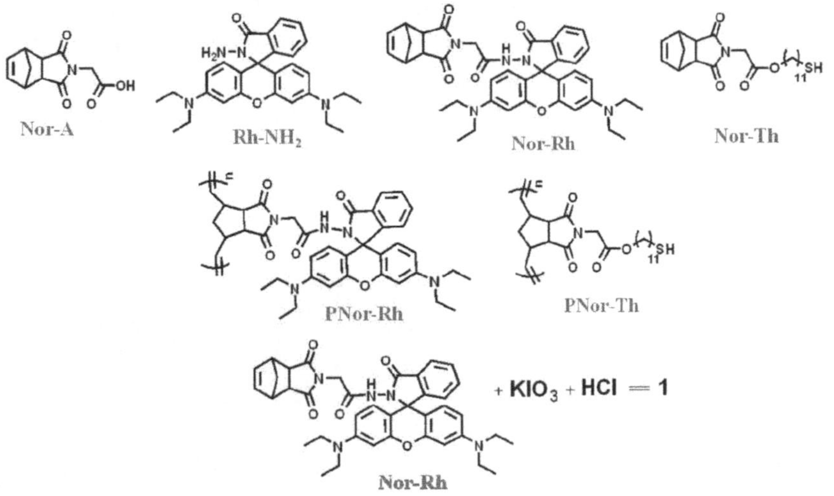

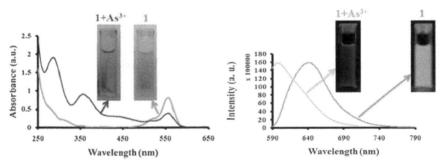

FIGURE 10-24 Schematic representation of As(III) sensor molecules tool box.

FIGURE 10-25 UV–visible and emission spectra of solution 1 (Nor-Rh, KIO$_3$, and HCl) before and after the addition of As(III) and it's color change under naked eye and a hand-held UV-lamp.

10.1.3. SENSING MECHANISM

$$2 \, AsO_2^- + 2 \, IO_3^- + 8 \, H^+ \longrightarrow I_2 + 2 \, AsO_3^- + 4 \, H_2O$$

Liberated iodine from the sensing component forming adduct to norbonene double bond simultaneously opening spirolactam ring. We hypothesized a possible halogen bonding formation (Figure 10-26) between the norbornene–I$_2$ adduct and the rhodamine carbonyl oxygen. To prove this hypothesis, ^{1}H NMR and IR spectroscopy experiments of Nor-Rh and 1 + As(III) are carried out. From ^{1}H NMR analysis, a new signal is observed at 3.6 ppm in 1 + As(III) solution (Figure 10-27a). This is attributed to the signal from the norbornene–I_2 adduct. Also the rhodamine aromatic protons are shifted down-field due to the spirolactam ring opening of the rhodamine motif. FT-IR spectroscopy showed a 12 cm^{-1} shift of the amide bond of the rhodamine motif due to the iodine interaction (Figure 10-27b) and therefore we got unique response through UV–visible and fluorescence spectroscopy.

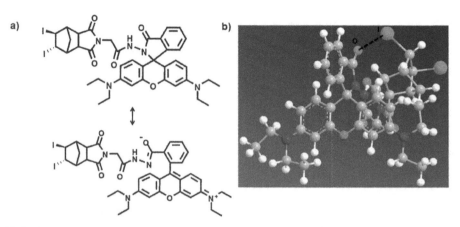

FIGURE 10-26 (a) Proposed chemical reaction of 1 + As(III). (b) Proposed halogen bonding between iodine and the carbonyl functionality of the rhodamine motif.

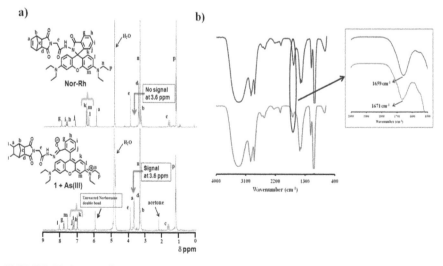

FIGURE 10-27 (a) ^{1}H NMR and (b) FT-IR spectrum of Nor-Rh and 1 + As(III).

To demonstrate the suitability of a practical device using this sensing system, paper strips are prepared by coating with a solution of PNor-Rh, KIO_3, and HCl and are then dried well. The paper strips, coated with the sensor component, are dipped into different concentrations of arsenic solution at the ppb level. It is observed that with increasing arsenic concentration, the intensity of the brown color increased. Interestingly, the change of color is instantaneous and prominent to the naked eye (Figure 10-28c). Despite the simplicity of this system it has an excellent detection limit (in ppb level) and appears to be very versatile.

A calibration curve is made from the UV–vis spectroscopic response of 1 (characteristic absorbance at 288) with different concentrations of As(III) (Figure 10-28a). Using the calibration curve the unknown concentration of the As(III) sample is identifed. The concentration of the unknown As(III) samples is simultaneously checked by using an ICP-MS instrument as well (Figure 10-28b). It is very exciting to note that both the results are in excellent agreement. The results also showed the advantage of our design over other analytical techniques including ICP-MS since no extra effort is needed for the sample preparation.

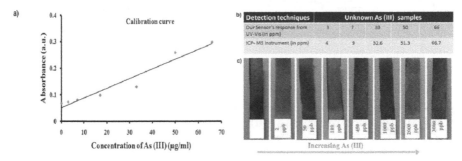

FIGURE 10-28 Comparisons of the efficiency of our sensing method to standard analytical methods. (a) A calibration curve is made from the UV–vis absorbance response of the sensing system with known concentrations of As(III). (b) A table summarizing the concentration of unknown As(III) samples obtained from our calibration curve as well as ICP-MS instrumentation. (c) Our paper strip coated with 1 shows the response based on the concentration of As(III).

Since As(III) has a specific affinity towards thiol functionality, a thiol based norbornene monomer (Nor-Th) and its polymer (PNor-Th) are synthesized to remove As(III) from the arsenic contaminated water. 1 mL of a known concentration of As(III) containing water (3223 ppb) is passed through alumina packed PNor-Th. The eluent is collected from the column and the presence of arsenic is measured using ICP-MS instrumentation. It is very interesting to note the As(III) content in the solution decreased from 3223 to 1.2 ppb (Figure 10-29). This procedure is repeated over five cycles and up until the fifth cycle the column's efficiency in trapping As(III) remained almost same.

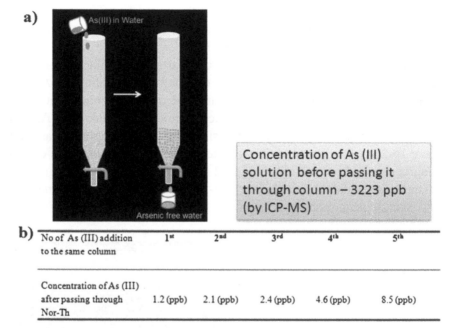

a)

Concentration of As (III) solution before passing it through column – 3223 ppb (by ICP-MS)

b)

No of As (III) addition to the same column	1st	2nd	3rd	4th	5th
Concentration of As (III) after passing through Nor-Th	1.2 (ppb)	2.1 (ppb)	2.4 (ppb)	4.6 (ppb)	8.5 (ppb)

FIGURE 10-29   (a) A cartoon representation of As(III) removal. (b) A table summarizing the ICP-MS values after passing the As(III) solution through the Nor-Th containing column.

The novel feature of this invention is that it is a redox based sensing where during the sensing event converting more poisonous As(III) to As(V) with liberation of iodine. This iodine is responsible for the adduct formation to norbornene double bond simultaneously spirolactam ring opening of the norbornene motif. To prove this, we measured the pH of solution 1 upon gradual addition of As(III). Interestingly, it is observed that the pH of the solution 1 with addition of As(III) increased from 1.34 to 4.23 (Figure 10-30). When the pH of the solution reached the value above 4, the color of the solution turned into brown. We attributed the observed increase in pH which is due to the consumption of acid for the redox reaction between As(III) and KIO_3 to get As(V) and I_2. As a result we got very unique response through UV and fluorescence spectroscopy.

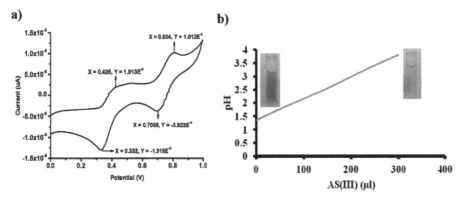

FIGURE 10-30 (a) Cyclic voltammogram of As(III) and 1. (b) Control experiment regarding pH change.

In conclusion, herein we have discussed polymer (PNor-Rh) coated paper strips could greatly simplify in-field detection of As(III) without the need for special equipment. This novel invention can sense poisonous metal arsenic in ppb level (detection limit 200 nM) very selectively, sensitively in aqueous environment.

PART B: NERVE GAS SENSING BY ROMP BASED POLYMERS

Recent toxic gas attack in Syria, which killed almost thousand people, reminds us again the necessity and urgency of continuing and developing nerve gas sensing research. The ease of production with available starting materials and their extreme toxicity never opts out the possibility of using this menace by the terrorists. In 1988, deadly nerve agent Sarin is used against the Kurds.[51] Again in 1995, Sarin is used to kill 12 people and injure 5000 during the terrorist attack on Tokyo's subway in Japan.[52,53] So continuing research for the detection of these odorless and colorless nerve agents as well as its destruction is always in need. Several scientists and their groups are working on this to find a way to sense nerve gas. If the presence of nerve can be detected before it started spreading, damage to the human life and environment will be less (Figure 10-31).

Rapid and severe effect on human and animal health made these nerve agents a deadly weapon. Highly reactive and volatile nerve agents can enter into body by inhalation as well as through skin. These nerve agents are series of organophosphates containing a labile leaving group attached to phosphate moiety, which made them highly reactive. Nerve agents can be divided into two types such as G-series and V-series. G-series nerve agents are sarin (GB), soman (GD), tabun (GA), and cyclosarin (GF) in the name of their discoverer, that is, German scientists. V-series nerve agents are VE, VG, VM, and VX. Whether G-series nerve agents hydrolyse rapidly, persistency of V-series nerve agents made them more toxic and dangerous. But for laboratory use, simulants of nerve agents of less toxicity are used. Dicyclohexyl chlorophosphate (DCP), diisopropyl fluorophosphate (DFP), and diphenylchlorophosphate (DPCP) are of similar reactivity and can be used instead of highly toxic nerve agents for experimental purpose. Here we have shown the chemical structures of nerve gases and their simulants.

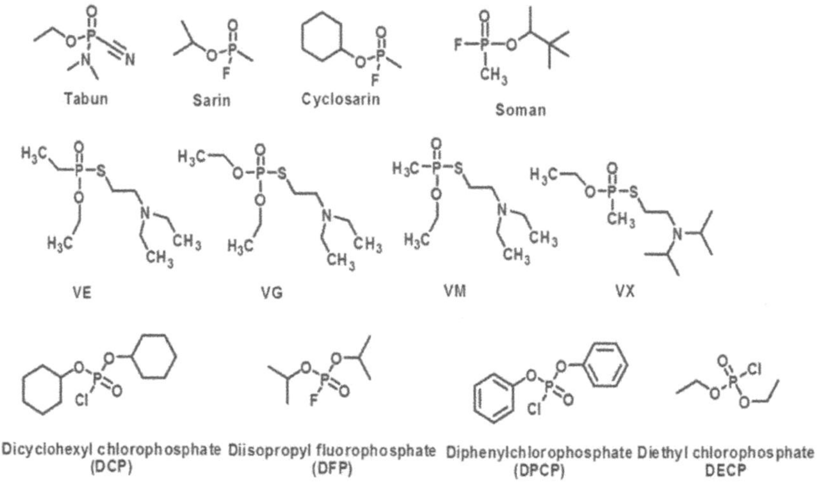

FIGURE 10-31 Chemical structures of G-series and V-series nerve agents and their mimics.

To look into the reactivity of the nerve agent first we need to know how our nervous system works. In our nervous system acetylcholine, a neurotransmitter, carries the signal from one organ to another. After delivering the signal, acetylcholinesterese, an enzyme destroys acetylcholine. This acetylcholinesterese consists of a serine residue having a hydroxyl group. When nerve agent enters into body it reacts with this serine residue and inhibits its reactivity, that is, destruction of acetylcholine. Hence, it leads to accumulation of acetylcholine at a particular region of organ and for this signal sending continues uninterruptedly. Muscles responding to the signal could not relax and in consequence, several physiological disorder even neuromuscular paralyses may happen which brings death.[54] Preliminary symptoms after immediate inhalation are respiratory problem along with anxiety, headache, and slurred speech. Depending upon the time of exposure the degree of effect varies.

Though a variety of detection methods for detection of Chemical Warfare Agent (CWAs) have been developed including enzymatic assays,[55] gas chromatography-mass spectrometry,[56] surface acoustic wave (SAW) devices,[57] molecular imprinting,[58] and flame photometric detectors.[59] However, all the methods mentioned above are not portable device for in-field use. Even all these methods have limitations like operational complexiblily, high cost, slow response time, low sensitivity, and selectivity. To overcome these difficulties optical change based sensors aiming for in-field application are designed and synthesized by several groups. Generating an optical event for sensing is the easiest way to follow by even an untrained person. Chromogenic and fluorogenic reagents for the sensing of nerve agents are designed and developed by various groups (Figure 10-32).

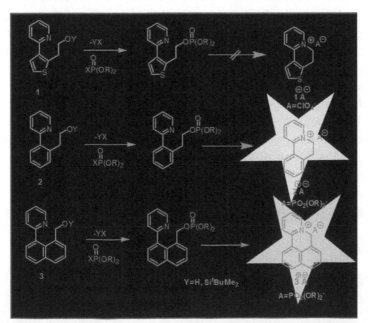

FIGURE 10-32   Structural demonstration of fluorescence emission through cyclization in presence of DFP by three compounds synthesised by Swager and Zhang.[60]

Zhang and Swager in 2003 developed a sensing technique of nerve gas sensing through intramolecular cyclization reaction that produce shift in absorption and emission spectra. Three molecules (**1, 2,** and **3**) are synthesized having one pyridine group bound to an aromatic system carrying an activated hydroxyl end.[60] In presence of organophosphates it formed phosphate ester followed by cyclization reaction to form a rigid molecule that eliminates rotation and produce change in its spectra. Following this principles it is observed that after forming organophosphate ester **1** is unable to cyclize in normal conditions due to unfavorable transition state structure. Molecule **2** undergone intramolecular cyclization by reacting with DFP, but quantum yield is too less due to unrestricted rotation of the pyridine-benzene single bond. Molecule **3** is synthesized by inclusion of naphthalene group to restrict the rotation. Reacting with DFP, it formed cyclized structure with enhanced emission and high quantum yield. It is proved that response of **3** is 17 times faster enough that molecule **2**.

Rebek and Dale developed fluorescent sensor for nerve gas by using PET process. They designed and synthesized four molecules having pyrene as dye at variable position from an amine group and having a primary hydroxyl group to react with organophosphate.[61] Fluorescence of the dye is quenched due to the PET process from the lone pair of the amine to the dye (Figure 10-33). As the length of the spacer (from $-CH_2-$ to $(-CH_2-)_4$) increases PET effect decreases. In presence of DFP the hydroxy group forms phosphate ester as an intermediate followed by immediate cyclization due to the nucleophilic attack of the amine to the ester. Lone pair of the amine being not available any more, the PET process stopped and generates fluorescence. Definitely molecule having one methelyne group shows higher fluorescence among all. When paper soaked with this compound is exposed to the vapor of DFP, blue emission is observed. Even by replacing the dye pyrene with coumarin, same "turn-on" response is noticed.

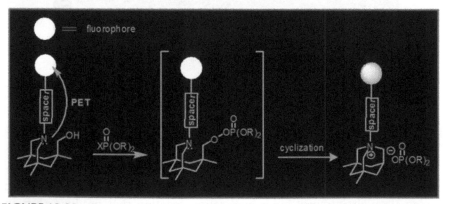

FIGURE 10-33 Detection of nerve gas agents by Rebek et al.[61] using photoinduce electron transfer.

A new and faster way of nerve gas detection technique is developed by using a supernucleophile as a sensor. A supernucleophile is one that contains an atom with an unshared pair of electrons adjacent to a nucleophilic centre, for example, oxime and hydrazone. This oxime based molecules react faster due to α-effect. It is shown that in basic media coumarin dye having oxime group emitted no fluorescence due to PET quenching from oxime to the dye.[62] On phosphorylation with DFP the PET deactivated and it resumed its emission (Figure 10-34a). Generally, this same principle of reactivity using supernucleophile is used for the antidote (e.g., 2-pyridinealdoxime; 2- PAM) of nerve gas poisoning.

To make them more effective a β-hydroxy group is introduced into oxime functionality by Dale and Rebek, which will lead to cyclization eliminating the phosphate ester to form isoxazole. Four oxime-based sensors are synthesized with the aromatic cores naphthalene, pyrene, coumarin, and pyridine and their reactivity is tested against nerve gas surrogate.[63] Their reaction with dicyclohexylchlorophosphate lead to formation of a cyclized structure called aryloxazole and enhances the fluorescence (Figure 10-34b). After comparing the fluorescence data it is seen that the response of naphthalene based sensor is the strongest and three times than the previously reported system D reported by same group. One advantage of this system is, it is water soluble which is the important criterion of nerve agent sensors.

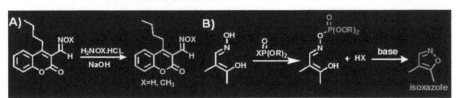

FIGURE 10-34 (a) Oxime based fluorometric sensing of nerve agent by Anslyn et al.[62] (b) β–hydroxy group containing oxime sensor for nerve gas by Dale and Rebek.[63]

Even micrometer and sub-micrometer fibers for sensing of nerve agents are also manufactured.[64] Pyrene imine molecule attached to norbornene backbone is synthesized and incorporated in polystyrene films as well as their fibers. The pyrene imine molecule acts as a "turn-on" system against sarin surrogate, SAS-Cl. A bright visible green emission is observed upon addition of SAS-Cl to the molecule E. These polystyrene based fibers carrying the sensor can be utilized as photonics as well as optical devices.

Our research is mainly focused on developing polymer based sensor systems. Most of the existing sensing systems are small molecules based.

It is expected that polymer based sensor materials will provide a better response than small molecules or monomers, as a polymer composed of several monomeric units. Aiming that way, we have designed our monomers that contain target binding group with a fluorophore as responsive moiety. Depending upon the target analyte, the binding group is varied and the fluorophore is chosen accordingly.

Generally, for metal binding, molecules containing electron donating group such as nitrogen, oxygen, and sulphur are used. Polymerization of these monomers is possible only when polymer backbone as well as polymerization catalyst will be compatible with these functional groups. In that case norbornene backbone is very helpful. It is easy to functionalize norbornene as it can carry two molecules at his two ends. As a polymerization technique we found ROMP of norbornene monomers using Grubbs catalyst is the best one. Also the efficiency, control, and tolerability of the ruthenium catalysts to a variety of pendant functional groups make this polymerization method a unique way for us. The most important property of the polymerization is its livingness. Random and block co-polymers can be synthesized easily. Also after polymerization, ruthenium bind to polymer backbone can be cleaved just by adding ethyl vinyl ether. In some cases, where monomers are very sensitive to the polymerization condition, post polymer modification also can be done.

Here, we have shown a cartoon representation to show our design towards a polymer based sensor system. A spacer is used between binding group and fluorophore that can control the electron transfer effect between this two. It can be observed that before binding to the analyte the fluorophore is not emitting. After binding it emits fluorescence. This is a "turn-on" system (Figure 10-35).

To develop a nerve agent sensing material terpyridine-Eu(III) complex system is designed, as it is known that lanthanide ions with terpy exhibits excellent emission properties.[65] Norbornene attached terpyridine (**NDT**) is synthesised and characterised. Surprisingly, it was found that **NDT** produces an unusual emission on excitation. This unusual emission along with terpy-Eu(III) lanthanide emission is utilized for sensing nerve agent surrogate. The toolbox showed here the target molecules, sensor molecules and the control molecules we developed during this work.

First of all, terpyridine is functionalized with 3-aminopropanol to produce amine functionalized terpyridine which is coupled to norbornene functionalized carboxylic acid using DCC. Crude product is purified through column chromatography. ^{1}H NMR, ^{13}C NMR, IR, and mass spectroscopy confirmed formation of **NDT** (Figure 10-36).

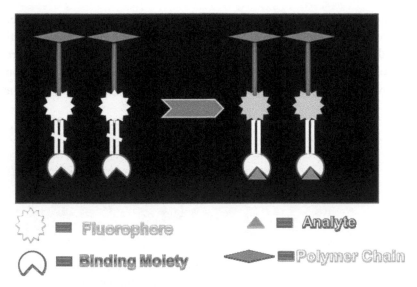

FIGURE 10-35   A general overview of the polymer based sensing scheme.

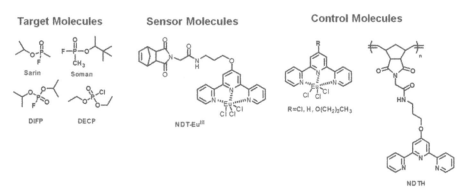

FIGURE 10-36   Toolbar consisting of target, sensor, and control molecules used for sensing.

NDT emerges an unusual blue emission under handheld UV lamp. The absorption and emission spectrum of NDT is recorded in methanol. Absorbance spectrum showed two maximum at 240 and 280 nm. NDT emits at 375, 450, and 480 nm when excited at 330 nm. Quantum yield of NDT is also measured by using quinine sulphate as standard and it is 0.22. Density functional theory calculations at the B3LYP/6-31G level are carried out to understand this unusual property. For the NDT molecule, the results showed that the HOMO wave function is located on the terpyridine while the LUMO wave function is located on the norbornene. It is found that the significant difference in the position of the HOMO and LUMO wave functions that prevented the nonradiative relaxation pathways is responsible for this property. Involvement of the norbornene double bond in this unusual emission is proved by formation of homopolymer (NDTH) of NDT (Figures 10-37 and 10-38).

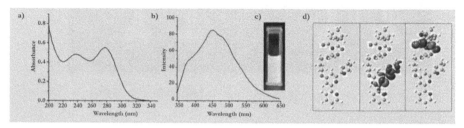

FIGURE 10-37   (a) UV–vis absorption. (b) Emission spectrum of NDT in methanol. (c) Image of NDT under UV light in methanol. (d) Optimized structure and molecular orbital amplitude plots of HOMO and LUMO energy levels of NDT molecule calculated with the use of B3LYP/6-31G basis set is shown.

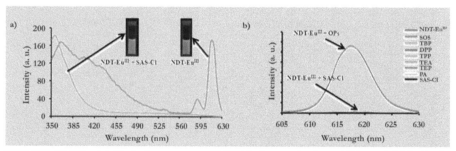

FIGURE 10-38   (a) Emission spectra of NDT before and after addition of SAS-Cl. (b) Plot of emission intensity at 620 nm upon addition of DIFP and other OPs, pinacoly methyl phosphonic acid (SOS), tributylphosphate (TBP), diphenylphosphate (DPP), triphenyl phosphate (TPP), triethylamine (TEA), triethylphosphate (TPP), and phosphoric acid (PA).

Surprisingly, NDTH did not show any unusual emission at all. This interesting observation strongly supported our hypothesis. The weak binding nature of lanthanide to terpyridine is utilized to sense highly reactive nerve agents. For laboratory use less toxic nerve agent mimic diethylchlorophosphate (DECP) is used. Response of **NDT**-Eu complex against surrogate is studied through fluorescence in methanol solvent. The characteristic **NDT**-Eu emission is stopped after SAS-Cl addition.

Under UV lamp it is seen that magenta emission changed to blue after immediate addition of DECP. The response is instantaneous and the lowest detection limit is 1.7 ppm, as determined. The sensor responded only to this DECP even in presence of other phosphate molecules. As explanation, it is proposed that the reactive DECP stopped the antenna effect between terpy nitrogens and Eu, which regenerate the blue emission of NDT.

10.2 OUR RESEARCH FOR NERVE AGENT SENSING

Nerve agent sensing research is still ongoing in our group to develop a colorimetric sensing technique based upon polymer. It is known that in spirolactum ring closed form of rhodamine it is colorless and nonfluorescent due to the lack of conjugation. When this spirolactum ring opens in contact with some metal or acid it shows color as well as fluorescent as the structure behaves as conjugated system (Figure 10-39). This phenomenon we would like to develop in polymer system. To that goal, norbornene attached rhodamine is prepared (Figure 10-40).

FIGURE 10-39 Spirolactum ring opening of rhodamine B induced by acid.

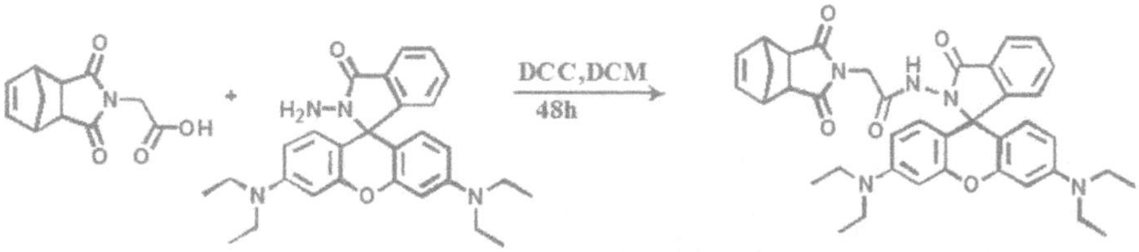

FIGURE 10-40 Formation of monomer Nor-Rh through coupling reaction.

Firstly, rhodamine B is converted to rhodamine amine following a reported procedure.[66] Norbornene acid is attached to rhodamine amine by

DCC coupling. Crude product is purified through precipitation of methanol. Formation of product is confirmed through NMR and mass spectroscopy.

For this monomer, we have tested its response against nerve agent surrogate and undoutedly it responded as a colorimetric as well as fluorometric sensor for nerve agent mimic DCP. It is observed that in UV spectrum Nor-Rh showed no absorbance. But with addition of DCP rhodamine ring opened form showed peak at 550 nm appeared and solution color changed from colorless to pink. In fluoroscence spectroscopy with addition of DCP new peak at 560 nm appeared and solution color is changed from colorless to yellow under UV lamp (Figure 10-41).

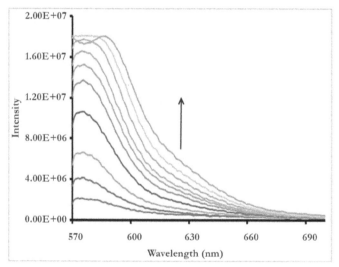

FIGURE 10-41   Emission specta of Nor-Rh with addition of DCP. Λ_{Ex}=550 nm.

To make the system more prominent and effective, FRET based sensor is planned to design. It is found that CdSe nanoparticles[67] emits in the same region where rhodamine ring opened form absorbs (Figure 10-42). This is the main principle of FRET to happen. Towards that goal we have deigned our scheme that is shown in Scheme 10-4. A thiol protected norbornene monomer containing long aliphatic chain is synthesised that could act as a nanoparticle capping agent after thiol deprotection. Another monomer norbornene attached n-hydroxysucccinimide is prepared for rhodamine attachment. Rhodamine amine is planned to attach through post polymer modification method, as in presence of ruthenium metal from Grubbs' catalyst rhodamine sprirolactum ring can be opened.[46]

Nor-NHS and Nor-TH monomer is synthesised and characterised. This two monomers are copolymerized in different batches to form random copolymers in DCM solvent. But it is found to be difficult to deprotect the thiol after copolymerisation. By different trial with different molecular weight polymers we failed to get the thiol containing polymers.

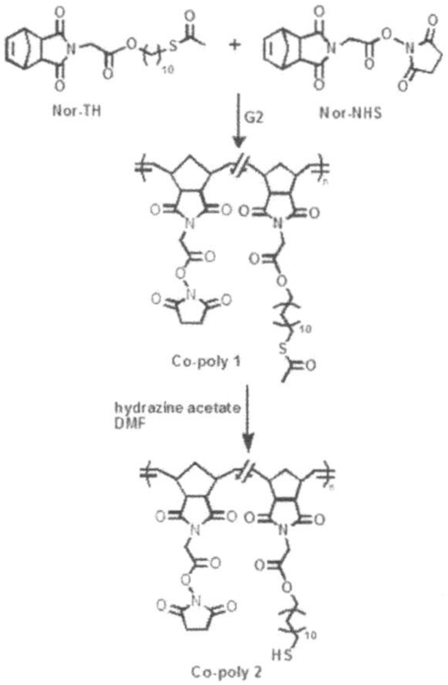

FIGURE 10-42 Schematic design of the copolymer synthesis for CdSe NPs and rhodamine together.

Our current research is involved in finding a better approach to produce copolymer having CdSe NPs and rhodamine together. We would like to develop it further for in-field application.

REFERENCES

1. Lakowicz, J. R. *Principles of Fluorescence Spectroscopy*, 2nd ed.; Kluwer/Plenum: New York, 1999.
2. Czarnik, A. W. *Fluorescent Chemoprobes for Ion and Molecule Recognition*; American Chemical Society: Washington, DC, 1993.
3. Valeur, B. *Molecular Fluorescence: Principles and Applications*; Wiley-VCH: Weinheim, 2001.
4. Kobayashi, H.; Ogawa, M.; Alford, R. Choyke, P. L.; Urano, Y. *Chem. Rev.* **2010,** *110*, 2620.
5. Tsien, R. Y. *Annu. Rev. Biochem.* **1998,** *67*, 509.

6. Terai, T.; Nagano, T. *Curr. Opin. Chem. Biol.* **2008**, *12*, 515.
7. Parker, D. *Coord. Chem. Rev.* **2000**, *205*, 109.
8. Gunnlaugsson, T.; Glynn, M.; Tocci, G. M.; Kruger, P. E.; Pfeffer, F. M. *Coord. Chem. Rev.* **2006**, *250*, 3094.
9. Shinoda, S. Tsukube, H. *Analyst* **2011**, *136*, 431.
10. Zhao, Q.; Li, F.; Huang, C. *Chem. Soc. Rev.* **2010**, *39*, 3007.
11. Thibon, A.; Pierre, V. C. *Anal. Bioanal. Chem.* **2009**, *394*, 107.
12. Valeur, B.; Leray, I. *Coord. Chem. Rev.* **2000**, *205*, 3.
13. de Silva, A. P.; Moody, T. S.; Wright, G. D. *Analyst* **2009**, *134*, 2385.
14. Jiang, P. J.; Guo, Z. J. *Coord. Chem. Rev.* **2004**, *248*, 205.
15. Carlson, H. J.; Campbell, R. E. *Curr. Opin. Biotechnol.* **2009**, *20*, 19.
16. Wu, J.; Liu, W.; Ge, J.; Zhang, H.; Wang, P. *Chem. Soc. Rev.* **2011**, *40*, 3483.
17. Grabowski, Z. R.; Rotkiewicz, K. *Chem. Rev.* **2003**, *103*, 3899.
18. Sumalekshmy, S.; Henary, M. M.; Siegel, N.; Lawson, P. V.; Wu, Y.; Schmidt, K.; Brédas, J.-L.; Perry, J. W.; Fahrni, C. J. *J. Am. Chem. Soc.* **2007**, *129*, 11888.
19. Clapp, A. R.; Medintz, I. L.; Mauro, J. M.; Fisher, B. R.; Bawendi, M. G.; Mattoussi, H. Protein Acceptors. *J. Am. Chem. Soc.* **2004**, *126*, 301–310.
20. Miyake, Y.; Togashi, H.; Tashiro, M.; Yamaguchi, H.; Oda, S.; Kudo, M.; Tanaka, Y.; Kondo, Y.; Sawa, R.; Fujimoto, T.; Machinami, T.; Ono, A. *J. Am. Chem. Soc.* **2006**, *128*, 2172–2173.
21. Ono, A.; Togashi, H. *Angew. Chem., Int. Ed.* **2004**, *43*, 4300–4302.
22. Zhou, Z.; Yu, M.; Yang, H.; Huang, K.; Li, F.; Yi, T.; Huang, C. *Chem. Commun.* **2008**, 3387.
23. Jun, M. E.; Roy, B.; Ahn, K. H. *Chem. Commun.* **2011**, *47*, 7583.
24. Dujols, V.; Ford, F.; Czarnik, A. W. *J. Am. Chem. Soc.* **1997**, *119*, 7386.
25. Chen, X.; Pradhan, T.; Wang, F.; Kim, J. S.; Yoon, J. *Chem. Rev.* **2012**, *112*, 1910.
26. Yang, Z.; She, M.; Yin, B.; Cui, J.; Zhang, Y.; Sun, W.; Li, J.; Shi, Z. J. *Org. Chem.* **2012**, *77*, 1143–1147.
27. Lu , C. Y.; Liu, Y. W.; Hung , P. J.; Wan, C. F.; Wu, A. T. *Inorg. Chem. Commun.* **2013**, *35*, 273–275.
28. Dwivedi, A. K.; Saikia, G.; Iyer, P. K. *J. Mater. Chem.* **2011**, *21*, 2502–2507.
29. Wu, X.; Xu, B.; Tong, H.; Wang, L. *Macromolecules* **2010**, *43* (21), 8917–8923.
30. Bhattacharya, S.; Rao, V. N.; Sarkar, S.; Shunmugam, R. *Nanoscale* **2012**, *4*, 6962–6966.
31. (a). Jane, A. M.; Matin, M. S.; Amy, T.; John, M. H.; Polly, A. N. *J. Natl. Cancer Inst.* **2006**, *98*, 869–873. (b). Violaine, V.; Dominique, L.; Philippe, H. *J. Toxicol. Environ. Health, Part A*, **2003**, *6*, 227–255. (c). Åkesson, A.; Julin, B.; Wolk, A. *Cancer Res.* **2008**, *68*, 6435–6441.
32. (a). Anthemidis, A. N.; Karapatouchas, C. P. *Microchim. Acta* **2008**, *160*, 455. (b). Kaya, G.; Yaman, M. *Talanta* **2008**, *75*, 1127. (c). Davis, A. C.; Calloway, C. P.; Jones, B. T. *Talanta* **2007**, *71*, 1144. (d). Huston, M. E.; Engleman, C.; Czarnik, A. W. *J. Am. Chem. Soc.* **1990**, *112*, 7054. (e). Bronson, R. T.; Michaelis, D. J.; Lamb, R. D.; Husseini, G. A.; Farnsworth, P. B.; Linford, M. R.; Izatt, R. M.; Bradshaw, J. S.; Savage, P. B. *Org. Lett.* **2005**, *7*, 1105. (f). Liu, W. M.; Xu, L. W.; Sheng, R. L.; Wang, P. F.; Li, H. P.; Wu, S. K. *Org. Lett.* **2007**, *9*, 3829. (g) Tang, X.; Peng, X.; Dou, W.; Mao, J.; Zheng, J.; Qin, W.; Liu, W.; Chang, J.; Yao, X. *Org. Lett.* **2008**, *10*, 3653.
33. (a). Charles, S.; Yunus, S.; Dubois, F.; Donckt, E. V. *Anal. Chim. Acta* **2001**, *440*, 37. (b). Gunnlaugsson, T.; Lee, T. C.; Parkesh, R. *Org. Lett.* **2003**, *5*, 4065. (c). Zhou, Y.; Xiao, Y.; Qian, X. *Tetrahedron Lett.* **2008**, *49*, 3380. (d). Li, H. ; Zhang, Y.; Wang, X.

Sens. Actuators B **2007,** *127,* 593. (e). Banerjee, S.; Kar, S.; Santra, S. *Chem. Commun.* **2008,** 3037. (f). Wang, H.-H.; Gan, Q.; Wang, X.-J.; Xue, L.; Liu, S.-H.; Jiang, H. *Org. Lett.* **2007,** *24,* 4995.

34. Sarkar, S.; Shunmugam, R. *ACS Appl. Mater. Interfaces* **2013,** *5,* 7379.
35. Bielawski, C. W.; Grubbs, R. H. *Prog. Polym. Sci.* **2007,** *32,* 1–29.
36. Lochow, C. F.; Miller, R. G. *J. Am. Chem. Soc.* **1976,** *98,* 1283–1285.
37. Smith, D.; Pentzer, E. B.; Nguyen, S. T. *Polymer Reviews* **2007,** *47,* 419–459.
38. Conrad, R. M.; Grubbs, R. H. *Angew. Chem. Int. Ed.* **2009,** *48,* 8328.
39. (a). Helal, A.; Rashid, M. H.; Choi, C. H.; Kim, H. S. *Tetrahedron* **2011,** *67,* 2794. (b). Goswami, S.; Sen, D.; Das, N. K. *Org. Lett.* **2010,** *12,* 4.
40. Maity, D.; Govindaraju T. *Chem. Eur. J.* **2011,** *17,* 1410.
41. Goswami, S.; Sen, D.; Das, N. K.; Fun, H. K.; Quah, C. K. *Chem. Commun.* **2011,** *47,* 9101.
42. Xu, Z.; Qian, X.; Cui, J. *Org. Lett.* **2005,** *7,* 3029.
43. Rey, N. A.; Howarth, O. W.; Pereira-Maia, E. C. *J. Inorg. Biochem.* **2004,** *98,* 1151 – 1159.
44. Burford, N.; Eelman, M. D.; Groom, K. *J. Inorg. Biochem.* **2005,** *99,* 1992–1997.
45. Krezel, A.; Lesniak, W.; Bojczuk, M. J.; Miyanarz, P.; Brasun, J.; Kozlowksi, H.; Bal, W. *Inorg. Biochem.* **2001,** *84,* 77–88.
46. Bhattacharya, S.; Sarkar, S.; Shunmugam, R. *J. Mater. Chem. A* **2013,** *1,* 8398–8405.
47. Pillai, A.; Sunita, G.; Gupta, V. K. *Anal. Chim. Acta* **2000,** *408,* 111.
48. Kalluri, R. K.; Arbneshi, T.; Khan, S. A.; Neely, A.; Candice P.; Varisli, B.; Washington, M.; McAfee, S.; Robinson, B.; Banerjee, S.; Singh, A. K.; Senapati D.; Ray, P. C. *Angew. Chem., Int. Ed.* **2009,** *48,* 1.
49. Ezeh, V. C.; Harrop, T. C. *Inorg. Chem.* **2012,** *51,* 1213.
50. Roy, S.; Palui, G.; Banerjee, A. *Nanoscale* **2012,** *4,* 2734–2740.
51. Romano, J. A.; Lukey, B. J.; Salem, H. *Chemistry, Pharmacology, Toxicology, and Therapeutics,* 2nd ed.; CRC Press: Boca Raton, 2007.
52. Tu, A. T. *Toxin Rev.* **2007,** *26,* 231.
53. Yang, Y. C.; Baker, J. A.; Ward, J. R. *Chem. Rev.* **1992,** *92,* 1729.
54. (a). Ikeda, A.; Shinkai, S. *Chem. Rev.* **1997,** *97,* 1713. (b). de Namor, A. F. D.; Cleverly, R. M.; Zapata-Ormachea, M. L. *Chem. Rev.* **1998,** *98,* 2495.
55. Russell, A. J.; Berberich, J. A.; Drevon, G. F.; Koepsel, R. R. *Annu. Rev. Biomed. Eng.* **2003,** *5,* 1.
56. Steiner, W. E.; Klopsch, S. J.; English, W. A.; Clowers, B. H.; Hill, H. H. *Anal. Chem.* **2005,** *77,* 4792.
57. (a). Yang, Y.; Hi, H.-F.; Thundat, T. *J. Am. Chem. Soc.* **2003,** *125,* 1124. (b). Thompson, H.; Hu, J.; Kaganove, S. N.; Keinath, S. E.; Keeley, D. L.; Dvornic, P. R. *Chem. Mater.* **2004,** *16,* 5357.
58. Jenkins, A. L.; Uy, O. M. *Anal. Commun.* **1997,** *34,* 221.
59. Kendler, S.; Zaltsman, A.; Frishman, G. *Instrum. Sci. Technol.* **2003,** *31,* 357–375.
60. Zhang, S.-W.; Swager, T. M. *J. Am. Chem. Soc.* **2003,** *125,* 3420–3421.
61. Dale, T. J.; Rebek, J. *J. Am. Chem. Soc.* **2006,** *128,* 4500.
62. Wallace, K. J.; Fagbemi, R. I.; Folmer-Andersen, F. J.; Morey, J.; Lynth, V. M.; Anslyn, E. V. *Chem. Commun.* **2006,** 3886.
63. Dale, T. J.; Rebek, J. *Angew. Chem., Int. Ed.* **2009,** *48,* 7850.
64. Rathfon, J. M.; AL-Badri, Z. M.; Shunmugam, R.; Berry, S. M.; Pabba, S.; Keynton, R. S.; Cohn, R. W.; Tew, G. N. *Adv. Funct. Mater.* **2009,** *19,* 689.

65. (a). Schubert, U. S.; Eschbaumer, C. *Angew. Chem.* **2002,** *114*, 3016; *Angew. Chem., Int. Ed.* **2002,** *41*, 2893. (b). Shunmugam, R.; Tew, G. N. *J. Am. Chem. Soc.* **2005,** *127*, 13567. (c). Shunmugam, R.; Tew, G. N. *J. Polym. Sci., Part A: Polym. Chem.* **2005,** *43*, 5831.
66. Du, J.; Fan, J.; Peng, X.; Sun, P.; Wang, J.; Li, H.; Sun, S. *Org. Lett.* **2010,** *12*, 476.
67. (a). Clapp, A. R.; Medintz, I. L.; Mauro, J. M.; Fisher, B. R.; Bawendi, M. G.; Mattoussi, H. *J. Am. Chem. Soc.* **2004,** *126*, 301. (b). Wang, S.; Meng, X.; Zhu, M. *Tetrahedron Letters* **2011,** *52*, 2840.

INDEX

Q

Milton Keynes UK
Ingram Content Group UK Ltd.
UKHW031138141024
449569UK00024B/1233